Polysaccharide Hydrogels

Polysaccharide Hydrogels

Characterization and Biomedical Applications

edited by
Pietro Matricardi
Franco Alhaique
Tommasina Coviello

Pan Stanford Publishing

Published by

Pan Stanford Publishing Pte. Ltd.
Penthouse Level, Suntec Tower 3
8 Temasek Boulevard
Singapore 038988

Email: editorial@panstanford.com
Web: www.panstanford.com

British Library Cataloguing-in-Publication Data
A catalogue record for this book is available from the British Library.

Polysaccharide Hydrogels: Characterization and Biomedical Applications

ISBN 978-981-4613-61-3 (Hardcover)
ISBN 978-981-4613-62-0 (eBook)

Printed in the USA

"The pursuit of science has often been compared to the scaling of mountains, high and not so high. But who amongst us can hope, even in imagination, to scale the Everest and reach its summit when the sky is blue and the air is still, and in the stillness of the air survey the entire Himalayan range in the dazzling white of the snow stretching to infinity? None of us can hope for a comparable vision of nature and of the universe around us. But there is nothing mean or lowly in standing in the valley below and awaiting the sun to rise over Kinchinjunga."

Truth and Beauty: Aesthetics and Motivations in Science (1987)
Subrahmanyan Chandrasekhar (Nobel Prize in Physics, 1983)

Contents

Preface

"What is a hydrogel?" "Sorry for taking your time Prof., but I need an explanation from you: what is the difference between a very viscous concentrated polymer solution and a hydrogel?" "Excuse me, I don't want to bother you, but I have also another question: how can I characterize a hydrogel?"

How many times polymer scientists, involved in the field of hydrogels, received these important and sometimes challenging questions? And how many times are we embarrassed while answering due to the difficulties in explaining the main features of this particular "state of matter" in a simple and straightforward manner?

"...the gel ...is easier to recognize than to define...". This is the famous statement that Dorothy Jordan Lloyd wrote in her book in 1926. Although since then much work has been carried out on hydrogels for their better characterization and definition, for many people and even for "experts" in this subject, this part of science still remains in the "dark side of the moon".

After a long scientific and professional activity, mainly devoted to polysaccharide hydrogel innovation, characterization, and evaluation of possible uses in the field of pharmaceutics and biomedical devices, we decided to summarize the state of the art in this discipline, trying, at the same time, to answer to the main questions that those who want to start a scientific approach to hydrogels and their applications in pharmacy and medicine must face.

The aim behind the preparation of this book is to offer students, scientists, and professionals a valid support that can help them find their way within the complex world of polysaccharide hydrogel characterization. In this book, we have selected topics, mainly related to the experimental aspects, that we encountered more frequently during our various studies. At the same time, we were forced to leave out several aspects that could also be considered as important topics, but, as it happens in most of the events of our life, selections and exclusions are needed because of mandatory limitations such as the size of this book.

While preparing this book, we asked for the assistance of many of our friends whom we met over the years during our studies, sharing with them competencies and receiving knowledge and friendship. We thank all of them for their willingness.

We would also like to thank Dr. Claudia Cencetti for her invaluable work in text revision.

Finally, we would like to thank all the colleagues that helped us in directing our steps within this discipline, since the beginning and over the years. In this sense we would like to remember in particular Prof. Vittorio Crescenzi, a mentor and a group leader in this field. It was under his guidance that, enlightened by his knowledge, we discovered the fascinating world of polysaccharide hydrogels.

Pietro Matricardi
Franco Alhaique
Tommasina Coviello
Autumn 2015

Chapter 1

Introduction

Franco Alhaique, Tommasina Coviello, and Pietro Matricardi
Department of Drug Chemistry and Technologies, University of Rome "La Sapienza", Piazzale Aldo Moro 5 00185, Rome, Italy
franco.alhaique@uniroma1.it

Among the numerous polymers that are actually used, or have been proposed, for the formulation of conventional and modified release dosage forms, natural polysaccharides and their derivatives are indeed the most employed and versatile polymeric materials. But it must be pointed out that polysaccharides show an even wider variety of biomedical applications, ranging from their intrinsic biological activity to their ability in living cell encapsulation, as well as from bone and cartilage repair to the preparation of friendly scaffolds for tissue engineering. Furthermore, remaining within quite close areas of interest, polysaccharides are often also present in cosmetic and personal care products [70], are used as food additives [125] and have even been proposed in formulations for paper and painting cleaning [86]. The peculiar and diversified properties of polysaccharides can be related to the different molecular weights and chemical composition that these macromolecules may have, as well as to the various type and number of reactive groups that can be present on each single unit. These properties allow numerous chemical modifications that can be

Polysaccharide Hydrogels: Characterization and Biomedical Applications
Edited by Pietro Matricardi, Franco Alhaique, and Tommasina Coviello

ISBN 978-981-4613-61-3 (Hardcover), 978-981-4613-62-0 (eBook)
www.panstanford.com

appropriately tailored according to the specific use that is assigned to these macromolecules. Polysaccharides are usually abundant (it is well known that cellulose is the most abundant organic compound on earth) and available from various renewable natural sources such as animals (e.g., chitin/chitosan, chondroitin), plants (e.g., pectin, guar gum), algae (e.g., alginates, agar, carrageenans), microbes, and fungi (e.g., dextran, xanthan gum, gellan, scleroglucan). Finally, but not less important, polysaccharides are, with only a few exceptions [103, 112], biocompatible and nontoxic products and can be classified as GRAS (generally recognized as safe).

Just as other synthetic and natural polymers, polysaccharides can form three-dimensional networks capable of imbibing large amounts of water or biological fluids, i.e., hydrogels [98]. The overall polymeric structure of such hydrogels can be chemically or physically based, according to the type of crosslinking among the macromolecules, but both chemical and physical crosslinks may also coexist in the same network [19]. Furthermore, polysaccharide hydrogels can be obtained with a single type of polymer or can be intercalated with other macromolecules (different polysaccharides, natural or synthetic polymers), leading to semi-interpenetrating (semi-IPN) or interpenetrating (IPN) networks which, most often, show physicochemical properties that can remarkably differ from those of the macromolecular constituents [84]. The ability of hydrogels to retain significant amounts of water makes them quite similar to natural living tissues, thus allowing an even wider range of applications for polysaccharides when used for the formation of hydrogel networks.

As evidenced by Lee at al. [72], the number of publications focused on hydrogels has increased exponentially during the last decade, and correspondingly the papers on drug, gene, and protein delivery from polysaccharide hydrogels as well as on tissue engineering and regenerative medicine based on such materials have followed the same trend with hundreds of research articles and numerous reviews written in recent years on this topic, and even whole issues of outstanding international journals dedicated to this subject. Consequently, the enormous amount of published data does not allow a complete comprehensive overview. Anyhow, the next sections of this chapter will give a glance to several specific

Table 1.1 Reviews focused on general aspects

Censi, R., Di Martino, P., Vermonden, T., and Hennink W. E. (2012). Hydrogels for protein delivery in tissue engineering, *J. Contr. Rel.*, **161**, pp. 680–692.
Coviello, T. Matricardi, P., Marianecci, C., and Alhaique F. (2007). Polysaccharide hydrogels for modified release formulations, *J. Contr. Rel.*, **119**, pp. 5–24.
Coviello, T. Matricardi, P., and Alhaique F. (2006). Drug delivery strategies using some polysaccharidic gels, *Expert Opin. Drug Deliv.*, **3**, 395–404.
Hamidi, M., Azadi, A., and Rafiei, P. (2008). Hydrogel nanoparticles in drug delivery, *Adv. Drug Deliv. Rev.*, **60**, pp. 1638–1649.
Hoffmann, A. S. (2012). Hydrogels for biomedical applications, *Adv. Drug Deliv. Rev.*, **64**, pp. 18–23.
Kim, J. K., Kim, H. J., Chung, J.-Y., Lee, J.-H., Young, S.-B., and Kim, Y.-H. (2014). Natural and synthetic biomaterials for controlled drug delivery, *Arch. Pharm. Res.*, **37**, pp. 60–68.
Lee, S. C., Kwon, K., and Park, K. (2013). Hydrogels for delivery of bioactive agents: a historical perspective, *Adv. Drug Deliv. Rev.*, **65**, pp. 17–20.
The whole issue of: Park, K., ed. (2001). Recent development in Hydrogels, *Adv. Drug Deliv. Rev.*, 48, pp. 1–166.
The whole issue of: Wang, Y., and Wang, P. G., eds. (2013). Polysaccharide-based systems in drug and gene delivery, *Adv. Drug Deliv. Rev.*, **65**, pp. 1121–1282.

biomedical applications of polysaccharide hydrogels, thus providing the reader with the basic information about specific topics and, at the same time, giving him or her the instruments needed to go deeper into the heart of the matter of more specific personal interest by means of the large number of reported references of research and review papers.

In the Tables 1.1 and 1.2 a preliminary, but of course partial, list of review papers on single polysaccharides and/or on general aspects of polysaccharide hydrogels suitable for biomedical applications is given.

1.1 Tissue Engineering and Regenerative Medicine

1.1.1 *Bone, Cartilage, and Skin*

Within the wide field of tissue engineering and regenerative medicine, scaffolds represent a fundamental aspect because they allow obtaining functional living implants from different types of cell cultures. The 3D porous structures needed for these purposes can

Table 1.2 Reviews focused on single polysaccharides

Balamurugan, M. (2012). Chitosan: a perfect polymer used in fabricating gene delivery and novel drug delivery systems, *Int. J. Pharm. Pharm. Sci.*, **4**(3), pp. 54–56.
Bernkop-Schnürch, A., and Dünnhaupt, S. (2012). Chitosan-based drug delivery systems, *Eur. J. Pharm. Biopharm.*, **81**, pp. 463–469.
Bidarra, S. J., Barrias, C. C., and Granja, P. L. (2014). Injectable alginate hydrogels for cell delivery in tissue engineering, *Acta Biomaterialia* **10**, pp. 1646–1662.
Coviello, T., Palleschi , A., Grassi, M., Matricardi , P., Bocchinfuso, G., and Alhaique, F. (2005). Scleroglucan: a versatile polysaccharide for modified drug delivery, *Molecules,* **10**, pp. 6–33.
Gao, C., Liu, M., Lü, S., Chen, C., Huang, Y., and Chen, Y. (2013). Preparation of sodium alginate hydrogel and its application in drug release, *Prog. Chem.*, **25**, pp. 1012–1022.
Gombotz, W. R., and Wee, S. F. (2012). Protein release from alginate matrices , *Adv. Drug Deliv. Rev.*, **64** suppl., pp. 194–205.
Hu, L., Sun, Y., and Wu, Y. (2013). Locust bean gum: a versatile biopolymer, *Nanoscale*, **5**, pp. 3103–3111.
Lee, K. Y., and Mooney, D.J. (2012). Alginate: properties and biomedical applications, *Adv. Polym. Sci.,* **37**, pp. 106–126.
Li, L., Ni, R., Shao, Y., and Mao, S. (2014). Carrageenan and its applications in drug delivery, *Carbohydr. Polym.*, **103**, pp. 1–11.
Matricardi, P., Di Meo, C., Coviello, T., and Alhaique, F. (2008). Recent advances and perspectives on coated alginate microspheres for modified drug delivery, *Exp. Opin. Drug Del.,* **5**, pp. 417–425.
Osmalek, T., Frelich, A., and Tasarek, S. (2014). Application of gellan gum in pharmacy and medicine, *Int. J. Pharm.*, **466**, pp. 328–340.
Prajapati, V. D. , Jani, G. K., and Khanda, S. M. (2013). Pullulan: an exopolysaccharide and its various applications, *Carbohydr. Polym.*, **95**, pp. 540–549.
Prajapati, V. D., Jani, G. K., Moradiya, N. G., Randeria, N. P., and Nagar, B. J. (2013). Advances in chitosan-based drug delivery vehicles, *Carbohydr. Polym.*, **94**, pp. 814–821.
Shukla, S. K., Mishra, A. K., Arotiba, O. A., and Mamba, B. B. (2013). Chitosan-based nanomaterials: a state-of-the-art review, *Int. J. Biol. Macromol.* **59**, pp. 46–58.
Xu, X., Jha, A. K., Harrington, D. A., Farach-Carson, M. C., and Jia, X. (2012). Hyaluronic acid-based hydrogels: from a natural polysaccharide to complex networks, *Soft Matter*, **8**, pp. 3280–3294.

be obtained using various polymeric materials, and polysaccharide hydrogels have been widely studied and proposed because of their versatility in obtaining external shapes that can be modelled according to specific needs and, at the same time, well-defined networks with appropriate porosity capable of hosting cells. In most cases the presence of appropriate growth factors, such as TGF-β, IGF, and FGF, is fundamental for the control of cellular behavior in bone and cartilage regeneration.

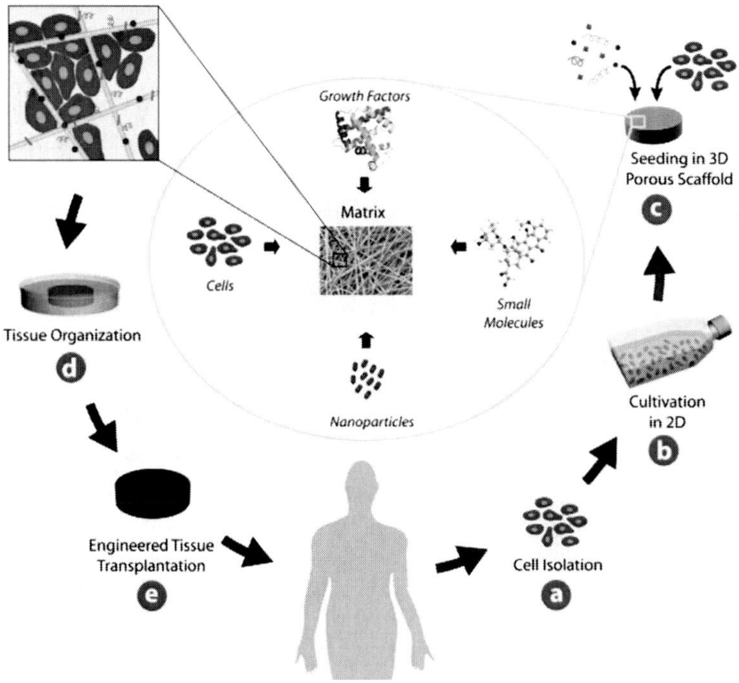

Figure 1.1 Scheme of the basic principles of tissue engineering. http://www.tau.ac.il/lifesci/departments/biotech/members/dvir/dvir.html).

Landers et al., by means of a rapid prototyping technique, developed in their laboratories [67] the use of hydrogels, obtained with agarose and alginic acid in the presence of calcium ions, for the formation of multilayered scaffolds on which human osteosarcoma cells and mouse connective tissue fibroblasts were seeded and grown [68, 69]. Alginate (ALG)-based hydrogels, covalently modified with RDG-containing peptides, were proposed for bone regeneration, with particular attention to the critical aspects related to scaffold degradation [3, 4, 128]. ALG hydrogels were also studied in the form of nanohydrogels by Lim et al. [75] in order to enhance osteogenic differentiation of human bone marrow stromal cells by the controlled release of a growth factor.

As expected, in all cases the method of scaffold fabrication, together with cell-dispensing technique, represents crucial factors in the optimization of cell viability and proliferation, as underlined in a recent paper by Lee et al. [71] where an innovative technique was proposed using again an ALG hydrogel.

Since 1975 Kosher and Church [60] evidenced the importance of glycosaminoglycans (GAGs) in chondrogenesis stimulation. Following this information, numerous research studies were carried out regarding the use of GAGs and GAG analogues for cartilage and bone regeneration, while Lu et al. [79] demonstrated that a significant increase in chondrocyte density occurred when a chitosan (CHT) solution was injected into the articular cavity of rats. Several papers report the use of CHT and of hyaluronic acid (HA), often in combination with other polymers, as scaffolding materials in cartilage engineering. Lindenhayn et al. [76] showed how the presence of HA in ALG beads is fundamental for human and porcine chondrocyte cultures. Lahiji et al. [64] reported that CHT is a biocompatible substrate suitable for human condrocytes as well as for osteoblasts. A wide overview, up to the beginning of this century, on the possible use of CHT-based biomaterials for these purposes was given by Suh and Matthew [134]. In more recent years, CHT was mainly used in combination with other polymers and/or inorganic materials, according to the type of cell regeneration which was required. CHT/Poly(L-lactic acid), CHT/HA, CHT/Alg/HA, and also CHT/pluronic acid as a thermosensitive scaffold [22, 45, 96, 156] have been proposed for cartilage repair, while, for bone repair, addition of calcium phosphate, hydroxyapatite, or even coralline biomaterials significantly improved the mechanical, and sometimes also adhesive, properties [40, 161].

Because of some drawbacks observed with HA, modifications were carried out for an improvement of its properties, such as covalent- and photo-crosslinkings, which increased with time mechanical strength and stability of the obtained materials. Swine chondrocytes within the pores of the novel hydrogels, obtained according to these approaches, were capable of producing neocartilage for several weeks [11]. Further improvements were obtained with HA ethyl and benzyl esters (HYAFF 7 and 11): it was demonstrated that such HA-based biomaterials, when used for cartilage repair, allow not only the expression of specific extracellular matrix molecules by human chondrocytes grown onto them, but also a downregulation of some catabolic factors and facilitate tissue repair. It was thus possible to suggest the use of these HA-derivative scaffolds in the treatment of early degenerative lesions in osteoarthritic patients [41–43]. Another strategy proposed for cartilage regeneration was based on the in situ formation of scaffolds obtained through the chemical synthesis of HA/benzoyl-cysteine derivatives. The results obtained showed, together with the suitable swelling properties of the new material, the chondrocyte proliferation within the in situ formed scaffold [97]. HA was also used in combination with hydroxyethyl-methacrylate-derivatized dextran in the particular form of a semi-interpenetrated network: it was shown that 3D bioprinted constructs of such material were suitable for condrocyte cell growth [84, 99].

To a lesser extent, numerous other polysaccharide hydrogels have been proposed for bone and cartilage regeneration, among them several starch- [38, 39, 83, 117, 127] and cellulose-based scaffolds [33, 143–145]. A wide survey is given in a review paper by Khan and Ahmad [56].

The studies on micro- and nanostructures increased significantly in the past years, and also regenerative medicine has been involved in this type of research. In this sense it is interesting to point out the quite relevant number of papers that appeared on the use of microenvironments for cell organization and orientation leading to micro-organ formation. A wide overview on this specific subject was very recently published by Verhulsel et al. [142].

Skin repair and wound healing is another important field related to tissue engineering. For these purposes, too, bio-inspired polysaccharide hydrogels have been widely studied. An ideal scaffold suitable for this type of applications should protect the wound from bacterial infections, prevent dehydration by controlling evaporative water loss, absorb exudates, allow gas diffusion, and provide an appropriate environment for cell proliferation, thus promoting skin repair. Again extracellular matrices prepared with GAGs and GAG analogues have been used for these purposes. Hydrogel films obtained from HA and chondroitin sulfate, chemically crosslinked with PEG-propiondialdehyde, were evaluated as bio-interactive wound dressings. Their swelling properties were studied and the in vivo efficacy of these films was evaluated on mouse models [59]. More recently a porous scaffold from HA, in combination with collagen and gelatine was tested, and in vivo results showed that such scaffold, together with a relevant water absorption capacity, significantly ameliorated wound healing, including decreasing neutrophil infiltrates and thickening newly generated skin compared to the group without treatments [146].

Also the use of CHT physical hydrogels was proposed for skin regeneration following third-degree burns [10]. It was demonstrated that after 100 days, the new tissue was quite similar to native skin, especially by its aesthetic aspect and its great flexibility. Another CHT physical gel promoting wound healing was proposed by Mayol et al. [85]. In this case the CHT powder was preliminarily thermally treated in autoclave and dissolved in acetic acid, which was then evaporated. Tested on mice, the obtained gel accelerated the healing process by reducing inflammatory markers and increasing re-epithelization parameters. Another polysaccharide proposed for the same purpose was pullulan [151, 152]: the authors suggest that pullulan hydrogel systems may prove beneficial for progenitor-cell-based approaches to skin regeneration, and it was evidenced that a hydrogel scaffold obtained with this polymer in combination with collagen was capable of improving early cutaneous wound healing.

Besides the cell proliferation promotion, as pointed out above, protection from bacterial infection and exudates absorption is an important feature in wound dressing, and some polysaccharide hydrogels possessing such properties have been proposed. One of the most recent approaches is based on a silver-loaded gellan/PVA/borax hydrogel, which showed a sustained release of silver and a significant antibacterial activity, also in comparison with a product on the market [15].

Although numerous improvements were obtained in recent years and several details about wound healing processes were elucidated, a recent editorial [135] clearly underlined how it is still a challenging problem to counteract all scar-forming factors during wound treatment.

1.1.2 *Cardiac Tissues*

Despite the recent advances, heart disease still kills more people than all other major diseases. Consequently, together with prevention, medication, and even heart transplants, a variety of materials are being studied for the repair and replacement of impaired heart tissues. Therefore, because of their properties and versatility, another important application of polysaccharide hydrogels is related to cardiac tissue regeneration. For this purpose, calcium-crosslinked biotin-labelled ALG was injected into the infarct seven days after anterior myocardial infarction in rats. The biomaterial was replaced within six weeks, the scar thickness was increased, and left ventricular systolic and diastolic dilatation and dysfunction were attenuated. Improvements were obtained also in the case of old myocardial infarction [66]. Infarct healing and repair was tested on larger animals [74] too, and it was also evidenced that peptide-modified ALG solutions were not as effective as the unmodified ALG [140]. On the other hand, Yu et al. [158, 159] demonstrated that RDG-modified ALG can improve myocardial functioning in a rat model of ischemic cardiomyopathy and that human mesenchymal stem cells, encapsulated within RDG-modified ALG microspheres, facilitated myocardial repair in rats. More recently, using a porous ALG scaffold bearing two peptides, a cardiac tissue which showed the striation and muscle fiber structure similar to that of a mature cardiac tissue was developed in vitro [119]. Together with ALG, other polysaccharide hydrogels were proposed for cardiac tissue regeneration; among them, a CHT scaffold in the form of nanofibers allowed to generate 3D cardiac tissue constructs [47].

1.1.3 *Ovaries and Oviducts*

Since ovarian cancer is the fifth leading cause of cancer deaths in women, numerous investigators utilized various approaches for culturing ovarian tissues or isolated follicles [30]. For this purpose also, polysaccharide hydrogels have been proposed, and, in particular, ALG hydrogels were used for three-dimensional ovarian organ cultures for the study of ovarian surface epithelial wound repair [49] and for the acquisition of further information about the origin of ovarian cancer and its progression from normal cells to tumors [58].

1.1.4 *Langerhans Cell Encapsulation and Insulin Delivery*

According to the International Diabetes Federation, diabetes caused 5.1 million deaths in 2013. At present 382 million people in the world have diabetes, and by 2035 this number will rise to 592 million. Thus, optimization of insulin delivery and Langerhans cell encapsulation can represent an important approach for the improvement of patients' quality of life as well as their life expectancy.

As pointed out earlier, owing to their similarity with natural extracellular matrices, hydrogels are attractive for biofabrication and allow cell encapsulation. Because of such features, hydrogels, usually in the presence growth factors such as vascular endothelial, fibroblast, or hepatocytic growth factors (VEGF, FGF, HGF), have been proposed for transplantation of pancreatic islets as a therapeutic modality for type 1 diabetes. For this purpose, ALG capsules were selected, often coated with polycation layers such as poly-L-lysine to obtain permselectivity [122] and an overview on the use of such polysaccharide for this specific application was published by Qi et al. [104]. Together with ALG, other polysaccharides were also proposed [31]. It was shown that agarose beads were capable of protecting islet xenografts from rejection and provide a microenvironment where the islets maintain and support their function in vivo [50]. CHT in combination with gellan [163] and pullulan/sulfonylurea [57] was also proposed. A wide survey on encapsulated islets for diabetes therapy was published very recently by Scharp and Marchetti [121] (Fig. 1.2).

Beside the above-reported use for Langerhans cell encapsulation and proliferation, polysaccharide hydrogels have been studied as effective and reliable insulin delivery systems for oral administration. Quite recently [150], insulin-loaded multilayered nanoparticles, formed by ALG and dextran sulphate nucleating around calcium and binding to poloxamer, stabilized by CHT and then coated with albumin, were tested on diabetic rats. The results that were

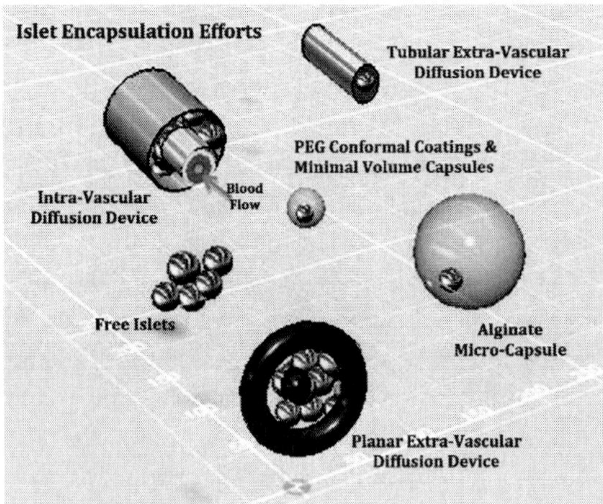

Figure 1.2 Schematic representation of various islet encapsulation efforts. Reprinted with permission from Ref. 121. Copyright © 2014, Elsevier.

obtained evidenced a reduced plasma glucose level up to 40% with a sustained hypoglycemic effect over 24 hours. CHT was also one of the components of poly(methacrylic acid)-CHT PEG microparticles, which were loaded with cyclodextrin complexed insulin. In this case also, a significant reduction in blood glucose levels was observed in vivo with Wistar rats [116]. Nanoparticles obtained with CHT prepared by the ion gelation method and based on the interaction between the positively charged groups of CHT and the negative ones of arabic gum showed, after an initial burst effect, a slow release of loaded insulin and a relevant mucoadhesion on excise rat jejunum [6]. Wider information on hydrogels for oral insulin delivery can be acquired from a recent review paper published by Chaturvedi et al. [16], which actually is mostly devoted to the possible use of CHT and its analogues.

1.2 Ocular Delivery

It is well known that numerous factors can limit the usefulness of ocular drug application. After an eyedrop instillation in the inferior fornix of the conjunctiva, the drug mixes with the lacrimal fluid, and drug contact time is affected by lacrimation, tear drainage, and turnover. A relevant number of investigations aimed to improve ocular drug contact time and drug delivery have been carried out in past decades, including the development of ointments, gels, liposome formulations, and various sustained and controlled-release substrates. According to these preliminary remarks, numerous studies have been devoted to the possible use of polysaccharide hydrogels in the field of ocular drug delivery, for both systemic and local applications. The use of gellan hydrogels for ophthalmic drug administration is reported in the chapter devoted to such polysaccharide, but several other polysaccharidic polymers have been proposed for ocular delivery. CHT-based hydrogels, owing to their adhesive and penetration-enhancing properties, have received much attention. Felt et al. [34] showed, by gamma scintigraphy, that the clearance of the tobramycin formulations labelled with 99mTc-DTPA was significantly delayed in the presence of CHT with respect to a commercial product. The effect of the polysaccharide molecular weight and concentration was also evaluated. More recently [14], a new copolymer, poly(*N*-isopropylacrylamide)–CHT, because of its thermosensitive in situ gel-forming properties, was investigated for ocular timolol maleate delivery. The results obtained from studies carried out on rabbits indicated that the new formulation had a stronger capacity to reduce the intraocular pressure when compared to conventional eye drops of same concentration over a period of 12 hours. A hybrid nanoparticulate hydrogel, again based on PNIPAAm and CHT, this time functionalized with acrylic acid, was tested as an innovative controlled release vehicle in ophthalmic drug delivery [7]. Rassu et al. [108] studied CHT microspheres prepared by spray drying and loaded with rokitamycin for the treatment of granulomatous amoebic keratitis; in this case both ocular and nasal applications were proposed as alternative routes for drug administration to the brain. HA/CHT nanoparticles, obtained by ionotropic gelation and intended for the delivery of genes to the cornea and conjunctiva, were studied by de la Fuente et al. [27]. The nanoparticles were loaded with either the model plasmid pEGFP or pβ-gal. Transfection and toxicological studies were conducted in human corneal epithelial and normal human conjunctival cell lines. The mechanism of internalization of the nanoparticles by the corneal and conjunctival cells was investigated by fluorescence confocal microscopy. Results showed that nanoparticles provided high transfection levels, without affecting cell viability, that they were internalized by fluid endocytosis, and that this endocytic process was mediated by the hyaluronan receptor CD44. The same authors published a wide overview on the use of CHT in ophthalmics [28]. Prolonged ocular delivery and enhanced residence times of antibiotics (gatifloxacin, cyclosporine A) were obtained using CHT and CHT/ALG nanoparticles [8, 90], and also a monoclonal antibody such as bevacizumab, used for the treatment of age-related macular degeneration, was proposed in a sustained-release formulation based on a crosslinked CHT and ALG hydrogel [154]. In order to reduce the risk, associated with glaucoma, of the onset and progression of intraocular pressure, dorzolamide-loaded CHT nanoparticles were prepared by ionotropic gelation. For the preparation of an in situ gel, the nanoparticles obtained were suspended in a sodium ALG solution, and the gamma scintigraphic study showed good corneal retention compared to a marketed formulation [54].

Current dry eye treatment includes delivering of HA, but besides this specific use, this polysaccharide was also proposed for ocular drug administration. Since noncrosslinked HA offers limited control over site retention of drugs, a recent paper describes the use of a chemically modified HA which was mixed with latanoprost-loaded liposomes and then crosslinked to give a nanocomposite hydrogel. Release studies showed longer sustained release of the drug from the composite hydrogel compared with liposomes or hydrogels alone [149].

At the end of this paragraph it seems necessary to underline that the treatment of posterior segment disease is to a significant extent limited by the difficulty in delivering effective doses of drugs to target tissues: intravitreal injections still provide the most used method that, nevertheless, shows important side effects and is not well tolerated by the patient. Actually, in spite of the significant improvements made in recent years, the delivery of therapeutic doses of drugs to the tissues in the posterior segment of the eye still remains an important and stimulating challenge, and nanohydrogels based on polysaccharides may represent an investigation that should be further developed.

1.3 Nasal Delivery

Intranasal administration is an alternative promising route for systematically acting drugs with poor bioavailability, and together with other polymers, polysaccharides have often been proposed for the formulation of hydrogels containing therapeutic agents with different pharmacological effects. The broad number of drugs that can be administered by such route (peptides/hormones, anesthetics, anti-emetics, corticosteroids, sedatives, antimigraine drugs, vaccines, drugs capable of overcoming the BBB, etc.) led to an extremely diversified literature on nasal delivery, and the number of publications is so wide that it is almost impossible to summarize the research that has been carried out on this topic. Therefore, it seemed appropriate to preliminarily report some recent review articles on this complex subject [130, 133, 136, 139, 147]. Anyhow, some significant results obtained in the last decade will be reported here. In order to demonstrate that high molecular weight substances are capable of crossing the nasal mucosa, gellan gel formation upon contact with the nasal mucosa was investigated for an evaluation of the increased and prolonged uptake of 3000 Da fluorescein dextran in comparison with a mannitol solution [52]. Using different concentrations of gellan, Cao et al. [12] prepared an ion-activated in situ gel system loaded with scopolamine HBr, and the antimotion sickness efficacy was investigated in rats and compared with subcutaneous and oral administrations of the same drug. The results obtained indicated that the new formulation was more effective than the references without nasal citotoxicity. The same authors [13], by means of an in situ gel formation based on gellan, showed that nasal administration of mometasone furoate, a corticosteroid used for relieving the symptoms of seasonal allergic rhinitis, was more effective than a common suspension of that drug. Again gellan gum and its in situ hydrogel formation were used to improve the antimigraine activity of sumatriptan succinate [36], while CHT microspheres were tested

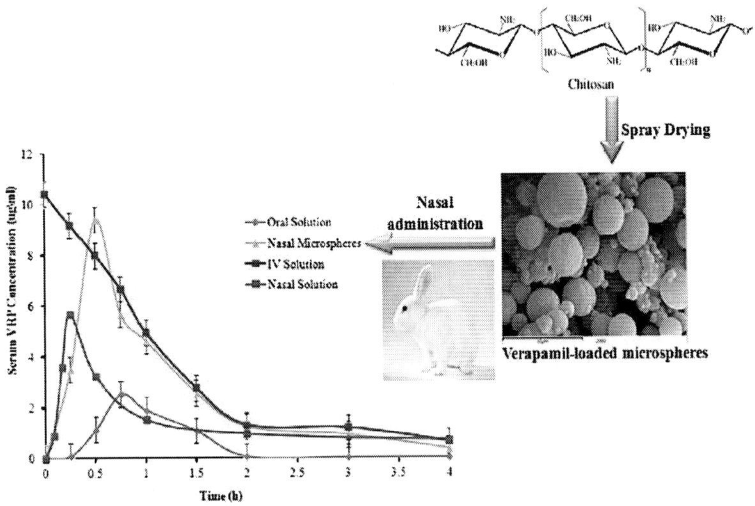

Figure 1.3 Verapamil-loaded chitosan microspheres and comparison of drug concentration–time profiles following administration of intravenous, oral, nasal solutions as well as nasal microspheres in rabbits. Reprinted with permission from Ref. 1. Copyright © 2014, Elsevier.

for the bioavailability enhancement of Verapamil HCl [1] (Fig. 1.3), and antigen/*N*-trimethylaminoethylmethacrylate CHT conjugates were tested for nasal immunization in order to prevent the diseases infected through the respiratory tract [78].

1.4 Buccal Administration

The oral cavity offers an attractive and easy route of drug administration which is not just limited to a local activity but is also significant for systemic delivery because of avoidance of drug degradation in the gastrointestinal tract and first-pass metabolism as well as patient acceptability. Although this route has been mainly restricted to small lipophilic molecules that readily cross the buccal mucosa, in recent years it became a significant means for successful systemic delivery of large molecules such as proteins and oligonucleotides.

Owing to their mucoadhesive properties, polysaccharide have often been proposed for buccal delivery, but in most cases these polymers, often in the form of thiolated derivatives to increase the interaction with mucin, were actually tested in the form of tablets and patches. Thus, such formulations cannot be considered real hydrogels, although, during swelling in the mouth environment, they have the appearance of strong and stable gels. Anyhow, several polysaccharide hydrogels have been proposed, throughout the years, as buccal drug delivery systems for both local/periodontal and systemic activity. Some examples of the possible applications of this type of formulations are reported below. Şenel et al. [123] demonstrated that a CHT gel containing chlorhexidine gluconate was capable of inhibiting the adhesion of *Candida albicans* to human buccal cells and showed antifungal activity. The same authors showed the enhancing effect of a CHT hydrogel on the transit of a peptide drug, the growth factor TGF-β [124], and again CHT was used for applications against a periodontal pathogen [48]. The anti-inflammatory activity of dexamethasone by buccal application of a CHT hydrogel was reported by Pignatello et al. [100], while the anesthetic activity of a mucoadhesive hydrogel based on CHT glutamate and lidocaine HCl was assessed in vivo [101]. Also insulin was proposed for buccal administration, and a pronounced hypoglycaemic effect by means of an insulin-loaded mucoadhesive CHT-EDTA hydrogel system was evidenced [21]. If, for the sake of brevity, only examples related to the use of CHT have been reported, other polysaccharides and derivatives have been studied for the same purposes. Among them, agarose, alginic acid, hyaluronic acid, guar gum, xanthan, cellulose and starch derivatives [91] as well as tamarind seed polysaccharide (xyloglucan), whose possible applications were recently reviewed [81].

1.5 Vaginal Delivery

Although somehow underutilized, vaginal delivery may have a wide number of possible applications for both local and systemic therapies [46]. The major advantages of this route include accessibility, good blood supply, the ability to bypass first-pass liver metabolism, and permeability to large molecular weight drugs. Actually, formulations for vaginal drug release can be used, or have been proposed, for several clinical and potential usages [26]: as antivirus agents (e.g., HIV prevention [113]), contraceptives [92], antibacterial/antifungal agents [9, 95], labour inducers [37], for hormone therapy as well as for vaccine and peptide/protein administration [110], and also against vaginal dryness [17], just to mention the most studied topics for practical utilization in this field. Together with suppositories, vaginal tablets and rings, gels and hydrogels have been also proposed for an appropriate vaginal distribution and retention of the dosage form, and polysaccharides were often employed for this purpose. As reported by Valenta [141] in her review on this topic, chitosan, cellulose derivatives, HA and its derivatives, pectins, carrageenan, tragacanth, and xanthan gums were used because of their gelling and bioadhesion properties. Also, starch was proposed, as in the case of insulin administration by means of starch microspheres, whose absorption was further increased by the presence of lysophosphatidylcholine [110]. Sometimes, to increase bioavailability, absorption enhancers were co-administered [111]. Together with bioadhesion, viscosity properties are important for this type of formulations, both for the ease of local administration and for the residence time. For this purpose, thermally sensitive gels capable of gelifying at physiological temperature have also been studied, and in a recent paper the better performances of a chitosan thermosensitive gel with respect to a poloxamer gel were evidenced [114]. Furthermore, since one of the problems related to vaginal administration is the modification of the rheological and mucoadhesive properties when diluted with the vaginal fluids that lead to a reduction of residence time, more recently a new hydrogel obtained with pluronics and hydroxypropylmethyl cellulose was proposed for this type of application. The tested thermosensitive gels, which showed a T_{gel} value near 30°C, kept their rheological and mucoadhesive properties after dilution with vaginal fluids [2]. Finally, it is interesting to point out that, besides drug delivery systems, placebo vaginal gels (i.e., without loaded drugs) may act as physical barriers to HIV during sexual transmission [65].

1.6 Colon Delivery

Targeted drug delivery into the colon is highly desirable both for the local treatment of a variety of bowel diseases and for the systemic delivery of drugs, such as proteins and peptides. Colon drug delivery systems must have specific formulations which are capable of protecting the drug en route to that segment of the intestine and, at the same time, avoid absorption in the stomach and in the small intestine. According to a review published several years ago [35], and to the more recent ones [44, 77, 115, 126, 131], it can be stated that polysaccharides have been proposed for colon drug delivery since more than 30 years. Consequently, the number of papers on hydrogels obtained from such macromolecules and targeted to that site of the gastrointestinal tract is extremely relevant, also because of the wide possibilities of applications and the large number of polysaccharides that have been and are being studied for this purpose. Just a couple of examples: in their review Jain et al. [51] report a list of 18 polysaccharides and several derivatives investigated for colon delivery, and a similar number, as well as numerous different types of polymer combinations, is reported by Shukla and Tiwari [126]. As already pointed out, together with the local treatment of inflammatory bowel diseases, ulcerative and Crohn's colitis, colorectal cancer, etc., the large intestine can be the site of absorption of high molecular weight drugs, thus allowing, if appropriately formulated, oral administration of this type of molecules. Together with the above-mentioned review papers, there are other recent publications that give overviews of such subjects, with specific attention to polysaccharides [18, 94, 153]. An up-to-date wide panorama of the possible use of the gel-forming Konjac glucomannan for colon targeting was published by Zhang et al. [160]. Thus, it appears more interesting here to focus mainly on some intriguing studies within this topic. First of all, it appears quite interesting to underline how numerous researches are still related to the improvement of "old" drugs such as 5-aminosalicilic acid (5ASA) and sulphasalazine (first introduced by Nanna Svartz in 1942 and which can actually be considered as a prodrug of 5ASA). Examples in this sense are given in several recent papers: Das and Pal [25] described a hydrogel based on hydroxypropyl

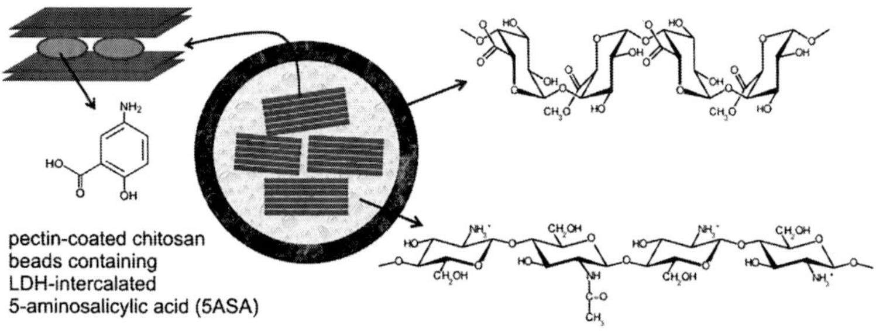

Figure 1.4 Scheme of a pectin-coated chitosan bead incorporating 5ASA intercalated in Mg_2Al layered double hydroxide, a new drug delivery system proposed by Ribeiro et al. (2014). Reprinted from Ref. 109 with permission from Elsevier.

methyl cellulose (HPMC) grafted with polyacrylamide for 5ASA colon delivery, while Ribeiro et al. [109] reported about pectin-coated CHT beads incorporating 5ASA intercalated in Mg_2Al layered double hydroxide as new colon targeted delivery systems (Fig. 1.4).

Anyhow, as stated before, numerous other drugs for different therapies can be orally administered and are capable of reaching the last part of the intestine by means of polysaccharide hydrogel formulations. Several anti-inflammatory agents were used for this purpose: prednisolone, which was encapsulated in pH-sensitive hydrogel beads prepared with ALG and ethylcellulose [106] or in a calcium-ALG CHT-coated matrix [5]; indomethacin, which was loaded in a hydrogel matrix based on Ca-ALG/chondroitin sulphate [32] or in pectin and charge-modified pectin beads [53]. Furthermore, a variation of the Pulsincap® device, using different polysaccharides, was proposed for ibuprofen colon release [155]. For the treatment of parasitic and bacterial infections, tinidazole within a CHT-sodium-ALG multiparticulate system, obtained by a dual ionic gelation method, was investigated [63]. Polysaccharide-gel-based colon cancer therapy was studied recently using 5-fluorouracil in CHT micro- [107] and nanospheres [102] or paclitaxel in chitin nanoparticles [132]. Protein delivery from polysaccharide gels was also studied: BSA, used as a model, was investigated by Zeng et al. [162], Cwp84-loaded pectin beads were tested for oral vaccination against Clostridium difficile [118], while a review reporting some examples of the use of polysaccharide gels for insulin oral admistration was recently published by Maroni et al. [82]. Numerous studies on delivery of probiotic bacteria by means of their entrapment in polysaccharide gels were also carried out [20]. Recently, CHT and thiolated CHT were investigated with the purpose of appropriately modifying the mucoadesion properties of the gel [17], while laccase-crosslinked arabinoxylan was proposed because of its capability of being degraded by bacteria of the intestinal microbiota [89].

1.7 Tumor Therapies

Drug targeting to tumor represents one of the main challenges of medical research [62], and within the numerous materials that have been proposed, polysaccharides are among the most studied macromolecules, often in the form of micro- and nanostructures. Improvements in the performance of well-known drugs have been recently made. Just a few examples: innovative cellulose hollow microspheres were prepared for daunorubicin modified release [87]; an HA microhydrogel prepared by a biotin-avidin system approach was studied for doxorubicin targeting [23]; for better targeting and sustained release to the gastrointestinal tract, methotrexate performances were improved by means of a chemical crosslinking carried out on a psyllium polysaccharide hydrogel [129]; bovine serum albumin oxidase, an enzyme capable of inducing cytotoxicity in cancer cell by converting polyamines and overexpressed in malignant cells into hydrogen peroxide, was studied for the treatment of melanoma [88]. Another interesting approach is a combination therapy based on a gellan nanohydrogel, an anti-inflammatory molecule (prednisolone chemically conjugated with the polysaccharide), and an anticancer drug (paclitaxel, whose solubility was dramatically increased within the hydrogel) [24] (Fig. 1.5).

Furthermore, a multiple presence (i.e., imaging and therapeutic agents) within the same structure is needed also for the combination of diagnostics with therapy (theranostics). HA nanoparticles for tumor-targeted imaging and drug release were studied by Yoon et al. [157] and a review on the use of this polysaccharide for theranostic

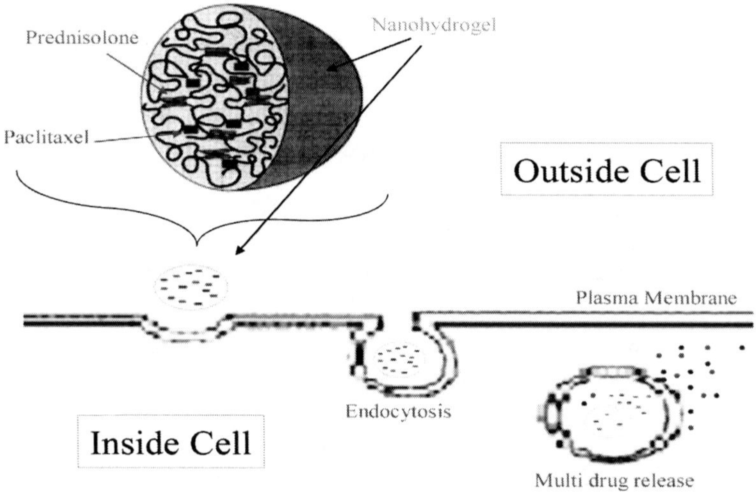

Figure 1.5 Schematic representation of a gellan nanogel system with prednisolone chemically linked to the polysaccharide and Paclitaxel physically loaded within the hydrophobic core of the nanostructure, according to the method proposed by D'Arrigo et al. [24].

applications was published very recently by Saravanakumar et al. [120]. Other polysaccharides were also proposed, such as glycol chitosan nanoparticles, which were capable of selectively accumulating at tumor site, thereby allowing in vivo visualization and potential therapy in cancer treatment [55, 73].

The ever-increasing interest in small interfering RNA (siRNA) as a powerful drug for gene silencing in cancer and in other severe pathologies, together with the limitations of its use due to degradation, poor cellular uptake, and rapid clearance, stimulated the development of innovative approaches for SiRNA protection and transport to the target [93, 138]. To overcome such limitations and improve SiRNA efficiency and specificity, several nanodevices using synthetic polymers and polysaccharides have been studied [137], and chitosan, because of its cationic nature accompanied by low toxicity and biocompatibility, was most often proposed for this type of application. A wide overview of this topic is given in the paper by Ragelle at al. [105], in which complexation and ionic gelation methods for the formulation of SiRNA-loaded nanoparticles are described together with the therapeutic applications in both local and systemic delivery. More recently, Deng et al. [29] proposed a co-delivery of another tumor suppressor (miR-34) and doxorubicin embedded in HA-chitosan nanoparticles for improved therapeutic treatment of breast cancer, in particular in case of resistance to conventional therapeutics.

1.8 Conclusions

A variety of naturally occurring polysaccharides and their derivatives have been successfully studied and practically utilized for the formulation of hydrogels suitable for a wide number of applications in the field of biomedicine. The versatility of the properties of such polysaccharide hydrogels allowed significant progresses in drug delivery and targeting, in theranostics, as well as in tissue engineering and regenerative medicine. Nanostructured hydrogels, which represent one of the most stimulating subjects for researchers from all over the world, certainly show great potential for future innovative results aimed at improving drug efficacy, life expectancy, and quality of life. Nevertheless, in spite of such developments, it should be also known that the passage to the market of all these new materials and formulations is, up to now, quite limited. Consequently, much work is still needed for fundamental studies to better understand and more appropriately tailor the physicochemical and biological properties of polysaccharide hydrogels as well as for further investigations on hydrogel systems capable of better fulfilling unmet medical and pharmaceutical needs.

References

1. Abdel Mouez, M., Zaki, N. M., Mansour, S., and Geneidi, A. S. (2014). Bioavailability enhancement of verapamil HCl via intranasal chitosan microspheres, *Eur. J. Pharm. Sci.*, **51**, pp. 59–66.
2. Aka-Any-Grah, A., Bouchemal, K., Koffi, A., Agnely, F., Zhang M., Djabourov, M., and Ponchel, G. (2010). Formulation of mucoadhesive vaginal hydrogels insensitive to dilution with vaginal fluids, *Eur. J. Pharm. Biopharm.*, **76**, pp. 296–303.

3. Alsberg, E., Anderson, K. W., Albeiruti, A., Franceschi, R. T., and Mooney, D. J. (2001). Cell-interactive alginate hydrogels for bone tissue engineering, *J. Dent. Res.*, **80**, pp. 2025–2029.
4. Alsberg, E., Kong, H. J., Hirano, Y., Smith, M. K., Albeiruti, A., and Mooney, D. J. (2003). Regulating bone formation via controlled scaffold degradation, *J. Dent. Res.* **82**, pp. 903–908.
5. Araujo, V., Gamboa, A., Caro, N., Abugoch, L., Gotteland, M., Valenzuela, F., Merchant, H. A., Basit, A. W., and Tapia, C. (2013). Release of prednisolone and inulin from a new calcium-alginate chitosan-coated matrix system for colonic delivery, *J. Pharm. Sci.*, **102**, pp. 2748–2759.
6. Avadi, M. R., Sadeghi, A. M. M., Dounighi, N. M., Dinarvand, R., Atyabi, F., and Rafiee-Tehrani, M. (2011). Ex vivo evaluation of insulin nanoparticles using chitosan and arabic gum, *ISRN Pharmaceutics*, Article ID 860109, pp. 1–6.
7. Barbu, E., Verestiuc, L., Iancu, M., Jatariu, A., Lungu, A., and Tsibouklis, J. (2009). Hybrid polymeric hydrogels for ocular drug delivery: nanoparticulate systems from copolymers of acrylic acid-functionalized chitosan and *N*-isopropylacrylamide or 2-hydroxyethyl methacrylate, *Nanotechnology*, **20**, 225108.
8. Basaran, E., Yenilmez, E., Berkman, M. S., Büyükköroğlu, G., and Yazan, Y. (2014). Chitosan nanoparticles for ocular delivery of cyclosporine A, *J. Microencapsulation*, **31**, pp. 49–57.
9. Bonferoni, M. C., Giunchedi, P., Scalia, S., Rossi, S., Sandri, G., and Caramella, C. (2006). Chitosan gels for the vaginal delivery of lactic acid: Relevance of formulation parameters to mucoadhesion and release mechanisms, *AAPS PharmSciTech*, **7**, Article no. 104.
10. Boucard, N., Viton, C., Agay, D., Mari, E., Roger, T., Chancerelle, Y., and Domard, A. (2007). The use of physical hydrogels of chitosan for skin regeneration following third-degree burns, *Biomaterials*, **28**, pp. 3478–3488.
11. Burdick, J. A., Chung, C., Jia, X., Randolph, M. A., and Langer, R. (2005). Controlled degradation and mechanical behavior of photopolymerized hyaluronic acid networks, *Biomacromolecules*, **6**, pp. 386–391.
12. Cao, S.-L., Zhang, Q.-Z., and Jiang, X.-G. (2007). Preparation of ion-activated *in situ* gel systems of scopolamine hydrobromide and evaluation of its antimotion sickness efficacy, *Acta Pharm. Sinica*, **28**, pp. 584–590.
13. Cao, S.-L., Ren, X.-W., Zhang, Q.-Z., Hen, E., Xu, F., Chen, J., Liu, L.-C., and Jiang, X.-G. (2009). In situ gel based on gellan as new carrier for nasal

administration of mometasone furoate, *Int. J. Pharm.*, **365**, pp. 109–115.

14. Cao, Y., Can, Z., Shen, W., Cheng, Z., Yu, L., and Ping, O. (2007). Poly(*N*-isopropylacrylamide)–chitosan as thermosensitive *in situ* gel-forming system for ocular drug delivery, *J. Contr. Rel.*, **120**, pp. 186–194.
15. Cencetti, C., Bellini, D., Pavesio, A., Senigaglia, D., Passariello, C., Virga, A., and Matricardi, P., (2012). Preparation and characterization of antimicrobial wound dressing based on silver, gellan, PVA and borax, *Carbohydr. Polym.*, **90**, pp. 1362–1370.
16. Chaturvedi, K., Ganguly, K., Nadagouda, M. N., and Aminabhavi, T. M. (2013). Polymeric hydrogels for oral insulin delivery, *J. Contr. Rel.*, *165*, pp. 129–138.
17. Chen, S., Cao, Y., Ferguson, L. R., Shu, Q., and Garg, S. (2013). Evaluation of mucoadhedive coatings of chitosan and thiolated chitosan for the colonic delivery of microencapsulated probiotic bacteria, *J. Microencapsulation*, **30**, pp. 103–115.
18. Chourasia, M. K., and Jain, S. K. (2004). Plysaccharides for colon targeted delivery, *Drug. Deliv.*, **11**, pp. 129–148.
19. Clark, A. H., and Ross-Murphy, S. B. (1987). Structural and mechanical properties of biopolymer gels, *Adv. Polym. Sci.* **83**, pp. 57–192.
20. Cook, M. T., Tzortzis, G., Charalampopulos, D., and Khutoryanskiy, V. V. (2012). Microencapsulation of probiotics for gastrointestinal delivery, *J. Contr. Rel.* **162**, pp. 56–67.
21. Cui, F., He, C., He, M., Tang, C., Yin, L., Quian, F., and Yin, C. (2009). Preparation and evaluation of chitosan-ethylenediaminetetracetic acid hydrogel films for the mucoadhesive transbuccal delivery of insulin, *J. Biomed. Mater. Res. Part A*, **89**, pp. 1063–1071.
22. Cui, Y. L., Qi, A. D., Liu, W. G., Wang, X. H., Wang, H., Ma, D. M., and Yao, K. D. (2003). Biomimetic surface modification of poly(L-lactic acid) with chitosan and its effects on articular chondrocytes in vitro, *Biomaterials*, 24, pp. 3859–3868.
23. Cui, Y., Li, Y., Duan, Q., and Kakuchi, T. (2013). Preparation of hyaluronic acid micro-hydrogel by biotin-avidin-specific bonding for doxorubicin-targeted delivery, *Appl. Biochem. Biotechnol.*, **169**, pp. 239–249.
24. D'Arrigo, G., Navarro, G., Di Meo, C., Matricardi, P., and Torchilin, V. (2014). Gellan gum nanohydrogel containing anti-inflammatory and anti-cancer drugs: a multi delivery system for a combination therapy in cancer treatment, *Eur. J. Pharm. Biopharm.*, **87**, pp. 208–216.

25. Das, R., and Pal, S. (2013). Hydroxypropyl methyl cellulose grafted with polyacrylamide: Application in controlled release of 5-amino salicylic acid, *Coll. Surf. B: Biointerf.* **110**, pp. 236–241.
26. das Neves, J., and Bahia, M. F. (2006). Gels as vaginal delivery systems, *Int. J. Pharm.*, **318**, pp. 1–14
27. de la Fuente, M., Seijo, B., and Alonso, M. J. (2008). Novel hyaluronic acid-chitosan nanoparticles for ocular gene therapy, *Invest. Ophthalmol. Vis. Sci.*, **49**, pp. 2016–2024
28. de la Fuente, M., Raviña, M., Paolicelli, P., Sanchez, A., Seijo, B., and Alonso, M. J. (2010). Chitosan-based nanostructures: a delivery platform for ocular therapeutics, *Adv. Drug. Deliv. Rev.*, **62**, pp. 100–117.
29. Deng, X., Cao, M., Zhang, J., Hu, K., Yin, Z., Zhou, Z., Xiao, X., Yang, Y., Sheng, W., Wu, Y., and Zeng, Y. (2014). Hyaluronic acid-chitosan nanoparticles for co-delivery of MiR-34a and doxorubicin in therapy against triple negative breast cancer, *Biomaterials*, **35**, pp. 4333–4334.
30. Devine, P. J., Rajapaksa, K. S., and Hoyer, P. B. (2002). In vitro ovarian tissue and organ culture, a review, *Front. Biosci.* **7**, pp. d1979.1989.
31. de Vos, P., Lazarjani, H. A., Poncelet, and Faas, M. M. (2014). Polymers in cell encapsulation from an enveloped cell perspective, *Adv. Drug. Deliv. Rev.*, **67–68**, pp. 15–34.
32. Fajardo, A.r., Silva, M. B., Lopes, L. C., Piai, J. F., Rubira, A. F., and Muniz, E. C. (2012). Hydogel based on an alginate-Ca^{2+}/chondroitin sulphate matrix as a potential colon-specific drug delivery system, *RSC Adv.* **2**, pp. 11095–11103.
33. Fang, B., Wan, Y.-Z., Tang, T.-T., Gao, C., and Dai, K.-R. (2009). Proliferation and osteoblastic differentiation of human bone marrow stromal cells on hydroxyapatite/bacterial cellulose nanocomposite scaffolds, *Tissue Eng. Part A*, 15, pp. 1091–1098.
34. Felt, O., Furrer, P., Mayer, J. M., Plazonnet, B., Buri, P., and Gurny, R. (1999). Topical use of chitosan in ophthalmology: tolerance assessment and evaluation of precorneal retention, *Int. J. Pharm.*, **180**, pp. 185–193.
35. Friend, D. R. (1991). Colon specific drug delivery, *Adv. Drug Deliv. Rev.* **7**, pp. 149–199.
36. Galgatte, U. C., Kumbhar, A. B., and Chaudhari, P. D. (2014). Development of *in situ* gel for nasal delivery: design, optimization, in vitro and in vivo evaluation, *Drug Deliv.* **21**, pp. 62–73.

37. Gauger, L. J. (1984). Hydroxyethylcellulose gel as a dinoprostone vehicle, *Am. J. Hosp. Pharm.* **41**, pp. 1761–1762.

38. Gomes, M. E., Reis, R. L., Cunha, A. M., Blitterswijk, C. A., and de Bruijn, J. D. (2001). Cytocompatibility and response of osteoblastic-like cells to starch-based polymers: effect of several additives and processing conditions, *Biomaterials*, **22**, pp. 1911–1917.

39. Gomes, M. E., Godinho, J. S., Tchalamov, D., Cunha, A. M., and Reis, R. L. (2002). Alternative tissue engineering scaffolds based on starch: processing methodologies, morphology, degradation and mechanical properties, *Mater. Sci.Eng. C*, **20**, pp. 19–26.

40. Gravel, M., Vago, R., and Tabrizian, M. (2006). Use of natural coralline biomaterials as reinforcing and gas-forming agent for developing novel hybrid biomatrices: microarchitectural and mechanical studies, *Tissue Eng.*, **12**, pp. 589–600.

41. Grigolo, B., Roseti, L., Fiorini, M., Fini, M., Giavaresi, G., Nicoli Aldini, N., Giardino, R., and Facchini, A. (2001). Transplantation of chondrocytes seeded on a hyaluronan derivative (Hyaff®-11) into cartilage defects in rabbits, *Biomaterials*, **22**, pp. 2417–2474.

42. Grigolo, B., Lisignoli, G., Piacentini, A., Fiorini, M., Gobbi, P., Mazzotti, G., Duca, M., Pavesio, A., and Andrea Facchini, A. (2002). Evidence for redifferentiation of human chondrocytes grown on a hyaluronan-based biomaterial (HYAFF® 11): molecular, immunohistochemical and ultrastructural analysis, *Biomaterials*, **23**, pp. 1187–1195

43. Grigolo, B., De Franceschi, L., Roseti, L., Cattini, L., and Facchini, A. (2005). Down regulation of degenerative cartilage molecules in chondrocytes grown on a hyaluronan-based scaffold, *Biomaterials*, **26**, pp. 5668–5676.

44. Gulbake, A., and Jain, S. K. (2012). Chitosan: a potential polymer for colon-specific drug delivery system, *Exp. Opin. Drug. Del.*, **9**, pp. 713–729.

45. Hsu, S-h., Whu, S. W., Hsieh, S.-C., Tsai, C.-L., Chen, D. C., and Tan, T.-S. (2004). Evaluation of Chitosan-alginate-hyaluronate complexes modified by an RGD-containing protein as tissue-engineering scaffolds for cartilage regeneration, *Artif. Organs*, **28**, pp. 693–703.

46. Hussain, A., and Ahsan, F. (2005). The vagina as a route for systemic drug delivery, *J. Contr. Rel.*, **103**, pp. 301–313.

47. Hussain, A., Collins, G., Yip, D., and Cho, C. H. (2013). Functional 3-D cardiac co-culture model using bioactive chitosan nanofiber scaffolds, *Biotechnol. Bioeng.* **110**, pp. 637–647.

48. İkinci, G., Şenel, S., Akincibay, H., Kaş, S., Ercis, S., Wilsom, C. G., and Hincal, A. A. (2002). Effect of chitosan on a periodontal pathogen *Porphyromonas gingivalis*, *Int. J. Pharm.* **235**, pp. 121–127.

49. Jackson, K. S., Inoue, K., Davis, D. A., Hilliard, T. S., and Burdette, J. E. (2009). Three-dimensional ovarian organ culture as a tool to study normal ovarian surface epithelial wound repair, *Endocrinology*, **150**, pp. 3921–3926.

50. Jain, K., Yang, H., Cai, B.-R., Haque, B., Hurvitz, A. I., Diehl, C., Miyata, T., Smith, B. H., Stenzkl, K., Suthanthiran, M., and Rubin, A. L. (1995). Retrievable, replaceable, macroencapsulated pancreatic islet xenografts long-term engraftment without immunosuppression, *Transplantation*, 59, pp. 319–324.

51. Jain, A., Gupta, Y., and Jain, S. K. (2007). Perspectives of biodegradable natural polysaccharides for site specific drug delivery to the colon, *J. Pharm. Pharmaceut. Sci.***10**, pp. 86–128.

52. Jansson, B., Hägerström, H., Fransén, N., Edsman, K., and Björk, E. (2005). The influence of gellan gum on the transfer of fluorescein dextran across rat nasal epithelium in vivo, *Eur. J. Pharm. Biopharm.* **59**, pp. 557–564.

53. Jung, J., Arnold, R. D., and Wicker, L. (2013). Pectin and charge modified pectin hydrogel beads as a colon-targeted drug delivery carrier, *Coll. Surf. B. Bionterf.*, **104**, pp. 116–121.

54. Kativar, S., Pandit, J., Mondal, R. S., Mishra, A. K., Chuttani, K., Aqil, M., Ali, A., and Sultana, Y. (2014). *In situ* gelling dorzolamide loaded chitosan nanoparticles for the treatment of glaucoma, *Carbohydr. Polym.*, **102**, pp. 117–124.

55. Key, J., Cooper, C., Kim, A. Y., Dhawan, D., Knapp, D. W., Kim, K., Park, J. H., Choi, K., Kwon, I. C., Park, K., and Leary, J. (2012). In vivo NIRF and MR dual-modality imaging using glycol chitosan nanoparticles, *J. Contr. Rel.* **163**, pp. 249–255.

56. Khan, F., and Ahmad, S. R. (2013). Polysaccharides and their derivatives for versatile tissue engineering application, *Macromol. Biosci.*, **13**, pp. 395–421.

57. Kim, S., Chae, S. Y., Na, K., Kim, S. W., and Bae, Y. H. (2003). Insulinotropic activity of sulfonylurea/pullulan conjugate in rat islet microcapsule, *Biomaterials*, **24**, pp. 4843–4851.

58. King, S. M., Quartuccio, S., Tyvette, S., Hilliard, T. S., Inoue, K., and Burdette, J. E. (2011) Alginate hydrogels for three-dimensional organ culture of ovaries and oviducts, *J. Vis. Exp.*, **52**, pp. 2804–2810.

59. Kirker, K. R., Luo, Y., Nielson, J. H., Shelby, J., and Prestwich, J. D. (2002). Glycosaminoglycan hydrogel films as bio-interactive dressings for wound healing, *Biomaterials* **23**, pp. 3661–3671.

60. Kosher, R. A., and Church, R. L. (1975). Stimulation of in vitro somite chondrogenesis by procollagen and collagen, *Nature*, **258**, pp. 327–330.

61. Kumar, S. P., Prathibha, D., Parthibarajan, R., and Rechal, C. R., (2012). Novel control delivery drug dlivery system, *Int. J. Pharm. Pharm. Sci.*, **4** (suppl. 1), pp. 22–29.

62. Kwon, K., Lee, S. C., Han, B., and Park, K. (2012). Analysis on the current status of targeted drug delivery to tumors, *J. Contr. Rel.* **164**, pp. 108–114.

63. Lanjhiyana, S. K., Bajpayee, P., Kesavan, K., Lanjhiyana, S., and Muthu, M. S. (2013). Chitosan-sodium alginate blended polyelectrolyte complexes as potential multiparticulate carrier system: colon-targeted delivery and gamma scintigraphic imaging, *Exp. Opin. Drug Del.* **10**, pp. 5–15.

64. Lahiji, A., Sohrabi, A., Hungerford, D. S., and Frondoza, C. G. (2000). Chitosan supports the expression of extracellular matrix proteins in human osteoblasts and chondrocytes, *J. Biomed. Mater. Res.* (*A*) **51**, pp. 586–595.

65. Lai, B. E., Geonotti, A. R., De Soto, M. G., Montefiori, D. C., and Katz, D. F. (2010). Semi-solid gels function as physical barriers to human immunodeficiency virus transport *in vitro*, *Antiviral Res.*, **88**, pp. 143–151.

66. Landa, N., Miller, L., Feinberg, M. S., Holbova, R., Shachar, M., Freeman, I., Cohen, S., and Leor, J. (2008). Effect of injectable alginate implant on cardiac remodeling and function after recent and old infarcts in rat, *Circulation*, **117**, pp. 1388–1396.

67. Landers, R., and Mulhaupt, R. (2000). Desktop manufacturing of complex objects, prototypes and biomedical scaffolds by means of computer-assisted design combined with computer-guided 3D plotting of polymers and reactive oligomers, *Macromol. Mater. Eng.*, **282**, pp. 17–21.

68. Landers, R., Hübner, U., Schmelzeisen, R., and Mülhaupt, R. (2002). Rapid prototyping of scaffolds derived from thermoreversible hydrogels and tailored for applications in tissue engineering, *Biomaterials*, **23**, pp. 4437–4447.

69. Landers, R., Pfister, A., Hübner, U., John., H., Schmelzeisen, R., and Mülhaupt, R. (2002). Fabrication of soft tissue engineering scaffolds

by means of rapid prototyping techniques, *J. Mater. Sci.*, **37**, pp. 3107–3116.

70. Lautenschläger, H. (2009). (Poly) Saccharides in cosmetic products-From Alginate to Xanthan Gum, *Kosmetische Praxis*, **4**, pp. 12–15.
71. Lee, H. J., Ahn, S. H., Chun, W., and Kim, G. H. (2014). Enhancement of cell viability by fabrication of macroscopic 3D hydrogel scaffolds using an innovative cell-dispensing technique supplemented by preosteoblast-laden micro-beads, *Carbohydr. Polym.*, **104**, pp. 191–198.
72. Lee, S. C., Kwon, K., and Park, K. (2013). Hydrogels for delivery of bioactive agents: a historical perspective, *Adv. Drug Deliv. Rev.* **65**, pp. 17–20.
73. Lee, S. J., Koo, H., Lee, D.-E., Min, S., Lee, S., Chen, X., Choi, Y., Leary, J. F., Park, K., Jeong, S. Y., Kwon, I. K., Kim, K., and Kuiwon, C. (2011). Tumor-homing photosensitizer-cojugated chitosan nanoparticles for synchronous photodynamic imaging and therapy based on cellular on/off system, *Biomaterials*, **32**, pp. 4021–4029.
74. Leor, J., Tuvia, S., Guetta, V., Manczur, F., Castel, D., Willenz, U., Petnehàzy, Ö., Landa, N., Feinberg, M. S., Konen, E., Goitein, O., Tsur-Gang, O., Shaul, M., Klapper, L., and Cohen, S. (2009). Intracoronary injection of in situ forming alginate hydrogel reverses left ventricular remodeling after myocardial infarction in swine, *J. Am. Coll. Cardiol.*, **54**, pp. 1014–1023.
75. Lim, H. J., Ghim, H. D., Choi, J. H., Chung, H. Y., and Lim, J. O. (2010). Controlled release of BMP-2 from alginate nanohydrogels emhanced osteogenic differentiation of human bone marrow stromal cells, *Macromol. Res.*, **18**, pp. 787–792.
76. Lindenhayn, K., Perka, C., Spitzer, R. S., Heilmann, H. H., Pommerening, K., Mennicke, J., and Sittinger, M. (1999). Retention of hyaluronic acid in alginate beads: aspects for *in vitro* cartilage engineering, *J. Biomed. Mater. Res.* (*A*) **44**, pp. 149–155.
77. Liu, L. S., Fishman, M. L., Kost, J., and Hicks, K. B. (2003). Pectin-based systems for colon-specific drug delivery via oral route, *Biomaterials*, **24**, pp. 3333–3343.
78. Liu, Q., Zhang, C., Zheng, X., Shao, X., Zhang, X., Zhang, Q., and Jiang, X. (2014). Preparation and evaluation of antigen/*N*-trimethylaminoethylmethacrylate chitosan conjugates for nasal immunization, *Vaccine*, **32**, pp. 2582–2590.
79. Lu, J. X, Prudhommeaux, F., Maunier, A., Sedel, L., and Guillemin, G. (1999). Effects of chitosan on rat knee cartilages, *Biomaterials*, **20**, pp. 1937–1944.

80. Madry, H., Rey-Rico, A., Venkatesan, J. K., Johnstone, B., and Cucchiarini, M. (2014). Transforming growth factor beta-releasing scaffolds for cartilage tissue engineering, *Tissue Eng. Part B: Reviews*, **20**, pp. 106–125.
81. Manchanda, R., Arora, S. C., and Manchanda, R. (2014). Tamarind seed polysaccharide and its modification–versatile pharmaceutical excipients: a review, *Int. J. Pharm. Tech. Res.*, **6**, pp. 412–420.
82. Maroni, A., Del Curto, M. D., Zema, L., Foppoli, A., and Gazzaniga, A. (2013). Film coating for oral colon delivery, *Int. J. Pharm*, **457**, pp. 372–395.
83. Marques, A. P., and Reis, R. L. (2005). Hydroxyapatite reinforcement of different starch-based polymers affects osteoblast-like cells adhesion/spreading and proliferation, *Mater. Sci.Eng. C*, **25**, pp. 215–229.
84. Matricardi, P., Di Meo, C., Coviello, T., Hennink, W. E., and Alhaique, F. (2013). Interpenetrating polymer networks polysaccharide hydrogels for drug delivery and tissue engineering, *Adv. Drug Deliv. Rev.*, **65**, pp. 1172–1187.
85. Mayol, L., De Stefano, D., Campani, V., De Falco, F., Ferrari, E., Cencetti, C., Matricardi, P., Maiuri, L., Carnuccio, R., Gallo, A., Maiuri, M. C., and De Rosa, G. (2014). Design and characterization of a chitosan physical gel promoting wound healing, *J. Mater. Sci. Mater. Med.* DOI 10.1007/s10856-014-5157-5.
86. Mazzuca, C., Micheli, L., Carbone, M., Basoli, F., Cervelli, E., Iannuccelli, S., Sotgiu, S., and Palleschi, A. (2014) Gellan hydrogel as a powerful tool in paper cleaning process: a detailed study, *J Colloid Interface Sci.*, **416**, pp. 205–211.
87. Metaxa, A. F., Efthimiadou, E. K., Boukos, N., and Kordas, G. (2012). Polysaccharides as a source of advanced materials: cellulose hollow microspheres for drug delivery in cancer therapy, *J. Coll. Interf. Sci.*, **384**, pp. 198–206.
88. Montanari, E., Capece, S., Di Meo, C., Meringolo, M., Coviello, T., Agostinelli, E., and Matricardi, P. (2013). Hyaluronic acid nanohydrogels as a useful tool for BSAO immobilization in the treatment of melanoma cancer cells, *Macromol. Biosci.*, **13**, pp. 1185–1194.
89. Morales-Ortega, A., Carvajal-Millan, E., Brown-Bojorquez, F., Rascón-Chu, A., Torres-Chavez, P., López-Franco, Y. L., Lizardi-Mendoza, J., Martinez-López, A. L., and Campa-Mada, A. C. (2014). Entrapment of probiotics in water extractable arabinoxylan gels: rheological and microstructural characterization, *Molecules*, **19**, pp. 3628–3637.

90. Motwani, S. K., Chopra, S., Talegaonkar, S., Kohli, K., Ahmad, F. J., and Khar, R. K. (2008). Chitosan–sodium alginate nanoparticles as submicroscopic reservoirs for ocular delivery: formulation, optimisation and *in vitro* characterisation, *Eur. J. Pharm. Biopharm.*, **68**, pp. 513–525.

91. Naskar, S., Roy, S. K., and Kuotsu, R. (2013). Drug delivery based on buccal adhesive systems: a review, *Int. J. Pharm. Bio Sci.*, **4**, pp. 240–256.

92. Owen, D. H., Dunmire, E. N., Pleny, A. M., and Katz, D. F. (1999). Factors influencing nonoxynol-9 permeation and bioactivity in cervical mucus, *J. Contr. Rel.*, **60** pp. 23–34.

93. Ozpolat, B., Sook, A. K., and Lopez-Berestein, G. (2010). Nanomedicine based approaches for the delivery of SiRNA in cancer, *J. Intern. Med.* **267**, pp. 44–53.

94. Pachuau, L., and Mazumder, B. (2013). Colonic drug delivery systems based on natural polysaccharides and their evaluation, *Mini-Rev. Med. Chem.* **13**, pp. 1982–1991.

95. Palmeira-De-Oliveira, A., Ribeiro, M. P., Palmeira-De-Oliveira, R., Gaspar, C., Costa-De-Oliveira, S., Correia, I. J., Pina Vaz, C., Martinez-De-Oliveira, J., Queiroz, J. A., and Rodrigues, A. G. (2010). Anti-Candida activity of a chitosan hydrogel: mechanism of action and cytotoxicity profile, *Gyn. Obst. Invest.*, **70**, pp. 322–327.

96. Park, K. M., Lee, S. Y., Joung, Y. K., Na, J. S., Lee, M. C., and Park, K. D. (2009). Thermosensitive chitosan–Pluronic hydrogel as an injectable cell delivery carrier for cartilage regeneration, *Acta Biomater.*, **5**, pp. 1956–1965.

97. Palumbo, F. S., Pitarresi, G., Fiorica C., Matricardi, P., Albanese, A., and Giammona, G. (2012). *In situ* forming hydrogels of new amino hyaluronic acid/benzoyl-cysteine derivatives as potential scaffolds for cartilage regeneration, *Soft Matter*, **8**, pp. 4918–4927.

98. Peppas, N. A., Bures, P., Leobandung, W., and Ichikawa, H. (2000). Hydrogels in pharmaceutical formulations, *Eur. J. Pharm. Biopharm.* **50**, pp. 27–46.

99. Pescosolido, L., Schuurman, W., Malda, J., Matricardi, P., Alhaique, F., Coviello, T., Van Weeren, P., Dhert, W., Hennink, W. E., and Vermonden, T. (2011). Hyaluronic acid and dextran based semi-IPN hydrogels as biomaterials for bioprinting, *Biomacromolecules*, 12, pp. 1831–1838.

100. Pignatello, R., Stancampiano, A. H. S., Ventura, C. A., and Puglisi, G. (2007). Dexamethasone sodium phosphate-loaded chitosan-based delivery systems for buccal application, *J. Drug Targeting.*, **15**, pp. 603–610.

101. Pignatello, R., Basile, L., and Puglisi, G. (2009). Chitosan glutamate hydrogels with local anesthetic activity for buccal application, *Drug Deliv.*, **16**, pp. 176–181.

102. Prados, J., Melguizo, C., Ortiz, R., Perazzoli, G., Cabeza, L., Àlvarez, P. J., Rodriguez-Serano, F., and Aranega, A. (2013). Colon cancer therapy: recent developments in nanomedicine to improve the efficacy of conventional chemotherapeutic drugs, *Anti-cancer Ag. Med. Chem.*, **13**, pp. 1204–1216.

103. Qi, H., Liu, X., Wang, K., Liu, D., Huang, L., Liu, S., and Zhang, Q. (2013). Subchronic toxicity study of ulvan from *Ulva pertusa* (Chlorophyta) in Wistar rats, *Food Chem. Toxicol.*, **62**, pp. 573–578.

104. Qi, M., McGarrigle, J. J., Wang, Y., Marchese, E., Bochenek, M. A., Vaca, P., Davis, M. E., Ahn, S. J., Schwartz, A., Strand, B., and Lacík, I. (2013). Transplantation of pancreatic islets immobilized in alginate-based microcapsules: from animal studies to clinical trials, *Micro Nanosys.*, **5**, pp. 186–193.

105. Ragelle, H., Vandermeulen, G., and Préat, V. (2013). Chitosan-based SiRNA delivery systems, *J. Contr. Rel.* **172**, pp. 207–218.

106. Rai, J. K., Sharma, U. K., Sharma, U., Agrahari, P. K., Rai, D., Kumar, R., and Singh, A. (2012). Prednisolone loaded pH sensitive hydrogel beads for colonic delivery, *Int. J. Pharm. Res.* **4**, pp. 94–98.

107. Raj, B. S., Shanthi, Nair, R. S., Samraj, P. I., and Vidya (2013). Formulation and evaluation of coated microspheres for colon targeting, *J. Appl. Pharm. Sci.*, **3 (S1)**, pp. S68–S74.

108. Rassu, G., Gavini, E., Jonassen, H., Zambito, Y., Fogli, S., Breschi, M. C., and Giunchedi, P. (2009). New chitosan derivatives for the preparation of rokitamycin microspheres designed for ocular or nasal administration, *J. Pharm. Sci.*, **98**, pp. 4852–4865.

109. Ribeiro, L. N. M., Alcântra, A. C. S., Darder, M., Aranda, P., Araújo-Moreira, F. M., and Ruiz-Hirtzky, E. (2014). Pectin-coated chitosan-LDH bionanocomposite beads as potential systems for colon-targeted drug delivery, *Int. J. Pharm.* **463**, pp. 1–9.

110. Richardson, J. L., and Illum, L. (1992). The vaginal route of peptide and protein drug delivery, *Adv. Drug. Deliv. Rev.*, **8**, pp. 341–366.

111. Richardson, J. L., Illum, L., and Thomas, N. W. (1992). Vaginal absorption of insulin in the rat: effect of penetration enhancers on insulin uptake and mucosal histology, *Pharm. Res.*, **9**, pp. 878–883.

112. Robbins, D. J., Moulton, J. E., and Booth, A. N. (1964). Subacute toxicity study of a microbial polysaccharide fed to dogs, *Food Cosmetics Toxicol.*, **2**, pp. 545–550.

113. Rohan, L. C., and Sassi, A. B. (2009). Vaginal drug delivery systems for HIV prevention, *AAPS J.*, **11**, pp. 78–87.

114. Rossi, S., Ferrari, F., Bonferoni, M. C., Sandri, G., Faccendini, A., Puccio, A., and Caramella, C. (2014). Comparison of poloxamer- and chitosan-based thermally sensitive gels for the treatment of vaginal mucositis, *Drug Develop. Ind. Pharm.*, **40**, pp. 352–360.

115. Rubinstein, A. (2005). Colonic drug delivery, *Drug Discovery Today: Technologies*, **2**, pp. 33–37.

116. Sajeesh, S., Bouchemal, K., Marsaud, V., Vauthier, C., and Sharma, C. P. (2010). Cyclodextrin complexed insulin encapsulated hydrogel microparticles: an oral delivery system for insulin, *J. Contr. Rel.*, **147**, pp. 377–384.

117. Sá-Lima, H., Caridade, S. G., Mano J. F., Rui, L., and Reis, R. L. (2010). Stimuli-responsive chitosan-starch injectable hydrogels combined with encapsulated adipose-derived stromal cells for articular cartilage regeneration, *Soft Matter*, **6**, pp. 5184–5195.

118. Sandolo, C., Péchiné, S., Le Monnier, A., Hoys, S., Janoir, C., Coviello, T., Alhaique, F., Collignon, A., Fattal, E., and Tsapis, N. (2011). Encapsulation of Cwp84 into pectin beads for oral vaccination against Clostridium difficile, *Eur. J. Pharm. Biopharm.*, **79**, pp. 566–573.

119. Sapir, Y., Kryukov, O., and Cohen, S. (2011). Integration of multiple cell-matrix interactions into alginate scaffolds for promoting cardiac tissue regeneration, *Biomaterials*, **32**, pp. 1838–1847.

120. Saravanakumar, G., Deepagan, V. G., Jayakumar, R., and Park, J. K. (2014). Hyaluronic acid-based conjugates for tumor-targeteg drug delivery and imaging. *J. Biomed. Nanotechnol.*, **10**, pp. 17–31.

121. Scharp, D. W., and Marchetti, P. (2014). Encapsulated islets for diabetes therapy: History, current progress, and critical issues requiring solution. *Adv. Drug Deliv. Rev.*, **67–68**, pp. 35–73.

122. Schneider, S., Feilen, P. J., Slotty, V., Kampfner, D., Preuss, S., Berger, S., Beyer, J., and Pommersheim, R. (2001). Multilayer capsules: a promising microencapsulation system for transplantation of pancreatic islets, *Biomaterials*, **22**, pp. 1961–1970.

123. Şenel, S., İkinci, G., Kaş, S., Yousefi-Rad, A., Sargon, M. F., and Hincal, A. A. (2000). Chitosan films and hydrogels of clorhexidine gluconate for oral mucosal delivery, *Int. J. Pharm.*, **193**, pp. 197–203.

124. Şenel, S., Kremer, M. J., Kaş, S., Wertz, P. W., Hincal, A. A., and Squirer, C. A. (2000). Enhancing effect of chitosan on peptide drug delivery across buccal mucosa, *Biomaterials*, **21**, pp. 2067–2071.

125. Shepherd, R., (2011). Polysaccharides, in *Food Additives Data Book*, 2nd ed. (Smith, J., and Hong-Shum, L., eds.). Wiley-Blackwell, Oxford, UK. DOI: 10.1002/9781444397741.ch10.
126. Shukla, R. K., and Tiwari, A. (2012). Carbohydrate polymers: applications and recent advances in delivering drugs to the colon, *Carbohydr. Polym.* **88**, pp. 399–416.
127. Silva, G. A., Coutinho, O. P., Ducheyne, P., Irving, M. Shapiro, I. M., Rui, L., and Reis, R. L. (2007). The effect of starch and starch-bioactive glass composite microparticles on the adhesion and expression of the osteoblastic phenotype of a bone cell line, *Biomaterials*, **28**, pp. 326–334
128. Simmons, C. A., Alsberg, E., Hsiong, S., Kim, W. J., and Mooney, D. J. (2004). Dual growth factor delivery and controlled scaffold degradation enhance in vivo bone formation by transplanted bone marrow stromal cells, *Bone*, **35**, pp. 562–569.
129. Singh, B., and Bala, R. (2014). Polysaccharide based hydrogels as controlled drug delivery system for GIT cancer, *Int. J. Biol. Macromol.*, **65**, pp. 524–533.
130. Singh, R. M., Kumar, A., and Pathak, K. (2013). Mucoadhesive *in situ* nasal gelling drug delivery systems for modulated drug delivery, *Exp. Opin. Drug. Del.*, **10**, pp. 115–130.
131. Sinha, V. R., and Kumria, R. (2001). Polysaccharides in colon-specific dug delivery, *Int. J. Pharm.*, **224**, pp. 19–38.
132. Smitha, K. T., Anitha, A., Furuike, T., Tamura, H., Nair, S. V., and Yajakumar, R. (2013). In vitro evaluation of paclitaxel loaded amorphous chitin nanoparticles for colon cancer drug delivery, *Coll. Surf. B: Biointerf.*, **104**, pp. 245–253.
133. Sreeja, C. N., Mable, S., and Anoop, K. R. (2014). *In situ* gel: an innovative approach for safe and sustained nasal drug delivery, *Int. J. Pharm.Sci. Rev. Res.*, **24**, pp. 1–7.
134. Suh, J. K. F., and Matthew, H. W. T. (2000). Application of chitosan-based polysaccharide biomaterials in cartilage tissue engineering: a review, *Biomaterials*, **21**, pp. 2589–2598.
135. Sun, G., Owens, D. M., and Mao, J. J. (2014). Scareless skin regeneration: are we there yet? *JSM Regen. Med.*, **2**, pp. 1007–1008.
136. Swamy, N. G. N., and Abbas, Z. (2012). Mucoadhesive *in situ* gels as nasal drug delivery systems: an overview, *Asian J. Pharm. Sci.*, **7**, pp. 168–180.
137. Tiera, M. J., Shi. Q., Barbosa, H. F. G., and Fernandes, J. C. (2013). Polymeric systems as nanodevices for SiRNA delivery, *Curr. Gene Ther.*, **13**, pp. 358–369.

138. Tokatlian, T., and Segura, T. (2010). SiRNAapplications in nanomedicine, *Wiley Interdiscip. Rev. Nanomed. Nanobiotechnol.*, **2**, pp. 305–315.

139. Touitou, E., and Illum, L., (eds.) (2013). *Drug Delivery Translational Res.*, **3**(1), pp. 1–109.

140. Tsur-Gang, O., Ruvinov, E., Landa, N., Holbova R., Micha, S., Feinberg, M. S., Leor J., and Cohen, S. (2009). The effects of peptide-based modification of alginate on left ventricular remodeling and function after myocardial infarction, *Biomaterials*, **30**, pp. 189–195.

141. Valenta, C. (2005). The use of mucoadhesive polymers in vaginal delivery, *Adv. Drug. Deliv. Rev.*, **57**, pp. 1692–1712.

142. Verhulsel, M., Vignes, M., Stéphanie Descroix, S., Malaquin, L., Vignjevic, D. M., and Viovy, J. L. (2014). A review of microfabrication and hydrogel engineering for micro-organs on chips, *Biomaterials*, **35**, pp. 1816–1832.

143. Vinatier, C., Magne, D., Weiss, P., Trojani, C., Rochet, N., Carle, G. F., Vignes-Colombeix, C., Chadjichristos, C., Galera, P., Daculsi, G., and Guicheux, J. (2005). A silanized hydroxypropyl methylcellulose hydrogel for the three-dimensional culture of chondrocytes, *Biomaterials*, **26**, pp. 6643–6651.

144. Vinatier, C., Magne, D., Moreau, A., Gauthier, O., Malard, O., Vignes-Colombeix, C., Daculsi, G., Weiss, P., and Guicheux, J. (2007). Engineering cartilage with human nasal chondrocytes and a silanized hydroxypropyl methylcellulose hydrogel, *J. Biomed. Mater. Res.*, **80A**, pp. 66–74.

145. Vinatier, C., Gauthier, O., Fatimi, A., Merceron, C., Masson, M., Moreau, A., Moreau, F., Fellah, B., Weiss P., and Guicheux, J. (2009). An injectable cellulose-based hydrogel for the transfer of autologous nasal chondrocytes in articular cartilage defects, *Biotechnol. Bioeng.*, **102**, pp. 1259–1267.

146. Wang, H.-M., Chou, Y.-T., Wen, Z.-H., Wang, Z.-R., Chen, C.-H., and Ho, M.-L. (2013). Novel biodegradable porous scaffold applied to skin regeneration, *PLoS ONE*, **8**, Article no. e56330.

147. Wang, X., Liu, G., Ma, J., Guo, S., Gao, L. Jia, Y., Li, X., and Zhang, Q. (2013). *In situ* gel-forming system: an attractive alternative for nasal drug delivery, *Crit. Rev. Ther. Drug Carrier Syst.*, **30**, pp. 411–434.

148. Wang, Y., and Wang, P. G. (eds). (2013). Polysaccharide-based systems in drug and gene delivery, *Adv. Drug Deliv. Rev.*, **65**, pp. 1121–1282.

149. Widjaja, L. K., Bora, M., Chan, P. N., Lipik, V., Wong, T. T., and Venkatraman, S. S. (2013). Hyaluronic acid-based nanocomposite

hydrogels for ocular drug delivery applications, *J. Biomed. Mater. Res. A*, **102**, pp. 3056–3065.

150. Woitiski, C. B., Neufeld, R. J., Veiga, F., Carvalho, R. A., and Figueiredo, I. V. (2010). Pharmacological effect of orally delivered insulin facilitated by multilayered stable nanoparticles, *Eur. J. Pharm. Sci.*, **41**, pp. 556–563.

151. Wong, V. W, Rustad, K. C., Glotzbach, J. P., Sorkin, M., Inayathullah, M., Longaker, M. T., Rajadas, J., and Gurtner, G. C. (2011). Pullulan hydrogels improve mesenchymal stem cell delivery into high-oxidative-stress wounds, *Macromol. Biosci.*, **11**, pp. 1458–1466.

152. Wong, V. W, Rustad, K. C., Galvez, M. G., Neofytou, E., Glotzbach, J. P., Januszyk, M., Major, M. R., Sorkin, M., Inayathullah, M., Longaker, M. T., Rajadas, J., and Gurtner, G. C. (2011). Engineered pullulan-collagen composite dermal hydrogels improve early cutaneous wound healing, *Tissue Eng. A.*, **17**, pp. 631–644.

153. Xiao, B., and Merlin, D. (2012). Oral colon-specific therapeutic approaches toward treatment of inflammatory bowel disease, *Exp. Opin Drug Del.*, **9**, pp. 1393–1407.

154. Xu, X., Weng, Y., Xu, L., and Chen, H. (2013). Sustained release of avastin® from polysaccharides cross-linked hydrogels for ocular drug delivery, *Int, J. Biol. Macromol.*, **60**, pp. 272–276.

155. Yalavarthi, P. R., Vulava, J., Vadlamudi, H. C., Balambhaigari, R. Y., and Nair, R. (2013). Modified pulsincap of ibuprofen-a novel approach for chronotherapy, *Curr, Drug Deliv.*, **10**, pp. 299–308.

156. Yamane, S., Iwasaki, N., Majima, T., Funakoshi, T., Masuko, T., Harada, K., Minami, A., Kenji Monde, K., and Nishimura, S. (2005). Feasibility of chitosan-based hyaluronic acid hybrid biomaterial for a novel scaffold in cartilage tissue engineering, *Biomaterials*, **26**, pp. 611–619.

157. Yoon, H. Y., Koo, H., Choi, K. Y., Lee, S. J., Kim, K., Kwon, I. C., Leary, J. F., Park, K., Yuk, S. H., Park, J. H., and Choi, K. (2012). Tumor-targeting hyaluronic acid nanoparticles for photodynamic imaging and therapy, *Biomaterials*, **33**, pp. 3980–3989.

158. Yu, J., Gu, Y., Du, K. T., Mihardja, S., Sievers, R. E., and Lee, R. J. (2009). The effect of injected RGD modified alginate on angiogenesis and left ventricular function in a chronic rat infarct model, *Biomaterials*, **30**, pp. 751–756.

159. Yu, J., Du, K. T., Fang, Q., Gu, Y., Mihardja, S. S., Sievers, R. E., Wu, J. C., and Lee, R. J. (2010). The use of human mesenchymal stem cells encapsulated in RGD modified alginate microspheres in the repair of myocardial infarction in the rat, *Biomaterials*, **31**, pp. 7012–7020.

160. Zhang, C., Chen, J. D., and Yang, F.-Q. (2014). Konjac glucomannan, a promising polysaccharide for OCDDS, *Carbohydr. Polym.* **104**, pp. 175–181.

161. Zhang, Y., Ni, M., Zhang, M., and Ratner, B. (2003). Calcium phosphate-chitosan composite scaffolds for bone tissue engineering, *Tissue Eng. Part A*, **9**, pp. 337–345.

162. Zheng, X. F., Lian, Q., and Lian Song, S. T. (2013). Chitosan-gelatin-pectin composite films from polyion-complex hydrogels, *Asian J. Chem.*, **25**, pp. 5363–5366.

163. Zhi, Z.-L., Liu, B., Jones, P. M., and Pickup, J. C. (2010). Polysaccharide multilayer nanoencapsulation of insulin-producing β-cells grown as pseudoislets for potential cellular delivery of insulin, *Biomacromolecules*, **11**, pp. 610–616.

Chapter 2

Hydrogels

Massimiliano Borgogna, Eleonora Marsich, Ivan Donati, Sergio Paoletti, and Andrea Travan

Department of Life Sciences, University of Trieste, Via Licio Giorgieri 5, I-34127 Trieste, Italy

biomat@units.it

2.1 Definition

Defining the concept of "hydrogel" is a very challenging task. Most of the times, hydrogels are framed with respect to their use, especially in the field of regenerative medicine. Along this line, hydrogels are materials for the treatment of wounds, for the design of contact lenses, or they are synthetic substitutes for the regeneration of natural soft tissues (e.g., cartilage).

Nowadays, a precise and comprehensive definition of hydrogel is still lacking from medical and pharmaceutical encyclopedias. Flory's definition of gels, including hydrogels, attests to their complexity; it is rigorous, but very difficult to apply on a routine basis [48].

Moreover, it should be pointed out that while, strictly speaking, the term *hydrogel* should necessarily imply the presence of water

Polysaccharide Hydrogels: Characterization and Biomedical Applications
Edited by Pietro Matricardi, Franco Alhaique, and Tommasina Coviello

ISBN 978-981-4613-61-3 (Hardcover), 978-981-4613-62-0 (eBook)
www.panstanford.com

(*hydro-*) as a component, it is now generally applied to gels containing any liquid as a solvent.

The modern history of hydrogels traces back to the fifties, with the synthesis of a semisolid 3D structure based on copolymers of 2-hydroxyethyl methacrylate and ethylene dimethacrylate [142]. Such systems have been proposed and largely used for soft contact lenses. Their commercial success has boosted interest for hydrogels.

In general terms and according to a commonly used definition, hydrogels are composed of an insoluble polymer which absorbs a large amount of solvent, or else it is simply a solvent-swollen polymer network. Another definition has been proposed by Hoffman and it pictures a hydrogel as a permanent or chemical gel stabilized by covalently crosslinked networks [66]. According to the definition of the *Webster New Twentieth Century Dictionary*, gel is a "jellylike substance formed by a colloidal solution in its solid phase," where jelly is defined as a "soft, resilient, partially transparent, semisolid, gelatinous food resulting from the cooling of fruit juice boiled with sugar, or of meat juice cooked down."

As early as 1861, Thomas Graham used the following words to address the problem of the definition of hydrogels: "While the rigidity of the crystalline structure shuts out external expressions, the softness of the gelatinous colloid partakes of fluidity, and enables the colloid to become a medium for liquid diffusion, like water itself" [55]. He further introduced the terms *hydrosol* and *hydrogel* for the liquid and gelatinous hydrates of silicic acid.

In 1926 Dorothy Jordan Lloyd wrote, "The colloid condition, the gel, is one which is easier to recognize than to define" [70]. Or, in other words, "if it looks Jello, it must be a gel" [131]. Jordan Lloyd recognized the difficulties hidden in framing a precise definition saying that "the limits between gel and sol, on the one hand, and gel and what may be termed curd, on the other, are not precise, but consist of a gradual change." She proceeded further defining that hydrogels "must be built up from two components, one of which is a liquid . . . , and the other of which, the gelling substance proper . . . is a solid, i.e., it can maintain its form under the stress of its own weight." Regarding hydrogels, she was convinced that "there is little doubt . . . that they all possess a solid phase."

Hermans expanded the definition proposed by Jordan Lloyd, describing hydrogels as being constituted by at least two components and exhibiting mechanical properties of the solid state, but in addition he added that the "dispersed component and the dispersion medium extend themselves continuously throughout the whole system" [64]. Bungenberg de Jong underlined the solid characteristic of the hydrogel saying that it is "a colloidal system of solid character, in which the colloidal particles somehow constitute a coherent structure, the latter being interpenetrated by a (usually liquid) system consisting in kinetic units smaller than colloidal particles" [13].

Flory and Stockmayer attempted to identify some criteria to define the hydrogel state and introduced concepts such as infinite network, three-dimensional structure, coherence, connectedness. Interestingly enough, in this definition there is no indication of the minimum number of components of the system required for hydrogel to be classified as such. Hence, according to Flory and Stockmayer definition, unswollen rubbers are considered hydrogels [49, 124, 125].

Ferry, in view of the inherent complexity of the definition of the hydrogel, proposed its phenomenological description as "substantially diluted system which exhibits no steady state flow" [45].

Along this line, Burchard and Ross-Murphy framed a definition of hydrogels stressing that they "possess at least one property which can stand as the operational definition of a gel; they possess a plateau in the real part of the complex modulus extending over an appreciable window of frequencies—i.e. they are, or can be coaxed under appropriate conditions to be, viscoelastic solids" [14].

A definition of hydrogels based on phenomenological characteristics has recently been given by Amdal et al. [1], who stated that hydrogels possess the following characteristics: (i) consisting of "two or more components, one of which is a liquid, present in substantial quantity"; (ii) they are "a soft, solid or solid-like material"; (iii) "solid-like gels are characterized by the absence of an equilibrium modulus, by a storage modulus, G', which exhibits a pronounced plateau extending to times at least of the order of seconds, and a loss modulus, G'', which is considerably smaller that the storage modulus in the plateau region." It is to be noted that, according to this definition also, entangled systems are considered as hydrogels [72].

Recently Rogovina et al. pointed out that the lack of equilibrium modulus is observed for physical gels but this does not hold for chemical gels. In addition, the invariability of the storage modulus over a sufficiently wide range of frequency should not be regarded as absolute. Therefore, Rogovina proposed that "a gel is an elastic solid (material) composed of at least two components, one of which (polymer) forms a physically or chemically bonded three-dimensional network that occurs in a medium of another component, a liquid, wherein the amount of the liquid is sufficient for ensuring the elastic properties of the gel" [107]. Moreover, "a physical gel presents one or two yield points at stresses above which network degradation and gel transition from the solid to the fluid state takes place" [107].

To complete this short recapitulation of the definition of hydrogels, it should be reminded that a philosophical definition of gels has been provided by Nijenhuis [130] and re-proposed by Nishinari [95]: "A gel is a gel, as long as one cannot prove that it is not a gel."

The term *hydrogel* has, nowadays, widened to include several related systems such as weak gel, quasi gel, temporary gel, microgel, pseudogel, heterogel, and nanogel [1, 14, 26, 42, 64, 112]. In addition, hydrogels devoid of the liquid component are often referred to with hydrogel-related terms such as aerogel and xerogels [1].

2.2 Mechanical Properties of Hydrogels

2.2.1 *Structure and Compression Resistance*

The process which leads to the formation of a hydrogel is called gelation. This is based on intermolecular reactions among segments of a linear polymer which increase its molecular weight to a point in which it becomes infinite. At this point, defined as the gelation point, the relaxation time becomes infinite, there is no flow of the system,

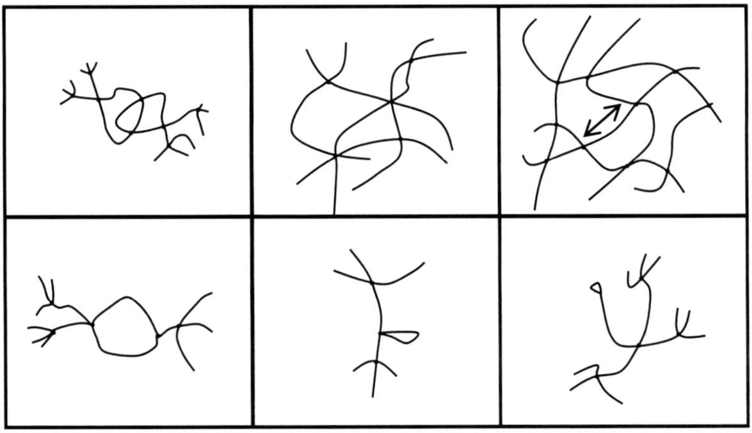

Figure 2.1 Possible hydrogel network structures. (a) Ideal network with tetrafunctional crosslinks. (b) Presence of multifunctional junctions. (c) Molecular entanglements. (d) Presence of unreacted functionalities (e, f) Presence of chain loops.

and the reticulated network spans the whole container in which it sits. The gel point, α_c, shows an inverse dependence from the degree of polymerization (DP) (Eq. 2.1) [47, 72].

$$\alpha_c \propto DP^{-1} \tag{2.1}$$

Structural and mechanical properties of hydrogels must be taken into consideration when selecting materials for specific applications in which mechanical stress could compromise the performance of the device. In general, these materials can be considered as highly hydrated constructs in which the polymeric chains are crosslinked by chemical or physical bonds; the network structure is composed of tetrafunctional crosslinks, multifunctional junctions, and molecular entanglements, which contribute to the mechanical properties of the hydrogel, at variance with unreacted functionalities and chain loops (Fig. 2.1).

In order to increase the mechanical resistance of hydrogels composed of more than one polymer, a multiple crosslinking process can be exploited (Fig. 2.2).

For several applications, it is important to evaluate the wet-state stability of hydrogels, which depends on the crosslinking density: the higher this density in the gel, the higher its mechanical stability.

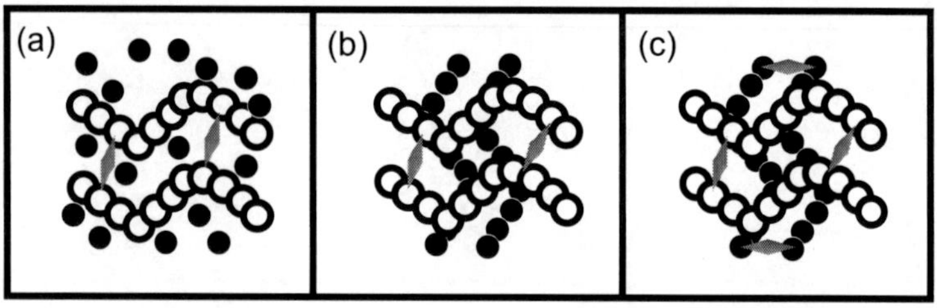

Figure 2.2 Schematic representation of a multiple crosslinking process. (a) Only one of the two species in solution is polymerized and crosslinked. (b) A second polymerization occurs leading to the formation of a semi-interpenetrated network. (c) A second crosslinking leads to the formation of a fully interpenetrated network.

Unconfined compression testing is a technique which allows to evaluate mechanical parameters such as compression modulus and ultimate strength of the hydrogels. This technique is complementary to rheological measurements that are employed to investigate the viscoelastic properties by measuring storage (G') and loss (G'') moduli.

Static compression tests apply an escalating compressive load until failure, or apply a specific load and hold it for a certain period, while dynamic compression tests involve cycling between two (or more) load conditions. Depending on the nature of the material, compression tests can help determine the material resistance, service life, or other critical performance characteristics.

When a hydrogel is placed under a compression force with a constant deformation rate and there is no limitation to the lateral expansion, a typical force–displacement curve is obtained (Fig. 2.3).

In these conditions, the hydrogel withstands the load by deforming itself up to a point where the applied stress becomes higher than the material strength, which results in sample failure. By referring to the geometrical characteristics of the sample (surface on which the force is applied, height of the specimen), the stress–strain curve of the hydrogel can be calculated. The compression modulus is represented by the slope of the initial linear part of the curve: the higher the slope, the higher the modulus, and the stiffer the hydrogel. The force increases in the nonlinear region until the maximum load is reached: at this point a crack is formed and starts

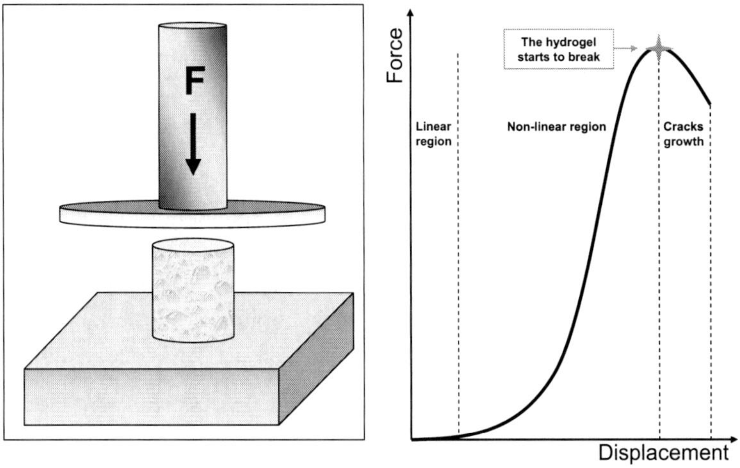

Figure 2.3 Compression testing of hydrogels and typical force–displacement curve for this class of materials.

to propagate within the structure. The slope of the curve in the third region indicates the rate at which hydrogel breaks apart: a mild slope corresponds to a slow and ductile breaking mode, while a sharp slope corresponds to a brittle fracture.

The mechanical characterization of hydrogels is of primary importance for several biomedical applications: for instance, hydrogels for tissue engineering should be able to offer sufficient mechanical support to cells; at the same time the stiffness of the support was shown to direct cellular activities like attachment, migration, and differentiation [62].

Considering the mechanical behavior, swollen hydrogels can be regarded as rubber-like systems. The relationship between the mechanical response and the gel network characteristics has been derived, according to the rubber elasticity theory, by means of a thermodynamic approach. This theoretical model assumes the mechanical response of the hydrogel upon application of a stress as deriving from the loss of configurational freedom of the (Gaussian) chains between the crosslinks in the network.

Assuming no volume variation upon application of the deformation, which is equivalent to saying that the macroscopic deformation on the gel is transmitted to the polymer chains that are constrained by the crosslinks, the shear-stress per unit area, τ, of an ideal rubber can be written as (Eq. 2.2):

$$\tau = \frac{\rho RT}{\overline{M}_c}\frac{\overline{r}_0^2}{\overline{r}_f^2}\left(\lambda - \frac{1}{\lambda^2}\right) \tag{2.2}$$

where λ is the extension ratio, ρ is the polymer density, and $\overline{M}_C$ is the average molecular weight between entanglements. r_0^2 and r_f^2 are the (squares of the) end-to-end distance in a real network and in the isolated chains, respectively. Equation 2.2 shows that the rubber elasticity theory predicts a nonlinear dependence of the shear-stress from the strain which is in accordance with experimental data only when λ is small (deformation <20%). Further, Eq. 2.2 neglects the existence of non-elastically active chains (loops) or loose ends. In ideal systems, all the molecules composing the rubber are crosslinked to form an assembly of mutually interconnected segments or network branches. However, in real systems several network imperfections are encountered. Among these, one can find interloping or entanglements of the chain, formation of closed loops, termination of the chains (loose ends). To take this into account, a correction to Eq. 2.2 is applied which reads (Eq. 2.3):

$$\tau = \frac{\rho RT}{\overline{M}_c}\frac{\overline{r}_0^2}{\overline{r}_f^2}\left(1 - \frac{2\overline{M}_c}{\overline{M}_n}\right)\left(\lambda - \frac{1}{\lambda^2}\right) \tag{2.3}$$

In this case, the shear modulus equals (Eq. 2.4):

$$G = \frac{\rho RT}{\overline{M}_c}\frac{\overline{r}_0^2}{\overline{r}_f^2}\left(1 - \frac{2\overline{M}_c}{\overline{M}_n}\right) \tag{2.4}$$

This expression, in the case of $\overline{M}_n >> \overline{M}_c$ and assuming that $\overline{r}_0^2/\overline{r}_f^2$ equals 1, reconciles with the well-known relation between the shear modulus, G, and the crosslink density, v_e, i.e., the moles of elastically active chains per unit volume (Eq. 2.5):

$$G = v_e RT = \frac{\rho RT}{\overline{M}_c} \tag{2.5}$$

Equation 2.5 was further modified by Flory introducing the so-called rubber-front factor, g, which accounts for deviations from the ideal rubber behavior (Eq. 2.6).

$$G = g\frac{\rho RT}{\overline{M}_c} \tag{2.6}$$

In these terms, the approach based on the rubber elasticity theory has been used to describe alginates hydrogels [38, 90, 134].

In the case of swollen networks, like hydrogels, the expression of the shear modulus has to be modified introducing the polymer volume fraction, Φ (Eq. 2.7):

$$G = \frac{\rho RT}{\overline{M}_c}\Phi^{1/2} \tag{2.7}$$

2.2.2 *Hydrogels with Improved Mechanical Behavior*

The design of hydrogels with adequate mechanical properties is of great importance in many application areas of soft materials. Highly swollen hydrogels normally do not possess high mechanical strength because of the lack of an efficient energy dissipation mechanism and irregular distribution of crosslinking points [11, 27].

Several attempts, such as topological gels and double network gels, have been made in recent years to design hydrogels with improved mechanical stability [129]. The techniques include adding polymeric particles into the gel network [149], incorporating carbon nanotubes [133] and inorganic clay [123], and forming (semi)-interpenetrating polymer networks [145] or double-network structures [54]. The use of toxic crosslinking agents represents a drawback in the case of hydrogels that are supposed to be in contact with human body.

Recently it has been shown that inducing a permanent deformation upon plastic compression leads to hydrogels with a large increase in strength; this mechanism is due to irreversible water exclusion and it was verified in the case of fibrin and collagen gels [5]. Haugh et al. demonstrated that the application for 30 minutes of unconfined compression to fibrin hydrogels led to a considerable increase of the compression modulus, which persists even after several weeks of incubation in cellular medium [62] (Fig. 2.4).

The dispersion of micro- and nano-sized particles and fibres in polymer networks is a technique that offers significant enhancements in the structural properties of hydrogels. Several reinforcements have been considered to increase stiffness and toughness. Recently, Zhou et al. developed a high-strength nanocomposite polyacrylamide (PAM) hydrogel containing chitosan nanofibres

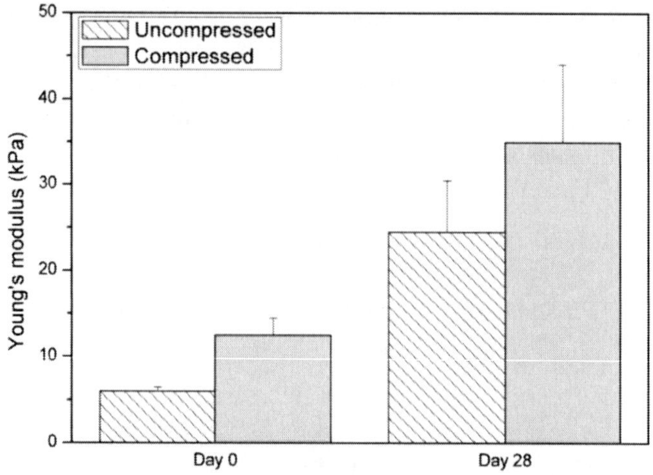

Figure 2.4 Plastic compression results in a stable increase in compressive modulus of fibrin hydrogels. Reprinted with permission from Ref. 62. Copyright © 2012, Elsevier.

[148]. Compression studies showed that the nanofibers efficiently distributed the load improving strength due to strong interfacial adhesion between PAM chains and the reinforcement which limited the growing of microcracks. In a similar study, Borges et al. prepared methacrylate-based hydrogels with nanofibrillated cellulose [9]; the low concentration of the fibers (maximum 1.6% wt) caused an eight-fold increase of the elastic modulus. Recently, Karasslan et al. [71] developed a poly(2-hydroxyethylmethacrylate) (PHEMA) matrix reinforced with cellulose whiskers resulting in hydrogels with enhanced toughness, increased viscoelasticity, and improved recovery behavior, with potential use in load-bearing biomedical applications. Turco et al. [135] found that alginate gels loaded with nano-hydroxyapatite (nHAp) showed an eight-fold increase of the compression strength of the material. Later, Chang et al. [22] introduced nHAp into chitin hydrogels obtaining a 10-fold increase of the same property for the chitin composite.

In general, the addition of micro- and nanofillers or the inducement of permanent deformations—to improve the mechanical properties of hydrogels—represent valuable strategies to tailor the resistance of the construct by engineering hydrogels with enhanced performance.

2.3 Interaction with Solvent: Swelling Properties

Due to their nature, hydrogels can absorb large quantities of solvent; in general these materials display considerable mechanical stability in their swollen state. Although their solvent content can be over 99%, hydrogels still retain the appearance and properties of solid materials [93].

The swelling capacity of a hydrogel depends on the space within the polymer network available for accommodating solvent molecules. In hydrogels, the solvent plays a dual role; it provides for a moist environment and controls the permeation through the network structure. A dried hydrogel, upon contact with a proper solvent, can swell immediately and absorb up to 1000 times its own weight. When a polysaccharide-based hydrogel is considered, the solvent used is typically water. Water in hydrogels can be operationally classified into four different states. Bound water is attached to the polymer chain by hydration of polar functional groups (hydrogen bonding) or ions and ionic groups (electrostriction). Interstitial water is not attached to the chains but is physically entrapped within the mesh. Free water is the outermost layer and can be removed by centrifugation or mechanical compression. Semibound water has intermediate properties between bound and free water. These water types can be identified by DSC or by NMR relaxometry measurements.

The swelling process depends on the rate at which solvent molecules can diffuse into the network structure and on the rate at which polymer chains relax. Thus, in the case of a highly crosslinked polymer, the swelling behavior is substantially governed by the diffusion mechanism, since polymer chains movements are limited by the high crosslink density.

The swelling properties of a hydrogel depend also on the polymer/water ratio; a low polymer/water ratio accounts for higher swelling capacity, while a high ratio increases the stability of the gel in the swollen state.

Swelling of a polysaccharide hydrogel is a consequence of polymer-water interactive forces which increase with the hydrophilic character of the macromolecules. When equilibrium conditions are attained for a hydrogel immersed in a fluid, two opposing forces take place: the polymer-solvent mixing contribution (thermodynamic) and the retractile forces which arise from the elastic response of the chains. In terms of total Gibbs free energy (ΔG_{total}), this is written as (Eq. 2.8):

$$\Delta G_{\text{total}} = \Delta G_{\text{mixing}} + \Delta G_{\text{elastic}} \tag{2.8}$$

In terms of chemical potential, it reads (Eq. 2.9)

$$\Delta \mu_{\text{total}} = \Delta \mu_{\text{mixing}} + \Delta \mu_{\text{elastic}} \tag{2.9}$$

The chemical potential of mixing of polymer chains and solvent is expressed by the Flory–Huggins solution theory as (Eq. 2.10) [47]:

$$\Delta \mu_{\text{mixing}} = RT \left[\ln (1 - \phi) + \phi + \chi \cdot \phi^2\right] \tag{2.10}$$

Here χ is the Flory–Huggins interactions parameter between solvent and the polymer chains and ϕ is the volume fraction of polymer in the gel.

At equilibrium, the chemical potential of the solvent outside and inside the hydrogel cancels out; the contributions for elastic retractile force equals the mixing term. Flory and Rehner have tackled the problem by evaluating the elastic retraction force by means of the theory of rubber elasticity and obtained an expression which allows determining the molecular weight between two crosslinks, $\overline{M}_{\text{c}}$ [50, 51, 101]. Peppas and Merril have extended the expression of Flory–Rehner for hydrogels prepared in the presence of solvent by considering changes in the elastic potential due to the solvent and proposed the following equation (Eq. 2.11) [102, 119]:

$$\frac{1}{\overline{M}_{\text{c}}} = \frac{2}{\overline{M}_n} - \frac{\left(\frac{\overline{v}}{V_1}\right)\left[\ln (1 - \phi_{\text{s}}) + \phi_{\text{s}} + \chi \phi_{\text{s}}^2\right]}{\left(\phi_{\text{s}}^{1/3} - \frac{\phi_{\text{s}}}{2}\right)} \tag{2.11}$$

where $\overline{M}_n$ is the molecular weight of the polymer chains in the absence of crosslinks, $\overline{v}$ the specific volume of the polymer, $v_{2,s}$ the polymer fraction in the swollen state, and V_1 the molar volume of water.

In the case of ionic hydrogels, an additional chemical potential term has to be taken into account arising from the contribution due to the charged groups in the network, $\Delta\mu_{\text{ionic}}$ (Eq. 2.12):

$$\Delta\mu_{\text{total}} = \Delta\mu_{\text{mixing}} + \Delta\mu_{\text{elastic}} + \Delta\mu_{\text{ionic}} \tag{2.12}$$

This was analytically derived by Slaughter and Peppas [119], and in the case of a hydrogel with anionic groups immersed in a solvent, it leads to (Eq. 2.13):

$$\frac{V_1}{4 \cdot I \cdot M_{\text{r}}}\left(\frac{v_{2,\text{s}}}{\bar{v}}\right)\left(\frac{K_{\text{a}}}{10^{-pH} + K_{\text{a}}}\right)^2 = \left[\ln\left(1 - v_{2,\text{s}}\right) + v_{2,\text{s}} + \chi_1 \cdot v_{2,\text{s}}^2\right] + \left(\frac{V_1}{\bar{v} \cdot \overline{M}_{\text{c}}}\right)\left(1 - \frac{2\overline{M}_{\text{c}}}{\overline{M}_n}\right) v_{2,\text{r}}\left[\left(\frac{v_{2,\text{s}}}{v_{2,\text{r}}}\right)^{1/3} - \left(\frac{v_{2,\text{s}}}{2v_{2,\text{r}}}\right)\right] \tag{2.13}$$

where I is the ionic strength used, K_{a} the dissociation constant for the acid group, M_{r} the molecular weight of the repeating unit, and $v_{2,\text{r}}$ the polymer volume fraction in the relaxed state. In this case, the 3D network is assumed to be composed of Gaussian chains: this represents a limitation for polysaccharide-based networks. Moreover, simple and generalized equations describing gel swelling in both salt-free and in high ionic concentration environments are lacking. Along this line, an equilibrium swelling equation was derived by Neufeld and coworkers accounting for the effect of non-ideal Gaussian chains and an ionic effect evaluated on the basis of the non-ideal Donnan equilibrium [19–21].

In general terms, the swelling tendency of a hydrogel increases with the content of ionic groups. This fact is due to the simultaneous increase of the number of counterions inside the gel, which produces an additional osmotic pressure which swells the gel [47]. Uptaken water due to counterion-driven osmotic swelling is essentially of the "free" type, while, in parallel, a nonnegligible amount of water is strongly "bound" around "hard" (e.g., sulfate) or "soft" (e.g., carboxylate) ionic groups, with a density higher than 1 [88]. From a general point of view, when only one type of charge is present, the ionic groups of the polymer repel each other, thus creating an expansion of the structure in aqueous solution allowing water absorption. The sum of these three components (polymer–water interaction, osmotic pressure, and electrostatic forces) induces hydrogel expansion upon water absorption; the network expansion

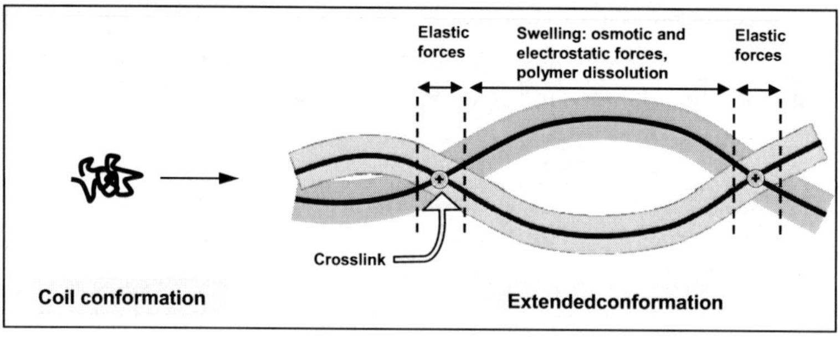

Figure 2.5 Swelling forces in hydrogels.

is counterbalanced by the elastic forces of the network crosslink junctions, which determines the equilibrium swelling capacity of the hydrogel (Fig. 2.5).

The osmotic pressure difference, Q_e, between the inside and outside of a charged hydrogel can be written as (Eq. 2.14) [94, 118]

$$Q_e \propto \alpha^{\frac{6}{5}} \cdot \nu_e^{\frac{4}{5}} \tag{2.14}$$

where α is the ionization degree. Along this line, the elastic modulus of the gel can be written as (Eq. 2.15)

$$E \propto \nu_e^{-\frac{8}{5}} \cdot \alpha^{-\frac{2}{5}} \tag{2.15}$$

The excess swelling with respect to the swelling of the corresponding non-ionic hydrogels can be reduced by increasing the salt concentration in the external solution, which decreases the difference of concentration of the counterions between inside and outside the hydrogel.

Durmaz et al. [41] studied the swelling behavior of ionic PAAm hydrogels of various charge densities in water and in aqueous sodium chloride solutions. The ionic co-monomer used in the hydrogel preparation was the sodium salt of 2-acrylamido-2-methylpropane sulfonic acid (AMPSNa), which dissociates completely over the entire pH range causing pH-independent swelling. The study pointed out that increasing the AMPSNa content from 0 to 80 mol% results in a 27-fold increase in the hydrogel volume in water, while in 1.0 M NaCl solution, the swelling ratio was almost independent on the ionic group content due to screening of charge interactions within the hydrogel.

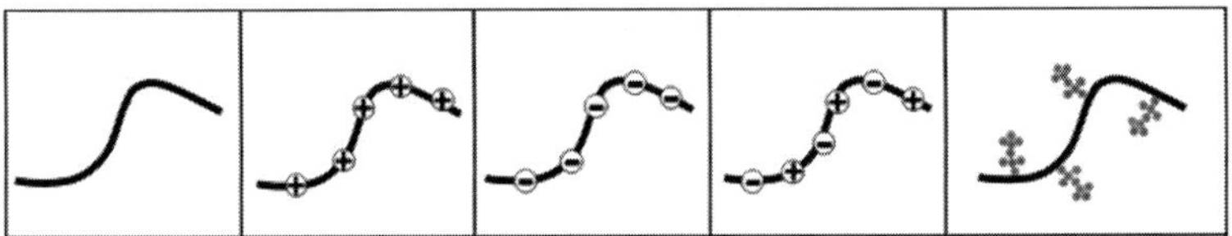

Non-ionic	Cationic	Anionic	Ampholytic	Hydrophobically modified
No pH dependency	Low pH favours repulsive forces	High pH favours repulsive forces	Low pH: cationic, high pH: anionic	Temperature increase favours aggregation of hydrophobic groups

Figure 2.6 Influence of pH on different hydrogel structures.

The swelling of ionic hydrogels depend on the pH of the aqueous media: due to chain repulsion, cationic hydrogels display increased swelling in acidic media while anionic hydrogels swell more in basic media. In the case of polymers with hydrophilic backbone and hydrophobic pendant groups, the aggregation in aqueous solution changes with temperature (Fig. 2.6).

The determination of the swelling equilibrium allows calculating the correlation length, d, as the linear distance between two adjacent crosslinks as (Eq. 2.16):

$$d = \varphi_{\mathrm{s}}^{-1/3} \cdot \left(\frac{2C_{\mathrm{n}}\overline{M}_{\mathrm{c}}}{M_{\mathrm{r}}} \right)^{1/2} \cdot l \tag{2.16}$$

where C_n is the Flory characteristic ratio and l the geometric length of the repeating unit. d is often used to describe the size of the pores in hydrogels and its evaluation, together with its dependence from the network parameters, allows tailoring the diffusion properties of the hydrogels.

From the experimental point of view, gel pore size can be determined by means of thermoporosimetry and by BSA diffusion. In the latter case, the diffusion through the gel (D_{gel}) of a model compound, the protein BSA, is related to its diffusion in water, $D_{\mathrm{H_2O}}$, according to the following equation:

$$D_{\mathrm{gel}} = D_{\mathrm{H_2O}} \cdot e^{\left[-\pi \frac{r_{\mathrm{H}}+r_{\mathrm{f}}}{d+2r_{\mathrm{f}}}\right]} \tag{2.17}$$

where r_{f} is the radius (cross section) of the polymer chain and r_{H} the hydrodynamic radius of BSA. Neuman et al. have confronted the theoretical predictions and the experimental results obtained from BSA diffusion and found a good agreement among the two [19].

2.4 Classification of the Gels

Several classifications of hydrogels have been proposed in the literature [95]. The first one was reported by Flory in 1974 [48] on the basis of structural criteria and included:

(i) Well-ordered lamellar structures, including gel mesophases, such as soap gels and inorganic gels formed from minerals and clays;
(ii) Covalent polymeric networks completely disordered, such as vulcanized rubber or polyacrylamide gels in which network is formed by interchain covalent crosslinks;
(iii) Polymer networks formed through physical aggregation; predominantly disordered, but with regions of local order, including gelatin, agar, carrageenan, methylcellulose gels where the junctions zones are believed to be formed by the aggregation of helices or stiff chains;
(iv) Particulate, disordered structures, which may include many gels of globular proteins where the network is formed by clusters of connected spheres.

In 1995, Keller discussed various gel structures in detail and illustrated two classes of hydrogels arising from liquid to crystalline phase transformation (crystallization) [73]:

(i) Junctions, bundles, or micelles, aggregated rigid chains which are connected by released chains;
(ii) The network elements are formed by fibrous crystals constituted by rigid or semiflexible chains that do not fold on crystallization.

Tanaka recently classified the gelation of hydrogels into six categories on the basis of the structure of the crosslinks [95, 128]:

(i) Simple gelation, where one single component of a polymer forms a gel by pairwise crosslinking in a solvent;

(ii) Gelation by multiple crosslinking, where more than three chains are linked in the junction point;
(iii) Gelation from two different polymers, which include interpenetrating networks, alternately crosslinked networks, and randomly mixed networks;
(iv) Gelation competing with hydration, and coexisting with hydration, where gelation is either blocked or accelerated by hydration;
(v) Gelation induced by polymer conformational transition such as coil–helix transition, coil–globule transition, and coil–rod transition;
(vi) Gelation coupled with liquid crystallization, where hydrogen-bonded mesogens form crosslinks.

It should be pointed out that all above definitions fail to clearly address important ionic hydrogels like calcium alginate. A classification of hydrogels has also been proposed on the basis of their application in the field of tissue engineering, and it comprises hydrogels as space-filling agents, as delivery vehicles for bioactive molecules, and as three-dimensional structures for tissue regeneration [40]. Space-filling agents are used in applications such as bulking and biological glues. At variance, hydrogels can be designed to encapsulate secretory cells to deliver specific drugs or hormones, such as in the case of Langerhans islets entrapped in alginate. Finally, a third application of hydrogels in tissue engineering envisages a three-dimensional network, to be transplanted into the body, to support cell growth and tissue integration.

Other classifications of hydrogels are based on optical properties of the networks or on thermal reversibility of the gel forming process. However, the most commonly used classification is based on the type of crosslinks and divides hydrogels into physical and chemical networks.

2.4.1 *Chemical Gels*

Chemical gels are based on the formation of covalent bonds among linear high-molecular-weight polymer chains. A chemical gel can be regarded as a true macromolecule with an infinite molecular weight and relaxation time. Chemical gels are thermo-irreversible and can be prepared according to two strategies: swelling of pre-crosslinked polymer or crosslinking of oligomers in a medium. Typically, chemical gels display higher mechanical resistance than physical ones and they further possess an equilibrium modulus of elasticity [72]. The mechanical breakage of the chemical hydrogels is caused by the cleavage of the covalent bonds induced by the application of an excessive stress.

Focusing on polysaccharide-based systems, several strategies have been exploited to form chemical hydrogels through covalent bonds [15]. Examples include disulfide bonds, photopolymerization reactions, oxidation of tyramine conjugates [79], thiol-ene couplings [89], and Diels–Alder reactions [127]. An excellent example of the versatility of polysaccharides for the formation of chemical hydrogels is given by hyaluronan (Fig. 2.7) [143].

Photopolymerization is a form of radical polymerization induced by the use of UV or visible light which leads to the formation of an in situ hydrogel. The key points of photo crosslinked hydrogels, photoinitiators and photopolymerizable groups have been recently reviewed and discussed [93, 138]. Among the photopolymerizable groups, acrylate and methacrylate moieties are the most commonly encountered. Photopolymerization has found numerous application in the biomedical field since Hubbell's work [115]. The main reason for that is the fast kinetics of the process and of the photocuring and its compatibility with laparoscopic and minimally invasive surgical techniques [12, 43, 139]. One of the drawbacks of this process is connected with the formation of radical species from photoinitiator decomposition. Another drawback of photopolymerized hydrogels is represented by the polymeric species formed during the process (poly(acrylates) and poly(methacrylates)) which are not biodegradable/bioerodable and their kidney clearance depends on the molecular weight. This approach has been applied to several polysaccharide based photocrosslinkable hydrogels. As an example, hydrogels based on methacrylated-dextran and methacrylated-dextran-HEMA-dimethylaminoethyl have been proposed. In addition, alginate and hyaluronan based photocrosslinkable hydrogles have been reported for protein delivery and tissue engineering. Photopolymerized methacrylate dextran, in combination with

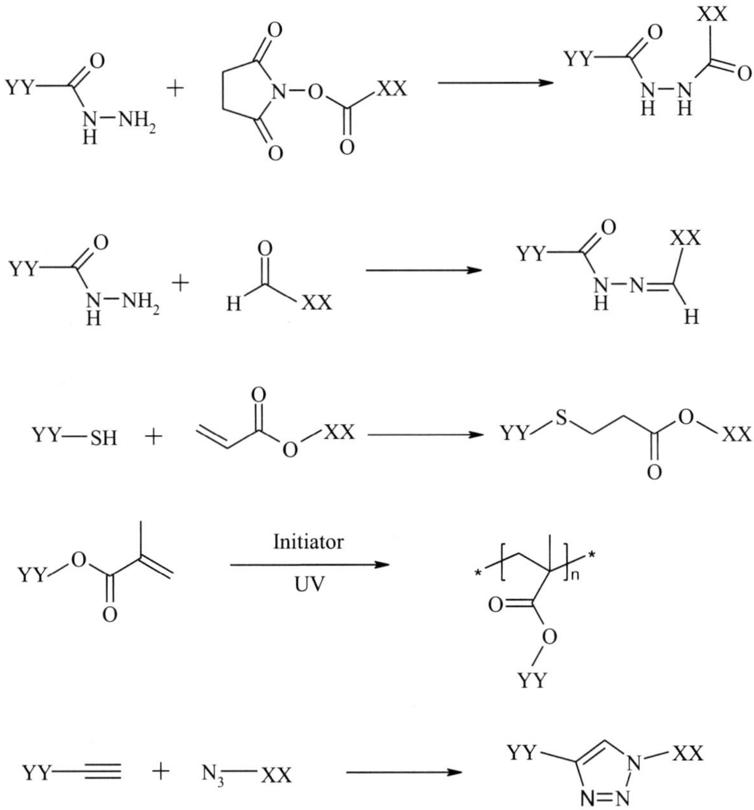

Figure 2.7 Examples of chemical reactions used for gel formation. XX and/or YY represent hyaluronan.

alginate, was shown to be an efficient method for preserving the enzymatic activity of horseradish peroxidase [103, 139]. In addition, photopolymerized alginate capsules, prepared starting from the polysaccharide bearing methacrylate moieties exclusively on mannuronic residues, was proven to be biocompatible with the encapsulation of human Langerhans islets and to account for a notable increase in swelling stability [109].

Another interesting approach leading to the formation of chemical hydrogels makes use of the well-known Michael's addition [139]. This chemical process is compatible with the aqueous medium and takes place at room temperature and at physiological pH values [86, 139]. A wide class of olefins and alkenes are available as Michael acceptor. Among these, acrylate esters, acrylonitrile, acrylamides, maleimides, alkyl methacrylates, cyanoacrylates, and vinyl sulfones are the most common ones. As Michael donors, apart from carbon-based nucleophiles, groups bearing heteroatoms like nitrogen and sulfur are of particular importance for the formation of hydrogels. In many gel-related applications, the higher selectivity of Michael acceptors for thiols than for amines is exploited [53]. By means of the Michael addition, hydrogels were prepared for load bearing applications as substitutes of cartilage and for tissue engineering purposes [82, 105, 140]. Hyaluronan modified with 2-aminoethyl methacrylate was reticulated using sulfhydryl groups in the crosslinker molecule and used for sustained protein delivery [65]. Along the same line, acrylated hyaluronan was reticulated through Michael's addition by means of a thiolated PEG and the obtained hydrogel, when bone morphogenetic protein-2 was incorporated within the network, showed a high in vivo mature bone formation [74]. Another example was based on a methacrylated hyaluronan and used thiol-containing peptides as crosslinkers; the gel thus obtained accounted for an elevated plasma concentration of EPO for 7 days [59, 60]. Thiolated hyaluronic acid was used for a Michael addition with a methacrylated triblock copolymer leading to an injectable hydrogel [17]. Similarly, a thiolated hyaluronan has been used to prepare injectable hydrogels with a hyperbranched methacrylate-bearing PEG system [39]. A nontoxic in situ forming scaffold based on thiolated glycol chitosan and vinyl sulfone-modified PEG was tested for chondrocyte culture [81]. Finally, a hyaluronan based hydrogel composed of aminoethyl methacrylated HA and thiolated HA was obtained through a Michael addition and it was reported to control the release of human growth hormone for up to 4 weeks [144].

A novel strategy for the preparation of chemical hydrogels is based on a native chemical ligation (Fig. 2.8).

This involves the reaction between a thioester and an N-terminal cysteine resulting eventually in a native peptide bond [33]. Native ligation has a high chemoselectivity and takes place under physiological conditions without toxic reagents or catalysts. Thus, it

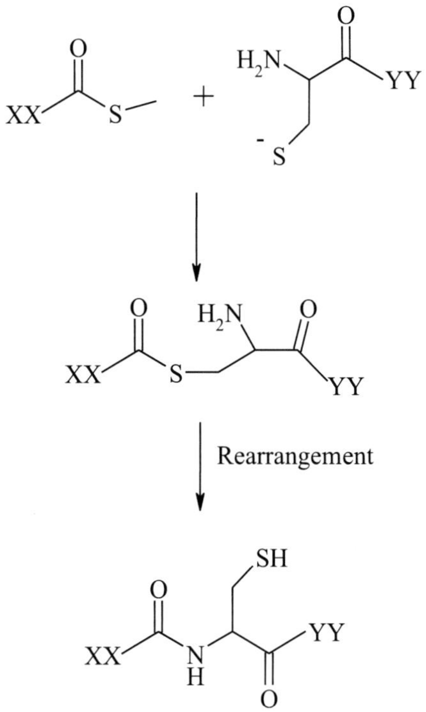

Figure 2.8 Schematic representation of the native chemical ligation for hydrogel formation.

is compatible with biomolecules like peptides and polysaccharides [67, 126]. Although examples of hydrogels formed by this approach are limited [126], it can be expected that, due to the high chemoselectivity, natural chemical ligation will be exploited for the synthesis of future generation hydrogels for drug delivery and tissue engineering.

A strategy that has become widely used for the preparation of chemical hydrogels is the so-called click chemistry, which is based on a chemical crosslink between azides and acetylenes [83]. This reticulation is catalyzed by Cu^+ and offers the great advantage of a high specificity and a quick kinetics without side reactions occurring with other functional groups on the hydrogel components. However, the toxicity of the catalyst represents a concern and limits the applicability of this technique to injectable hydrogels; its replacement presents some inconveniences such as slow reaction kinetics, although some work along this line is being carried on [78]. Click chemistry represents an attractive way to produce novel hydrogels, and several examples based on natural components have been reported [137]. A chitosan-based hydrogel has been produced for the preparation of immunonanoparticles [78]. Hyaluronan has been modified with either azide or alkyne groups and the click ligation catalyzed by Cu^+ led to the formation of a hydrogel which showed interesting properties as to the release of doxorubicine [28]. Alkyne-functionalized guar gum was reticulated using the Cu^+ catalyzed click reaction with a diazido-poly[(ethylene glycol)-co-(propylene glycol)] to obtain a thermosensitive hydrogel suitable for the temperature-dependent drug release [132].

An emerging and alternative strategy in the field of chemical hydrogel formation is represented by the enzyme catalyzed crosslinking reactions [92]. The mild reaction conditions, compatible with the physiological environment, which allow for its use in the presence of cells and tissues, represent the main advantage of this approach. Among the most commonly used enzymes one finds transglutaminase(s) (TGAse), which catalyze the formation of covalent bonds between a free amine group from a lysine and the γ-carboxamide group of a glutamine [29], tyrosinase, a copper-containing enzyme that catalyzes the oxidation of phenols into activated quinones [24], lysyl oxidase, transforming primary amines of lysines to aldehydes which react further to crosslink the hydrogel and peroxidases, which catalyze the conjugation of phenol derivatives in the presence of hydrogen peroxide [69, 114]. Figure 2.9 reports a few examples of enzyme catalyzed formation of chemical hydrogels.

2.4.2 *Physical Gels*

The term *physical gels* has been introduced by de Gennes [34] to describe the semisolid systems where polymer chains are "physically," rather than chemically, reticulated by crosslinks which have small but finite energy and/or finite lifetime [72]. Several biopolymer gels are based on noncovalent physical crosslinks which

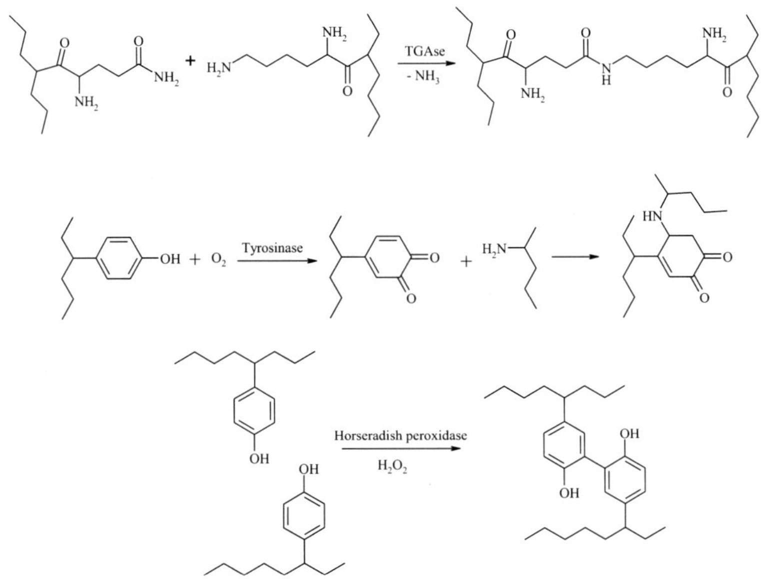

Figure 2.9 Examples of enzyme based reticulation for chemical hydrogel formation.

are formed by ordered secondary structures, called junction zones [72]. Indeed, with the advent of the sliding gels proposed by de Gennes or slide-ring gels of Okumura–Ito, the term physical gel has been extended to systems in which crosslinking points are composed by topological constraints [35, 97].

Physical network are formed by polymers containing two kind of groups: those strongly interacting with the solvent and those strongly interacting with one another [107]. These latter groups aggregate in network junction on the molecular scale. Network junctions result from (i) entanglements between macromolecules, (ii) ionic and ionic-coordination bonding, and (iii) interaction of helices [10, 112, 117, 122]. Associated junctions do not dissolve at all in commonly used solvents and, when they are formed by large and rigid crystallites, they give rise to physical hydrogels behaving like chemical gels [107]. Indeed, the dissolution of a physical gel into a solvent is prevented by physical interactions between different polymer chains [63]. From the thermodynamic point of view, physical hydrogels are systems with incomplete phase separation or one-phase or two-phase systems with a lower or upper critical solution temperature. Ross-Murphy has introduced the discrimination between "weak" and "strong" networks [72, 111]. While both types of physical gels respond as solids at small deformations, the strong gels behave as solids also at large deformation whereas the weak hydrogels (structured fluids) flow as liquids at large deformation.

The physical description of these hydrogels is complicated by the fact that the number and position of the crosslinks fluctuate with time and temperature [72]. Moreover, the junctions forming the three-dimensional network often undergo a large lateral aggregation to form bundles. These supramolecular arrangements have consequences on the mechanical response of the hydrogel, and it is clear that the application of the classic statistical approach (rubber elasticity) for their description requires numerous assumptions and simplifications. Indeed, the use of a front factor to adapt the rubber elasticity theory to polysaccharide-based physical gels is, in these cases, quite common.

Physical crosslinking can be obtained through hydrophobic interaction, ionic interactions, hydrogen bonding, host-guest interactions or their combination [139]. All these interactions are reversible and can be disrupted by changes in physical–chemical conditions or applications of stress [110]. There has been an increasing interest in physical-hydrogels because a careful selection of conditions like pH, temperature, and ion concentration can lead to the formation of a hydrogel without requiring chemical crosslinking agents; this improves the biocompatibility of the construct.

Several biopolymer hydrogels are based on noncovalent physical crosslinking which is formed through different mechanisms [72].

2.4.2.1 Thermal gelation

Thermally crosslinked gels are obtained upon cooling a hot solution of the biopolymer. In several cases, but invariably concerning semirigid polysaccharides, a decrease of temperature energetically favors the attainment of an ordered conformation (at least on a chain length scale of 10^1 to 10^2 nm) at the expense of a very moderate loss of configurational entropy of the semiflexible polymer. As a result, two or more macromolecules align along the chain axis (either parallel or antiparallel) bringing about the formation of an interchain junction. The simultaneous build up of such contacts produce the formation of the physical gel junctions, interconnected by more or less disordered (albeit well-hydrated) chain stretches. Agarose and carrageenans are examples of such gelling polysaccharides: they both are algal galactans, but the latter ones bear variable amounts of sulfate ester groups. In the case of agarose, the cooling of a solution of the polysaccharide induces the aggregation into fiber bundles through extensive hydrogen bonding [3]. For agarose, both double-helix [2, 96] and single-helix [52, 57] models have been proposed. In the case of carrageenans, extensive experimental evidence indicates that the fundamental ordered conformation of the associating stretches is a single helical one, for both κ- [8, 136] and ι- [6], followed by the formation of bundles of aggregated chains [7]. Bacterial polysaccharides like xanthan, gellan, welan and rhamsan also undergo gelation upon cooling aqueous solutions, driven by a disorder-to-order conformational transition of the polysaccharide chain [106].

van der Waals interactions in some cases may accompany the attainment of an ordered (interchain) conformation. (Highly) methylated pectins, when treated in slightly acidic medium, lead to gel formation upon decrease of temperature. This process has been largely exploited by the food industry.

Other polysaccharide derivatives, such as methyl- or hydroxypropyl methyl cellulose, undergo an inverse sol–gel transition in water giving rise to gel formation upon heating of the polymer solution, due to the prevailing effect of hydrophobic interactions [113]. For such polymers, geometrical constraints related to chain ordering and alignment are much less relevant.

In all these cases, the hydrogels are thermoreversible; gel formation can be reversed by varying the temperature.

2.4.2.2 Ionic gelation

In the case of ion-driven gelation, the formation of a hydrogel structure occurs upon treatment of an ionic polysaccharide with mono-, di-, or trivalent counterions. Several examples of ionic gelation exist for polysaccharides of which alginate is probably the most well known [36]. In this case, the gelling ion, typically Ca^{2+}, Ba^{2+}, and Sr^{2+}, bind to stretches of guluronic acid residues within the polysaccharide chain [91]. This leads to the formation of junctions, described according to the egg-box model [56, 120], which physically hold together the polysaccharide chains in a 3D continuum. In a much similar way, galacturonic acid can form hydrogels upon treatment with the same divalent ions [76].

Ionic gelation can also occur in the case of polycations with an anion as the crosslinker. Specifically, the ionic interaction between chitosan (polycation) and the trivalent negatively charged glycerol phosphate was shown to induce hydrogel formation [147].

κ- and ι-carrageenan represent examples of physical gel obtained by combination of both thermal treatment and ionic bonding. In particular, upon cooling a hot κ-carrageenan solution, a conformational coil–helix transition takes place and in the presence of monovalent ions, such as K^+ and Cs^+ [30], single helices aggregate in the gel junction zones; a very similar behavior is shown by ι-carrageenan, but for a higher affinity for alkaline-earth metal ions. Much like agarose, but at variance with Me^{2+}/polyuronates gels, carrageenan gels are thermoreversible.

2.4.2.3 Complex coacervation

Physical hydrogels can be obtained by means of complex coacervation between a polycation and a polyanion. The electrostatic interactions among oppositely charged polymers lead to hydrogel formation. Gel formation depends on parameters like pH and ionic strength. Coacervation of chitosan and xanthan represent an example of this type of physical hydrogel [58].

2.4.2.4 Hydrophobic interactions

Hydrophobic association is another driving force leading to gel formation. The hydrophobic modification of polysaccharide can lead to micelles, nanoparticles, microspheres, liposomes and hydrogels [61].

This is the case of HA samples produced by FAB - Fidia Advanced Biopolymers (HYADD-3 and HYADD-4) where the polysaccharide has been modified through an amidation reaction inserting, on its backbone, dodecyl or hexadecyl side chains. Due to the presence of hydrophobic interactions, formation of physical gel takes place and the product has been proposed as an injectable matrix for reconstruction of soft tissues [46, 104]. Alginate was modified with dodecyl side chains to undergo a sol–gel transition upon treatment with sodium chloride [80, 99]. In the case of chitosan, the modification of the polysaccharide backbone with hydrophobic groups from laurate to stearate also leads to hydrogel formation through hydrophobic interactions [18, 85].

2.4.3 *Semi-interpenetrating Polymer Networks (SIPN) and Interpenetrating Polymer Networks (IPN)*

Semi-interpenetrating polymer networks (SIPNs) can be described as systems of polymers entangled within polymeric networks. The official IUPAC definition says: "A polymer comprising one or more networks and one or more linear or branched polymer(s) characterized by the penetration on a molecular scale of at least one of the networks by at least some of the linear or branched macromolecules" [68].

If compared with typical immiscible polymer blends, the polymers of a SIPN are more intimately mixed and mechanically interlocked, yielding to high mechanical performance and greater miscibility [44]. Such systems are based both on natural and synthetic polymers, and also on their combination.

SIPNs and IPNs can be prepared following two main synthetic pathways, namely in situ and sequential synthesis [25, 121], as reported in Fig. 2.10. In the first case, all the components are mixed before starting any crosslinking (or polymerization for synthetic networks) reaction. For the IPNs case, the formation of the two networks may be simultaneous or sequential. The conditions should be carefully controlled, in order to avoid crossed reactions. In the second case, the sequential synthesis, firstly a polymer network is obtained and then filled (impregnation step) with another polymer or with the components (or precursors) of the second network.

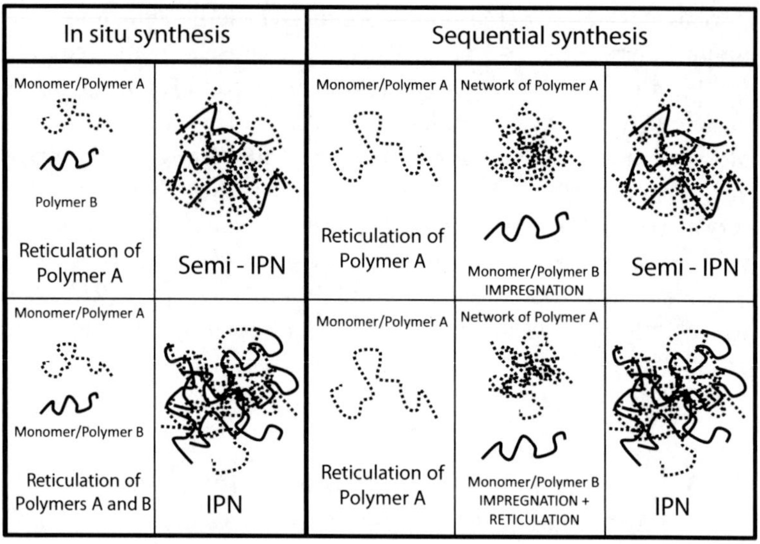

Figure 2.10 Schematic in situ and sequential synthesis for semi-IPN and IPN. Adapted from Ref. 121.

SIPNs have been widely exploited for the preparation of multiphase systems for human applications, in particular in the pharmaceutical (drug delivery) and biomedical (tissue engineering) fields. The use of SIPNs enables for the preparation of structures characterized by an overall homogeneity, in which polymers with different properties are combined, giving also the possibility to complement the deficiencies of either of the two [4, 25]. SIPNs have been proposed as systems able to overcome some hydrogel disadvantages, as the low mechanical strength [32]. Thus, natural polymers characterized either by high solubility and degradability in physiological conditions, or by poor mechanical properties, are generally combined with other polymers (natural or synthetic) [77]. On the other side, the introduction of a biopolymer into a synthetic matrix improves the biological response of a material, often exhibiting interesting swelling and mechanical properties, tunable with the biopolymer content [77].

In the following section some examples of the several SIPNs proposed in the literature are reported.

SIPN hydrogels based on linear hydroxypropyl cellulose (HPC) and crosslinked poly[(*N*-tert-butylacrylamide)-co-acrylamide] [poly(TBA-co-AAm)] have been described. In this system TBA is a temperature sensitive component, AAm improves the mechanical strength of PTBA hydrogels, and HPC has both hydrophilic properties and good biocompatibility. The resulting material presents good mechanical strength, temperature sensitivity, and enhanced biocompatibility. Highly hydrophilic and pH-sensitive SIPNs hydrogels have been also prepared through the combination of a poly(acrylic acid) network with bis-acrylamide (BIS) as crosslinker and potassium persulphate (KPS) as initiator in the presence of carboxymethyl cellulose (CMC). This SIPN can be used as vehicle for controlled delivery of ciprofloxacine, dependent on pH, ion strength, and temperature [23]. SIPN microspheres of gelatin and carboxymethyl cellulose for controlled release of drugs (such as ketorolac tromethamine) have been also proposed [108].

The marine polysaccharides alginate and chitosan are widely employed for the preparation of SIPNs, combined together, with other natural polymers, or with synthetic ones.

Multiphase alginate-based systems have been prepared in combination with a poly(2-hydroxyethylmethacrylate-co-2-methacryloxyethyltrimethylammonium) (poly(HEMA-co-METAC)) matrix in a semi-interpenetrating structure. The system obtained was biocompatible, appropriate for cell adhesion and proliferation, and displayed improved and tunable swelling and mechanical properties. In another similar study, SIPNs based on the same synthetic polymers and bioactive extra cellular matrix components (such as hyaluronic acid and chondroitin sulfate) have been described [75, 77]. Gelatin-alginate SIPNs hydrogels have been proposed for controlled delivery of small molecules [75]. Also pH-sensitive SIPNs superabsorbent hydrogels based on alginate-poly(acrylate) networks and linear poly(vinylpyrrolidone) (PVP) have been described. Superabsorbent hydrogels are moderately crosslinked polymer networks capable to retain high amounts of liquids, also under certain heat or pressure conditions. Due to such unique properties, superior to conventional absorbents, these hydrogels find potential application in several field, such as agriculture, hygiene, wastewater treatment, sealing material, and drug delivery systems [141].

On the chitosan side, in situ gelling double networks of oxidized dextrans and thiolated chitosan were proposed for healing purposes, showing faster gelation kinetics if compared to auto-gelling thiolated chitosan hydrogels [16, 146]. SIPNs based on chitosan and poly(methacryloyl glycylglycine) [poly(MAGG)] have been developed to create in vitro cell culture models to be used as platform for cell pathophysiology investigations. Moreover, poly(MAGG), being negatively charged, can form polymer complexes with chitosan. Thus, poly(MAGG) shows several applications as macromolecular component in drug delivery formulations [32]. Several chitosan-based SIPNs prepared with polyether, silk, PEO, and PVP have been reviewed [4].

The combination of both alginate and chitosan is also widely exploited. For example native polymers have been used to prepare alginate-chitosan SIPNs scaffolds by freeze-drying process for improved cartilage tissue engineering [116]. Recently a bioactive scaffold has been obtained from a blend of alginate and a lactose-modified chitosan (chitlac) [84]. The calcium-based gelation of the alginate resulted in the formation of SIPN hydrogel beads: the alginate 3D supporting structure was combined with the bioactivity of the chitosan derivative, which provides interactions with porcine articular chondrocytes [37]. The hydrogel showed improved mechanical properties (bead stability), thanks to the formation of soluble complexes between the (oppositely) charged polysaccharides, which give ionic extra-contributions to the hydrogel formation, resulting in a structural stabilization. Moreover, biochemical and biological studies demonstrated the capability of the system to maintain chondrocyte phenotype, and to significantly stimulate and promote cell growth and proliferation.

Interpenetrating polymer network (IPN) hydrogels are composed of two crosslinked and interpenetrated polymer 3D structures. Each component will thus be reticulated either physically or chemically and the final hydrogel entangles the two networks and benefits of enhanced mechanical and physical properties deriving from the synergistic combination of the two structures. In IPNs, neither chemical nor strong physical interactions are taking place between the two networks. The IUPAC definition is "A polymer comprising two or more networks which are at least partially interlaced on a molecular scale but not covalently bonded to each other and cannot be separated unless chemical bonds are broken" [68].

In the latest years, IPNs have shown a high potential in medical and pharmaceutical applications, especially in the field of drug delivery [31, 87]. In one of the recent applications proposed for IPN, an alginate-HA interpenetrated structure has been based on the physical gelation of alginate, driven by the presence of Ca^{2+}, and the chemical gelation of a methacrylated hyaluronan which takes place upon treatment with a UV light [31]. Along a similar path, HEMA-derivatized dextran and alginate were used to prepare a IPN based on the combination of a physical (for alginate) and chemical (for modified dextran) reticulation [87]. In another example, an interpenetrated matrix between alginate and gelatin was formed through two different physical gelation processes, i.e., ionic for the presence of Ca^{2+} and thermal with the formation of hydrogen bonds [75].

References

1. Almdal, K., Dyre, J., Hvidt, S., and Kramer, O. (1993). Towards a phenomenological definition of the term "gel," *Polym. Gels Netw.*, **1**, pp. 5–17.
2. Arnott, S., Fulmer, A., Scott, W. E., Dea, I. C. M., Moorhouse, R., and Rees, D. A. (1974). The agarose double helix and its function in agarose gel structure, *J. Mol. Biol.*, **90**, pp. 269–284.
3. Aymard, P., Martin, D. R., Plucknett, K., Foster, T. J., Clark, A. H., and Norton, I. T. (2001). Influence of thermal history on the structural and mechanical properties of agarose gels, *Biopolymers*, **59**, pp. 131–144.
4. Bhattarai, N., Gunn, J., and Zhang, M. (2010). Chitosan-based hydrogels for controlled, localized drug delivery, *Adv. Drug Deliver. Rev.*, **62**, pp. 83–99.
5. Bitar, M., Salih, V., Brown, R. A., and Nazhat, S. N. (2007). Effect of multiple unconfined compression on cellular dense collagen scaffolds for bone tissue engineering, *J. Mater. Sci. Mater. Med.*, **18**, pp. 237–244.

6. Bongaerts, K., Paoletti, S., Denef, B., Vanneste, K., and Cuppo, F. R. H. (2000). Light scattering investigation of ι-carrageenan aqueous solutions. Concentration dependence of association, *Macromolecules*, **33**, pp. 8709–8719.
7. Bongaerts, K., Reynaers, H., Zanetti, F., and Paoletti, S. (1999). Equilibrium and nonequilibrium association processes of κ-carrageenan in aqueous salt solutions, *Macromolecules*, **32**, pp. 683–689.
8. Bongaerts, K., Reynaers, H., Zanetti, F., and Paoletti, S. (2000). On the molar mass of κ-carrageenan in the course of conformational transition from the disordered to the fundamental ordered from, *Macromolecules*, **32**, pp. 675–682.
9. Borges, A. C., Eyholzer, C., Duc, F., Bourban, P. E., Tingaut, P., Zimmermann, T., Pioletti, D. P., and Månson, J. A. (2011). Nanofibrillated cellulose composite hydrogel for the replacement of the nucleus pulposus, *Acta Biomater.*, **7**, pp. 3412–3421.
10. Braudo, E. E., and Plashchina, I. G. (1995). *Food Macromolecules and Colloids*, The Royal Society of Chemistry, Cambridge.
11. Brown, H. R. (2007). A model of the fracture of double network gels, *Macromolecules*, **40**, pp. 3815–3818.
12. Bryant, S. J., Nuttelman, C. R., and Anseth, K. S. (2000). Cytocompatibility of UV and visible light photoinitiating systems on cultured NIH/3T3 fibroblasts in vitro, *J. Biomat. Sci. Polym. E.*, **11**, pp. 439–457.
13. Bungenberg de Jong, H. G. (1949). A survey of the study objects in this volume, in *Colloid Science*, Kruyt, H. R., ed., Elsevier, Amsterdam, pp. 1–18.
14. Burchard, W., and Ross-Murphy, S. B. (1990). Introduction: physical gels from synthetic and biological macromolecules, in *Physical Networks: Polymers and Gels*, Burchard, W., and Ross-Murphy, S. B., eds., Elsevier, London, UK, pp. 1–14.
15. Burdick, J. A., and Prestwich, G. D. (2011). Hyaluronic acid hydrogels for biomedical applications, *Adv. Mater.*, **23**, pp. H41-H56.
16. Censi, R., di Martino, P., Vermonden, T., and Hennink, W. E. (2012). Hydrogels for protein delivery in tissue engineering, *J. Control. Rel.*, **161**, pp. 680–692.
17. Censi, R., Fieten, P. J., di Martino, P., Hennink, W. E., and Vermonden, T. (2010). In situ forming hydrogels by tandem thermal gelling and michael addition reaction between thermosensitive triblock copolymers and thiolated hyaluronan, *Macromolecules*, **43**, pp. 5771–5778.

18. Cerchiara, T., Luppi, B., Bigucci, F., Orienti, I., and Zecchi, V. (2002). Physically cross-linked chitosan hydrogels as topical vehicles for hydrophilic drugs, *J. Pharm. Pharmacol.*, **54**, pp. 1453–1459.

19. Chan, A. W., and Neufeld, R. J. (2009). Modeling the controllable pH-responsive swelling and pore size of networked alginate based biomaterials, *Biomaterials*, **30**, pp. 6119–6129.

20. Chan, A. W., Whitney, R. A., and Neufeld, R. J. (2008). Kinetic controlled synthesis of pH-responsive network alginate, *Biomacromolecules*, **9**, pp. 2536–2545.

21. Chan, A. W., Whitney, R. A., and Neufeld, R. J. (2009). Semisynthesis of a controlled stimuli-responsive alginate hydrogel, *Biomacromolecules*, **10**, pp. 609–616.

22. Chang, C., Peng, N., He, M., Teramoto, Y., Nishio, Y., and Zhang, L. (2013). Fabrication and properties of chitin/hydroxyapatite hybrid hydrogels as scaffold nano-materials, *Carbohyd. Polym.*, **91**, pp. 7–13.

23. Chang, C., and Zhang, L. (2011). Cellulose-based hydrogels: present status and application prospects, *Carbohyd. Polym.*, **84**, pp. 40–53.

24. Chen, T., Embree, H. D., Brown, E. M., Taylor, M. M., and Payne, G. F. (2003). Enzyme-catalyzed gel formation of gelatin and chitosan: potential for in situ applications, *Biomaterials*, **24**, pp. 2831–2841.

25. Chikh, L., Delhorbe, V., and Fichet, O. (2011). (Semi)Interpenetrating polymer networks as fuel cell membranes, *J. Membrane Sci.*, **368**, pp. 1–17.

26. Clark, A., and Ross-Murphy, S. (1987). Structural and mechanical properties of biopolymer gels, in *Biopolymers*, Springer, Berlin/Heidelberg, pp. 57–192.

27. Cohen, Y., Ramon, O., Kopelman, I. J., and Mizrahi, S. (1992). Characterization of inhomogeneous polyacrylamide hydrogels, *J. Polym. Sci. B Polym. Phys.*, **30**, pp. 1055–1067.

28. Crescenzi, V., Cornelio, L., Di Meo, C., Nardecchia, S., and Lamanna, R. (2007). Novel hydrogels via click chemistry: synthesis and potential biomedical applications, *Biomacromolecules*, **8**, pp. 1844–1850.

29. Crescenzi, V., Francescangeli, A., and Taglienti, A. (2002). New gelatin-based hydrogels via enzymatic networking, *Biomacromolecules*, **3**, pp. 1384–1391.

30. Cuppo, F., Reynaers, H., and Paoletti, S. (2002). Association of κ-carrageenan induced by Cs^+ ions in iodide aqueous solution: a light scattering study, *Macromolecules*, **35**, pp. 539–547.

31. D'Arrigo, G., Di Meo, C., Geissler, E., Coviello, T., Alhaique, F., and Matricardi, P. (2012). Hyaluronic acid methacrylate derivatives and calcium alginate interpenetrated hydrogel networks for biomedical applications: physico-chemical characterization and protein release, *Colloid. Polym. Sci.*, **290**, pp. 1575–1582.
32. Dash, M., Ferri, M., and Chiellini, F. (2012). Synthesis and characterization of semi-interpenetrating polymer network hydrogel based on chitosan and poly(methacryloylglycylglycine), *Mater. Chem. Phys.*, **135**, pp. 1070–1076.
33. Dawson, P. E., Muir, T. W., Clark-Lewis, I., and Kent, S. B. (1994). Synthesis of proteins by native chemical ligation, *Nature*, **266**, pp. 776–779.
34. de Gennes, P. G. (1979). *Scaling Concepts in Polymer Physics*, Cornell University Press, Ithaca, NY.
35. de Gennes, P. G. (1999). Sliding gels, *Phys. A*, **271**, pp. 231–237.
36. Donati, I., and Paoletti, S. (2009). Material properties of alginates, in *Alginates: Biology and Applications*, Rhem, B. H. A., Springer-Verlag, Berlin/Heidelberg, pp. 1–53.
37. Donati, I., Stredanska, S., Silvestrini, G., Vetere, A., Marcon, P., Marsich, E., Mozetic, P., Gamini, A., Paoletti, S., and Vittur, F. (2005). The aggregation of pig articular chondrocyte and synthesis of extracellular matrix by a lactose-modified chitosan, *Biomaterials*, **26**, pp. 987–998.
38. Donati, I., Mørch, Y. A., Strand, B. L., Skjåk-Bræk, G., and Paoletti, S. (2009). Effect of elongation of alternating sequences on swelling behavior and large deformation properties of natural alginate gels, *J. Phys. Chem. B*, **113**, pp. 12916–12922.
39. Dong, Y., Saeed, A. O., Hassan, W., Keigher, C., Zheng, Y., Tai, H., Pandit, A., and Wang, W. (2012). "One-step" preparation of thiol-Ene clickable PEG-based thermoresponsive hyperbranched copolymer for in situ crosslinking hybrid hydrogel, *Macromol. Rapid Commun.*, **33**, pp. 120–126.
40. Drury, J. L., and Mooney, D. J. (2003). Hydrogels for tissue engineering: scaffold design variables and applications, *Biomaterials*, **24**, pp. 4337–4351.
41. Durmaz, S., and Okay, O. (2000). Acrylamide/2-acrylamido-2-methylpropane sulfonic acid sodium salt-based hydrogels: synthesis and characterization, *Polymer*, **41**, pp. 3693–3704.
42. Eicke, H.-F., Quellet, C., and Xu, G. (1990). Atypical gels: examples of polymer networks in microemulsions, in *Physical Networks: Polymers*

and Gels, Burchard, W., and Ross-Murphy, S. B., eds., Elsevier Applied Science, London, UK, pp. 169–184.

43. Elisseeff, J., Anseth, K., Sims, D., McIntosh, W., Randolph, M., and Langer, R. (1999). Transdermal photopolymerization for minimally invasive implantation, *P. Natl. Acad. Sci. USA*, **96**, pp. 3104–3107.
44. Fang, C., Julius, D., Tay, S. W., Hong, L., and Lee, J. Y. (2013). Preparation of semi-interpenetrating polymer networks with adjustable mesh width and hydrophobicity, *Polymer*, **54**, pp. 134–142.
45. Ferry, J. D. (1961). *Viscoelastic Properties of Polymers*, Wiley, New York.
46. Finelli, I., Chiessi, E., Galesso, D., Renier, D., and Paradossi, G. (2011). A new viscosupplement based on partially hydrophobic hyaluronic acid: a comparative study, *Biorheology*, **48**, pp. 263–275.
47. Flory, P. J. (1953). *Principles of Polymer Chemistry*, Oxford University Press, Ithaca.
48. Flory, P. J. (1974). Introductory lecture, *Faraday Discuss. Chem. Soc.*, **57**, pp. 7–18.
49. Flory, P. J. (1941). Molecular size distribution in three dimensional polymers. I. Gelation, *J. Am. Chem. Soc.*, **63**, pp. 3083–3090.
50. Flory, P. J., and Rehner, J. (1943). Statistical mechanics of cross-linked polymer networks I. Rubberlike elasticity, *J. Chem. Phys.*, **11**, pp. 512–520.
51. Flory, P. J., and Rehner, J. (1943). Statistical mechanics of cross-linked polymer networks II. Swelling, *J. Chem. Phys.*, **11**, pp. 521–526.
52. Foord, S. A., and Atkins, E. D. Y. (1989). New x-ray diffraction results from agarose: extended single helix structures and implications for gelation mechanism, *Biopolymers*, **28**, pp. 1345–1365.
53. Friedman, M., Cavins, J. F., and Wall, J. S. (1965). Relative nucleophilic reactivities of amino groups and mercaptide ions in addition reactions with α,β-unsaturated compounds, *J. Am. Chem. Soc.*, **87**, pp. 3672–3682.
54. Gong, J. P., Katsuyama, Y., Kurokawa, T., and Osada, Y. (2003). Double-network hydrogels with extremely high mechanical strength, *Adv. Mater.*, **15**, pp. 1155–1158.
55. Graham, T. (1861). Liquid diffusion applied to analysis, *Philos. Tr. R. Soc. Lond.*, **151**, pp. 183–224.
56. Grant, G. T., Morris, E. R., Rees, D. A., Smith, P. J. C., and Thom, D. (1973). Biological interactions between polysaccharides and divalent cations: the egg-box model, *FEBS Lett.*, **32**, pp. 195–198.

57. Guenet, J. M., Brûlet, A., and Rochas, C. (1993). Agarose chain conformation in the sol state by neutron scattering, *Int. J. Biol. Macromol.*, **15**, pp. 131–132.

58. Gulrez, S. K. H., Al-Assaf, S., and Phillips, G. O. (2011). Hydrogels: methods of preparation, characterisation and applications, in *Progress in Molecular and Environmental Bioengineereing: From Analysis and Modeling to Technology Applications*, Carpi, A., ed., InTech, 117–150.

59. Hahn, S. K., Oh, E. J., Miyamoto, H., and Shimobouji, T. (2006). Sustained release formulation of erythropoietin using hyaluronic acid hydrogels crosslinked by Michael addition, *Int. J. Pharm.*, **322**, pp. 44–51.

60. Hahn, S. K., Park, J. K., Tomimatsu, T., and Shimoboji, T. (2007). Synthesis and degradation test of hyaluronic acid hydrogels, *Int. J. Biol. Macromol.*, **40**, pp. 374–380.

61. Hassani, L. N., Hendra, F., and Bouchemal, K. (2012). Auto-associative amphiphilic polysaccharides as drug delivery systems, *Drug Discov. Today*, **17**, pp. 608–614.

62. Haugh, M. G., Thorpe, S. D., Vinardell, T., Buckley, C. T., and Kelly, D. J. (2012). The application of plastic compression to modulate fibrin hydrogel mechanical properties, *J. Mech. Behav. Biomed. Mater*, **16**, pp. 66–72.

63. Hennink, W. E., and Nostrum, C. F. (2002). Novel crosslinking methods to design hydrogels, *Adv. Drug Discov. Rev.*, **54**, pp. 13–36.

64. Hermans, P. H. (1949). Gels, in *Colloid Science*, Kruyt, H. R., ed., Elsevier, Amsterdam, pp. 483–651.

65. Hirakura, T., Yasugi, K., Nemoto, T., Sato, M., Shimoboji, T., Aso, Y., Morimoto, N., and Akiyoshi, K. (2010). Hybrid hyaluronan hydrogel encapsulating nanogel as a protein nanocarrier: new system for sustained delivery of protein with a chaperone-like function, *J. Controlled Rel.*, **142**, pp. 483–489.

66. Hoffman, A. S. (2002). Hydrogels for biomedical applications, *Adv. Drug Deliv. Rev.*, **54**, pp. 3–12.

67. Hu, B. H., Su, J., and Messersmith, P. B. (2009). Hydrogels cross-linked by native chemical ligation, *Biomacromolecules*, **10**, pp. 2194–2200.

68. Jenkins, A. D., Kratochvìl, P., Stepto, R. F. T., and Suter, U. W. (1996). Glossary of basic terms in polymer science, *Pure Appl. Chem.*, **68**, pp. 2287–2311.

69. Jin, R., Moreira Teixeira, L. S., Dijkstra, P. J., Karperien, M., van Blitterswijk, C. A., Zhong, Z. Y., and Feijen, J. (2009). Injectable chitosan-based hydrogels for cartilage tissue engineering, *Biomaterials*, **30**, pp. 2544–2551.

70. Jordan-Lloyd, D. (1926). In *Colloid Chemistry*, Alexander, J., ed., The Chemical Catalog Co., New York, pp. 767–782.

71. Karaaslan, M. A., Tshabalala, M. A., Yelle, D. J., and Buschle-Diller, G. (2011). Nanoreinforced biocompatible hydrogels from wood hemicelluloses and cellulose whiskers, *Carbohyd. Polym.*, **86**, pp. 192–201.

72. Kavanagh, G. M., and Ross-Murphy, S. B. (1998). Rheological characterisation of polymer gels, *Prog. Polym. Sci.*, **23**, pp. 533–562.

73. Keller, A. (1995). Introductory lecture. Aspects of polymer gels, *Faraday Discuss.*, **101**, pp. 1–49.

74. Kim, J., Kim, I. S., Cho, T. H., Lee, K. B., Hwang, S. J., Tae, G., Noh, I., Lee, S. H., Park, Y., and Sun, K. (2007). Bone regeneration using hyaluronic acid-based hydrogel with bone morphogenic protein-2 and human mesenchymal stem cells, *Biomaterials*, **28**, pp. 1830–1837.

75. Klak, M.-C., Lefebvre, E., Rémy, L., Agniel, R., Picard, J., Giraudier, S., and Larreta-Garde, V. (2013). Gelatin-alginate gels and their enzymatic modifications: controlling the delivery of small molecules, *Macromol. Biosci.*, **13**, 687–695

76. Kohn, R. (1975). Ion binding on polyuronates: alginate and pectin, *Pure Appl. Chem.*, **42**, pp. 371–397.

77. La Gatta, A., Schiraldi, C., D'Agostino, A., Papa, A., and De Rosa, M. (2012). Properties of newly-synthesized cationic semi-interpenetrating hydrogels containing either hyaluronan or chondroitin sulfate in a methacrylic matrix, *J. Funct. Biomater.*, **3**, pp. 225–238.

78. Lallana, E., Fernandez-Megia, E., and Riguera, R. (2009). Surpassing the use of copper in the click functionalization of polymeric nanostructures: a strain-promoted approach, *J. Am. Chem. Soc.*, **131**, pp. 5748–5750.

79. Lee, F., Chung, J. E., and Kurisawa, M. (2009). An injectable hyaluronic acid–tyramine hydrogel system for protein delivery, *J. Control. Release*, **134**, pp. 186–193.

80. Leonard, M., Rastello de Boisseson, M., Hubert, P., and Dellacherie, E. (2004). Production of microspheres based on hydrophobically associating alginate derivatives by dispersion/gelation in aqueous sodium chloride solutions, *J. Biomed. Mater. Res. A*, **68A**, pp. 335–342.

81. Li, F., Ba, Q., Niu, S., Guo, Y., Duan, Y., Zhao, P., Lin, C., and Sun, J. (2012). In-situ forming biodegradable glycol chitosan-based hydrogels: Synthesis, characterization, and chondrocyte culture, *Mater. Sci. Eng. C*, **32**, pp. 2017–2025.

82. Lutolf, M. P., and Hubbell, J. A. (2003). Synthesis and physicochemical characterization of end-linked poly(ethylene glycol)-co-peptide hydro-

gels formed by Michael-type addition, *Biomacromolecules*, **4**, pp. 713–722.

83. Malkoch, M., Vestberg, R., Gupta, N., Mespouille, L., Dubois, P., Mason, A. F., Hedrick, J. L., Liao, Q., Frank, C. W., Kingsbury, K., and Hawker, C. J. (2006). Synthesis of well-defined hydrogel networks using Click chemistry, *Chem. Commun.*, pp. 2774–2776.
84. Marsich, E., Borgogna, M., Donati, I., Mozetic, P., Strand, B. L., Salvador, S. G., Vittur, F., and Paoletti, S. (2008). Alginate/lactose-modified chitosan hydrogels: a bioactive biomaterial for chondrocyte encapsulation, *J. Biomed. Mater. Res. A*, **84A**, pp. 364–376.
85. Martin, L., Wilson, C. G., Koosha, F., Tetley, L., Gray, A. I., Senel, S., and Uchegbu, I. F. (2002). The release of model macromolecules may be controlled by the hydrophobicity of palmitoyl glycol chitosan hydrogels, *J. Controlled Rel.*, **80**, pp. 87–100.
86. Mather, B. D., Viswanathan, K., Miller, K. M., and Long, T. E. (2006). Michael addition reactions in macromolecular design for emerging technologies, *Prog. Polym. Sci.*, **31**, pp. 487–531.
87. Matricardi, P., Pontoriero, M., Coviello, T., Casadei, M. A., and Alhaique, F. (2008). In situ cross-linkable novel alginate-dextran methacrylate IPN hydrogels for biomedical applications: Mechanical characterization and drug delivery properties, *Biomacromolecules*, **9**, pp. 2014–2020.
88. Mentré, P., and Hui Bon Hoa, G. (2001). Effects of high hydrostatic pressures on living cells: a consequence of the properties of macromolecules and macromolecule-associated water, *Int. Rev. Cytol.*, **201**, pp. 1–84.
89. Mergy, J., Fournier, A., Hachet, E., and Auzély-Velty, R. (2012). Modification of polysaccharides via thiol-ene chemistry: a versatile route to functional biomaterials, *J. Polym. Sci. A Polym. Chem.*, **50**, pp. 4019–4028.
90. Moe, S. T., Elgsaeter, A., Skjåk-Bræk, G., and Smidsrød, O. (1993). A new approach for estimating the crosslink density of covalently crosslinked ionic polysaccharide gels, *Carbohyd. Polym.*, **20**, pp. 263–268.
91. Mørch, Y. A., Donati, I., Strand, B. L., and Skjåk-Bræk, G. (2006). Effect of Ca^{2+}, Ba^{2+}, and Sr^{2+} on alginate microbeads, *Biomacromolecules*, **7**, pp. 1471–1480.
92. Moreira Teixeira, L. S., Feijen, J., van Blitterswijk, C. A., Dijkstra, P. J., and Karperien, M. (2012). Enzyme-catalyzed crosslinkable hydrogels: Emerging strategies for tissue engineering, *Biomaterials*, **33**, pp. 1281–1290.

93. Nguyen, K. T., and West, J. L. (2002). Photopolymerizable hydrogels for tissue engineering applications, *Biomaterials*, **23**, pp. 4307–4314.

94. Nisato, G., Schosseler, F., and Candau, S. J. (1996). Swelling equilibrium properties of partially charged gels: the effect of salt on the shear modulus, *Polym. Gels Netw.*, **4**, pp. 481–498.

95. Nishinari, K. (2009). Some thoughts on the definition of a gel, in *Gels: Structures, Properties, and Functions* Tokita, M., and Nishinari, K., eds., Springer, Berlin/Heidelberg, 87–94.

96. Norton, I. T., Goodall, D. M., Austen, K. R. J., Morris, E. R., and Rees, D. A. (1986). Dynamics of molecular organization in agarose sulphate, *Biopolymers*, **25**, pp. 1009–1029.

97. Okumura, Y., and Ito, K. (2001). The polyrotaxane gel: a topological gel by figure-of-eight cross-links, *Adv. Mater.*, **13**, pp. 485–487.

98. Omidian, H., and Park, K. (2010). Introduction to hydrogels, in *Biomedical Applications of Hydrogels Handbook*, Ottenbrite, R. M., Ark, K., and Okano, T., eds., Springer, New York, pp. 1–16.

99. Pawar, S. N., and Edgar, K. J. (2012). Alginate derivatization: a review of chemistry, properties and applications, *Biomaterials*, **33**, pp. 3279–3305.

100. Peppas, N. A. (2004). Hydrogels, in *Biomaterials Science: An Introduction to Materials in Medicine* Ratner, B. D., Hoffman, A. S., Schoen, F. J., and Lemons, J. E., eds., Elsevier Academic Press, San Diego, California, pp. 100–107.

101. Peppas, N. A., Hilt, J. Z., Khademhosseini, A., and Langer, R. (2006). Hydrogels in biology and medicine: from molecular principles to bionanotechnology, *Adv. Mater.*, **18**, pp. 1345–1360.

102. Peppas, N. A., and Merrill, E. W. (1977). Crosslinked poly(vinyl alcohol) hydrogels as swollen elastic networks, *J. Appl. Polym. Sci.*, **21**, pp. 1763–1770.

103. Pescosolido, L., Vermonden, T., Malda, J., Censi, R., Dhert, W. J. A., Alhaique, F., Hennink, W. E., and Matricardi, P. (2011). In situ forming IPN hydrogels of calcium alginate and dextran-HEMA for biomedical applications, *Acta Biomater.*, **7**, pp. 1627–1633.

104. Rhodes, N. P. (2007). Inflammatory signals in the development of tissue-engineered soft tissue, *Biomaterials*, **28**, pp. 5131–5136.

105. Rizzi, S. C., and Hubbell, J. A. (2005). Recombinant protein-co-PEG networks as cell-adhesive and proteolytically degradable hydrogel matrixes. Part I: Development and physicochemical characteristics, *Biomacromolecules*, **6**, pp. 1226–1238.

106. Robinson, G., Manning, C. E., and Morris, E. R. (1991). Conformation and physical properties of the bacterial polysaccharides gellan, welan and rhamsan, in *Food Polymers, Gels and Colloids*, Special Publication No. 82, Dickinson, E., ed., Royal Society of Chemistry, Cambridge, pp. 22–33.

107. Rogovina, L. Z., Vasil'ev, V. G., and Braudo, E. E. (2008). Definition of the concept of polymer gel, *Polym. Sci. Ser. C*, **50**, pp. 85–92.

108. Rokhade, A. P., Agnihotri, S. A., Patil, S. A., Mallikarjuna, N. N., Kulkarni, P. V., and Aminabhavi, T. M. (2006). Semi-interpenetrating polymer network microspheres of gelatin and sodium carboxymethyl cellulose for controlled release of ketorolac tromethamine, *Carbohyd. Polym.*, **65**, pp. 243–252.

109. Rokstad, A. M., Donati, I., Borgogna, M., Oberholzer, J., Strand, B. L., Espevik, T., and Skjåk-Bræk, G. (2006). Cell-compatible covalently reinforced beads obtained from a chemoenzymatically engineered alginate, *Biomaterials*, **27**, pp. 4726–4737.

110. Rosiak, J. M., and Yoshii, F. (1999). Hydrogels and their medical applications, *Nucl. Instrum. Meth. B*, **151**, pp. 56–64.

111. Ross-Murphy, S. B. (1984). Rheological methods, in *Biophysical Methods in Food Research*, Chan, H. W. S., ed., Blackwell Scientific, Oxford, pp. 138–199.

112. Ross-Murphy, S. B. (1991). Physical gelation of synthetic and biological macromolecules, in *Polymer Gels: Fundamentals and Biomedical Applications* De Rossi, D., Kajiwara, K., Osada, Y., and Yamauchi, A., eds., Plenum Press, New York, pp. 21–39.

113. Ruel-Gariépy, E., and Leroux, J. C. (2004). In situ-forming hydrogels–review of temperature-sensitive systems, *Eur. J. Pharm. Biopharm.*, **58**, pp. 409–426.

114. Sakai, S., Ito, S., Ogushi, Y., Hashimoto, I., Hosoda, N., Sawae, Y., and Kawakami, K. (2009). Enzymatically fabricated and degradable microcapsules for production of multicellular spheroids with well-defined diameters of less than 150 μm, *Biomaterials*, **30**, pp. 5937–5942.

115. Sawhney, A. S., Pathak, C. P., and Hubbell, J. A. (1993). Bioerodible hydrogels based on photopolymerized poly(ethylene glycol)-co-poly(alpha-hydroxy acid) diacrylate macromers, *Macromolecules*, **26**, pp. 581–587.

116. Seda Tigli, R., and Gümüsderelioglu, M. (2009). Evaluation of alginate-chitosan semi IPNs as cartilage scaffolds, *J. Mater. Sci. Mater. Med.*, **20**, pp. 699–709.

117. Shenoy, S. L., Painter, P. C., and Coleman, M. M. (1999). The swelling and collapse of hydrogen bonded polymer gels, *Polymer*, **40**, pp. 4853–4863.

118. Skouri, R., Schosseler, F., Munch, J. P., and Candau, S. J. (1995). Swelling and elastic properties of polyelectrolyte gels, *Macromolecules*, **28**, pp. 197–210.

119. Slaughter, B. V., Khurshid, S. S., Fisher, O. Z., Khademhosseini, A., and Peppas, N. A. (2009). Hydrogels in regenerative medicine, *Adv. Mater.*, **21**, pp. 3307–3329.

120. Smidsrød, O. (1974). Molecular basis of some physical properties of alginates in the gel state, *Faraday Discuss. Chem. Soc.*, **57**, pp. 263–274.

121. Sperling, L. H. (1994). Interpenetrating polymer networks: an overview, in *Interpenetrating Polymer Networks*, Klempner, D., Sperling, L. H., and Utracki, L. A., eds., American Chemical Society, Washington, DC, pp. 3–38.

122. Stadler, R. (1988). Association phenomena in macromolecular systems: influence of macromolecular constraints on the complex formation, *Macromolecules*, **21**, pp. 121–126.

123. Stefanescu, E. A., Stefanescu, C., Daly, W. H., Schmidt, G., and Negulescu, I. I. (2008). Hybrid polymer–clay nanocomposites: A mechanical study on gels and multilayered films, *Polymer*, **49**, pp. 3785–3794.

124. Stockmayer, W. H. (1944). Theory of molecular size distribution and gel formation in branched polymers II. General cross linking, *J. Chem. Phys.*, **12**, pp. 125–131.

125. Stockmayer, W. H. (1952). Molecular distribution in condensation polymers, *J. Polym. Sci.*, **9**, pp. 69–71.

126. Su, J., Hu, B. H., Lowe, J., Kaufman, D. B., and Messersmith, P. B. (2010). Anti-inflammatory peptide-functionalized hydrogels for insulin-secreting cell encapsulation, *Biomaterials*, **31**, pp. 308–314.

127. Tan, H., Rubin, J. P., and Marra, K. G. (2011). Direct synthesis of biodegradable polysaccharide derivative hydrogels through aqueous Diels–Alder chemistry, *Macromol. Rapid Commun.*, **32**, pp. 905–911.

128. Tanaka, F. (2006). Theory of molecular association and thermoreversible gelation, in *Materials with Molecular Gels: Self-Assembled Fibrillar Netwoks*, Weiss, R. G., and Terech, P. eds., Springer, Dordrecht, pp. 17–78.

129. Tanaka, Y., Gong, J. P., and Osada, Y. (2005). Novel hydrogels with excellent mechanical performance, *Prog. Polym. Sci.*, **30**, pp. 1–9.

130. te Nijenhuis, K. (1997). Introduction, in *Thermoreversible Networks* Springer, Berlin/Heidelberg, pp. 1–12.

131. Terech, P., and Weiss, R. G. (1997). Low molecular mass gelators of organic liquids and the properties of their gels, *Chem. Rev.*, **97**, pp. 3133–3160.

132. Tizzotti, M., Creuzet, C., Labeau, M.-P., Hamaide, T., Boisson, F., Drockenmuller, E., Charlot, A., and Fleury, E. (2010). Synthesis of temperature responsive biohybrid guar-based grafted copolymers by click chemistry, *Macromolecules*, **43**, pp. 6843–6852.

133. Tong, X., Zheng, J., Lu, Y., Zhang, Z., and Cheng, H. (2007). Swelling and mechanical behaviors of carbon nanotube/poly(vinyl alcohol) hybrid hydrogels, *Mater. Lett.*, **61**, pp. 1704–1706.

134. Turco, G., Donati, I., Grassi, M., Marchioli, G., Lapasin, R., and Paoletti, S. (2011). Mechanical spectroscopy and relaxometry on alginate hydrogels: a comparative analysis for structural characterization and network mesh size determination, *Biomacromolecules*, **12**, pp. 1272–1282.

135. Turco, G., Marsich, E., Bellomo, F., Semeraro, S., Donati, I., Brun, F., Grandolfo, M., Accardo, A., and Paoletti, S. (2009). Alginate/hydroxyapatite biocomposite for bone ingrowth: a trabecular structure with high and isotropic connectivity, *Biomacromolecules*, **10**, pp. 1575–1583.

136. Ueda, K., Itoh, M., Matsuzaki, Y., Ochiai, H., and Imamura, A. (1998). Observation of the molecular weight change during the helix-coil transition of κ-carrageenan measured by the SEC-LALLS method, *Macromolecules*, **31**, pp. 675–680.

137. Uliniuc, A., Popa, M., Hamaide, T., and Dobromir, M. (2012). New approaches in hydrogel synthesis: click chemistry—a review, *Cell. Chem. Technol.*, **46**, pp. 1–11.

138. Van Tomme, S. R., Storm, G., and Hennink, W. E. (2008). In situ gelling hydrogels for pharmaceutical and biomedical applications, *Int. J. Pharm.*, **355**, pp. 1–18.

139. Vermonden, T., Censi, R., and Hennink, W. E. (2012). Hydrogels for protein delivery, *Chem. Rev.*, **112**, pp. 2853–2888.

140. Vernon, B., Tirelli, N., Bächi, T., Haldimann, D., and Hubbell, J. A. (2003). Water-borne, in situ crosslinked biomaterials from phase-segregated precursors, *J. Biomed. Mater. Res. A*, **64A**, pp. 447–456.

141. Wang, W., and Wang, A. (2010). Synthesis and swelling properties of pH-sensitive semi-IPN superabsorbent hydrogels based on sodium alginate-g-poly(sodium acrylate) and polyvinylpyrrolidone, *Carbohyd. Polym.*, **80**, pp. 1028–1036.

142. Wichterle, O., and Lim, D. (1960). Hydrophilic gels for biological use, *Nature*, **185**, pp. 117–118.

143. Xu, X., Jha, A. K., Harrington, D. A., Farach-Carson, M. C., and Jia, X. (2012). Hyaluronic acid-based hydrogels: from a natural polysaccharide to complex networks, *Soft Matter*, **8**, pp. 3280–3294.

144. Yang, J. A., Kim, H., Park, K., and Hahn, S. K. (2011). Molecular design of hyaluronic acid hydrogel networks for long-term controlled delivery of human growth hormone, *Soft Matter*, **7**, pp. 868–870.

145. Yin, L., Fei, L., Cui, F., Tang, C., and Yin, C. (2007). Superporous hydrogels containing poly(acrylic acid-co-acrylamide)/O-carboxymethyl chitosan interpenetrating polymer networks, *Biomaterials*, **28**, pp. 1258–1266.

146. Zhang, H., Qadeer, A., and Chen, W. (2011). In situ gelable interpenetrating double network hydrogel formulated from binary components: Thiolated chitosan and oxidized dextran, *Biomacromolecules*, **12**, pp. 1428–1437.

147. Zhao, Q. S., Ji, Q. X., Xing, K., Li, X. Y., Liu, C. S., and Chen, X. (2009). Preparation and characterization of novel porous hydrogel films based on chitosan and glycerolphosphate, *Carbohyd. Polym.*, **76**, pp. 410–416.

148. Zhou, C., and Wu, Q. (2011). A novel polyacrylamide nanocomposite hydrogel reinforced with natural chitosan nanofibers, *Colloids Surfaces B: Biointerfaces*, **84**, pp. 155–162.

149. Zhou, C., Wu, Q., Yue, Y., and Zhang, Q. (2011). Application of rod-shaped cellulose nanocrystals in polyacrylamide hydrogels, *J. Colloid Interf. Sci.*, **353**, pp. 116–123.

Chapter 3

Rheological Characterization of Hydrogels

Romano Lapasin
Department of Engineering and Architecture, University of Trieste, Piazzale Europa, Trieste, 34127, Italy
romano.lapasin@di3.units.it

3.1 Introduction

Rheology deals with complex systems whose deformation behavior and flow properties cannot be described by the classical approaches belonging to continuum mechanics, i.e., the elasticity theory and fluid dynamics.

Complex systems (typically polymeric and disperse systems) possess molecular or structural length scales much larger than atomic and then they differ from simple liquid and solids whose mechanical behavior is quite simpler and can be described by linear constitutive equations (the Newton law or the Hooke equation) and hence characterized by a single property (viscosity or elastic modulus), whose value does not depend on flow intensity or deformation amplitude, respectively. Complex systems display

Polysaccharide Hydrogels: Characterization and Biomedical Applications
Edited by Pietro Matricardi, Franco Alhaique, and Tommasina Coviello

ISBN 978-981-4613-61-3 (Hardcover), 978-981-4613-62-0 (eBook)
www.panstanford.com

mechanical responses that are in a certain sense intermediate between those of liquids and solids because they contain both viscous and elastic components, albeit definitely more complex than in simple systems. Their behavior is viscoelastic and nonlinear, depending on both type and magnitude of the field conditions (i.e., local stresses, strains, or strain rates) as well as on their duration and variation over time. Consequently, diverse material functions must be used in the place of simple properties to fully characterize the viscoelastic responses even under simple (shear or extensional) kinematic conditions. Moreover, for every viscoelastic system the relative balance between viscous and elastic components depends on the timescale at which the system is probed. It is properly weighed through the Deborah number (De), a dimensionless quantity defined as the ratio between the relaxation time of the material, λ, and a time constant, Λ, which characterizes the kinematic conditions. At high De ($\lambda/\Lambda \ll 1$) the system behaves like an elastic solid, while its response is liquid-like at low De ($\gg 1$). Far from asymptotic conditions, more or less sophisticated constitutive equations, generally written in integral form, are necessary to describe thoroughly and carefully the nonlinear viscoelastic behavior exhibited by various complex systems, since it normally depends on the whole kinematic history previously experienced by the material. Most of these models are phenomenological in nature and then experimental tests must be performed to check their validity and determine their adjustable parameters. Consequently, the rheological characterization plays a fundamental role in the evaluation and modeling of the macroscopic behavior exhibited by complex systems under flow and deformation conditions.

Hydrogels are undoubtedly complex materials and their rheological characterization represents an important tool also to investigate their structural characteristics, even better if combined with other techniques, to individuate the conditions required for their existence and formation (see sol–gel transition), to examine effects caused by changes in chemical and environmental parameters, as well as to evaluate their potential use and performance in various biomedical applications.

Hydrogels are hydrophilic three-dimensional polymer networks that, in spite of the large amount of water or biological fluids normally entrapped, can equally display soft and rubbery consistency, thus closely resembling living tissues in their physical properties. Moreover, the high water content also contributes to their excellent biocompatibility. Up to now, numerous polymers have been proposed for the preparation of hydrogels suitable for constructing three-dimensional scaffolding in tissue engineering, drug delivery systems, cell encapsulation devices, and other biomedical applications [17, 37, 47, 59, 79, 81, 82, 86, 101, 115, 117, 134]. Among them, polysaccharides and proteins present a number of advantages over the synthetic polymers which were initially employed in the field of pharmaceutics, since they are abundant and readily available from renewable sources as well as through recombinant DNA techniques.

The category of hydrogels comprises a wide variety of systems which differ not only in their primary polymer structure but even more in their supramolecular structure because of the different interchain bonds and mechanisms leading to its formation. Chemical hydrogels are water-swollen networks of hydrophilic homopolymers or copolymers linked by covalent bonds, while physical hydrogels derive from weaker crosslinks due to chain entanglements, association bonds including hydrogen bonds, ionic bonds or strong van der Waals interactions between chains, hydrophobic interactions, or crystallites bringing together two or more macromolecular chains. It must be underlined that such noncovalent intermolecular interactions and associations are often combined together in hydrogel formation. Moreover, physical gelation is usually a complex mechanism due to the transient reversible nature of physical network junctions and to different processes presiding over the gel formation, typically conformational changes like coil–helix transition and unfolding or dissociation of helical structures driven by temperature changes in polysaccharide, protein, and DNA systems.

The degree of connectivity of chemically crosslinked networks and the consequent mechanical properties are essentially ruled by the crosslink functionality and density. The strong covalent bonds cannot be broken by thermal motion of the constituent molecules, so that the topological features of a chemical gel are preserved as it is prepared. Conversely, the binding energy of physical bonds is of the same order of thermal energy and hence physical crosslinks can be reversibly formed and destroyed by a change in temperature. The average lifetime of the physical crosslinks is relatively short, even shorter than the time of observation, while the topological structure of chemical networks looks to be frozen in time. The structural features of physical gels may be very complex, often hierarchical as is the case of several biopolymer systems, thus implying a spatially heterogeneous and wide distribution of crosslink densities and junction lifetimes. Moreover, they can be appreciably modified by even small changes of parameters such as pH, temperature, or ion addition and through mechanical energy imparted during and after preparation. Thus, it can be more easily understood how quite different rheological properties are exhibited by diverse hydrogels and also by the same system under different experimental conditions and after different thermal and mechanical histories. A wide arc of viscoelastic behaviors variously located between those typical of concentrated polymer solutions and elastic solids is covered, and this explains why hydrogels are increasingly finding their applications in diverse areas, especially in the field of pharmaceutical devices and biomedical systems, where different rheological properties are required for the various employments, during their preparation and use as well as for their in situ performance.

Indeed, preformed hydrogels are suitable candidates for injectable therapeutic delivery vehicles provided that they display shear thinning flow behavior upon the application of a proper shear stress and rapidly self-heal into solids once the stress is removed, so recovering the initial gel properties as much an soon as possible. In wound healing and other topical uses, formulations should be able to retain an adequately high viscosity in order to remain on the target area for prolonged periods of time after their application.

Alternatively, gel formation can be promoted in situ through different possible mechanisms based on solvent exchange, UV irradiation, ionic crosslinking, pH change, and temperature modulation. In these cases, both gelation kinetics and final viscoelastic properties are decisive to evaluate the polymer suitability for the specific application.

For ophthalmic uses, in situ forming hydrogels may offer several advantages over conventional eye drops, mainly due to their prolonged corneal contact time. Polymer solutions should have optimum viscosity for easy instillation into the eye and undergo rapid sol–gel transition at specific pH, ions, and temperature of the eyeball. After administration, the ocular residence time depends upon the rheological properties of the formed hydrogel which must withstand shearing forces acting during and between blinking.

Appropriate viscoelastic and mucoadhesive properties are essential for hydrogel systems destined to drug delivery in mucosal areas such as intestine, stomach, vaginal tract, ocular mucosa, and buccal and nasal cavity, since they control the drug release and retain the preparation at the absorption site. Both properties can be suitably assessed through rheological measurements.

In the field of tissue engineering hydrogel scaffolds can serve as space filling agents, tissue barriers and bioadhesives, as delivery vehicles for bioactive molecules that encourage the natural reparative process, to encapsulate and deliver cells, and to provide structural integrity and bulk for cellular organization and morphogenesis control. The polymeric hydrogel becomes an attractive scaffolding material when its structural, chemical, and mechanical properties can be tailored to mimic those of the natural extracellular matrix (ECM), keeping the target organ in mind in order to develop a suitable tissue replacement. This may be a challenging task since, at least in principle, hydrogel mechanical properties should match all the specific ECM properties including tensile strength, failure strain, viscoelastic behavior, and compressibility.

Viscosupplementation is a symptomatic treatment of moderate osteoarthritis in articular joints, where lubricating and cushioning properties of the synovial fluid may be sensibly reduced. The injection of hydrogels, administered intra-articularly in a single or multiple injection therapies, may serve to restore desirable viscoelastic behavior and thus recreate the intra-articular joint homeostasis.

These brief and partial notes on biomedical uses of hydrogels reveal that rheological investigations can play an important role in assessing the suitability of polymeric systems candidate for a specific application keeping into account the thermomechanical history which should be experienced along the whole route leading to their in situ performance. This may signify kinematic conditions of diverse type (shear, extensional or compressional, hybrid) and different intensity, from very low strains to high strain rates, depending on the boundary conditions and geometry of the deformation or flow region and on the intensity of the driving forces or velocities. Correspondingly, also dynamic conditions are diverse with different spatial distributions and values of tangential and/or normal stresses throughout the system. Indeed, the rheological characterization can provide profitable results on condition that experimental conditions of laboratory tests are equivalent to those encountered by the system during its preparation and use.

When the rheological investigations are addressed to analyze the structural characteristics of formed hydrogels, to examine the conditions required for their formation or to study their gelation rate, the characterization of the linear viscoelastic properties over a wide range of experimental times or frequencies represents an almost obligatory task. Nonetheless, the mechanical responses of a system in nonlinear regime (large strains and/or strain rates) may be equally interesting, particularly when they help to check its same belonging to or its location within the wide and variegated category of gels.

All these objectives are usually pursued with experimental methods which are macrorheological and then apt to characterize the macroscopic behavior of the systems, even if interesting perspectives have been increasingly offered by microrheological approaches, in particular to investigate the viscoelastic properties on local micro- or nanoscales.

3.2 Instruments and Procedures for the Rheological Characterization

As discussed above, a wide gamma of rheological behaviors can be displayed by polymeric hydrogels in virtue of their diverse structural features resulting from different interchain bonds and gel formation mechanisms. Consequently, numerous are the experimental devices and techniques at disposal for the rheological characterization, in particular of physical gels. For the sake of brevity, hereafter attention will mainly be focussed on fundamental instruments and experimental procedures which are more often employed for investigating the shear behavior of various soft materials and complex systems, including non-Newtonian fluids, without neglecting more specific devices and mechanical tests apt to evaluate the mechanical properties of elastic bodies.

The experimental investigations are usually addressed to characterize the shear properties, to a lesser extent the extensional ones. They can be carried out with rotational rheometers under continuous or oscillatory shear conditions, thus covering very wide intervals of strain (and stress) amplitudes and strain rates, from linear viscoelastic regime to medium-high shear conditions (of the order of 10^3 s^{-1}). The extension to very high shear conditions necessarily implies the use of capillary rheometers.

Each rotational rheometer can normally be equipped with different measuring devices (cone/plate, parallel plates, coaxial cylinders) of different dimensions whose selection is guided by the same rheological responses of the sample under investigation (i.e., the magnitude of both controlled and measured quantities) and by its structural features as well. Indeed, the use of cone/plate geometries, commonly employed for polymer solutions, may be questionable in the case of several hydrogels and other complex systems since the conical gap may be too narrow in comparison to the dimensions of the individual structural elements (gel microdomains, chain or particle aggregates, etc.). They can be conveniently replaced by parallel plate devices, whose use is more flexible, because the gap can be adjusted in accordance with specific exigencies as for disk-shaped samples of preformed self-sustaining gels and pourable physical gels. Moreover, the wall slip phenomenon due to the formation of a depletion layer at the boundary can be minimized or eliminated by resorting to parallel plates with serrated or roughened plate surfaces. Edge effects caused by water evaporation are easily accommodated by using a humidity chamber or applying a thin film of silicone oil around the free surface of the aqueous sample.

Rotational rheometers differ essentially in the input function used, i.e., either stress or strain controlled. In the former case a controlled torque is applied to the mobile part of the measuring device through a low inertia motor, while its resulting angular displacement or speed is measured. In the latter one controlled kinematic conditions are applied by the motor, while the resulting torque is measured by a transducer. Independently of rheometer type (controlled stress or controlled rate), the local tangential stresses and shear rates (or strains) can be derived from the torque value and the angular velocity (or displacement), respectively, keeping into account the geometrical characteristics of the selected measuring device.

Simple experimental procedures can be used for characterizing the rheological behavior in shear conditions in order to examine separately the time- and shear-dependence of relevant material functions.

Particularly interesting are continuous and oscillatory shear tests performed at sufficiently low strain, shear rate, or stress values since they are suitable to reveal the mechanical properties of the system in equilibrium structural states close to rest conditions. This means to operate within the linear viscoelastic regime, where the tangential and normal stresses are linearly proportional to the corresponding strain or shear rate, and hence, all the material functions are independent on the strain, shear rate or stress values; here, the viscosity $\eta (\eta = \tau/\gamma)$ and the shear modulus G $(G = \tau/\gamma)$ obey the Newton law and the Hooke equation, respectively.

3.2.1 *Oscillatory Shear Tests*

Dynamic tests are carried out with rotational rheometers by imposing a sinusoidal angular displacement or torque to one of the two boundary surfaces. When they are performed under small amplitude oscillatory shear (SAOS) conditions, the time variations of stress and strain are both sinusoidal and the maximum amplitudes of the shear stress and strain (strain rate) are linearly proportional.

Let us consider the unsteady motion of a fluid within a cone/plate device, when the upper cone undergoes a controlled sinusoidal motion of small amplitude at constant frequency f. The strain and shear rate distributions inside the sample are uniformly described by the following relationships:

$$\gamma = \gamma^0 \sin(\omega t) \tag{3.1}$$

$$\dot{\gamma} = \gamma^0 \omega \cos(\omega t) = \dot{\gamma}^0 \cos(\omega t) \tag{3.2}$$

where ω is the angular frequency or pulsation ($\omega = 2\pi f$), γ^0 and $\dot{\gamma}^0$ are the maximum strain and shear rate values during oscillating motion, respectively. The oscillatory stress response has the same frequency, but it is shifted with respect to the strain:

$$\tau = \tau^0 \sin(\omega t + \delta) \tag{3.3}$$

where τ^0 and δ are the maximum stress and the phase shift, respectively.

The shear stress response can be split into two parts as follows:

$$\tau = \tau^0 \cos\delta \cdot \sin\omega t + \tau^0 \sin\delta \cdot \cos\omega t \tag{3.4}$$

The first term is in phase with strain and, hence, represents the elastic contribution, whereas the viscous one is given by the second term, which is in phase with shear rate. In fact, in the two limiting cases $\delta = 0$ and $\delta = \pi/2$, we get the purely elastic and viscous behaviors, respectively:

$$\tau = \tau^0 \sin\omega t = G\gamma^0 \sin\omega t \quad (\text{for } \delta = 0) \tag{3.5}$$

$$\tau = \tau^0 \cos\omega t = \eta\dot{\gamma}^0 \cos\omega t \quad (\text{for } \delta = \pi/2) \tag{3.6}$$

Accordingly, Eq. 3.4 can be rewritten in two equivalent forms, by introducing two pairs of viscoelastic quantities, G', G'' and η', η'':

$$\tau = G'\gamma^0 \cdot \sin\omega t + G''\gamma^0 \cdot \cos\omega t \tag{3.7}$$

$$\tau = \eta''\dot{\gamma}^0 \cdot \sin\omega t + \eta'\dot{\gamma}^0 \cdot \cos\omega t \tag{3.8}$$

G' and G'' are the storage modulus and the loss modulus, respectively, and normally vary with applied frequency ω. They represent the elastic and viscous contributions to the properties of the fluid, and are related to the corresponding viscosity functions η' and η'' through

$$G' = \eta''\omega, \quad G'' = \eta'\omega \tag{3.9}$$

For a Newtonian fluid, the dynamic viscosity η' coincides with the shear viscosity η and G' reduces to zero, whereas for a purely elastic solid we have $G'' = 0$ and the storage modulus is equal to the shear modulus G. Both the function pairs can be treated as real and imaginary parts of complex quantities, which measure the magnitude of the total response of the fluid. The complex viscosity η^* is defined by

$$\eta^* = \eta' - i\,\eta'' \tag{3.10}$$

and, hence, we obtain the expression of the complex modulus G^*:

$$\mathrm{G}^* = i\,\omega\eta^* = \mathrm{G}' + i\,\mathrm{G}'' \tag{3.11}$$

Its magnitude $|G^*|$ corresponds to the ratio between the maximum stress τ^0 and the maximum strain γ^0. It does not depends on γ^0 and then the complex stress response is linearly proportional to the applied strain, provided that the oscillation amplitude is small enough, within the linear viscoelastic regime.

3.2.1.1 Oscillatory strain and stress sweep tests

The extension of the linear viscoelastic regime can be defined through strain (or stress) sweep tests at constant frequency. At low γ^0 (or τ^0) values both moduli are constant regardless of oscillation amplitude, whereas they become strain-dependent and test conditions loose their nondestructive character above a critical strain value γ^0_{cr}. For ordinary polymer solutions and chemical gels the linear viscoelastic regime is extended up to relatively high strain values, while γ^0_{cr} may be quite low, below 10^{-2} or even less, for other complex systems such as physical gels and dispersions. Different strain-dependent behaviors are displayed by polymer solutions and gels in the nonlinear region owing to their different structural characteristics, mainly the strength and density of the interchain interactions and links. Typical profiles are illustrated in Fig. 3.1.

In polymer solutions, increasing strain causes a progressive decrease of both moduli, more pronounced for G' than for G'' (case a). In both concentrated solutions and polymer melts such strain thinning is essentially ruled by topological constraints for chain reptation within the temporary polymer network.

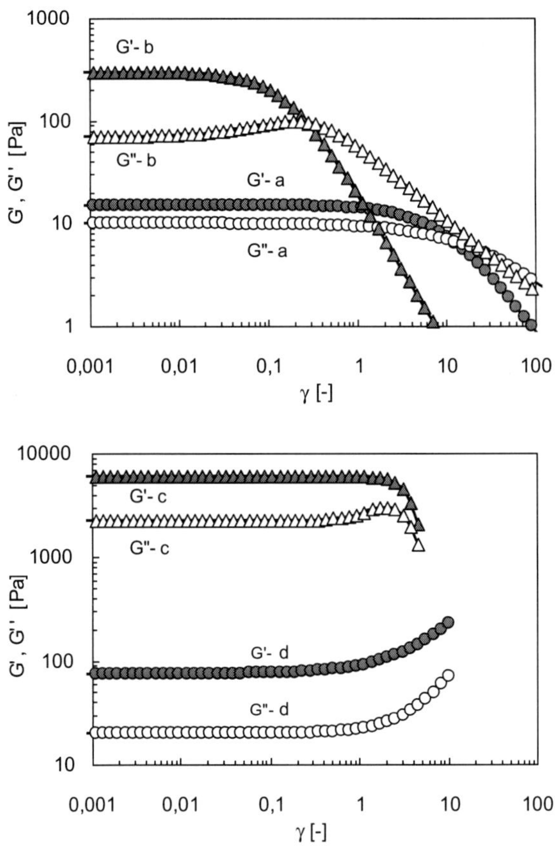

Figure 3.1 Typical strain dependences of viscoelastic moduli: (a) CMCA solution [55], (b) crosslinked CMCA microgel dispersion [55], (c) agar gel, and (d) crosslinked poly(4-vinylpyridine) network [132] (profiles obtained from experimental data fitting through Eq. 3.12). Adapted with permission from Ref. 132. Copyright © 2011, American Chemical Society.

A quite different pattern is often detected in the case of physical gels: G' decreases monotonically while G'' first increases and then decreases, so displaying a weak strain overshoot (case b). This behavior is mainly due to breakage and recovery of weak physical bonds linking adjacent chains and has been often observed for several other structured systems, as for concentrated dispersions, where the deformation-induced changes in the material's microstructure mainly concern adjacent particles or clusters and their more or less dense spatial arrangement. At higher strains a crossover condition ($G' = G''$) is attained after which the viscous component prevails.

In case of stronger physical gels an abrupt decay, mainly of the elastic component, occurs within a narrow stress range beyond the nonlinear viscoelastic region: increasing strains lead to sample rupture instead of more or less homogeneous viscous flow conditions (case c). The initial linear response can be followed by strain hardening before sample fracture (case d), resembling that shown by chemically crosslinked gels. Following the proposed mechanisms for polymeric networks, strain hardening is ascribable to strain-stiffening network components or strain-induced structural conversion of intrachain cross-linking to interchain crosslinking [132]. Strain hardening is often conjugated with highly elastic properties, and then it can be better examined through mechanical tests normally employed to determine elastic modulus, ultimate strength, and strain to failure, as in the case of interpenetrated network hydrogels, microgel-reinforced hydrogels and nanostructured polyelectrolyte hydrogels [38, 40, 70, 122, 130].

The strain dependence of both moduli can be described satisfactorily in all the cases examined by the following damping function

$$G' = G'_0 \frac{1 + a'\gamma}{1 + b'\gamma^{n'}} \quad G'' = G''_0 \frac{1 + a''\gamma}{1 + b''\gamma^{n''}} \tag{3.12}$$

where G'_0 and G''_0 are the limiting values of the storage modulus and loss modulus in the linear viscoelastic regime, respectively, whereas a', b', n', a'', b'', and n'' are adjustable parameters [54].

It must be noted that the viscoelastic response of the system becomes more complex under large amplitude oscillatory shear (LAOS) since it contains higher odd harmonics, as it emerges from Fourier transform analysis. Therefore, the results cannot be significantly expressed through only two quantities, typically G' and G'', and the linear viscoelasticity analysis looses its validity. Other approaches must be followed in order to thoroughly exploit also LAOS responses, that can play an important role for examining the structural characteristics and provide an additional rheological fingerprint of complex systems [42, 43].

3.2.1.2 Oscillatory frequency sweep test

The linear viscoelastic behavior of a system is fully characterized if we know the frequency dependence of the dynamic moduli, G' and G'', or of $|G^*|$ and loss tangent ($\tan\delta = G''/G'$) or of other two relevant material functions. Obviously, the frequency sweep test must be performed keeping the oscillation amplitude within the linear viscoelasticity border individuated through stress or strain sweep tests.

The frequency dependence of G' and G'' is normally plotted in a log–log diagram so giving the mechanical spectrum of the system, almost an identity card of its linear viscoelastic behavior, where the weights of the elastic and viscous components as well as the relaxation modes are accurately recorded. This means that mechanical spectra obtained from small amplitude oscillatory tests represent a powerful tool to distinguish different structural responses and to analyze the structural conditions of a system, offering a basis for molecular and structural interpretations. Indeed, the wide variety of linear viscoelastic behaviors exhibited by differently structured polymer systems, typically physical hydrogels, lies between those of entangled solutions and chemical crosslinked gels, where either topological constraints or covalent bonds are the only interchain links, respectively. Figure 3.2 shows four representative examples of mechanical spectra.

Different regions can be distinguished in the mechanical spectra of entanglements polymer solutions along the frequency (time) scale: in the low-frequency region the viscous component prevails over the elastic one, tending to simple scaling laws in the terminal frequency region ($G' \approx \omega^2$, $G'' \approx \omega$ for $\omega \rightarrow 0$) (case A). In such conditions the solution behaves as a viscous liquid. As ω increases, G' increases more rapidly with respect to G'', so that for sufficiently high polymer concentrations, a crossover of the $G'(\omega)$ and $G''(\omega)$ curves can be detected within the experimental window usually accessible with commercial rheometers.

A frequency region then appears where G'' is smaller than G' and may show a minimum, while a more or less extended plateau can be noted in the $G'(\omega)$ curve. Here, the observation time scales become shorter than the lifetime of the topological constraints

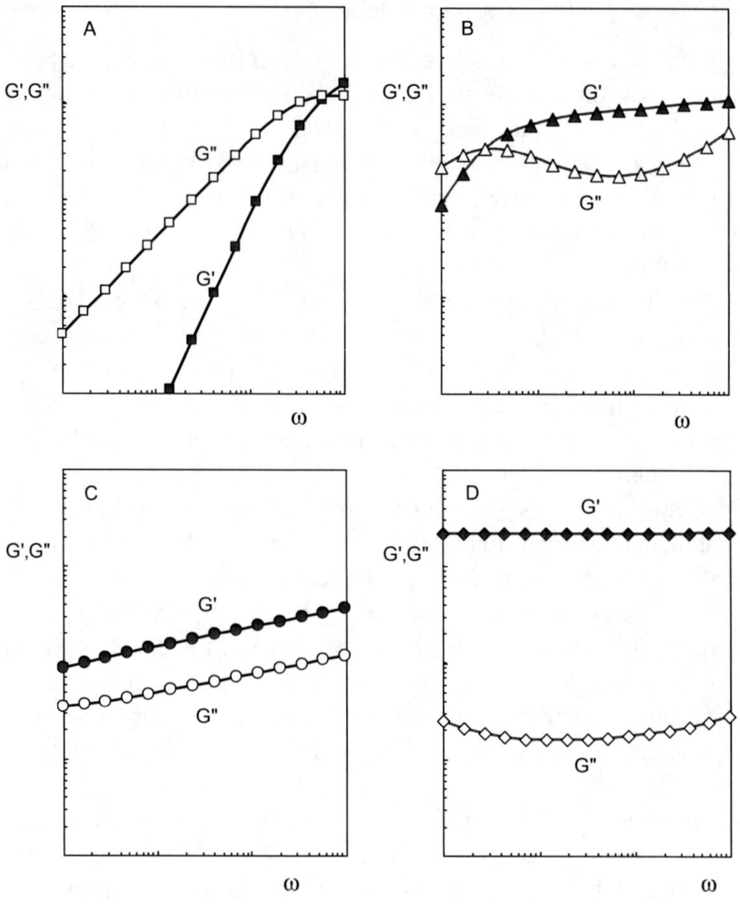

Figure 3.2 Typical mechanical spectra: (A) polymer solution, (B) polymer solution, (C) weak gel, and (D) strong gel.

entanglements (De > 1) and the solution behaves as a "pseudogel" [49] (case B).

In several other systems, polymer chains are physically crosslinked owing to hydrogen bonding, crystallizing segments, hydrophobic or ionic interactions that cooperate to produce intermolecular connections of different strength, lifetime, and spatial distribution. In the presence of two or more polymers, different structures, ranging from separated phases to interpenetrating networks, can be generated in virtue of different interchain connections, intramolecular interactions, and their interplay. Accordingly, all these systems exhibit diverse mechanical spectra of various shape and complexity, which cover the whole range between those of entangled solutions and chemical gels. Among them, physical gels represent a variegated category whose extension depends on the adopted mechanical terms of belonging. This argument is strictly related to sol–gel transition criteria that will be treated hereafter. Weak gels represent a special class of physical gels whose linear viscoelastic behavior is close to the sol–gel transition (case C).

In the case of chemical and strong physical gels (case D), crosslinked chains form a continuous three-dimensional network pervading the system from wall to wall, whose elastic character increases with increasing crosslink density. For covalently crosslinked gels G' is much higher than G'' and almost independent of frequency, tending to an equilibrium value G_e for $\omega \rightarrow 0$. Such systems actually are single macromolecules and behave as elastic solids.

Frequency sweep tests have proven to be reliable guidelines for assessing the sol–gel transition and for investigating several aspects related to hydrogels structural conditions and biomedical applications. For the sake of brevity, mention is made here only of some investigations concerning the effects related to pH or temperature variations, ion addition, polymer blending, derivatization and crosslinking parameters, interactions with drug molecules or other components, mucoadhesivity, and previous mechanical history [3, 5, 6, 10, 12, 15, 16, 18, 25, 26, 31–34, 36, 41, 51, 53, 59, 61, 67, 71, 83, 84, 90, 91, 96–98, 100, 102, 108, 114, 123, 131, 138].

Figure 3.3 reports two paradigmatic examples of gelation produced by temperature changes: (A) on cooling for a xanthan-locust bean gum blend (cold set gel), and (B) on heating for a hydroxypropyl methylcellulose (heat set gel).

Figure 3.4 shows the progressive changes in the mechanical spectra of scleroglucan solutions with increasing polymer concentration (C) and the sol–gel transition produced by an increasing content of crosslinked polymer in amidated carboxymethylcellulose blends composed of soluble CMCA and crosslinked CMCA (D).

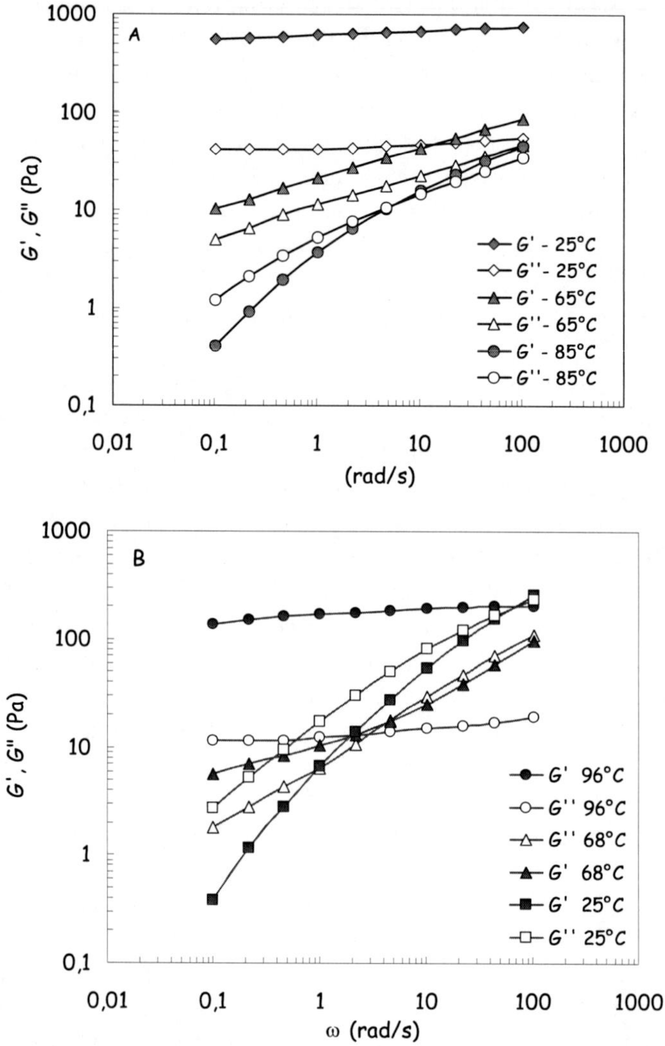

Figure 3.3 Transition from sol to gel behavior induced by temperature changes: (A) on cooling for 1:1 xanthan-locust bean gum blend (1% w/w, KCl 20 mM), and (B) on heating for hydroxypropyl methylcellulose K4M (3% w/w) [85].

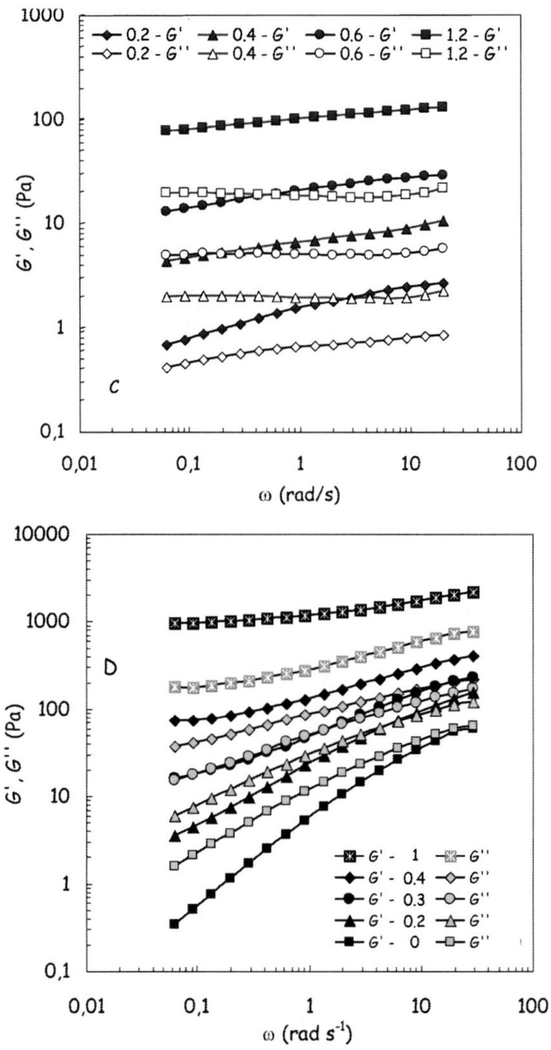

Figure 3.4 Mechanical spectra of scleroglucan solutions with different polymer concentrations (0.2 ÷ 1.2% wt/wt) (C) [4] and amidated carboxymethylcellulose (CMCA) systems (soluble CMCA [system 0], crosslinked CMCA [system 1]) and their blends with different gel fraction [0.2 ÷ 0.4]) (D) [55]. Reprinted with permission from Ref. 4. Copyright © 2005, Elsevier.

Mechanical spectra represent a canonical benchmark for assessing constitutive equations which appear to be valid candidates to provide a thorough description of the linear viscoelastic behavior. Several models have been collected in literature, starting from the classical mechanical models, resulting from parallel and/or series combinations of viscous and elastic elements, as in the Maxwell and the Voigt–Kelvin models that are their archetypes. For the sake of brevity, mention is made only of the Generalized Maxwell model: the parallel combination of N Maxwell elements gives the following expressions for the frequency dependence of viscoelastic moduli:

$$G' = \sum_{i=1}^{N} G_{\mathrm{i}} \frac{(\lambda_i \omega)^2}{1+(\lambda_i \omega)^2} \qquad G'' = \sum_{i=1}^{N} G_{\mathrm{i}} \frac{\lambda_i \omega}{1+(\lambda_i \omega)^2} \tag{3.13}$$

Thus, the linear viscoelastic behavior can be described in an equivalent and more compact manner through the discrete set of N pairs G_i-λ_i, which represents the relaxation time spectrum of the system. The differences between the $G_i(\lambda_i)$ profiles displayed by polymer solutions and various physical gels generally regard the longest relaxation modes, whose relative weights become more and more important with increasing distance from the sol behavior. This appears quite evident by comparing the relaxation time spectra of polymer blends with increasing content of crosslinked polymer, as illustrated in Fig. 3.5.

In the case of strong viscoelastic gels a better descriptive capacity is provided by the Weichert model in virtue of an additional elastic term (equilibrium modulus, G_{e}) appearing in the $G'(\omega)$ expression derived from the generalized Maxwell model.

Various models based on transient network theory have been suggested for describing and interpreting the rheological responses of physically crosslinked networks in which junctions are sufficiently weak to break and recombine in thermal fluctuations. They have been successfully applied to associating polymers, containing stickers along their backbone or only at each of their two ends [44, 45, 66, 105, 106, 113, 121, 124]. An extension to more complex physical gels should necessarily imply a substantial revision of the basic assumptions concerning the structural features of the network.

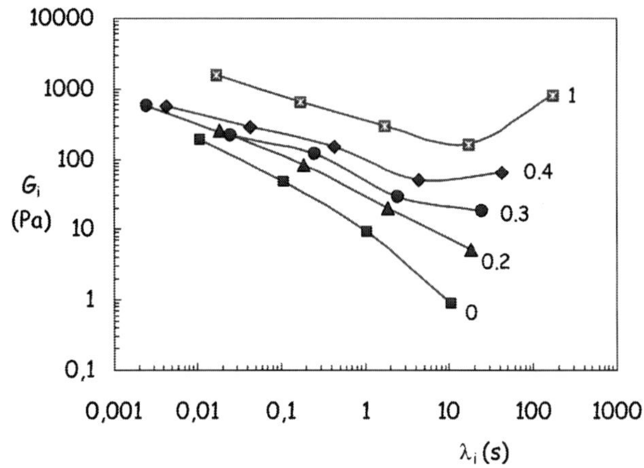

Figure 3.5 Relaxation time spectra of amidated carboxymethylcellulose (CMCA) systems (soluble CMCA (system 0), crosslinked CMCA (system 1) and their blends with different gel fraction (0.2 ÷ 0.4)) (D) [55].

Among phenomenological approaches, fractional viscoelastic models are particularly appealing for the variety of behaviors they are able to describe with satisfactory accuracy, without resorting to many adjustable parameters. More than 50 years ago, Scott-Blair suggested that the stress response of viscoelastic systems could be related to non-integer strain derivatives, in the place of strain or strain rate as usual for Hookean solids and Newtonian liquids [95]. Starting from such pioneering idea several constitutive equations with fractional integro-differential operators have been increasingly used for describing the linear viscoelastic behavior of polymeric systems. More details on this subject can be found in [30, 56, 62, 63].

Mechanical spectra are usually determined through isothermal tests performed in a frequency range whose amplitude may be determined by instrumental inertial effects and time duration limits at high and low ω, respectively. In principle, the experimental frequency window could be enlarged by applying the time–temperature superposition principle, even beyond the usual limits of commercial rheometers [28]. Accordingly, the mechanical spectra experimentally determined at different temperatures can be superposed only by shifting data along the frequency axis in accordance with the WLF or Arrhenius equations. Such a procedure can be applied to thermorheologically simple systems, typically polymer solutions and melts above the glass transition temperature, because the structural rearrangements and relaxation processes involved in deformation and flow do not change qualitatively and, even more importantly, their rates decrease with decreasing temperature in the same way. These inherent requirements often make questionable or, at least, difficult its use for hydrogels without violating the fundamental shifting criteria.

Conversely, if the system rapidly evolves at constant temperature owing to chemical crosslinking reaction or physical processes such as chain association or degradation, isothermal frequency sweeps are no longer suitable to determine the mechanical spectrum. They can be replaced by multiwave tests consisting of the simultaneous application of multiple oscillations of different frequencies. The total strain is the sum of the strains produced by all the individual oscillations acting independently and it must not exceed the limit of the linear viscoelastic region; the resulting complex wave response of the system is analyzed using a suitable data treatment. Multiwave tests repeated at different time intervals provide the same results derivable from multiple time sweep tests, but with a significant reduction in analysis time.

3.2.1.3 Oscillatory time sweep tests and temperature sweep tests

Oscillatory time sweep tests are carried out at constant amplitude and frequency of oscillation (typically, 1 Hz). They are important tools for testing systems undergoing macro- or microstructural re-arrangement with time since they directly illustrate how a material changes with time. The temporal evolution of the mechanical spectrum can be constructed from the linear viscoelastic moduli detected at equal times in time sweeps performed at different frequencies. This is particularly useful to individuate the sol–gel transition for a system undergoing chemical crosslinking in isothermal conditions as well as in a thermal gelation induced by heating or cooling. In the latter case, multiple temperature sweeps must be carried out with the same cooling or heating rate, since the thermal history may affect significantly the experimental results. During gelation both moduli, particularly G', often undergo a dramatic increase, even by more orders of magnitude, and the critical strain marking the border of the linear viscoelastic region may change sensibly. Consequently, problems may derive from using the same measuring device during the test: they concern instrumental sensitivity in the initial phase and data meaningfulness, especially near the sol–gel transition, where measurements may overcome the border of the linear region, so producing destructive effects on forming network.

For thermogelling systems a limited number of isothermal frequency sweeps may be a better solution than multiple temperature sweeps, since each isothermal test can be carried out with an appropriate device suitable for the relevant viscoelastic responses, obviously provided that stationary thermal and structural conditions are attained before test beginning.

Single time or temperature sweep tests have been frequently used to individuate the gelling time or temperature from the crossover point of G' and G'' curves in accordance with the empirical Tung–Dynes criterion [112]. An example concerning the thermal gelation for Pluronic–alginate blends is reported in Fig. 3.6. However, it must be underlined that the sol–gel transition does not

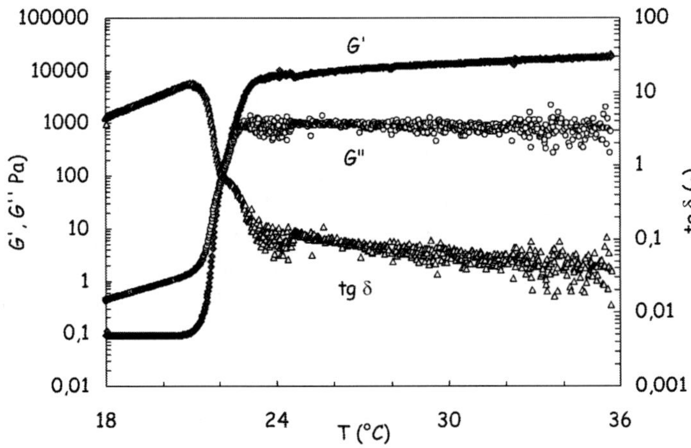

Figure 3.6 Temperature variation of elastic modulus G', loss modulus G'' and loss tangent for an aqueous Pluronic PF127 20% w/w system. Reprinted with permission from Ref. 33. Copyright © 2006, Elsevier.

necessarily coincide with the crossover condition as it was firstly demonstrated for chemically crosslinking polymers [8,9]. Another empirical method proposed for determining the gel point is based on the reduction in sample strain at the onset of gelation, via monitoring the strain–time curve at constant frequency [6].

In addition to the heating–cooling cycles traditionally used to test the reversibility of thermally induced gelation, temperature swing tests made of consecutive cycles are useful to assess the thermal stability of formed hydrogels in view of their uses [60].

3.2.2 *Oscillatory Tests and Rheological Criteria for Sol–Gel Transition*

Leaving apart all the definitions suggested for the gel state in structural terms, attention can now be profitably focused on the rheological criteria which have been proposed to individuate the sol–gel transition and then also establish the mechanical borders for the whole class of gel behaviors, since they are essentially based on the linear viscoelastic responses emerging from oscillatory tests. Several reviews are available on this subject [2, 49, 56, 78, 87, 89, 109, 110].

In accordance with the theoretical work of Flory [29], at the gel point the effective molecular weight of a crosslinking polymer goes to infinity and connectivity lines begins to extend over the entire sample volume. Consequently, the gel point should coincide with the divergence of several physical properties, including the zero shear viscosity ($\eta_0 \rightarrow \infty$), and longest relaxation time, tending to infinity, and with the equilibrium modulus G_e equal to zero. The sol–gel transition may be interpreted as a percolation threshold and typical scaling laws derived from the percolation theory can be applied to the dependence of η_0 and G_e on the distance to the gel point, below and above it, respectively [48].

This classical criterion has been successfully applied to the vast majority of chemically crosslinked polymers by performing combined measurements of the viscosity and the modulus up to and beyond the gel point. In the pre-gel state, the system is subjected to shear flow and its viscosity is checked until the stress reaches the upper limit of the instrument or the material breaks, whereas, for characterization beyond the gel point, the system is subjected to strain and the steady state modulus is measured during its growth with increasing extent of gelation. However, it must be noted that an accurate determination of the gel point through this procedure is impracticable, since the transition condition is defined through a singular behavior that is not accessible to these experiments except by extrapolation. Moreover, the shear flow measurements in the pre-gel state are enormously complicated in the case of physical gelation because of the transient nature of the resulting structures.

The criterion proposed by Tung and Dynes [112] has been frequently used to individuate the sol–gel transition in the course of crosslinking processes produced by covalent bonds under isothermal conditions or by physical interchain associations promoted by temperature changes. Accordingly, the gel point is identified with the incipient prevalence of the elastic component over the viscous one, i.e., with the crossover of G' and G'' curves determined through an oscillatory time (or temperature) sweep test. Such an empirical method is undoubtedly appealing for its simplicity and easy applicability since it requires only a single test at a given frequency, even if problems may arise in keeping test conditions within the linear viscoelastic regime. However, it does not account for the frequency dependence of the viscoelastic quantities: indeed, tests carried out at different ω values may yield different gelation times or temperatures. Furthermore, the crossover of G' and G'' is observable also in the absence of chemical crosslinks or stronger physical interchain interactions, as for polymer solutions, where only chain entanglements are present [126].

Starting from the results obtained for chemical crosslinking of polydimethylsiloxane [127], Chambon and Winter proposed a more rigorous criterion for sol–gel transition [8, 9], that fulfills the basic principle of Flory's theory and comprehends the Tung–Dynes method as a limiting case. Accordingly, at the gel point the dynamic moduli are parallel with the same power-law frequency-scaling ($G'(\omega) \sim G''(\omega) \sim \omega^{\mathrm{n}}$) over the entire ω range. The phase lag δ is independent of frequency and equal to $\pi n/2$. Correspondingly, a power law relaxation behavior is valid over the entire time domain:

$$G(t) = S\,t^{-n} \tag{3.14}$$

where the gel strength factor S and the relaxation exponent n are the parameters of the incipient gel network. The n value depends on the network connectivity. For stoichiometrically balanced chemical gels $n = 0.5$. In this case, G'' is equal to G', so finding the simple Tung–Dynes criterion. In the case of crosslinker deficiency, n values are larger and G'' prevails over G'.

The power law behavior and the relaxation exponent can be interpreted in the light of fractal geometry analysis, by assuming self-similarity of the polymeric network [72, 73, 119, 128], and then the fractal dimension, d_f, which characterizes the compactness of the polymer clusters can be derived. If hydrogels are treated as closely packed fractal clusters, scaling models, originally developed for flocculated dispersions [99, 118, 129], can be used to define the fractal characteristic of the hydrogel structure in more detail, by correlating the elastic modulus and the critical strain of the linear viscoelastic region with the polymer concentration [102]. The different character (strong-link or weak-link) of the hydrogel structure depends on the strength of the interfloc links relative to that of intrafloc links.

The Winter–Chambon method has been often used to determine the sol–gel transition in both chemical and physical gelation processes or to individuate the threshold conditions for gel state in polymeric systems as well as in colloidal dispersions, usually through the independence of tan δ on frequency. Congruency in G' and G'' has been generally assessed over larger (up to 3 decades) frequency intervals in chemical gelation than in physical ones [12, 25, 53, 64, 71, 98, 100, 104, 131]. The validation of the method may be questionable, if limited to quite narrow frequency intervals, and troublesome in the presence of more components and higher structural heterogeneity, when the mechanical spectra show more complex profile, as is the case with polymer blends and colloidal particles or gel microdomains dispersed in a polymeric solution or gel.

This method implies different experimental procedures and data sampling techniques, depending on the peculiar nature of the gel. For chemical gels the frequency dependence of the dynamic moduli could be analyzed after having stopped the crosslinking reaction at different degrees of conversion while, for thermosetting

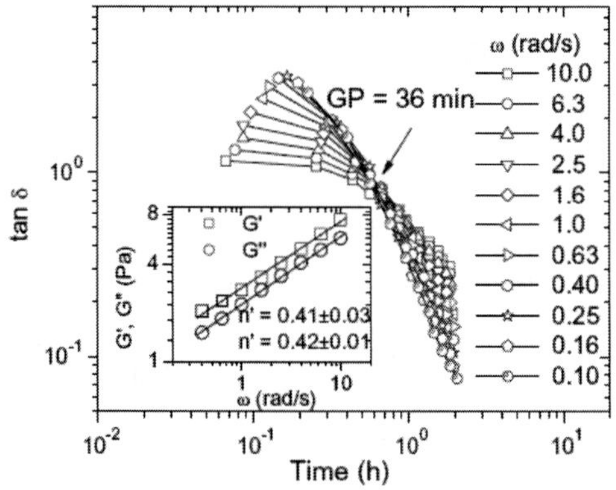

Figure 3.7 Loss tangent as a function of time for the frequencies indicated during the crosslinking process of 0.5 wt% of sodium hyaluronate with the Ugi condensation reaction with 0.13 wt% crosslinker agent. The gel point (GP) is indicated. The inset displays plot of G' and G'' versus frequency for the system at the gel point. Reprinted with permission from Ref. 64. Copyright © 2007, Elsevier.

physical gels, it can be carried out on samples kept at different temperature levels until the corresponding steady state is attained. Alternatively, a discrete series of time (or temperature) sweeps can be performed at different frequencies to provide an almost equivalent data set. The time of gelation can be identified at the point where a frequency-independent value of the loss tangent is observed, and the same scaling law exponent n is valid for both G' and G'', as illustrated in Fig. 3.7 for a cross-linking process of sodium hyaluronate.

However, it must noted that the investigation of the gelation of weak physical gels by classical rheometry may be hampered by the fact that physical bonds are shear sensitive and weak physical gels show a significant decrease of the critical strain close to the gel point. In the light of the Winter–Chambon method, several polysaccharides and other polymeric systems, usually termed as weak gels, can be also designated as critical gels since their rheological behavior is close to the sol–gel transition. Significant elastic components at low strains are conjugated with homogeneous flow properties at higher shearing conditions and this makes the difference from strong gels, which undergo rupture instead of flow, when subjected to sufficiently high stresses. A typical example of the peculiar properties exhibited by weak physical gels is reported in [31].

According to the criterion for gelation presented by Djabourov [21], if the value of the storage modulus as the frequency of oscillation approaches zero or relaxation time tends to infinity, is assumed as a measure of the gel rigidity G_e, then the first point at which G_e is nonzero can be associated with the onset of an infinite spanning network or the gel point. This criterion resembles the original Flory's proposal and is consistent with the Winter–Chambon method in that the modulus relaxation, $G(t)$, follows a power law at the gel point, while just beyond the gel point $G(t)$ it will tend to a finite value at long times. Its application is difficult due to experimental constraints associated with accessing low frequency data or long relaxation times.

Also the phenomenological definition of the term *gel* proposed by Almdal et al. [2] is based on the linear viscoelastic moduli: it should be limited to systems whose mechanical spectra are characterized by a pronounced plateau of the $G'(\omega)$ curve, extending to times at least of the order of seconds, and $G''(\omega)$ values which are considerably lower than the corresponding $G'(\omega)$ values in the plateau region.

Graphical methods constitutes an additional tool to evaluate how much the rheological behavior of structured polymeric systems is near or far from that exhibited by ordinary solutions. In the Cole–Cole diagram the real and imaginary components of the complex viscosity $|\eta^*|$ (or complex modulus $|G^*|$) are plotted against each other on linear scales. Deviations from the behavior of polymer solutions are revealed by η'' vs. η' curves whose profiles differ not only from the classical half-circle predicted by the Maxwell model but also from a more frequent skewed arc shape [76]. Almost linear traces are typical of several physical weak gels; for chemically crosslinking systems they indicate that the structural conditions attained during the process are close to the sol–gel transition in conformity with the Winter–Chambon criterion. Figure 3.8 shows how the contribution of the crosslinked polymer to the elastic character of CMCA blends can be highlighted through the Cole–Cole plot.

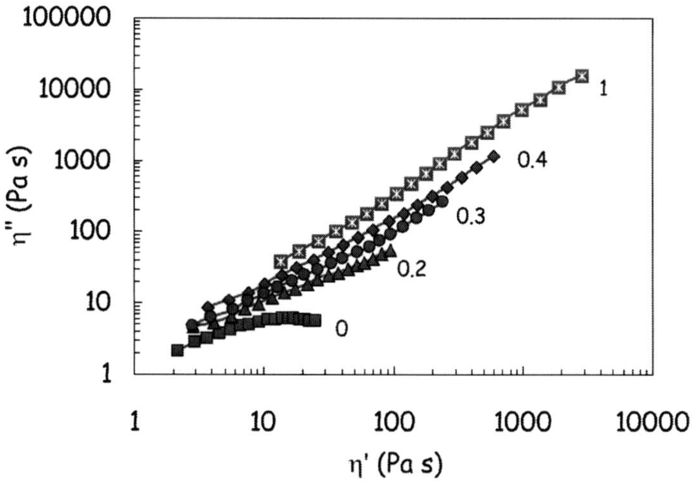

Figure 3.8 Cole–Cole plot of amidated carboxymethylcellulose (CMCA) systems (soluble CMCA [system 0], crosslinked CMCA [system 1], and their blends with different gel fraction [0.2 ÷ 0.4]) [55].

Many polymeric melts and concentrated polymer solutions obey the empirical Cox–Merz rule, which states that the complex viscosity and the steady shear viscosity must be equal at equal values of frequency and shear rate, respectively. It means that the curves of $|\eta^*|$ (ω) and $\eta(\dot{\gamma})$ can be superposed if plotted in the same graph. Thus, a surprising but powerful tool is offered for predicting the nonlinear flow behavior from linear viscoelastic properties and vice versa, provided that it holds, and, as with every rule, exceptions are found also in polymer solutions. The Cox–Merz rule fails for complex fluids whose microstructure undergo considerable alterations under nonlinear shear conditions. This happens also for weak physical gels, which are capable to display flow behavior: a significant distance may separate the two curves of $|\eta^*|$ (ω) and $\eta(\dot{\gamma})$ in the Cox–Merz plot. As it can be seen in Fig. 3.9, the Cox–Merz rule holds for CMCA solution, while it fails for crosslinked CMCA hydrogel and even for CMCA blends with relatively low gel fraction. However, it must underlined that the rule failure can represent a behavioral clue rather than a decisive proof of the structural diversity of a polymeric system from ordinary solutions.

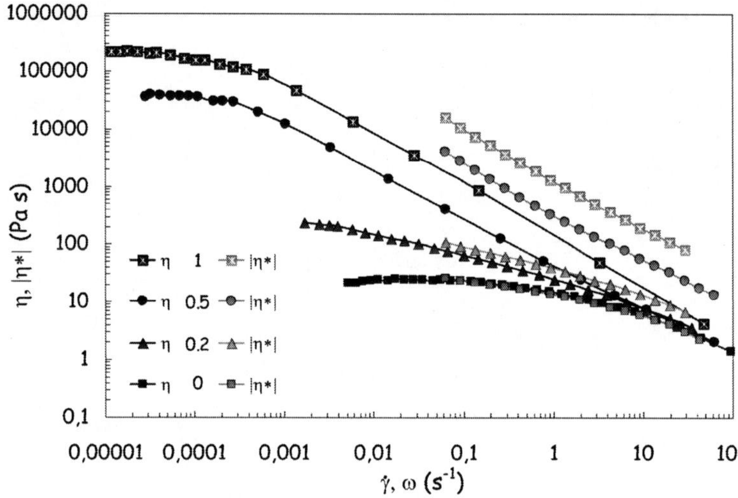

Figure 3.9 Cox–Merz diagram: comparison of the shear rate dependence of the steady viscosity η and the frequency dependence of the complex viscosity $|\eta^*|$ for amidated carboxymethylcellulose (CMCA) systems (soluble CMCA [system 0], crosslinked CMCA [system 1] and blends with different gel fractions [systems 0.2 and 0.5]) [55].

3.2.3 *Steady Shear Tests*

Several experimental procedures can be applied for characterizing the rheological behavior under unidirectional shear deformation or flow. They are made of one or more segments at controlled stress or strain or shear rate, whose value is kept constant or continuously increases or decreases with time.

3.2.3.1 Relaxation tests

Relaxation experiments can be performed by monitoring the stress response of fluids after cessation of steady shear flow or, more interestingly for various materials, when a single step strain is applied. In the latter tests the tangential stress remains constant for elastic solids, while it falls to zero immediately for Newtonian liquids. Viscoelastic systems generally exhibit stress decays $\tau(t)$ of exponential type, tending asymptotically to zero or to an equilibrium value. The relevant material function is the shear relaxation modulus

$G(t,\ \gamma^0)$ defined as $\tau(t,\ \gamma^0)/\gamma^0$. For sufficiently long times and several polymeric systems, it can be factorized into a function of time, the linear relaxation modulus $G(t)$, and a function of the applied strain γ^0, the damping function $h(\gamma^0)$, tending to 1 for $\gamma^0 \rightarrow 0$, i.e., $G(t, \gamma^0) = G(t)h(\gamma^0)$.

$G(t)$ plays a fundamental role in deriving any viscoelastic response in linear conditions because it corresponds to the kernel of the Boltzmann causal integral equation which defines how the stress state, expressed by the components of the extra-stress tensor, $\tau_{ij}(t)$, depends on the earlier kinematic history, $\dot{\gamma}_{ij}(t)$:

$$\tau_{ij}(t) = -\int_{-\infty}^{t} G\left(t - t'\right)\dot{\gamma}_{ij}\left(t'\right) dt' \tag{3.15}$$

In the case of the Weichert model, the relaxation modulus $G(t)$ is given by

$$G(t) = G_{\text{e}} + \sum_{i=1}^{N} G_{\text{i}}\exp(-t/\lambda_i) \tag{3.16}$$

where the residual asymptotic value of the exponential decay is given by the pure elastic component represented by the equilibrium modulus G_{e}.

The damping function $h(\gamma^0)$ measures the nonlinearity effects and can be determined experimentally by the vertical shift required to superpose the relaxation modulus data at high strains with a reference curve obtained in the linear viscoelastic region. For ordinary polymer solutions and melts its profile recalls the strain dependence of G' obtained from oscillatory tests.

Hydrogels and other structured polymeric systems may exhibit more complex relaxation responses. In particular, power law relaxation modes ($G(t) = St^{-n}$) are displayed by hydrogels whose structural conditions can be considered close to the sol–gel transition in conformity to the Winter–Chambon criterion [87, 102].

3.2.3.2 Creep tests

The creep test consists of measuring the time dependent strain resulting from the instantaneous application of a steady stress τ^0.

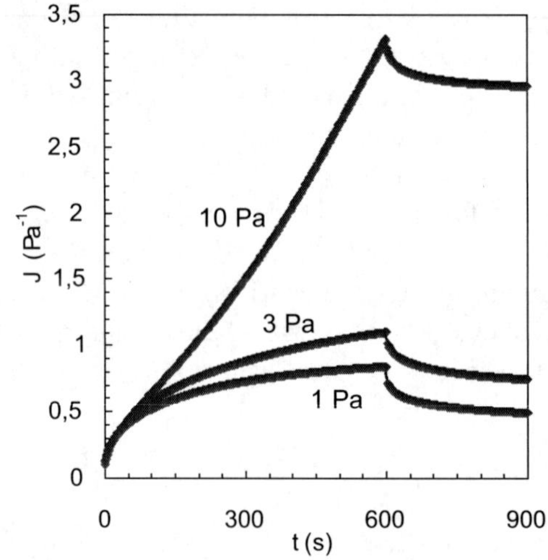

Figure 3.10 Creep-recovery tests for crosslinked CMCA hydrogel (2.5 wt%) in linear (1 Pa), nonlinear (3 Pa), and strongly nonlinear (10 Pa) conditions [55].

It is often followed by a recovery phase starting from a complete stress removal. For simple liquids strain linearly increases with time during creep test, remaining constant in the following recovery segment, while it remains constant and then instantaneously relaxes to zero for elastic solids. Viscoelastic systems display diverse nonlinear responses resulting from different balances between viscous and elastic components. The relevant material function is the creep compliance $J(t, \tau^0)$ defined as the ratio $\gamma(t, \tau^0)/\tau^0$. At sufficiently low τ^0 values it depends only on time and then a single curve $J(t)$ describes the creep behavior in linear viscoelasticity regime. An example of typical $J(t, \tau^0)$ profiles is given in Fig. 3.10.

The linear creep response can be often described with satisfactory approximation by the Burgers model resulting from the series combination of two elementary "dash-pot" models, which give the simplest representations of the viscoelastic behaviors, typical of liquids (the Maxwell model) and solids (the Voigt–Kelvin model) (see Fig. 3.11). The following expression is derived for creep

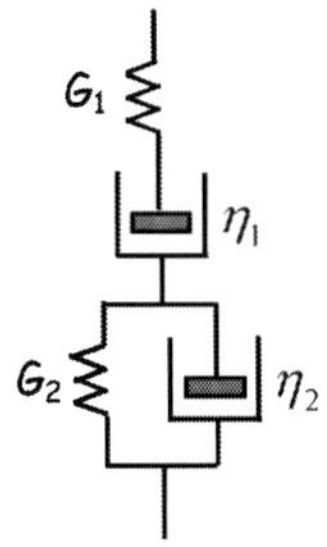

Figure 3.11 The Burgers model.

behavior:

$$J(t) = \frac{1}{G_1} + \frac{t}{\eta_1} + \frac{1}{G_2}\left(1 - \exp\left(-\frac{G_2}{\eta_2}t\right)\right)$$
$$= J_1\left(1 + \frac{t}{\lambda_1}\right) + J_2\left(1 - \exp\left(-\frac{t}{\lambda_2}\right)\right) \quad (3.17)$$

where (G_1, η_1) and (G_2, η_2) are the two parameter pairs of the Maxwell and the Voigt–Kelvin models, respectively. Their specific contributions are explicitly shown in the second version of the equation and can be differently balanced through the relative weight of J_1 and J_2 values.

The duration of creep tests should be sufficiently long to achieve a reliable definition of the asymptotic behavior, in order to distinguish a viscoelastic solid behavior from a liquid one, i.e., the attainment of an equilibrium strain or a steady strain rate, respectively. In the former case the asymptotic creep compliance can be compared to the inverse of the equilibrium gel modulus G_e determined from dynamic oscillatory tests, while in the latter one η_1 measures the steady viscous behavior at low shear conditions.

$J(t)$ profiles may be more complex or display opposite curvature (strain hardening); in other cases they are simpler as for systems close to sol–gel transition that exhibit a power law dependence ($J \approx t^n$), similar to that displayed in stress relaxation test.

$J(t)$ and $G(t)$ are intrinsically related since creep and relaxation are both manifestations of the same molecular mechanisms. However, even if simple reciprocal correlations can be established between initial and final values of both quantities, in general

$J(t) \neq 1/G(t)$. In particular, the relaxation response moves toward its equilibrium value more quickly than does the creep response.

The implicit correlation between $J(t)$ and $G(t)$ is

$$\int_0^t G(t')\, J(t-t')\, dt' = t \tag{3.18}$$

The interconversion between $J(t)$ and $G(t)$ can be done with several numerical methods as reviewed in [137], also resorting to the fractional viscoelastic models and their reciprocal interconversion [48, 63, 137].

In nonlinear conditions, as τ^0 increases, $J(t, \tau^0)$ curves no longer overlap but progressively shift towards higher values, often maintaining similar profiles. For rigid gels, nonlinear viscoelastic responses are abruptly interrupted by sample fracture after sufficiently long time. For physical weak gels and other structured systems, the shape of $J(t)$ curve may significantly change if the applied stress exceeds a critical yield value: its curvature changes in a narrow strain (and time) range since a rapid transition to appreciable shear flow takes place. This is typical of plastic systems whose rheological behavior abruptly changes within a narrow stress range, which can be individuated also through creep tests.

To this end, it must be underlined that the yield stress is a time-dependent property which has been defined in diverse manners also in relation to the criterion suggested for its determination. Consequently, different yield stress values can be obtained for different mechanical histories and experimental procedures, usually based on steady shear conditions [78, 111], even if also oscillatory strain or stress sweep tests can be used. In this case, plotting the in-phase stress component $\tau' = G'\gamma_0$ against the strain amplitude γ^0, the yield stress can be associated with the maximum of the resulting curve [133].

3.2.3.3 Stepwise sequences and ramps

Other experimental procedures aimed at characterizing the nonlinear behavior and performable with controlled stress rheometers consist of ascending stepwise sequences of constant stress segments (multicreep) and increasing stress ramps. The strain and/or strain rate responses generally depend also on time duration of stress segments or ramp. For sufficiently long segment times multicreep tests are suitable to define the steady shear-dependent behavior of physical gels and other viscoelastic fluids as well. A single test can make possible to cross the entire stress interval ranging from linear conditions to high shear flow or to sample fracture in strong gels. In the case of polymer solutions and other simpler shear thinning liquids, the results are equivalent to those obtained from shear rate ramps or ascending stepwise sequence of constant shear rate segments.

Several physical hydrogels and also associated polymer solutions exhibit marked shear thinning or apparently plastic behaviors, which are characterized by high viscosity values at low shear with a more or less accessible Newtonian plateau, followed by an extended power law region where viscosity markedly decreases with increasing shear rate or stress. An accurate description of the flow behavior is generally given by a truncated version of the Carreau–Yasuda equation:

$$\eta = \eta_0 \left(1 + (\lambda\dot{\gamma})^a\right)^{(n-1)/a} \tag{3.19}$$

where η_0 is the zero-shear rate viscosity and λ is a characteristic time, while the parameters n and a govern the transition from the Newtonian plateau to the power law region.

The distinction between plastic behavior and marked shear thinning is somewhat questionable since it depends on how large is the viscosity variation occurring in a narrow stress interval and how narrow is such an interval. When increasing stress ramps or stepwise stress sequences are applied to plastic systems, a dramatic viscosity drop of three or more orders of magnitude takes place in a very narrow stress range. Figure 3.12 shows how an increase in polymer concentration reflects into a transition from shear thinning to plastic behavior in scleroglucan systems [4].

Shear thinning and plastic behaviors can be associated to different structural processes, among which the breakup of interchain connections represents for physical hydrogels the predominant mechanism induced by increasing stresses and strains within nonlinear conditions.

Reversibility of shear-induced mechanisms and rheological behaviors are related to intrinsic structural characteristics (strength,

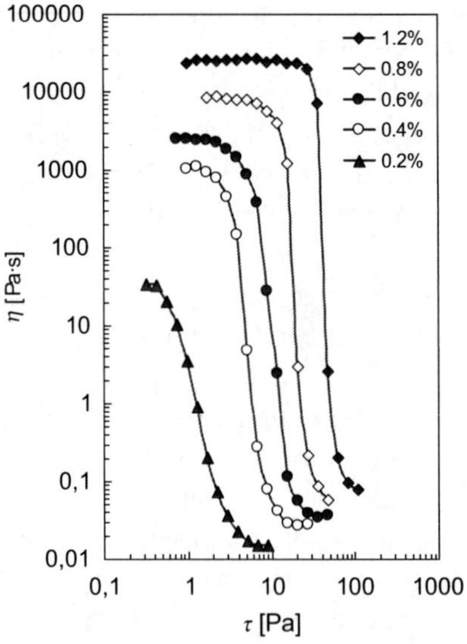

Figure 3.12 Steady viscosity versus shear stress for scleroglucan solutions with different polymer concentrations (0.2 ÷ 1.2% wt/wt). Reprinted with permission from Ref. 4. Copyright © 2005, Elsevier.

density, and distribution of intermolecular links) and mechanical history previously experienced in the non linear region.

Thixotropy is the term coined to designate the reversible time-dependent behavior associated with structural breakdown and buildup processes produced by changes in shear flow conditions. Nevertheless, it must be underlined that even structural degradation or weakening processes resulting into irreversible viscosity decays with time have been erroneously classified as thixotropic. Physical hydrogels may display thixotropic properties, which is normally conjugated with viscoelastic components and this makes more difficult their characterization [56].

Various procedures can be used to examine the time-dependent viscous behavior and its reversibility under flow conditions as well as the structural viscoelastic recovery at rest conditions.

Consecutive ascending/descending continuous stress or shear rate ramps (triangular procedures) are the experimental tests frequently preferred in industrial labs, mainly for their easy

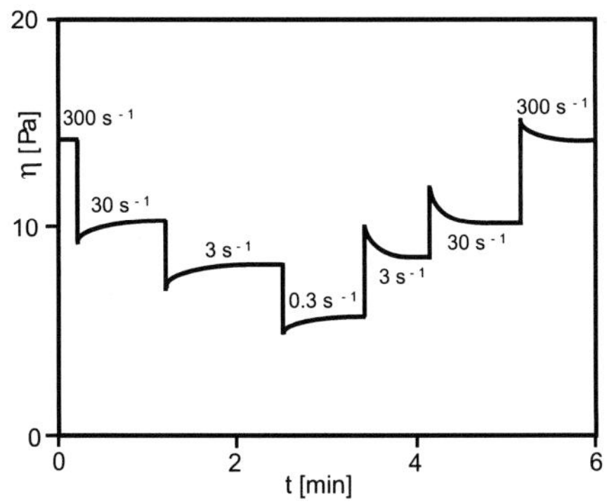

Figure 3.13 Shear stress transients in a stepwise test (descending-ascending sequence of constant shear rate values) for 0.75 wt% scleroglucan solution. Reprinted with permission from Ref. 35. Copyright © 1996, Elsevier.

application and short times required, even if the obtained results are usually dependent on test parameters (maximum shear rate or stress, cycle time) and previous mechanical history and then cannot be interpreted in an easy and unequivocal way.

Stepwise procedures composed of consecutive segments at constant shear rate (or stress) values are preferable, particularly for thixotropy characterization, to assess reversibility and kinetics of transient stress (or shear rate) responses. An example is reported in Fig. 3.13.

Oscillatory stress (or strain) sweep cycles at constant frequency serve to check the reversibility of the linear viscoelastic behavior, when the sample is subjected to large amplitude oscillations. As for triangular tests previously mentioned, the results obtained in the backward sweep are affected by the maximum oscillation amplitude reached in the upward segment.

The capacity of structural recovery after nonlinear shear conditions can be evaluated through alternate procedures composed of consecutive segments at higher and lower oscillatory constant stress (or strain) (SAOS-LAOS-SAOS sequences) or mixed procedures with consecutive SAOS and continuous shear segments at constant oscillation amplitude and shear rate, respectively. The latter ones may be useful to evaluate how much and quickly injectable hydrogels recover their viscoelastic properties after their application. An alternative procedure is given by frequency sweeps (replaced by multiwave tests for high recovery rates) performed before and after continuous shearing in nonlinear conditions [5].

3.3 Instruments and Procedures for Mechanical Characterization

In addition to the shearing tests previously illustrated, several other techniques typically used for highly elastic materials are available to examine the mechanical behavior of hydrogels, especially in view of tissue engineering applications [1].

Uniaxial tensile tests involve the application of tensile forces to strip-shaped samples placed between two clamps and stretched at constant extension rates in order to determine the stress–strain behavior over the entire range, from the initial linear region up to the ultimate failure (rupture) of the sample. Thus, several mechanical properties are derived including Young's modulus (defined as the ratio of tensile stress to tensile strain), yield strength and elongation at yield (marking the onset of plasticity region), and ultimate tensile strength and elongation at rupture. Recent examples of the application of tensile tests to the study of strain hardening of hydrogels are reported in [76, 130].

Extensiometry can serve also to study the viscoelastic properties of a hydrogel by elongating the sample to a preset length and examining the stress relaxation response over time at constant strain. Consecutive stress-relaxation segments performed after strain increments applied at controlled strain rate can help in determining reliable stress–strain data in the presence of appreciable time-dependent viscoelastic properties [14].

Tensile strength measurement is one of the most common methods used to evaluate the mucoadhesion of polymers thanks to its simplicity and because allowing ex vivo studies.

Another important property of hydrogels is fracture toughness, that measures the resistance to failure and the associated rate-dependent energy dissipation and plays a conditioning role for their performance as engineered tissues or drug delivery systems. Alternatively, tearing tests can be used to determine the fracture energy G, defined as the energy required to create a unit area of fracture surface in a sample gel. The two arms of a trouser-shaped piece are cramped and pulled apart at a constant velocity and the tearing force F is recorded. G is calculated by dividing the average of F during tearing by the width of the gel samples.

Both tensile and tearing tests are commonly used for highly elastic gels, whereas problems may arise in the case of soft physical hydrogels, starting from sample shaping and clamping. Difficulties of applying conventional fracture mechanics schemes can be attributed to intrinsic differences in fracture behavior. Unlike chemical hydrogels, physical gels do not fracture by chain scission but by viscous pull-out of whole chains from the network via plastic yielding of the crosslinks and this accounts for the rate dependence of their fracture behavior. Quasi-static fracture toughness can be evaluated by both the J-integral and EWF methods provided the effects of loading rate are investigated and accounted for [52].

Unlike extensiometry, uniaxial compression tests do not limit the hydrogel geometry to strips or rings even though they do require a flat surface. Unconfined compressive testing is usually carried out placing a cylindrical specimen between two smooth impermeable platens by compressing it. The mechanical properties of the hydrogel can be calculated from the pressure applied to the hydrogel surface and the variations in the thickness of the compressed specimen. Limitations of this approach come from bulging of the hydrogel under compression and difficulties in applying pressure evenly. Bulging can be overcome through confined compressive tests that are carried out in a confining chamber where the sample is loaded by a permeable piston.

Fundamental parameters derived from compression tests are Young's modulus, derived from the initial slope of the stress–strain curve and the compressive strength corresponding to sample fracture, although it must underlined that the accuracy of modulus values may be compromised by friction or buckling effects in short or long samples, respectively [46]. Using the Mooney–Rivlin model the elastic response of the hydrogels can be described quantitatively under both compression and elongation modes so determining the elastic modulus and providing basic information for the estimation of the effective network crosslink density [80]. An example of the use of compression tests for studying the large strain behavior of nanostructured polyelectrolyte hydrogels is reported in [70].

The large deformation behavior can be profitably tested in uniaxial compression and tension by performing loading/unloading cycles at different strain rates and up to different maximum strains [40,122]. Peculiar responses are given by double-network (DN) hydrogels, a special class of interpenetrated polymer networks with distinctly different chain lengths, that are promising scaffold candidates for regenerating load-bearing tissues and delivering drugs just in virtue of their anomalous high toughness (fracture energy over 1000 Jm^{-2}) [38, 122]. DN gels are the toughest hydrogels with a high modulus, even as tough as load-bearing cartilages and filled rubbers. Mullins effect and large hysteresis have been observed in the first loading–unloading cycle but negligible in the following cycles, if the deformation remains below the maximum value of the first cycle. No substantial recovery of their virgin behavior occurs when the deformed hydrogels are left to rest and the hysteretic dissipation has negligible dependence on deformation rate. All these phenomenological aspects can be attributed to the breaking of polymer chains and crosslinks in the shorter-chain network under deformation. A mechanistically motivated model has been developed to account for the large deformation of interpenetrating polymer networks, in particular the special Mullins effects, high fracture toughness and necking instability of DN hydrogels [136].

Another method used to examine the mechanical characteristics of hydrogels is the bulge test [1]. The sample is inflated through a window in the substrate by an applied pressure and the resulting displacement is measured using a CCD camera or a laser. A finite element model is then used to analyze the data.

Indentation tests has found increasing applications in examination of viscoelastic as well as elastic materials [1]. This technique implies indenting a sample at a single point to a preset displacement depth and measuring the reaction force required to cause the indentation, usually through a force transducer connected to the indenter. The force–displacement curve is used to calculate the elastic modulus of the material. Keeping constant the indentation depth of the sample, stress relaxation data can be obtained.

Light load indentation devices are employed to measure the adhesion properties of hydrogels by performing a loading–unloading experiment [84]. At the maximum loading, a pause time is applied before the unloading. At the end of the indentation test, the maximum adhesion force (pull-off force) required to break the contact between the indenter and the sample is measured.

Tensile (detachment) tests are employed also for the assessment of mucoadhesive properties of dosage forms, which are fundamental to the development of novel drug delivery systems. Accordingly, the mucoadhesive performance is generally evaluated from combining the maximal detachment force and the total work of adhesion (area under the detachment curve) [50].

Nondestructive techniques involving spherical indentation (long-focal microscopy based or optical coherence tomography [OCT] based) have been developed to monitor the mechanical properties of hydrogels over time under cell culture conditions [1]. Other advanced indentation techniques allow to measure the mechanical properties of hydrogels on microscale or nanoscale at different points on the material surface [24, 80].

These last citations introduce to recent developments in the field of experimental techniques addressed to the structural and rheological characterization of soft complex materials. In several hydrogels and even more in biological tissues, their structural complexity and heterogeneity reflect into local differences of the mechanical and rheological properties. High stiffness regions often exist beside regions having much lower values for the elastic modulus. These local differences cannot be detected by bulk measurements, such as tensile, compressive, and dynamic tests, whereas the local mechanical properties rather than the bulk properties can be of crucial importance for hydrogels design and applications in tissue engineering and drug delivery. Moreover, the above mentioned techniques and procedures may affect the structural integrity of the sample or imply its destruction.

Consequently, numerous non-invasive and nondestructive methods with high spatial resolution have been proposed to overcome the limits of the traditional mechanical and macrorheological techniques. They are grouped into the broadening category of "microrheology" approaches.

3.4 Microrheology

Microrheology generally exploits the motion of colloidal probe particles embedded in a material to measure the viscosity or viscoelastic moduli at the micrometer or submicrometer level. The "micro" prefix refers to both the probe size and the very small sample volume, typically of the order of a microliter. There are two broad classes of microrheology techniques that are distinguished by the driving force of the probe motion. In passive microrheology no external force is applied and the particle displacements are simply due to thermal energy (Brownian motion), while active microrheology approaches make use of electric or magnetic fields, or micromechanical forces to generate local stresses, stronger than those due to thermal forces.

Several advantages are offered by microrheology over traditional bulk rheological and mechanical measurements. Small sample volumes and short acquisition times (of the order of seconds) are required. Local probes can be suitably employed for spatially heterogeneous systems as well as for nonconventional geometries (thin films, biological cells, membranes). Passive techniques are usefully applied to analyze systems, whereas active techniques can extend the measurable range to relatively stiff materials with high elasticity. Linear viscoelastic responses can be measured over a wide frequency range up to 10^6 Hz, using high-frequency techniques, primarily DSW and optical tweezers. Conversely, these methods are computationally intensive and difficulties can arise in studying very stiff or viscous materials and nonlinear responses, since very large stresses may be required to obtain detectable strains.

Extensive reviews of the microrheology techniques and principles are reported in [13, 19, 20, 92, 94, 103, 120, 125].

3.4.1 *Passive Microrheology*

In passive microrheology, the thermal motion of the embedded probe particle is measured and the generalized Stokes–Einstein relation (GSER) is used to relate the mean-squared displacement (MSD) of the particle to the creep compliance $J(t)$

$$J(t) = 3\pi a \langle \Delta r^2(t) \rangle / 2kT \tag{3.20}$$

where k is the Boltzmann constant, T is the absolute temperature, and a is the probe radius. This imply that the material must be sufficiently soft in order for embedded particles to move detectably with only kT of energy. Results always lie within the linear viscoelastic regime because there is no external stress applied. Accordingly, the creep compliance J can be converted to other viscoelastic functions, such as the storage and loss moduli, $G'(\omega)$ and $G''(\omega)$. The GSER application to data treatment is valid in a very large frequency range (typically, 10 Hz–100 kHz) whose lower and upper limits are dictated by the appearance of significant compressional modes and inertial effects, respectively [23].

The MSD values are calculated from particle tracking methods that directly image the particle position as a function of time using laser detection schemes or obtained from ensemble-averaged light scattering experiments with a video (fluorescence or bright field) microscopy setup.

One-particle microrheology is very sensitive to the local environment of the embedded bead, and the measured responses reflect those of the bulk only if the probe size is larger than the length scale of heterogeneity in the sample. Two-particle microrheology may probe bulk responses even if individual particles do not, since it measures the cross-correlated motion of pairs of tracer particles whose distance becomes the length scale being probed in the place of the individual bead radius.

Multiparticle tracking is particularly apt to investigate materials that are heterogeneous at the length scales of the beads and to describe their bulk response. It has proven useful to measure the critical gelation of peptide and polyacrylamide solutions [57, 116]. Utilizing the connectivity-based scaling principles of time-cure superposition, the MSD values are collapsed onto master curves before and after the gel point, so determining the gel point, critical relaxation exponent and dynamic scaling exponents. Most importantly, this analytical procedure offers a means for characterizing the gelation process of weak or fast-forming gels that could previously not be monitored using traditional bulk rheology.

Significantly high throughput and resolution can be achieved with the μ^2 rheology method, resulting from the combination of multiparticle tracking and microfluidics. A rapid and efficient screening of gelation conditions is practicable with little material over a large composition space, as was proven for poly(ethylene glycol)–heparin hydrogels [93].

Also dynamic light scattering (DLS) is inherently concerned with large ensembles of particles. It is a bulk average technique, capable of providing better averaging and statistical accuracy but unable to resolve spatial heterogeneity. Its extension to opaque systems is known as diffusive wave spectroscopy (DWS) and offers the additional advantage of enlarging the accessible frequency range to much higher frequencies, that are inaccessible for traditional rheological measurements.

Passive microrheology can be profitable used to characterize gelation processes, especially when they are characterized by small initial moduli and fragile structures that are easily compromised in a typical stress controlled bulk rheometer.

Other non-invasive methods, like magnetic resonance elastography (MRE) and NMR relaxation tests have been recently used to characterize the elastic properties of tissues and gel samples. MRE obtains information about the stiffness of tissue by assessing the propagation of mechanical waves through the tissue with a special magnetic resonance imaging technique [65, 69]. The NMR relaxation rates of diffusants in peptide hydrogels have shown a linear dependency on the shear modulus of the hydrogels, so opening the door for the mechanical characterization of gels and tissues using this NMR technique [27].

3.4.2 *Active Microrheology*

Active microrheology methods allow the possibility of applying large stresses to stiff materials in order to obtain detectable strains and to extend the rheological analysis beyond the linear regime. In the magnetic tweezers techniques strong magnets are used to manipulate embedded paramagnetic particles and the particle displacements produced by the applied forces are measured with video microscopy. In addition to viscosity measurements performed by applying constant force, pulse and oscillatory forces can be applied to determine creep responses and frequency-dependent viscoelastic moduli, respectively. Bulk properties are measures in homogeneous systems provided the probe particles are larger than all length scales of the material, whereas local responses are recorder in heterogeneous samples.

Highly focused light beams are used in optical tweezers techniques to capture and move a single dielectric particle, so applying local stresses to the surrounding sample according to constant or oscillating force modes. The resulting particle displacements are recorded with direct imaging methods or laser-based detection schemes and interpreted in terms of viscoelastic response. Higher frequencies are usually attained with laser detection tweezers techniques than with video-based magnetic particle methods.

X-ray photon correlation spectroscopy (XPCS) has emerged as a powerful technique for investigating slow, nanometer-scale structural dynamics in materials. XPCS has several advantages in tracking colloidal probe motion over more widely used microrheology methods, since the small length scales and long time scales accessible with XPCS make it naturally suited for highly viscous, very stiff materials [58].

Atomic force microscopy (AFM) techniques can be used not only for imaging the topography of surfaces, but also for measuring local elasticity and viscoelasticity of thin samples and surfaces since they are sensitive to the force required to indent the surface [24]. AFM measurements can be conducted on DN gels to determine the local Young's moduli immediately below the fracture surfaces and below the usual molded surfaces, and compare the local modulus with bulk Young's moduli measured before and after the yielding deformation [7, 107].

Convergences and discrepancies between the results obtained from linear and nonlinear microrheology, and macroscopic rheology are inevitably expected for hydrogels and other complex fluids, where there is an irreducible, macromolecular level of granularity at a length scale more or less comparable to that of the probe particle.

Because of this, each microrheology experiment is, unavoidably, a study of both the rheological properties of a complex fluid and of meso-scale granularity, even if the increasing use of two-probe active microrheology may prove a significant advance in disentangling rheology from granularity of the complex fluid, and therefore yield a better understanding of both. Moreover, the inhomogeneous strain field around the probe differs qualitatively from local conditions in macroscopic shear rheology and this raises additional intrinsic difficulties. In comparing the results obtained from two techniques. Consequently, microrheology and bulk rheology must be considered as complementary tools for investigating the rheological and structural properties of complex fluids such as the hydrogels are and for extending the experimental window on viscoelastic behavior along both scales of frequency and viscoelastic moduli [11].

References

1. Ahearne, M., Yang, Y., and Liu, K.-K. (2008). *Topics in Tissue Engineering* **4**, ed. Ashammakhi, N., Reis, R., and Chiellini, F., Chapter 12, "Mechanical characterisation of hydrogels for tissue engineering in topics in tissue engineering" (Expertissues e-books, Oulu, Finland), pp. 1–16.
2. Almdal, K., Dyre, J., Hvidt, S., and Kramer, O. (1993). Towards a phenomenological definition of the term "gel," *Polym. Gels Netw.*, **1**, pp. 5–17.
3. Antunes, F. E., Coppola, L., Rossi, C. O., and Ranieri, G. A. (2008). Gelation of charged bio-nanocompartments induced by associative and non-associative polysaccharides, *Colloids Surf. B*, **66**, pp. 134–140.
4. Bais, D., Trevisan, A., Lapasin, R., Partal, P., and Gallegos, C. (2005). Rheological characterization of polysaccharide–surfactant matrices for cosmetic O/W emulsions, *J. Colloid Interface Sci.*, **290**, pp. 546–556
5. Barbucci, R., Leone, G., and Lamponi, S. (2006). Thixotropy property of hydrogels to evaluate the cell growing on the inside of the material bulk (Amber effect), *J. Biomed. Mater. Res. B*, **76**, pp. 33–40.
6. Bonino, C. A., Samorezov, J. E., Jeon, O., Alsberg, E., and Khan, S. A. (2011). Real-time in situ rheology of alginate hydrogel photocrosslinking, *Soft Matter*, **7**, pp. 11510–11517.
7. Brandl, F., Sommer, F., and Goepferich, A. (2007). Rational design of hydrogels for tissue engineering: impact of physical factors on cell behavior, *Biomaterials*, **28**, pp. 134–146.
8. Chambon, F., and Winter, H. H. (1985). Stopping of cross-linking reaction in a PDMS polymer at the gel point, *Polym. Bull.*, **13**, pp. 499–503.

9. Chambon, F., and Winter, H. H. (1987). Linear viscoelasticity at the gel point of a crosslinking PDMS with imbalanced stoichiometry, *J. Rheol.*, **31**, pp. 683–697.
10. Chang, C., Duan, B., and Zhang, L. (2009). Fabrication and characterization of novel macroporous cellulose-alginate hydrogels, *Polymer*, **50**, pp. 5467–5473.
11. Chen., D. T. N., Wen, Q., Janmey, P. A., Crocker J. C., and Yodh, A. G. (2010). Rheology of soft materials, *Ann. Rev. Condens. Matter Phys.*, **1**, pp. 301–322.
12. Choudhary, S., White, J. C., Stoppel, W. L., Roberts, S. C., and Bhatia, S. R. (2011). Gelation behavior of polysaccharide-based interpenetrating polymer network (IPN) hydrogels, *Rheol. Acta*, **50**, pp. 39–52.
13. Cicuta, P., and Donald, A. M. (2007). Microrheology: a review of the method and applications, *Soft Matter*, **3**, pp. 1449–1455.
14. Cloyd, J. M., Malhotra, N. R., Weng, L., Chen, W., Mauck, R. L., and Elliott, D. M. (2007). Material properties in unconfined compression of human nucleus pulposus, injectable hyaluronic acid-based hydrogels and tissue engineering scaffolds, *Eur. Spine J.*, **16**, pp. 1892–1898.
15. Copetti, G., Grassi, M., Lapasin, R., and Pricl, S. (1997). Synergistic gelation of xanthan gum with locust bean gum: a rheological investigation, *Glycoconjugate J.*, **14**, pp. 951–961.
16. Coviello, T., Coluzzi, G., Palleschi, A., Grassi, M., Santucci, E., and Alhaique, F. (2003). Structural and rheological characterization of scleroglucan/borax hydrogel for drug delivery, *Int. J. Biol. Macromol.*, **32**, pp. 83–92.
17. Coviello, T., Matricardi, P., Marianecci, C., and Alhaique, F. (2007). Polysaccharide hydrogels for modified release formulations, *J. Controlled Release*, **119**, pp. 5–24.
18. Crescenzi, V., Cornelio, L., Di Meo, C., Nardecchia, S., and Lamanna, R. (2007). Novel hydrogels via click chemistry: synthesis and potential biomedical applications, *Biomacromolecules*, **8**, pp. 1844–1850.
19. De Puit, R. J., Khair, A. S., and Squires, T. M. (2011). A theoretical bridge between linear and nonlinear microrheology, *Phys. Fluids*, **23**, 063102, pp. 1–13.
20. De Vlaminck, I., and Dekker, C. (2012). Recent advances in magnetic tweezers, *Annu. Rev. Biophys.*, **41**, pp. 453–472.
21. Djabourov, M. (1991). Gelation. A review, *Polym. Int.*, **25**, pp. 135–143.

22. Donald, A. M. (2010). Developments in characterizing soft matter, *MRS Bull.*, **35**, pp. 702–707.

23. Dooley, K., Kim, Y. H., Lu, H. D., Tu, R., and Banta, S. (2012). Engineering of an environmentally responsive beta roll peptide for use as a calcium-dependent cross-linking domain for peptide hydrogel formation, *Biomacromolecules*, **13**, pp. 1758–1764.

24. Drira, Z., and Yadavalli, V. K. (2013). Nanomechanical measurements of polyethylene glycol hydrogels using atomic force microscopy, *J. Mech. Behav. Biomed. Mater.*, **18** , pp. 20–28

25. Fatimi, A., Axelos, M. A. V., Tassin, J. F., and Weiss, P. (2008). Rheological characterization of self-hardening hydrogel for tissue engineering applications: gel point determination and viscoelastic properties, *Macromol. Symp.*, **266**, pp. 12–16.

26. Fatimi, A, Tassin, J.-F., Turczyn, R., Axelos, M. A. V., and Weiss, P. (2009). Gelation studies of a cellulose-based biohydrogel: the influence of pH, temperature and sterilization, *Acta Biomater.*, **5**, pp. 3423–3432.

27. Feng, Y., Taraban, M., and Yu, B. (2011). Linear dependency of NMR relaxation rates on shear modulus in hydrogels, *Soft Matter*, **7**, pp. 9890–9893.

28. Ferry, J. D. (1980). *Viscoelastic properties of polymers* (Wiley, New York, NY).

29. Flory, P. J. (1953). *Principles of polymer chemistry* (Cornell University Press, Ithaca, NY).

30. Friedrich, Chr., Schiessel, H., and Blumen, A. (1999). Constitutive behavior modeling and fractional derivatives, *Rheology Series*, **8**, pp. 429–466.

31. Gloria, A., Borzacchiello, A., Causa, F., and Ambrosio, L. (2012). Rheological characterization of hyaluronic acid derivatives as injectable materials toward nucleus pulposus regeneration, *J. Biomater. Appl.*, **26**, pp. 745–759.

32. Gong, Z., Yang, Y., Ren, Q., Chen, X., and Shao, Z. (2012). Injectable thixotropic hydrogel comprising regenerated silk fibroin and hydroxypropylcellulose, *Soft Matter*, **8**, pp. 2875–2883.

33. Grassi, G., Crevatin, A., Farra, R., Guarnieri, G., Pascotto, A., Rehimers, B., Lapasin, R., and Grassi, M. (2006). Rheological properties of aqueous Pluronic-alginate systems containing liposomes, *J. Colloid Interface Sci.*, **301**, pp. 282–290.

34. Grassi, M., Lapasin, R., Coviello, T., Matricardi, P., Di Meo, C., and Alhaique, F. (2009). Scleroglucan/borax/drug hydrogels: structure

characterisation by means of rheological and diffusion experiments, *Carbohyd. Polym.*, **78**, pp. 377–383.

35. Grassi, M., Lapasin, R., and Pricl, S. (1996). A study of the rheological behaviour of scleroglucan weak gels systems, *Carbohydr. Pol.,* **29**, pp. 169–181.
36. Hägerström, H., Paulsson, M., and Edsman, K. (2000). Evaluation of mucoadhesion for two polyelectrolyte gels in simulated physiological conditions using a rheological method, *Eur. J. Pharm. Sci.*, **9**, pp. 301–309.
37. Haidera, M., Megeeda, Z., and Ghandehari, H. (2004). Genetically engineered polymers: status and prospects for controlled release, *J. Controlled Release*, **95**, pp. 1–26.
38. Haque, M. A., Kurokawa, T., and Gong, J.P. (2012). Super tough double network hydrogels and their application as biomaterials, *Polymer,* **53**, pp. 1805–1822.
39. Harden, J. L., and Viasnoff, V. (2001). Recent advances in DWS-based micro-rheology, *Curr. Opin. Colloid Interface Sci.*, **6**, pp. 438–445.
40. Hu, J., Kurokawa, T., Nakajima, T., Sun, T. L., Suekama, T., Wu, Z. L., Liang, S. M., and Gong, J. P. (2012). High fracture efficiency and stress concentration phenomenon for microgel-reinforced hydrogels based on double-network principle, *Macromolecules*, **45**, pp. 9445–9451.
41. Hyland, L. L., Taraban, M. B., Feng, Y., Hammouda, B., and Yu, Y. B. (2012). Viscoelastic properties and nanoscale structures of composite oligopeptide-polysaccharide hydrogels, *Biopolymers*, **97**, pp. 177–188.
42. Hyun, K., Kim, S. K., Ahn, K. H., and Lee, S. J. (2002). Large amplitude oscillatory shear as a way to classify the complex fluids, *J. Non-Newtonian Fluid Mech.*, 107, pp. 51–65.
43. Hyun, K., Wilhelm, M., Klein, C. O., Cho, K. S., Nam, J. G., Ahn, K. H., Lee, S. J., Ewoldt, R. H., and McKinley, G. H. (2011). A review of nonlinear oscillatory shear tests: analysis and application of large amplitude oscillatory shear (LAOS), *Progr. Polym. Sci.*, **36**, pp. 1697–1753.
44. Indei, T. (2007). Rheological study of transient networks with junctions of limited multiplicity, *J. Chem. Phys.*, **127**, 144904; *ibidem*, 127, 144905.
45. Indei, T., and Tanaka, F. (2004). Rheological study of transient polymer networks crosslinked by two-component associative groups—Inversion of the gel skeletal structure, *J. Rheol.*, **48**, pp. 641–661.

46. Johnson, B., Bauer, J. M., Niedermaier, D. J., Crone, W. C., and Beebe, D. J. (2004). Experimental techniques for mechanical characterization of hydrogels at the microscale, *Exp. Mech.*, **44**, pp. 21–28.
47. Jonker, A. M., Löwik, D. W. P. M., and Van Hest, J. C. M. (2012), Peptide- and protein-based hydrogels, *Chem. Mater.*, **24**, pp. 759–773.
48. Katicha, S. W., and Flintsch, G. W. (2012). Fractional viscoelastic models: master curve construction, interconversion, and numerical approximation, *Rheol. Acta*, **51**, pp. 675–689.
49. Kavanagh, G.M., and Ross-Murphy, S. B. (1998). Rheological characterization of polymer gels, *Prog. Polym. Sci.*, **23**, pp. 533–562.
50. Khutoryanskiy, V. V. (2011). Advances in mucoadhesion and mucoadhesive polymers, *Macromol. Biosci.*, **11**, pp. 748–764.
51. Koop, H. S., Da-lozzo, E. J., de Freitas, R. A., Franco, C. R. C., Mitchell, D. A., and Silveira, J. L. M. (2012). Rheological characterization of a xanthan-galactomannan hydrogel loaded with lipophilic substances, *J. Pharm. Sci.*, **101**, pp. 2457–2467.
52. Kwon, H. J., Rogalsky, A. D., and Kim, D.-W. (2011). On the measurement of fracture toughness of soft biogel, *Polym. Eng. Sci.*, **51**, pp. 1078–1086.
53. Lack, S., Dulong, V., Le Cerf, D., Picton, L., Argillier, J. F., and Muller, G. (2004). Hydrogels based on pullulan crosslinked with sodium trimetaphosphate (STMP): Rheological study, *Polym. Bull.*, **52**, pp. 429–436.
54. Lapasin, R., Grassi, M., and Coceani, N. (2001). Effects of polymer addition on the rheology of O/W microemulsions, *Rheol. Acta*, **40**, pp. 185–192.
55. Lapasin, R., Suerz, M., Grassi, M., Segatti, F., Fusi, S., Mercuri, D., and De Conti, G. (2012). Rheological characterization of an injectable polymeric blend for viscosupplementation treatment, *Proc. 16th Intl Congress on Rheology*, pp. 446.
56. Lapasin, R., and Pricl, S. (1995). *Rheology of Industrial Polysaccharides: Theory and Applications* (Blackie Academic and Professional) London.
57. Larsen, T. H., and Furst, E. M. (2008). Microrheology of the liquid-solid transition during gelation, *Phys. Rev. Lett.*, **100**, 146001.
58. Leheny, R. L. (2012). XPCS: nanoscale motion and rheology, *Curr. Opin. Colloid Interface Sci.*, **17**, pp. 3–12.
59. Li, Y., Rodrigues, J., and Tomás, H. (2012). Injectable and biodegradable hydrogels: gelation, biodegradation and biomedical applications, *Chem. Soc. Rev.*, **41**, pp. 2193–2221.

60. Li, L., Thangamathesvaran, P. M., Yue, C. Y., Tam, K. C., Hu, X., and Lam, Y. C. (2001). Gel network structure of methylcellulose in water, *Langmuir*, **17**, pp. 8062–8068.

61. Magnin, D., Lefebvre, J., Chornet, E., and Dumitriu, S. (2004). Physicochemical and structural characterization of a polyionic matrix of interest in biotechnology, in the pharmaceutical and biomedical fields, *Carbohydr. Polym.*, **55**, pp. 437–453.

62. Mainardi, F. (2012). Short survey: an historical perspective on fractional calculus in linear viscoelasticity, *Fract. Calc. Appl. Anal.*, **15**, pp. 712–717.

63. Mainardi, F., and Spada, G. (2011). Creep, relaxation and viscosity properties for basic fractional models in rheology, *Eur. Phys. J. Special Topics*, **193**, pp. 133–160

64. Maleki, A., Kjøniksen, A.-L., and Nyström, B. (2007). Characterization of the chemical degradation of hyaluronic acid during chemical gelation in the presence of different cross-linker agents, *Carbohydr. Res.*, **342**, pp. 2776–2792.

65. Mariappan, Y. K., Glaser, K. J., and Ehman, R. L. (2010). Magnetic resonance elastography: a review, *Clin. Anat.*, **23**, pp. 497–511.

66. Marrucci, G., Bhargava, S., and Cooper, S. L. (1993). Models of shear-thickening behavior in physically cross-linked networks, *Macromolecules*, **26**, pp. 6483–6488.

67. Matricardi, P., Pontoriero, M., Coviello, T., Casadei, M. A., and Alhaique, F. (2008). In situ cross-linkable novel alginate-dextran methacrylate IPN hydrogels for biomedical applications: mechanical characterization and drug delivery properties, *Biomacromolecules*, **9**, pp. 2014–2020.

68. Michon, C., Cuvelier, G., and Launay, B. (1993). Concentration dependence of the critical viscoelastic properties of gelatin at the gel point, *Rheol. Acta*, **32**, pp. 94–103.

69. Minton, J.A., Iravani, A., and Yousefi, A.-M. (2012). Improving the homogeneity of tissue-mimicking cryogel phantoms for medical imaging, *Medical Physics* , **39**, pp. 6796–6807.

70. Miquelard-Garnier, G., Hourdet, D., and Creton, C. (2009). Large strain behaviour of nanostructured polyelectrolyte hydrogels, *Polymer*, **50**, pp. 481–490.

71. Moura, M. J., Figueiredo, M. M., and Gil, M. H. (2007). Rheological study of genipin cross-linked chitosan hydrogels, *Biomacromolecules*, **8**, pp. 3823–3829.

72. Muthukumar, M. (1989). Screening effect on viscoelasticity near the gel point, *Macromolecules*, **22**, pp. 4656–4658.

73. Muthukumar, M., and Winter, H. H. (1986). Fractal dimension of a cross-linking polymer at the gel point, *Macromolecules*, **19**, pp. 1284–1285.

74. Myung, D., Koh, W., Ko, J., Hu, Y., Carrasco, M., Noolandi, J., Ta, C. N., and Frank, C. W. (2007). Biomimetic strain hardening in interpenetrating polymer network hydrogels, *Polymer*, **48**, pp. 5376–5387.

75. Nakajima, T., Sun, T. L., Suekama, T., Wu, Z. L., Liang, S. M., and Gong, J. P. (2012), High fracture efficiency and stress concentration phenomenon for microgel-reinforced hydrogels based on double-network principle, *Macromolecules*, **45**, pp. 9445–9451.

76. Nakamura, K., Tanaka, Y., and Sakurai, M. (1996). Dynamic mechanical properties of aqueous gellan solutions in the sol-gel transition region, *Carbohydr. Polym.*, **30**, pp. 101–108.

77. Nguyen, Q. D., and Boger, D. V. (1992). Measuring the flow properties of yield stress fluids, *Ann. Rev. Fluid Mech.*, **24**, pp. 47–88.

78. Nishinari, K. (2009). *Progress in Colloid and Polymer Science* **136**, ed. Tokita, M., and Nishinari K., "Some thoughts on the definition of a gel" (Springer-Verlag Berlin), pp. 87–94.

79. Nishinari, K., Zhang, H., and Ikeda, S. (2000). Hydrocolloid gels of polysaccharides and proteins, *Curr. Opin. Colloid Interface Sci.*, **5**, pp. 195–201.

80. Oliver, W. C., and Pharr, G. M. (1992). An improved technique for determining hardness and elastic modulus using load and displacement sensing indentation experiments, *J. Mater. Res.*, **7**, pp. 1564–1583.

81. Overstreet, D. J., Dutta, D., Stabenfeldt, S. E., and Vernon, B. L. (2012). Injectable hydrogels, *J. Polym. Sci., Part B: Polym. Phys.*, **50**, pp. 881–903.

82. Peppas, N. A., Bures, P., Leobandung, W., and Ichikawa, H. (2000). Hydrogels in pharmaceutical formulations, *Eur. J. Pharm. Biopharm.*, **50**, pp. 27–46.

83. Pescosolido, L., Schuurman, W., Malda, J., Matricardi, P., Alhaique, F., Coviello, T., van Weeren, P. R., Dhert, W. J. A., Hennink, W. E., and Vermonden, T. (2011). Hyaluronic acid and dextran-based semi-IPN hydrogels as biomaterials for bioprinting, *Biomacromolecules*, **12**, pp. 1831–1838.

84. Pliszczak, D., Bordes, C., Bourgeois, S., Marote, P., Zahouani, H., Tupin, S., Mattei, C. P., and Lantéri, P. (2012). Mucoadhesion evaluation of

polysaccharide gels for vaginal application by using rheological and indentation measurements, *Colloids Surf. B*, **92**, pp. 168–174.

85. Pricl, S., Lapasin, R., and Grassi, M. (1997). Application of fractional viscoelastic models to polysaccharide systems, *Rheologie des materiaux pour les industries agro-alimentaires, cosmetiques et pharmaceutiques : 32eme Colloque annuel du Groupe francais de rheologie, Cahiers de rhéologie*, **15** (4), pp. 448–457.
86. Rinaudo, M. (2008). Main properties and current applications of some polysaccharides as biomaterials, *Polym. Int.*, **57**, pp. 397–430.
87. Rodd, A. B., Cooper-White, J., Dunstan, D. E., and Boger, D. V. (2001). Gel point studies for chemically modified biopolymer networks using small amplitude oscillatory rheometry, *Polymer*, **42**, pp. 185–198.
88. Rodd, A. B., Dunstan, D. E,. Ross-Murphy, S. B., and Boger, D. V. (2001). Dependence of linear viscoelastic critical strain and stress values on extent of gelation for a thermally activated gelling system, *Rheol. Acta*, **40**, pp. 23–30.
89. Ross-Murphy, S. B. (1994). Rheological characterization of polymer gels and networks, *Polym. Gels Netw.*, **2**, pp. 229–237.
90. Sandolo, C., Bulone, D., Mangione, M. R., Margheritelli, S., Di Meo, C., Alhaique, F., Matricardi, P., and Coviello, T. (2010). Synergistic interaction of locust bean gum and xanthan investigated by rheology and light scattering, *Carbohydr. Polym.*, **82**, pp. 733–741.
91. Sandolo, C., Matricardi, P., Alhaique, F., and Coviello, T. (2009). Effect of temperature and cross-linking density on rheology of chemical cross-linked guar gum at the gel point, *Food Hydrocolloid*, **23**, pp. 210–220.
92. Scheffold, F., Romer, S., Cardinaux, F., Bissig, H., Stradner, A., Rojas-Ochoa, L. F., Trappe, V., and Schurtenberger, P. (2004). New trends in optical microrheology of complex fluids and gels, *Prog. Coll. Pol. Sci.*, **123**, pp. 141–146.
93. Schultz, K. M., Bayles, A. V., Baldwin, A. D., Kiick, K. L., and Furst, E. M. (2011). Rapid, high resolution screening of biomaterial hydrogelators by μ^2 rheology, *Biomacromolecules*, **12**, pp. 4178–4182.
94. Schultz, K. M., and Furst, E. M. (2012). Microrheology of biomaterial hydrogelators, *Soft Matter*, **8**, pp. 6198–6205.
95. Scott-Blair, G. M. (1949). *Survey of general and applied rheology* (Pitman, London).
96. Seal, B. L., and Panitch, A. (2006). Physical matrices stabilized by enzymatically sensitive covalent crosslinks, *Acta Biomater.*, **2**, pp. 241–251.

97. Serrero, A., Trombotto, S., Bayon, Y., Gravagna, P., Montanari, S., and David, L. (2011). Polysaccharide-based adhesive for biomedical applications: correlation between rheological behavior and adhesion, *Biomacromolecules*, **12**, pp. 1556–1566.
98. Serrero, A., Trombotto, S., Cassagnau, P., Bayon, Y., Gravagna, P., Montanari, S., and David, L. (2010). Polysaccharide gels based on chitosan and modified starch: structural characterization and linear viscoelastic behaviour, *Biomacromolecules*, **11**, pp. 1534–1543.
99. Shih, W.-H., Shih, W. Y., Kim, S.-I., Liu, J., and Aksay, I. A. (1990). Scaling behavior of the elastic properties of colloidal gels, *Phys. Rev. A*, **42**, pp. 4772–4779.
100. Silioc, C., Maleki, A., Zhu, K., Kjøniksen, A.-L., and Nyström, B. (2007). Effect of hydrophobic modification on rheological and swelling features during chemical gelation of aqueous polysaccharides, *Biomacromolecules*, **8**, pp. 719–728.
101. Slaughter, B. V., Khurshid, S.S., Fisher, O. Z., Khademhosseini, A., and Peppas, N. A. (2009). Hydrogels in regenerative medicine, *Adv. Mater.*, **21**, pp. 3307–3329.
102. Song, F., Zhang, L.-M., Shi, J.-F., Li, N.-N., Yang, C., and Yan, L. (2010). Using hydrophilic polysaccharide to modify supramolecular hydrogel from a low-molecular-mass gelator, *Mat. Sci. Eng. C*, **30**, pp. 804–811.
103. Squires, T. M., and Mason, T. G. (2010). Fluid Mechanics of Microrheology, *Annu. Rev. Fluid Mech.*, **42**, pp. 413–438.
104. Stang Holst, P., Kjøniksen, A.-L., Bu, H., Sande, S. A., and Nyström, B. (2006). Rheological properties of pH-induced association and gelation of pectin, *Polym. Bull.*, **56**, pp. 239–246.
105. Tanaka, F., and Edwards, S. F. (1992). Viscoelastic properties of physically crosslinked networks. 1. Transient network theory, *Macromolecules*, **25**, pp. 1516–1523.
106. Tanaka, F., and Edwards, S. F. (1992). Viscoelastic properties of physically crosslinked networks. *J. Non-Newtonian Fluid Mech.*, **43**, pp. 247–271; *ibidem.*, **43**, pp. 273–288; *ibidem*, **43**, pp. 289–309.
107. Tanaka, Y., Kawauchi, Y., Kurokawa, T., Furukawa, H., Okajima, T., and Gong, J. P. (2008). Localized yielding around crack tips of double-network gels, *Macromol. Rapid Comm.*, **29**, pp. 1514–1520.
108. Tang, Y.-F., Du, Y.-M., Hu, X.-W., Shi, X.-W., and Kennedy, J. F. (2007). Rheological characterisation of a novel thermosensitive

chitosan/poly(vinyl alcohol) blend hydrogel, *Carbohydr. Polym.*, **67**, pp. 491–499.

109. te Nijenhuis, K. (1990). *Physical Networks*, ed. Burchard, W., and Ross-Murphy S. B., "Viscoelastic properties of thermoreversible gels" (Elsevier, Applied Science, London), pp. 15–33.

110. te Nijenhuis, K., and Winter, H. H. (1992). Mechanical properties at the gel point of a crystallizing poly(vinyl chloride) solution, *Macromolecules*, **22**, pp. 411–414.

111. Tiu, C., Guo, J., and Uhlherr, P. H. T. (2006). Yielding behaviour of viscoplastic materials, *J. Ind. Eng. Chem.*, **12**, pp. 653–662.

112. Tung, C.-Y. M., and Dyne, P. J. (1982). Relationship between viscoelastic properties and gelation in thermosetting systems, *J. Appl. Polym. Sci.*, **27**, pp. 569–574.

113. Vaccaro A., and Marrucci, G. (2000). A model for the nonlinear rheology of associating polymers, *J. Non-Newtonian Fluid Mech.*, **92**, pp. 261–273.

114. Vallée, F., Müller, C., Durand, A., Schimchowitsch, S., Dellacherie, E., Kelche, C., Cassel, J. C., and Leonard, M. (2009). Synthesis and rheological properties of hydrogels based on amphiphilic alginate-amide derivatives, *Carbohydr. Res.*, **344**, pp. 223–228.

115. Van Vlierberghe, S., Dubruel, P., and Schacht, E. (2011). Biopolymer-based hydrogels as scaffolds for tissue engineering applications: a review, *Biomacromolecules*, **12**, pp. 1387–1408.

116. Veerman, C., Rajagopal, K., Palla, C. S., Pochan, D. J., Schneider, J. P., and Furst, E. M. (2006). Gelation kinetics of β-hairpin peptide hydrogel networks, *Macromolecules*, **39**, pp. 6608–6614.

117. Vermonden, T., Censi, R., and Hennink, W. E. (2012). Hydrogels for protein delivery, *Chem. Rev.*, **112**, pp. 2853–2888.

118. Vignati, E., Piazza, R., Visintin, R. F. G., Lapasin, R., D'Antona, P., and Lockhart, T. P. (2005), Wax crystallization and aggregation in a model crude oil, *J. Phys.: Condens. Matter*, **17**, pp. 3651–3660.

119. Vilgis, T. A., and Winter, H. H. (1988). Mechanical self-similarity of polymers during chemical gelation, *Colloid Polym. Sci.*, **266**, pp. 494–500.

120. Waigh, T. A. (2005). Microrheology of complex fluids, *Rep. Prog. Phys.*, **68**, pp. 685–742.

121. Wang, S. Q. (1992). Transient network theory for shear-thickening fluids and physically crosslinked networks, *Macromolecules*, **25**, pp. 7003–7010.

122. Webber, R. E., Creton, C., Brown, H. R., and Gong, J. P. (2007). Large strain hysteresis and Mullins effect of tough double-network hydrogels, *Macromolecules,* **40**, pp. 2919–2927.

123. Weng, L. Chen, X., and Chen, W. (2007). Rheological characterization of in situ crosslinkable hydrogels formulated from oxidized dextran and N-carboxyethyl chitosan, *Biomacromolecules,* **8**, pp. 1109–1115.

124. Wientjes, R. H. W., Jongschaap, R. J. J., Duits, M. H. G., and Mellema, J. (1999). A new transient network model for associative polymer networks, *J. Rheol.,* **43**, pp. 375–391.

125. Wilson, L. G., and Poon, W. C. K. (2011). Small-world rheology: an introduction to probe-based active microrheology, *Phys. Chem. Chem. Phys.,* **13**, pp. 10617–10630.

126. Winter, H. H. (1987). Can the gel point of a cross-linking polymer be detected by the G'-G'' crossover ?, *Polym. Eng. Sci.,* **27**, pp. 1698–1702.

127. Winter, H. H., and Chambon, F. (1986). Analysis of linear viscoelasticity of a crosslinking polymer at the gel point, *J. Rheol.,* **30**, pp. 367–382.

128. Winter, H. H., and Mours, M. (1997). Rheology of polymers near liquid-solid transitions, *Adv. Polym. Sci.,* **134**, pp. 164–234.

129. Wu,H., and Morbidelli, M. (2001). A model relating structures of colloidal gels to their elastic properties, *Langmuir,* **17**, pp. 1030–1036.

130. Xia, Z., Patchan, M., Maranchi, J., Elisseeff, J., and Trexler, M. (2013). Determination of crosslinking density of hydrogels prepared from microcrystalline cellulose, *J Appl. Polym. Sci.* **127**, pp. 4537–4541.

131. Xing, H., Nierling, W., and Lechner, M. D. (2004). A feasible method to determine the critical DS for modified polysaccharide derivatives' gelation by gel point, tg, *Carbohydr. Polym.,* **56**, pp. 397–402.

132. Xu, D., and Craig, S. L. (2011). Strain hardening and strain softening of reversibly cross-linked supramolecular polymer networks. *Macromolecules,* **44**, pp. 7478–7488.

133. Yamaguchi, N., Chae, B.-S., Zhang, L., Kiick, K. L., and Furst, E. M. (2005). Rheological characterization of polysaccharide-poly(ethylene glycol) star copolymer hydrogels, *Biomacromolecules,* **6**, pp. 1931–1940.

134. Yan, C., and Pochan, D. J. (2010). Rheological properties of peptide-based hydrogels for biomedical and other applications, *Chem. Soc. Rev.,* **39**, pp. 3528–3540.

135. Yang, M. C., Scriven, L. E., and Macosko, C. W. (1986). Some rheological measurements on magnetic iron oxide suspensions in silicone oil, *J. Rheol.,* **30**, pp. 1015–1029.

136. Zhao, X. (2012). A theory for large deformation and damage of interpenetrating polymer networks, *J. Mech. Phys. Solids*, **60**, pp. 319–332.

137. Zhu, Y., Sun, L., and Xu, H. (2011). L-curve based Tikhonov's regularization method for determining relaxation modulus from creep test, *J. Appl. Mech., Trans. ASME*, **78**, 031002

138. Zuidema, J. M., Pap, M. M., Jaroch, D. B., Morrison, F. A., and Gilbert, R. J. (2011). Fabrication and characterization of tunable polysaccharide hydrogel blends for neural repair, *Acta Biomater.*, **7**, pp. 1634–1643.

Chapter 4

Hydrogel Mesh Size Evaluation

Mario Grassi,[a] Rossella Farra,[a] Simona Maria Fiorentino,[a] Gabriele Grassi,[b] and Barbara Dapas[b]

[a] *Department of Engineering and Architecture, Trieste University, Piazzale Europa 1, Trieste, Italy I-34127*
[b] *Department of Life Sciences, Cattinara University Hospital, Strada di Fiume 447, Trieste, Italy I-34149,*
mariog@dicamp.univ.trieste.it, ggrassi@units.it

The importance of hydrogels in the biomedical and biotechnological fields induced researchers to experimentally and theoretically study their properties. Among them, one of the most important is represented by the mesh size distribution of their polymeric network. Indeed, this characteristic heavily rules the mass exchange processes between the hydrogel and the surrounding. The aim of this chapter is to present and discuss some techniques devoted to the estimation of the polymeric network mesh size. In particular, attention will be focused on rheology, low-field NMR, crioporosimetry, and release tests.

4.1 Introduction

The demand for more and more improved performances requests for a very refined designing process of polymer-based systems.

Polysaccharide Hydrogels: Characterization and Biomedical Applications
Edited by Pietro Matricardi, Franco Alhaique, and Tommasina Coviello

ISBN 978-981-4613-61-3 (Hardcover), 978-981-4613-62-0 (eBook)
www.panstanford.com

In particular, in the case of hydrogels, one of the most important designing parameter is the mesh size distribution of the polymeric network. Indeed, it strongly regulates the mass transport properties of substances that leave the polymeric network or that come inside the polymeric network from the surrounding environment. In other words, the polymeric network is fundamental for both the drug release and the uptake processes. While the first one is typical of delivery systems [18], the second one regards, for example, the polymer-based implantable systems such as the immunoisolant membranes [24]. Indeed, the aim of these membranes is to encapsulate not human pancreatic cells, devoted to the insulin production and to protect them from the attack of the big, external human immunoglobulins. Conversely, these membranes must allow the delivery of insulin and the income of small substances, such as sugar and oxygen, fundamental for pancreatic cells survival.

As many common approaches devoted to the determination of system "nano-porosity," such as nitrogen absorption and mercury porosimetry, cannot be applied to hydrogels due to the presence of the swelling agent (water) in the polymeric network, researchers devoted their efforts to study alternative approaches. Among them, we can mention transmission electron microscopy [41], small-angle X-ray scattering [2], dynamic laser scattering [40], swelling equilibrium tests [31], rheology [8], crioporosimetry [32], low-field NMR [32], and release tests [19]. As a detailed description of all of them would lead far away, in the light of their simplicity, in this chapter we prefer to focus attention on the last four approaches.

The rheological approach implies the determination of hydrogel shear modulus, G, by fitting the frequency sweep data according to the generalized Maxwell model [23]. On the basis of G knowledge, Flory theory [13] and the equivalent network theory [37] allow to determine the crosslink density and, thus, the average mesh size. Crioporosimetry, instead, relies on the reduction of ice melting temperature and enthalpy with crystal size [6]. Indeed, in hydrogels, due to the presence of the polymeric network, water crystallizes as nanocrystals whose size is strictly related to the polymeric mesh size distribution. Low-field NMR is based on the effect of solid surfaces (here represented by the surface of the polymeric chains) on the relaxation process of the protons (they are characterized by a permanent dipole) belonging to the water filling the polymeric network [9]. This means that the protons relaxation behavior can be related to the network mesh size distribution. Finally, release tests, aimed at determining the drug diffusion coefficient in the polymeric network, allow the estimation of the average mesh size recurring to the well-known equation of Peppas [31]. This equation, based on the free volume theory of Duda and Vrentas [43], relates the ratio of the drug diffusion coefficient in the polymeric network and in the swelling solvent (water) with polymer volume fraction, drug radius, and mesh size.

4.2 Rheology

The rheological approach is essentially based on the rubber elasticity theory developed by Flory [13]. According to Flory, the polymeric network can be seen as an ensemble of different chains joint together by punctual junctions, named crosslinks. Moreover, crosslinks represent fixed points of the structure where four different chain ends meet. The use of this theory for biopolymer gels, whose macromolecular characteristics, such as flexibility, are far from those exhibited by rubbers, has been repeatedly questioned. However, recent results have shown that very stiff biopolymers might give rise to networks which are suitably described by Flory approach [15]. This holds when small deformations are considered, that is, under linear stress–strain relationship (linear viscoelastic region). Upon exceeding this threshold, not only the entropic characteristics of the network are important, but also the enthalpic ones need to be taken into account. Thus, for stiff polymers and in the linear viscoelastic range, a hydrogel can be assimilated to a rubber network hosting in its meshes the swelling fluid (water).

On the basis of this physical frame, it can be demonstrated that the network crosslink density ρ_x, defined as the moles of junctions between different polymeric chains per gel unit volume, is linked to the hydrogel shear modulus G according to the following relation:

$$\rho_x = G/RT \tag{4.1}$$

where R is the universal gas constant and T is absolute temperature. G can be computed [29] as the sum of the elastic contributions (G_i) pertaining to each element of the generalized Maxwell model (parallel combination of Maxwell elements, each one constituted by a spring and a dashpot in series) describing the hydrogel mechanical spectrum [23]:

$$G' = \sum_{i=1}^{n} G_i \frac{(\lambda_i \omega)^2}{1 + (\lambda_i \omega)^2}; \qquad G_i = \eta_i / \lambda_i \tag{4.2}$$

$$G'' = \sum_{i=1}^{n} G_i \frac{\omega \lambda_i}{1 + (\lambda_i \omega)^2}; \tag{4.3}$$

where G' and G'' are, respectively, the storage and loss moduli, $\omega = 2\pi f$ (f = solicitation frequency), n is the number of Maxwell elements considered, while η_i and λ_i represent, respectively, the dashpot viscosity and the relaxation time of the ith Maxwell element. The simultaneous fitting of Eqs. 4.2 and 4.3 to the experimental G' and G'' vs. ω data allows the determination of the unknown fitting parameters. As, usually [23], the relaxation times (λ_i) are scaled by a factor 10, model parameters are $1 + n$ (i.e., $\lambda_1 + G_i$). The choice of the proper number, n, of Maxwell elements to be considered relies on a statistical procedure [11] according to which the optimal n is that minimizing the product $\chi^2(1 + n)$. χ^2 is the sum of the squared errors deriving from the simultaneous Eq. 4.2 and 4.3 fitting to the experimental G' and G'' trend.

The link between the crosslink density and the average mesh size, ξ, is established by the equivalent network theory [37]. Starting from the evidence that, in most cases, a detailed description of a real polymeric network is rather complicated, if not impossible, this theory suggests replacing the real network topology by and idealized one, sharing the same average ρ_x (Fig. 4.1). In addition, the idealized network is made up by a collection of identical spheres centered around each crosslink. Remembering the definition of crosslink density (moles of crosslinks per hydrogel unit volume), it turns out that the volume competing to each crosslink in the real network ($1/(N_A \rho_x)$) equals the volume of each sphere as the two networks share the same ρ_x. Thus, the relation between ξ and ρ_x

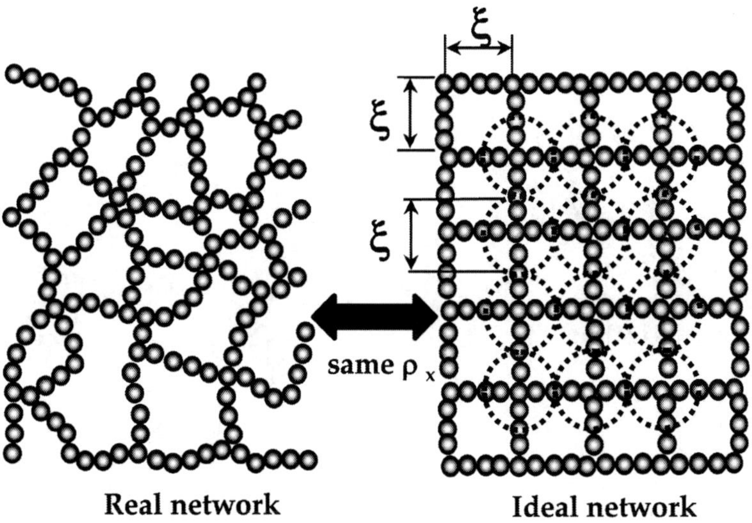

Figure 4.1 Schematic representation of a real polymeric network (gray spheres indicate rigid chain segments) and the ideal one according to the equivalent network theory. ρ_x and ξ indicate, respectively, network crosslink density and ideal network mesh size. Reprinted with permission from Ref. 37. Copyright © 1991, Elsevier.

reads

$$\frac{4}{3}\pi\left(\frac{\xi}{2}\right)^3 = \frac{1}{\rho_x N_A} \quad => \quad \xi = \sqrt[3]{6/\pi\rho_x N_A} \tag{4.4}$$

where N_A is the Avogadro number.

4.3 Crioporosimetry

Crioporosimetry relies on the triple point dependence on the dimension of the solid phase (ice crystal), as theoretically demonstrated by Brun and coworkers [6]. The Brun theory assumes different mathematical forms depending on the ice crystals shape (dictated by the polymeric mesh geometry where water solidification takes place) and boundary conditions (water in excess with respect to nanopores volume or not), as shown in Fig. 4.2:

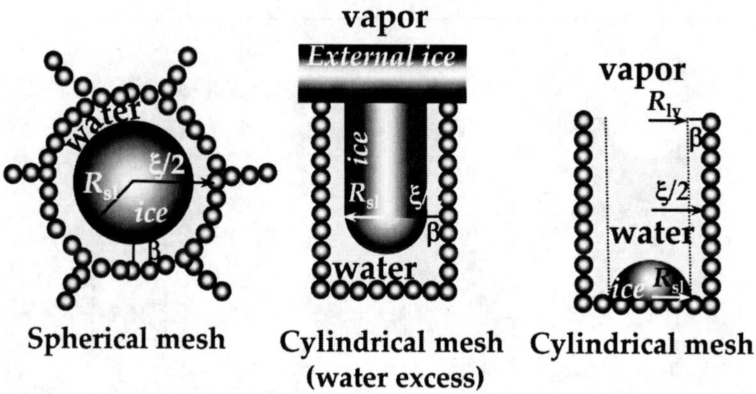

Figure 4.2 Ice crystals shape in relation to different polymeric mesh geometries (gray spheres indicate rigid chain segments). R_{sl} and R_{lv} indicate, respectively, the solid (ice)–liquid (water) and the liquid (water)–vapor interface radius of curvature, ξ is the mesh size, and β is the constant thickness of the non-freezable water layer adsorbed on the mesh wall.

cylindrical mesh with water excess, spherical mesh:

$$\int_{T_0}^{T_R} \frac{\Delta h_R}{T} dT = -\frac{2}{R_{sl}} \frac{\gamma_{sl}}{\rho_s} \tag{4.5}$$

cylindrical mesh with no water excess:

$$\int_{T_0}^{T_R} \frac{\Delta h_R}{T} dT = -\frac{2}{R_{sl}} \left(\frac{\gamma_{sl}}{\rho_s} - \gamma_{lv} \left(\frac{1}{\rho_s} - \frac{1}{\rho_l} \right) \right) \tag{4.6}$$

where T_0 represents the melting temperature of an infinitely large ice crystal (0°C), T_R and Δh_R are, respectively, ice melting temperature and specific melting enthalpy for a crystal characterized by a solid–liquid interface of curvature radius equal to R_{sl}, γ_{sl} and γ_{lv} are the ice–water and water–vapor interface tensions, and ρ_s and ρ_l are ice and water density, respectively.

In order to simultaneously take into account both situations, Eqs. 4.5 and 4.6 can be rewritten as

$$\int_{T_0}^{T_R} \frac{\Delta h_R}{T} dT = -\left(X \frac{2}{R_{sl}} \frac{\gamma_{sl}}{\rho_s} + (1-X) \frac{2}{R_{sl}} \left(\frac{\gamma_{sl}}{\rho_s} - \gamma_{lv} \left(\frac{1}{\rho_s} - \frac{1}{\rho_l} \right) \right) \right) \tag{4.7}$$

Although empirical, this choice of weighing the contribute of Eqs. 4.5 and 4.6 is coherent with a common approach adopted for the characterization of meshes that are neither spherical nor cylindrical [22]. Indeed, identifying the mesh shape factor, Z, as the power expressing the mesh volume dependence on mesh radius (3 for spherical meshes and 2 for cylindrical meshes), it is possible to connect Z to our weigh X according to the relation $Z = 2^*(1 - X) + 3^*X$. For $X = 1$ and 0, spherical ($Z = 3$) and cylindrical ($Z = 2$) meshes can be found, respectively. For $0 < X < 1$, Z varies between 2 and 3 and it represents meshes with geometrical characteristics in between those of cylinders and spheres. It is also important to point out that, due to the similar values of ice and water density (whatever temperature), the right hand side terms of Eqs. 4.5 and 4.6 do not differ too much each other. The numerical solution of Eq. 4.7 requires the knowledge of the Δh_R dependence on melting temperature and curvature radius. For this purpose, a classical thermodynamic relation can be used [46]:

$$\Delta h_R = \Delta h - \frac{3}{R_{sl}}\left(\frac{\gamma_{sv}}{\rho_s} - \frac{\gamma_{lv}}{\rho_l}\right) - \int_{T_R}^{T_0} \left(c_{ps} - c_{pl}\right) dT \tag{4.8}$$

where Δh is the specific ice melting enthalpy for an infinitely large crystal (the corresponding solid–liquid interface is flat; $R_{sl} \approx \infty$; $T_R = 0°C$), while c_{ps} and c_{pl} are, respectively, the ice and water specific heat capacities. The ρ_s, ρ_l, c_{ps}, and c_{pl} temperature dependence can be expressed by

$$\rho_s(\text{g/cm}^3) = 0.917^*(1.032 - 1.170^*T(\text{K})^*10^{-4}) \quad [21] \tag{4.9}$$

$$\rho_l(\text{g/cm}^3) = -7.1114 + 0.0882^*T(\text{K}) - 3.1959^*T^2(\text{K})^*10^{-4} + 3.8649^*T^3(\text{K})^*10^{-7} \quad [21] \tag{4.10}$$

$$c_{ps}[(\text{J/g}°\text{C}) = 2.114^*(1 + 373.7^*T(°\text{C})^*10^{-5}) \quad [7] \tag{4.11}$$

$$c_{pl}(\text{J/g}°\text{C}) = 4.222^*(1 - 54^*T(°\text{C})^*10^{-5}) \quad [7] \tag{4.12}$$

The γ_{sl}, γ_{lv}, and γ_{sv} dependence on R_{sl} can be described according to the Tolman approach [41]:

$$\frac{\gamma}{\gamma_\infty} = \left(1 + \frac{2\delta}{r}\right)^{-1} \tag{4.13}$$

where γ_∞ and γ are, respectively, the surface tension competing to a flat surface (infinite curvature radius) and a surface of curvature radius r (now identifiable with R_{sl}), while δ is the Tolmann length whose order of magnitude corresponds to the effective water molecule diameter [35] d_m and it is usually assumed [34] to be $d_m/3$ (in the case of water, $\delta \approx 0.0681$ nm). The numerical solution [32] of Eqs. 4.7–4.13 allows determining the Δh_R dependence on T_R as well as the T_R derivative with respect to R_{sl} (dT_R/dR_{sl}), fundamental functions for the determination of the mesh size distribution [22] dV_m/dR_m:

$$\frac{dV_m}{dR_m} = \frac{dV_s}{dR_m}\left(\frac{R_m}{R_m - \beta}\right)^Z = \frac{dT_R}{dR_{sl}}\frac{Q^\circ}{v\Delta h_R \rho_s}\left(\frac{R_m}{R_m - \beta}\right)^Z \tag{4.14}$$

where V_m is the meshes volume, R_m is the mesh radius ($= \xi/2$), V_s is the volume of the ice nanocrystals trapped inside the meshes, β is the constant thickness of the non-freezable water layer adsorbed on the mesh wall [21], v is the differential scanning calorimeter (DSC) heating speed, Q° is the DSC signal (W), and Z is the mesh shape factor. It is important to underline that the presence of a layer of non-freezable water adhering to the mesh walls makes the mesh volume (V_m) greater than the nanocrystals' volume (V_s) and, at the same time, it makes the mesh radius (R_m) equal to the crystal radius (R_{sl}) plus the thickness β (β being radius independent, $dR_{sl} = dR_m$). On the basis of Eq. 4.14, it is possible to determine the probability P of finding a mesh of diameter 2^*R_m:

$$P = dV_m \Bigg/ \int_{2R_{min}}^{2R_{max}} \left(\frac{dV_m}{d(2R_m)}\right) d(2R_m) \tag{4.15}$$

where R_{min} and R_{max} are, respectively, the minimum and maximum values for R_m ($= \xi/2$) (they indicate the distribution wideness).

The determination of β, essential for the P estimation, requires an iterative procedure [22, 32]. This procedure relies on the evidence that the total amount of water constituting the hydrogel (W_t) is given by the sum of the water amount that can freeze inside the hydrogel (W_f), the amount that cannot freeze inside the hydrogel due to the interaction with mesh wall (W_{nf}), and the amount of water that is out of the polymeric network (bulk or excess water, W_e). While W_t can be estimated by knowing the hydrogel mass and polymer concentration, W_f can be evaluated by measuring the amount of water melting below $0°C$:

$$W_f = \int_{T_R}^{0°C} \frac{Q°}{v\Delta h_R} dT \tag{4.16}$$

In addition, W_e can be evaluated measuring the amount of water melting for $T \geq 0°C$:

$$W_e = \int_{0°C}^{T>0°C} \frac{Q°}{v\Delta h_R} dT \tag{4.17}$$

Thus, the amount of non-freezable water (W_{nf}) can be calculated as follows:

$$W_{nf} = W_t - W_f - W_e \tag{4.18}$$

Consequently, mesh volume (V_m) can be calculated as the sum of freezable (V_f) and non-freezable (V_{nf}) water:

$$V_m = \frac{W_{nf}}{\rho_s(0°C)} + V_f \quad \text{where} \quad V_f = \int_{T_R}^{0°C} \frac{Q°}{v\Delta h_R \rho_s(T)} dT \tag{4.19}$$

Now, an iterative procedure is started assuming $\beta = 0$ and calculating the mesh volume V_{mc} as follows:

$$V_{mc} = \int_{R_{min}(T_R<0°C)}^{R_{max}(T=0°C)} \frac{dT_R}{dR_{sl}} \frac{Q°}{v\Delta h_R \rho_s} \left(\frac{R_m}{R_m - \beta}\right)^Z dR_{sl} \tag{4.20}$$

When the relative difference $\Delta V = |V_{mc} - V_m|/V_m$ is lower than a fixed tolerance ε, the assumed value for β is correct. Hence, the procedure is repeated by increasing β of $\Delta\beta$ up to convergence. As β is usually bigger than 0.5 nm [22], a prudential choice guaranteeing the convergence of the procedure consists in setting $\Delta\beta = 0.01$ nm [32]. This setting is usually compatible with the choice of $\varepsilon = 10^{-5}$.

4.4 Low-Field NMR

The nuclear magnetic resonance (NMR) approach relies on the effect of an external magnetic field on the magnetic moments of atoms characterized by an odd number of protons and/or electrons [20]. When the external magnetic field is absent, the orientation of the magnetic moments belonging to all the present atoms is, obviously, random and their vector sum (magnetization **M**) is zero. When, on the contrary, an external and constant magnetic field ($\mathbf{B_0}$) is applied, the majority of the magnetic moments tend to orient according to the $\mathbf{B_0}$ direction and against it. Thus, along the $\mathbf{B_0}$ direction (z = axial direction) the axial component of the global magnetization (M_z) will be greater than 0. Conversely, the **M** component in the *xy* plane (M_{xy}), normal to *z*, is zero. Indeed, the magnetic moments are not static, but they rotate around *z* with an inclination α and with a rotation frequency named the Larmor frequency ν_0 ($= \gamma_R{}^* B_0$; γ_R = gyromagnetic ratio). Although the rotating magnetic moments share the same rotation frequency, their phase is random and, consequently, the sum of their components in the *xy* plane is zero ($M_{xy} = 0$). If a magnetic field B_1, normal to B_0 (and much more intense than B_0), is suddenly applied, the majority of the magnetic moments will tend to orient in the B_1 direction and against it. Accordingly, for the same reasons explained in the case of the B_0 setup, the **M** component in the *xy* plane (M_{xy}) will be greater than 0, while the axial **M** component (M_z) will be vanishing. As soon as B_1 is removed, obviously, the majority of the magnetic moments will come back to the original B_0 direction and opposite it. This process, called *relaxation*, implies the progressive reduction of M_{xy} and the progressive increase of M_z. While in an ideal homogeneous magnetic field the time variation of M_{xy} and M_z is only due to *relaxation*, in real situations M_{xy} and M_z variation depends also on magnetic field inhomogeneity effects which cause different magnetic moments in the sample to precess at different rates. This inhomogeneous dephasing can be removed by applying, after a time interval equal to τ, a 180° inversion field that inverts the magnetization vectors (Hann Spin echo sequence: 90°– τ – 180°) [20].

The characteristic time of M_{xy} relaxation is called spin–spin, or transversal, relaxation time (T_2) while the characteristic time of M_z increase is called spin-lattice, or longitudinal, relaxation time (T_1). In the simplest situation, the relaxation process of M_{xy} and M_z is exponential and only one value of T_2 and T_1 fully characterizes the entire process. However, this is not always the case. Indeed, not only the two relaxation times depend on the atoms type, on temperature, and on the magnetic field strength, but they are also influenced by the surrounding. In particular, we refer to the effect of solid surfaces on the relaxation of atoms as it happens in the case of the protons of water trapped in a porous solid material. As it will be explained more in detail later, the relaxation time of protons near the solid surface is faster than that of protons not directly faced to the solid surface [14]. Thus, in principle, the relaxation of water protons trapped in a hole is the sum of two contributes: the relaxation of protons near the surface and the relaxation of protons far from the surface. A similar situation occurs in hydrogels, where the solid surface is that of the polymer chains. Thus, in general, the M_{xy} reduction process can be represented by a sum of exponential decays [9]:

$$\frac{M_{xy}\left(t>0\right)}{M_{xy}\left(t=0\right)}=I\left(t\right)=\sum_{i=1}^{m}A_{\mathrm{i}}e^{-t/T_{2\mathrm{i}}} \tag{4.21}$$

where t is time and A_{i} are the pre-exponential factors (dimensionless) proportional to the number of protons relaxing with the relaxation time $T_{2\mathrm{i}}$. Obviously, Eq. 4.21 holds also in the case of a mixture composed by n different liquids each one characterized by its own $T_{2\mathrm{i}}$. The determination of A_{i}, $T_{2\mathrm{i}}$, and m is achieved by fitting Eq. 4.21 to experimental $I(t)$ values. The statistically most probable value of m is that minimizing the product $\chi^{2}*(2m)$, where χ^{2} is the sum of the squared errors and $2m$ represents the number of fitting parameters of Eq. 4.21 [11]. The m (A_{i} - $T_{2\mathrm{i}}$) couples represent the so-called discrete relaxation times spectrum. In order to get the continuum relaxation time spectrum ($m \rightarrow \infty$), Eq. 4.21 has to be written in integral form:

$$I\left(t\right)=\int_{T_2^{\min}}^{T_2^{\max}} a\left(T_2\right)e^{\left(-t/T_2\right)}\mathrm{d}T_2 \tag{4.22}$$

where $T_2^{\min}$–$T_2^{\max}$ indicates the range of the T_2 distribution and $a(T_2)$ is the unknown amplitude of the spectral component at relaxation time T_2. $a(T_2)$ determination requires Eq. 4.22 to fit into the experimental $I(t)$ values. At this purpose, Eq. 4.22 can be discretized according to the criterion of Whittal and MacKay [45]:

$$I\left(t\right)=\int_{T_2^{\min}}^{T_2^{\max}} a\left(T_2\right)e^{\left(-t/T_2\right)}\mathrm{d}T_2\approx\sum_{i=1}^{N}a_{\mathrm{i}}\left(T_2^{\mathrm{i}}\right)e^{\left(-t/T_{2\mathrm{i}}\right)}\left(T_2^{\mathrm{i}+1}-T_2^{\mathrm{i}}\right)$$
$$=\sum_{i=1}^{N}A_{\mathrm{i}}\left(T_2^{\mathrm{i}}\right)e^{\left(-t/T_{2\mathrm{i}}\right)} \tag{4.23}$$

where the range of the T_2 distribution ($T_2^{\min}$–$T_2^{\max}$) is logarithmically subdivided into N parts (~200). Because of the noise disturbing the $I(t)$ experimental data, the fitting procedure must not minimize the χ^2 statistic, but a smoothed version (χ_s^2):

$$\chi_s^2=\sum_{i=1}^{N}\left(\frac{I_s\left(t_i\right)-I\left(t_i\right)}{\sigma_i}\right)^2+\mu\sum_{i=1}^{N-2}\left|A_{i+2}-2A_{i+1}+A_i\right|^2 \tag{4.24}$$

where σ_i is ith datum standard deviation and μ is the smoothing term weight (second summation in Eq. 4.24) proposed by Provencher [33]. Although different criteria can be followed for the μ determination, the strategy adopted by Wang [44] can be considered. According to this approach, Eq. 4.24 has to be repeatedly fitted to experimental data assuming increasing μ values. As the function Ln(χ_s) vs. Ln(μ) is linearly increasing up to a clear slope increase (curve heel), Wang suggests to take as optimal the μ value occurring just after the heel. Alternatively, Whittal [45] suggests, lacking further information, to choose the μ value ensuring that $\chi^2 \approx N$, where χ^2 is expressed by Eq. 4.24 with $\mu = 0$. Indeed, in this case, each datum is misfit by about one standard deviation.

4.4.1 *Relation between the Relaxation Time and the Mesh Diameter*

As anticipated in the previous paragraph, the interactions with the polymeric chains surface render the relaxation time of protons belonging to water molecules near the polymeric surface faster than that of bulk water protons [9]. Accordingly, in heterogeneous systems, the average relaxation time of protons will depend on the ratio between system volume and system solid surface, as demonstrated by Brownstein and Tarr [5] in the case of solid porous systems. These authors studied the problem assuming that the time (t) evolution of the magnetization density, $\rho(t)$, can be described by the following diffusion equation:

$$\frac{\partial \rho}{\partial t} = D\nabla^2 \rho - \theta\rho \quad M\,(\mathrm{t}) = \int\limits_{\mathrm{V}} \rho\,(r,t) \tag{4.25}$$

where D is the self-diffusion coefficient of water molecules, θ is a parameter accounting for losses in magnetization density occurring in the bulk, M is the total magnetization of the sample (proportional to the NMR signal), $\mathbf{r}$ is the position vector, and V is the system volume. Equation 4.25 is solved assuming that initially the magnetization density $\rho(\mathbf{r}, 0)$ is uniform and equal to $M(0)/V$, and on the solid surfaces of the system (pores wall) the following boundary condition holds:

$$(Dn \bullet \nabla\rho + \mathcal{M}\rho)|_S = 0 \tag{4.26}$$

where $\mathbf{n}$ is the surface normal vector, S is the solid surface (pores wall) and $\mathcal{M}$ (length/time) is an empirical parameter (relaxivity) accounting for the effect of surface on proton relaxations. Chui [9] adapted this theoretical approach for gels, building up the so-called fiber-cell model (Fig. 4.3).

In this theory, the polymeric network is considered as a unique long fiber, of radius R_f, surrounded by an annular region (defined the *pore*, in the terminology of Brownstein and Tarr [5]) of internal radius R_f and external radius R_c. R_c is defined as the distance from the fiber symmetry axis to the cylindrical surface where the gradient of the magnetization density zeroes ($\nabla\rho = 0$) (the cylinder of radius R_c is termed *cell*). Accordingly, Eq. 4.25 is solved by applying the condition $\nabla\rho = 0$ in R_c and the condition expressed by Eq. 4.26 on the fiber (polymer chain) surface, i.e., in R_f. As polymer gels are diluted systems (polymer volume fraction $\phi \leq 0.1$), it is supposed that the fiber content in the *cell* coincides with ϕ. Accordingly, R_c can be expressed by

$$\phi = \frac{\pi R_\mathrm{f}^2 L}{\pi R_\mathrm{c}^2 L} = \left(\frac{R_\mathrm{f}}{R_\mathrm{c}}\right)^2 \Rightarrow \frac{R_\mathrm{c}}{R_\mathrm{f}} = \frac{1}{\sqrt{\phi}} \tag{4.27}$$

where L is the total fiber (= polymer chains) length. On the basis of this frame, the pore hydraulic radius (R_h) can be deduced:

$$R_\mathrm{h} = \frac{2V_\mathrm{p}}{S} = 2\frac{\pi\left(R_\mathrm{c}^2 - R_\mathrm{f}^2\right)L}{L2\pi R_\mathrm{f}} = R_\mathrm{f}\left(\frac{1-\phi}{\phi}\right) \tag{4.28}$$

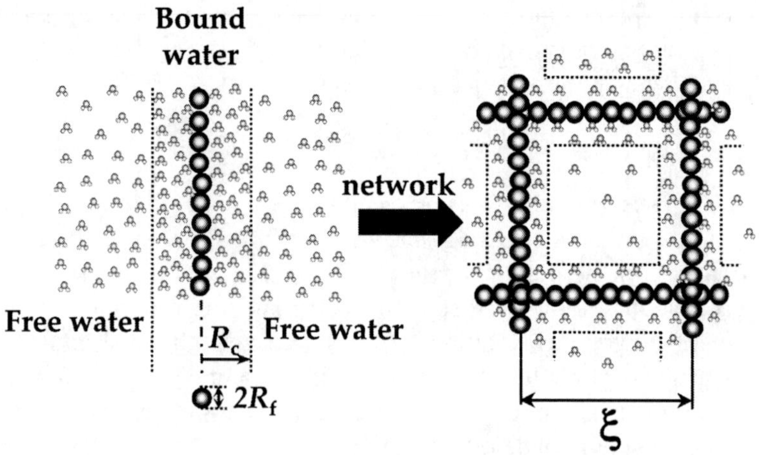

Figure 4.3 Fiber-cell model. The relaxation time of protons belonging to water molecules (light gray ensembles) near the polymeric chain surface (bound water) is lower than that competing to free water protons. R_f is the polymeric chain radius (dark gray spheres indicate rigid chain segments), ξ is the mesh size, while R_c indicates the radial position where the magnetization gradient zeroes.

where V_p is pore volume. The relation between R_h and the polymeric network mesh size (ξ) can be found recalling the Scherer theory [36]. This author assumed that the polymeric network can be represented as an ensemble of cylinders of radius R_f (i.e., polymeric chains) intersecting in a regular, cubic array. Accordingly, he demonstrated that the hydraulic radius of the cubic cell depends on ξ according to the following relation:

$$R_h = \frac{2V_p}{S} = 2\frac{(1-\phi)}{\left(6\pi - 24\sqrt{2}\,(R_f/\xi)\right)(R_f/\xi)^2} R_f \quad (4.29)$$

where $(R_f/\xi) \leq 0.5$. Starting from Eq. 4.29 it is possible to derive two simple expressions relating R_h to R_f and ξ to R_f. Indeed, defining x the ratio R_f/ξ, according to Sherer, we have:

$$\phi = 3\pi x^2 - 8\sqrt{2}x^3 \quad (4.30)$$

$$R_h = 2\frac{(1-\phi)}{\left(6\pi x^2 - 24\sqrt{2}x^3\right)} R_f \quad (4.31)$$

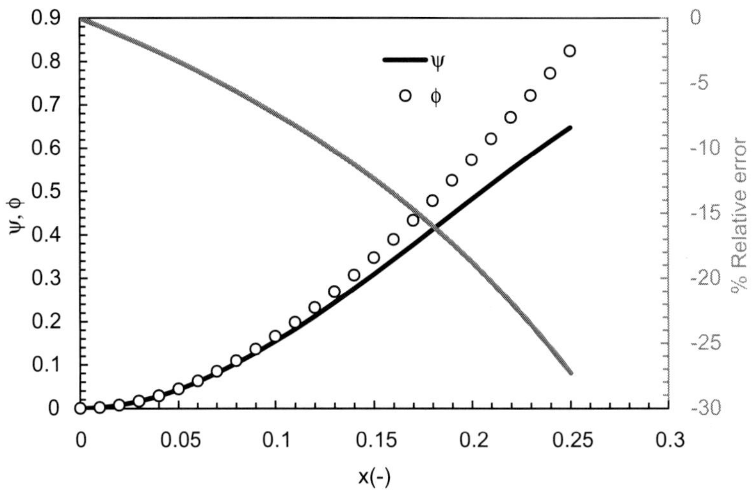

Figure 4.4 Comparison between functions ψ (solid black line, Eq. 4.32) and 2ϕ (open circles, Eq. 4.30). Gray solid line indicates the % relative error between ψ and ϕ (see right hand vertical axis).

In the limit $x \rightarrow 0$ (i.e., $\phi \rightarrow 0$), it is easy to see that the denominator appearing in Eq. 4.31 can be approximated by 2ϕ:

$$\psi = 6\pi x^2 - 24\sqrt{2}x^3 = 6\pi x^2 - 16\sqrt{2}x^3 - 8\sqrt{2}x^3 \quad (4.32)$$

$$\psi \approx 6\pi x^2 - 16\sqrt{2}x^3 = 2\phi \quad (4.33)$$

Figure 4.4 shows the comparison between ψ and ϕ for different x values. It can be seen that for $x \leq 0.1$, the error involved in the substitution of ψ by 2ϕ is lower than 8%. It can also be noted that $x \leq 0.1$ corresponds to $\phi \leq 0.16$.

Thus, Eq. 4.29 can be approximated by

$$R_\mathrm{h} \approx \frac{(1-\phi)}{\phi} R_\mathrm{f} \quad (4.34)$$

Now let's focus attention on the simplified relation between ξ and R_f. Assuming $z = 1/x$ and $y = R_\mathrm{f}/R_\mathrm{h}$, Eq. 4.29 can be rearranged to give

$$R_\mathrm{h}\left\{z^3\left(2y^2\right) - z6\pi\left(y^2 + y\right) + 8\sqrt{2}y\left(3 + 2y\right)\right\} = 0 \quad (4.35)$$

In the limit $y \rightarrow 0$, the cubic equation contained in Eq. 4.35 can be approximated by

$$z\left\{z^2\left(2y^2\right) - 6\pi\left(y\right)\right\} = 0 \quad (4.36)$$

Apart from the trivial and un-physical solution $z = 0$, the only admissible solution of Eq. 4.36 is

$$z = \sqrt{\frac{3\pi}{y}} \tag{4.37}$$

Figure 4.5 makes clear that Eq. 4.37 is a good approximation of the only physically admissible solution of Eq. 4.35 in the entire ϕ range explored.

Figure 4.5 is built by fixing y in the range 0.01–0.25, solving both Eq. 4.35 and Eq. 4.36 and calculating ϕ according to Eq. 4.30. Thus, Eq. 4.29 can be approximated by

$$\xi \overset{\text{Eq.(4.37)}}{\approx} R_\text{f}\sqrt{\frac{3\pi}{y}} \overset{\text{Eq.(4.34)}}{\approx} R_\text{f}\sqrt{3\pi\frac{(1-\phi)}{\phi}} \tag{4.38}$$

The error involved in adopting this simplified expression in place of the exact one (Eq. 4.29 or Eq. 4.35) is lower than 2.25% in the ϕ range considered $0 \leq \phi \leq 0.23$.

Equation 4.34 and Eq. 4.38 are fundamental for the practical application of the fiber-cell model. Indeed, Chui [9] demonstrated

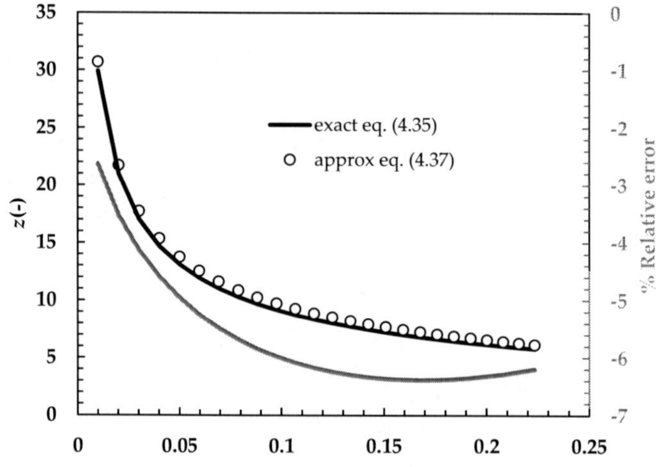

Figure 4.5 Comparison between the exact solution of Eq. 4.35 (solid black line) and the solution of Eq. 4.36 (open circles, Eq. 4.37). The solid gray line indicates the % relative error between the two solutions (right hand vertical axis).

that when the mobility of the water molecules, identifiable with D, is high compared to the rate of magnetization loss (identifiable with $\mathcal{M}R_c$; fast-diffusion regime: $\mathcal{M}R_c/D <<1$), the following relation holds:

$$\left\langle \frac{1}{T_2} \right\rangle = \frac{1}{T_{2\mathrm{H_2O}}} + 2\left\langle \frac{\mathcal{M}}{R_\mathrm{h}} \right\rangle = \frac{1}{T_{2\mathrm{H_2O}}} + 2\frac{\langle \mathcal{M} \rangle}{R_\mathrm{h}} \tag{4.39}$$

where $<1/T_2>$is the average value of the inverse of the relaxation time of the protons belonging to the water molecules trapped within the polymeric network of the gel, $T_{2\mathrm{H2O}}$ is the relaxation time of the protons of the bulk water (i.e., protons of the free water, whose relaxation is not affected by the presence of the polymeric chains), and $\langle \mathcal{M} \rangle$ is the $\mathcal{M}$ value averaged on the whole gel volume. In the case of a real polymeric network made up by cubic cells (polymeric mesh) of different dimensions (ξ_i), i.e., of different hydraulic radii R_{hi}, Eq. 4.39 becomes

$$\frac{1}{T_{2i}} = \frac{1}{T_{2\mathrm{H_2O}}} + 2\frac{\mathcal{M}_i}{R_{\mathrm{h}i}} = \frac{1}{T_{2\mathrm{H_2O}}} + 2\frac{\langle \mathcal{M} \rangle}{R_{\mathrm{h}i}} \tag{4.40}$$

where $T_{2\mathrm{i}}$ and $\mathcal{M}_i$ are, respectively, the relaxation time and the $\mathcal{M}$ value of the water molecules protons trapped in cubic cells (polymeric mesh) of size ξ_i (or hydraulic radius R_{hi}). Equation 4.40 is based on the assumption that M_i does not depend on mesh size ξ_i and $\mathcal{M}_i = \langle \mathcal{M} \rangle$ for every i. Thus, knowing R_h (Eq. 4.34), $\langle 1/T_2 \rangle$ (determinable by Eq. 4.21 or Eq. 4.22 data fitting) and $T_{2\mathrm{H2O}}$, Eq. 4.39 allows the determination of $\langle \mathcal{M} \rangle$. In addition, knowing $\langle \mathcal{M} \rangle$ and $T_{2\mathrm{i}}$ (see Eq. 4.21), together with Eq. 4.40, it is possible to evaluate ξ_i for each class of cubic cells. Finally, relying on the knowledge of $\langle \mathcal{M} \rangle$ and $a(T_2)$ (Eq. 4.22), it is possible to determine the continuous ξ distribution $a(\xi)$. Indeed, the combination of Eqs. 4.39 and 4.40 allows to find the relation between $\mathrm{d}\xi$ and $\mathrm{d}T_2$:

$$\xi = \frac{2\langle \mathcal{M} \rangle}{\sqrt{\frac{1-\phi}{\phi 3\pi}}} \bigg/ \left(\frac{1}{T_2} - \frac{1}{T_{2\mathrm{H_2O}}} \right) \quad \mathrm{d}\xi = \frac{2\langle \mathcal{M} \rangle}{\sqrt{\frac{1-\phi}{\phi 3\pi}}} \left(\frac{T_{2\mathrm{H_2O}}}{T_{2\mathrm{H_2O}} - T_2} \right)^2 \mathrm{d}T_2 \tag{4.41}$$

Consequently, $a(\xi)$ will be

$$a(\xi) = \frac{dT_2}{d\xi} a(T_2) = \frac{\sqrt{\frac{1-\phi}{\phi 3\pi}}}{2\langle \mathcal{M} \rangle} \left(\frac{T_{2\mathrm{H_2O}} - T_2}{T_{2\mathrm{H_2O}}} \right)^2 a(T_2) \tag{4.42}$$

Thus, the probability P(ξ) of finding a mesh of size ξ inside the polymeric network is

$$\mathrm{P}(\xi) = \frac{a(\xi)\,\mathrm{d}\xi}{\int_{\xi_{\min}}^{\xi_{\max}} a(\xi)\,\mathrm{d}\xi} \tag{4.43}$$

We would like to conclude this section remembering that we speak of low-field NMR as the magnetic field contained between 0.37 T and 2.43 T, while the more traditional high-field NMR implies magnetic fields lying in the 7.5–37 T interval.

4.5 Release Tests

This approach relies on the reduced mobility of a probe molecule inside a polymeric network due to the hindering effect exerted by the chains [3]. Indeed, in the case of negligible convection, network swelling, erosion, and probe molecule interaction with polymeric chains [18], probe molecule diffusion is essentially ruled by the presence of chains. In particular, in the case of a polymeric network, mesh diameter plays a key role in ruling probe movements. Probe mobility is represented by its diffusion coefficient D_p (dimensionally an areal velocity) that is sometimes referred to as *Fickian* diffusion coefficient as it is related to movements induced by the presence of a probe concentration gradient [12]. On the contrary, the *tracer* diffusion coefficient refers to the mobility of a probe molecule in an environment where no concentration gradient exists, i.e., probe movements are dictated only by Brownian motion. Indeed, if the probe molecule is present in traces, its concentration is almost vanishing as well as its concentration gradient. The *tracer* diffusion coefficient is also called *self-diffusion* coefficient as the absence of concentration gradient takes place also in a pure liquid where the molecules' movements are, again, due to only Brownian motion. In the case of diluted solution, the difference between the *Fickian* and the *tracer* diffusion coefficient becomes negligible [12].

The relation between D_p and the mean mesh diameter ξ is given by the Peppas and Merryl model [31]:

$$\frac{D_p}{D_0} = \left(1 - \frac{2r_s}{\xi}\right)\exp\left(-y_p\frac{\phi}{1-\phi}\right) \tag{4.44}$$

which combines the free volume theory [43] with the assumption that the probability of a solute of radius r_s having to pass through an opening of diameter ξ is linearly dependent on the ratio $2r_s/\xi$. In Eq. 4.44 D_0 indicates the probe (*Fickian*) diffusion coefficient in the pure solvent, ϕ is the polymer volume fraction in the gel, and y_p represents the ratio of the critical volume required for a successful translational movement of the probe molecule and the average free volume per molecule of the swelling medium. Although y_p depends on many factors, the authors suggest that when r_s is expected to be much smaller than ξ, it can be approximately assumed as equal to one. Thus, known D_0, r_s, and ϕ, the experimental measurement of D_p allows the determination of ξ according to Eq. 4.44. Although D_p can be measured by different methods such as the high-field NMR [4], the most common way of proceeding is to experimentally record the release of a probe molecule from a gel matrix [18]. The release curve is then fitted by a mathematical model relying on Fick'law [18]:

$$\frac{\partial C}{\partial t} = \nabla \left(D_p \nabla C \right) \quad (4.45)$$

$$V_r C_r(t) = V C_0 - \iiint\limits_V C(X, Y, Z, t) dV \quad (4.46)$$

where t is time, C is probe concentration in the generic position (X, Y, Z) of the releasing gel matrix of volume V, V_r and C_r are, respectively, the volume and the probe concentration of the release environment, ∇ is the nabla operator, and C_0 is the initial, uniform probe concentration in the gel matrix. While Eq. 4.45 expresses a mass balance on the probe at the microscale, Eq. 4.46 represents a mass balance on the probe at the macroscale. Indeed, it states that at each time, the amount of probe in the release environment $(V_r C_r)$ is equal to the difference between the initial amount of probe in the gel matrix (VC_0) less the amount still present in the gel matrix. Equation 4.45 has to be solved with the following initial and boundary conditions:

initial conditions

$$C(X, Y, Z) = C_0 \quad (4.47)$$

boundary conditions

$$C(\text{gel surface}) = k_p C_r \quad (4.48)$$

where k_p is the probe partition coefficient between the gel and the environmental release fluid. Equation 4.47 sets a uniform concentration in the gel at the beginning ($t = 0$), while Eq. 4.48 assumes that at the gel-release environment interface, the equilibrium condition for what concerns probe concentration holds. In order to determine D_p, Eqs. 4.45 and 4.46 have to be simultaneously solved (according to a numerical method [30]) and fitted to the experimental C_r trend, usually assuming $k_p = 1$.

4.6 Case Study

The final part of this chapter is devoted to the presentation and discussion of the results descending from the application of the different approaches previously illustrated.

4.6.1 *Alginates*

Alginate refers to a family of unbranched polysaccharides isolated from brawn seaweeds. They are composed of 1-4 linked α-D-mannuronic acid (M) and β-L-guluronic acid (G) arranged in a blockwise pattern with homopolymeric regions of M (M blocks) and G (G blocks) residues interspersed by regions of alternating structure (MG blocks) [42]. Alginate has largely been used in biotechnological and industrial applications for its gel-forming properties which were described in detail around the 1970s [17, 38]. Crosslinking between different chains implies the formation of junctions composed by two opposing G blocks which form a cavity that can accommodate divalent cations such as Ca^{2+} (egg-box). Abrami [1] and D'Agostino [10] studied a gel composed by 2% alginate (% mass fraction, 70% G content) and crosslinked by contact with a Ca^{2+} aqueous solution (5 g/l) for five minutes. The obtained film underwent a rheological characterization (stress and frequency sweep tests) to get the shear modulus G (62,500 Pa). Then, the crosslink density, ρ_x, was estimated according to a modified form of Eq. 4.1:

$$\rho_x = G/\eta RT \tag{4.49}$$

where η is called *front factor* and it coincides with the number of G blocks involved in each junction zone [42]. Indeed, the theory on which Eq. 4.1 relies assumes that crosslinks are punctual and not zonal as happens in alginate. Assuming $\eta \approx 14$ [42], ρ_x turns out to be $1.7*10^{-6}$ mol/cm^3. Thus, Eq. 4.4 yields $\xi \approx 12$ nm. This value is around one-half of that descending from the LNMR approach (Eq. 4.38, $\phi = 0.011$; $R_f = 0.8$ nm [3]). In addition, LNNMR approach allows to conclude that three different classes of meshes exist: $\xi_1 = 122$ nm ($A_1 = 12\%$), $\xi_2 = 27$ nm ($A_1 = 56\%$), and $\xi_3 = 15$ nm ($A_1 = 32\%$). It is evident that the last one is very close to the rheological estimation, and this seems to suggest that the rheological behavior is mainly affected by small meshes. The reliability of low-field NMR approach is ensured by the very law value of $\mathcal{M}R_c/D$ that is around 10^{-6}.

4.6.2 *Alginate–Methacrylated Dextran*

In order to improve the potentiality and the versatility of hydrogels, interpenetrated polymeric networks (IPN) can have an important role. Indeed, the presence of, at least, two networks can infer properties that would not be otherwise met by using one single polymer [26, 39]. In this direction, recently, Pescosolido and coworkers [32] moved to study an IPN made up of alginate (% mass fraction in the final system = 3%) and methacrylated dextran (MD) (MW 40,000, methacrylation degree = 5%; % mass fraction in the final system = 5%). Hydrogel formation occurred due to alginate crosslinking by Ca^{2+} ($CaCl_2$ solution 0.025 M plus NaCl 0.385 M) and MD crosslinking by a mercury UV lamp irradiation (wave length range 350–450 nm, light intensity 20 mW/cm^2) for 10 minutes. The final polymer volume fraction was approximately 0.046. The resulting IPN was characterized by means of rheology, low-field NMR, cryoporosimetry, and release tests (25°C). This characterization led to the conclusion that the average mesh size is 25.2 nm, 44.5 nm, 11.2 nm, and 18.5 nm according to rheology ($G = 489$ Pa), cryoporosimetry, low-field NMR ($R_f = 0.8$ nm for both alginate and methacrylated dextrane [3]), and release tests (myoglobin: $y_p = 1$; $r_{sM} = 2.1$ nm; $D_p = 1.05*10^{-10}$ m^2/s, $D_0 = 1.2*10^{-10}$ m^2/s) approach, respectively. Although all these estimates do not differ too much from each other, we can see that while rheology and release tests yield very similar conclusions, low-field NMR and cryoporosimetry provide, respectively, an estimation that is approximately one-half and double with respect to the that competing with the rheology-release test approaches. The fact that cryoporimetry provides the higher estimation could be connected to polymeric mesh widening upon water solidification. Anyway, considering the complexity of an IPN, it seems that all methods led to a similar estimation of the average mesh size.

4.6.3 *Scleroglucan*

Scleroglucan (Sclg), a water-soluble polysaccharide produced by fungi of the genus Sclerotium, consists of a main chain of (1-3)-linked β-D-glucopyranosyl units bearing, every third unit, a single β-D-glucopyranosyl unit linked (1-6). It is known that scleroglucan assumes a triple-stranded helical conformation in aqueous solution and a single-coiled disordered conformation in methylsulphoxide or at high pH values (NaOH $>$ 0.2 M) [27]. Due to its peculiar properties, Sclg was extensively used for various commercial applications such as secondary oil recovery, ceramic glazes, food, paints, and cosmetics [16], and it was also considered for modified/sustained release formulations [28]. More recently, by addition of borax to a Sclg solution, a new stronger gel was obtained by Grassi and coworkers [19]. This hydrogel ($\phi = 0.0045$) was characterized, at 37°C, according to rheology and release tests. Three different probe molecules were used: theophylline, $r_{sT} = 0.37$ nm and $D_0 = 8.2*10^{-10}$ m^2/s; vitamin B$_{12}$, $r_{sB12} = 0.85$ nm and $D_0 = 3.8*10^{-10}$ m^2/s; myoglobin, $r_{sM} = 2.1$ nm and $D_0 = 1.3*10^{-10}$ m^2/s. The rheological analysis provided $\xi = 66$ nm, 67 nm, and 73 nm in the case of hydrogels containing ($C_0 = 4$ mg/ml) theophylline ($G = 28.7$ Pa, $\rho_x = 1.1*10^{-8}$ mol/cm^3), B$_{12}$ ($G = 25.4$ Pa, $\rho_x = 1.0*10^{-8}$ mol/cm^3), and myoglobin ($G = 20$ Pa, $\rho_x = 8*10^{-9}$ mol/cm^3), respectively. The release tests, assuming $y_p = 1$ in Eq. 4.44, yielded $\xi = 65$ nm, 57 nm, and 65 nm in the case of hydrogels containing theophylline ($D_p = 8.07*10^{-10}$ m^2/s), B$_{12}$ ($D_p = 3.67*10^{-10}$ m^2/s), and myoglobin ($D_p = 1.21*10^{-10}$ m^2/s). It can be seen that the agreement is very good but the error associated with the estimation of the ratio D_p/D_0 (ranging between 10% and 15%) makes this evaluation a little bit delicate. These estimations are substantially confirmed by the low-field NMR analysis (Eq. 4.38, $R_f = 2.15$ nm, $\phi = 0.0045$), yielding to $\xi = 100$ nm.

4.7 Conclusions

This chapter, focusing the attention on four different approaches aimed at estimating the mesh size of a polymeric network, clearly evidenced the difficulties involved in this task. Indeed, the common drawback of these and other methods relies in the necessity of referring to an idealized network structure sharing with the real one some important characteristics such as the crosslink density. However, as a matter of fact, this is the only way we have to extract some mean network properties that would be otherwise impossible to determine due to the highly complex topology of real network (for this purpose it is sufficient to think about TEM pictures of hydrogel structure). In this sense none of them can be labeled the best one or the worst. Depending on the real network topology, the best approach is that of referring to the ideal network that better approaches the real one. As, a priori, it is virtually impossible to discern which is the best approximation of the real network, our suggestion is to consider more than one theoretical approach for the estimation of the mean mesh size. Indeed, the comparison among the results coming from different approaches can help a lot in individuating the most suitable. In addition, it is important to remember that the use of release tests can be very delicate as the error associated with the estimation of the ratio D_p/D_0 can be up to 15%, which implies a considerable variation in the ξ estimation, especially for what concerns hydrogels characterized by a very low polymer concentration (≤ 0.01). In the case of crioporosimetry, we feel that a certain overestimation of mesh size takes place due to a possible network deformation upon water freezing. Finally, we would like to mention the possibility of merging different techniques to get the mesh size distribution as done, for example, by Turco [42] and Pescosolido [32] who estimated the mean mesh size according to rheology and then considered low-field NMR to get the distribution. Indeed, low-field NMR, contrarily to the other approaches, has the drawback of a purely theoretical estimation of the mean mesh size.

Acknowledgments

Financial support from 'Fondo PRIN 2010–2011 (20109PLMH2), Fondazione Cassa di Risparmio of Trieste, and Fondazione Benefica Kathleen Foreman Casali of Trieste is acknowledged.

References

1. Abrami, M. (2013). Blend polimerici per la prevenzione dell'in-stent restenosis mediante gel paving. Master's degree thesis, Department of Engineering and Architecture, Trieste University.
2. Amamoto, Y., Kikuchi, M., Masunaga, H., Ogawa, H., Sasaki, S., Otsuka, H., and Takahara, A. (2011). Mesh-size control and functionalization of reorganizable chemical gels by monomer insertion into their cross-linking points. *Polym. Chem.*, **2**, pp. 957–962.
3. Amsden, B. (1998). Solute diffusion within hydrogels. Mechanisms and models. *Macromolecules*, **31**, pp. 8382–8395.
4. Aparicio, C. G., Garrido, I. Q., and Garrido, L. (2012). Diffusion of small molecules in a chitosan/water gel determined by proton localised NMR spectroscopy. *J. Coll. Inter. Sci.*, **368**, 14–20.
5. Brownstein, K. R., and Tarr, C. E. (1979). Importance of classical diffusion in NMR studies of water in biological cells. *Phys. Rev. A*, **19**, 2446–2453.
6. Brun, M., Lallemand, A., Quinson, J. F., and Eyraud, C. (1973). Changement d'etat liquide-solide dans les milieux poreux. *J. De Chimie Physique*, **70**, pp. 979–989.
7. Brun, M., Lallemand, A., Quinson, J.-F., and Eyraud, C. (1977). A new method for the simultaneous determination of the size and the shape of pores: the thermoporometry. *Thermochim. Acta*, **21**, pp. 59–88.
8. Cappelli, A., Galeazzi, S., Giuliani, G., Anzini, M., Grassi, M., Lapasin, R., Grassi, G., Farra, R., Dapas, B., Aggravi, M., Donati, A., Zetta, L., Boccia, A. C., Bertini, F., Samperi, F., and Vomero, S. (2009). Synthesis and spontaneous polymerization of oligo(ethylene glycol)-conjugated

benzofulvene macromonomers. A polymer brush forming a physical hydrogel. *Macromolecules*, **42**, pp. 2368–2378.

9. Chui, M. M., Phillips, R. J., McCarthy, M. J. (1995). Measurement of the porous microstructure of hydrogels by nuclear magnetic resonance. *J. Coll. Inter. Sci.*, **174**, pp. 336–344.
10. D'Agostino, I., Milcovich, G., Asaro, F., Fiorentino, S., Lapasin, R., Grassi, G., and Grassi, M. (2012). Caratterizzazione reologico/NMR di gel composti da alginato-pluronico per il rilascio di farmaci basati sul DNA. *Atti del XII Convegno dell'Associazione Italiana di Reologia-SIR*, Ustica (PA), Italy, pp. 9–14.
11. Draper, N.R., and Smith, H. (1966). *Applied Regression Analysis.* (John Wiley & Sons, Inc., New York, USA).
12. Fierro, D., Scharnagl, N., Emmler, T., Boschetti-de-Fierro, A., and Abetz, V. (2011). Experimental determination of self-diffusivities through a polymer network for single components in a mixture. *J. Membr. Sci.*, **384**, 63–71.
13. Flory, P.J. (1953). *Principles of Polymer Chemistry.* (Cornell University Press, Ithaca USA).
14. Gallegos, D. P., Munn, K., Smith, D. M., and Stermer, D. L. (1987). A NMR technique for the analysis of pore structure: application oto materials with well-defined pore structure. *J. Coll. Inter. Sci.*, **119**, pp. 127–140.
15. Gardel, M. L., Shin, J. H., MacKintosh, F. C., Mahadevan, L., Matsudaira, P., and Weitz, D. A. (2004). Elastic behavior of cross-linked and bundled actin networks. *Science.* **304**, pp. 1301–1305.
16. Giavasis, I., Harvey, L. M., and McNeil, B. (2002). Scleroglucan. In S. De Baets, E. J. Vandamme, and A. Steinbuchel (eds.). *Biopolymers, Polysaccharides II* (Vol. 6, pp. 37–60). (Wiley-VCH, Weinheim).
17. Grant, G. T., Morris, E. R., Rees, D. A., Smith, P. J. C., and Thom, D. (1973). Biological interactions between polysaccharides and divalent cations: the egg-box model. *FEBS Lett.* **32**, pp. 195–198.
18. Grassi, M., Grassi G., Lapasin R., and Colombo, I. (2007) *Understanding Drug Release and Absorption Mechanisms: A Physical and Mathematical Approach* (CRC Press, Boca Raton, USA).
19. Grassi, M., Lapasin, R., Coviello, T., Matricardi, P., Di Meo, C., and Alhaique, F. (2009). Scleroglucan/borax/drug hydrogels: structure characterisation by means of rheological and diffusion experiments. *Carb. Polym.*, **78**, pp. 377–383.
20. Hore, P. J. (1995). *Nuclear Magnetic Resonance* (Oxford University Press, New York, USA).

21. Ishikiriyama, K., Sakamoto, A., Todoki, M., Tayama, T., Tanaka, K., and Kobayashi, T. (1995). Pore size distribution measurements of polymer hydrogel membranes for artificial kidney using differential scanning calorimetry. *Thermochim. Acta*, **267**, pp. 169–180.
22. Ishikiriyama, K., Todoki, M., and Motomura, K. (1995). Pore size distribution (PSD) measurements of silica gels by means of differential scanning calorimtery. *J. Colloid Interface Sci.*, **171**, pp. 92–102.
23. Lapasin, R., and Pricl, S. (1995). *Rheology of Industrial Polysaccharides: Theory and Applications* (Chapman & Hall, London, GB).
24. Leoni, L., Desai, T. A., and Burke, J. E. (2004). Micromachined biocapsules for cell-based sensing and delivery, *Adv. Drug Del. Rev.*, **56**, pp. 211–229.
25. Lustig, S. R., and Peppas, N. A. (1988). Solute diffusion in swollen membranes. 9. Scaling laws for solute diffusion in gels. *J. Appl. Polym. Sci.*, **36**, pp. 735–747.
26. Marsano, E., Bianchi, E., and Viscardi, A. (2004). Stimuli responsive gels based on interpenetrating network of hydroxyl propylcellulose and poly(*N*-isopropylacrylamide). *Polymer*, **45**, 157–163.
27. Norisuye, T., Yanaki, T., and Fujita, H. (1980). Triple helix of a Schizophyllum commune polysaccharide in aqueous solution. *J. Polym. Sci. Part B - Polymer Physics*, **18**, pp. 547–558.
28. Palleschi, A., Coviello, T., Bocchinfuso, G., and Alhaique, F. (2006). Investigation on a new scleroglucan/borax hydrogel: structure and drug release. *Int. J. Pharm.*, **322**, pp. 13–21.
29. Pasut, E., Toffanin, R., Voinovich, D., Pedersini, C., Murano, E., and Grassi, M. (2008). Mechanical and diffusive properties of homogeneous alginate gels in form of particles and cylinders. *J. Biomed. Mat. Res.: Part A*, **87A**, pp. 819–824.
30. Patankar, S. V. (1990). *Numerical Heat Transfer and Fluid Flow* (Hemisphere Publishing, New York, USA).
31. Peppas, N. A., and Merrill, E. W. (1977). Crosslinked poly(vinyl alcohol) hydrogels as swollen elastic networks. *J. Appl. Polym. Sci.*, **21**, pp. 1763–1770.
32. Pescosolido, L., Feruglio, L., Farra, R., Fiorentino, S. M., Colombo, I., Coviello, T., Matricardi, P., Hennink, W. E., Tina Vermonden, T., and Grassi, M. (2012). Mesh size distribution determination of interpenetrating polymer network hydrogels. *Soft Matter*, **8**, pp. 7708–7715.
33. Provencher, S. W. (1982). A constrained regularization method for inverting data represented by linear algebraic or integral equations. *Comput. Phys. Comm.*, **27**, pp. 213–227.

34. Rowlinson, J. S., and Widom, B. (2003). *Molecular Theory of Capillarity* (Clarendon Press, Oxford, GB).
35. Samsonov, V. M., Sdobnyakov, N. Y., and Bazulev, A. N. (2004). Size deoendence of the surface tension and the problem of Gibbs thermodynamics extension to nanosystems. *Colloids Surf. A*, **239**, pp. 113–117.
36. Scherer, G. W. (1994). Hydraulic radius and mesh size of gels. *J. Sol-Gel Sci. Technol.*, **1**, pp. 285–291.
37. Schurz J. (1991). Rheology of polymer solutions of the network type. *Progr. Polym. Sci.*, **16**, pp. 1–53.
38. Smidsrød, O. (1974). Molecular basis for some physical properties of alginates in the gel state. *Faraday Discuss.* **57**, pp. 263–274.
39. Sperling, L. H. (1994). Interpenetrating polymer networks: an overview. In *Interpenetrating Polymer Networks*. Klempner, D., Sperling, L. H., Utracki, L. A. (eds.), Advances in Chemistry, **239** (American Chemical Society, Washington, DC), chap. 1, pp. 3–38.
40. Sumitomo, K., 1 Koichi Mayumi, K., Yokoyama, H., Sakai, Y., Minamikawa, H., Masuda, M., Shimizu, T., Ito, K., and Yamaguchi1, Y. (2009). Dynamic light-scattering measurement of sieving polymer solutions for protein separation on SDS CE. *Electrophoresis*, **30**, pp. 3607–3612.
41. Tolman, R. C. (1949). The effect of droplet size on surface tension. *J. Chem. Phys.*, **17**, pp. 333–337.
42. Turco, G., Donati, I., Grassi, M., Marchioli, G., Lapasin, R., and Paoletti, S. (2011). Mechanical spectroscopy and relaxometry on alginate hydrogels: a comparative analysis for structural characterization and network mesh size determination. *Biomacromolecules*, **12**, pp. 1272–1282.
43. Vrentas, J. S., Duda, J. L., Ju, S. T., and Liu, H. T. (1982). Prediction of diffusion coefficients for polymer-solvent systems. *AIChe J.*, **28**, pp. 279–285.
44. Wang, X., and Ni, Q. J. (2003). Determination of of cortical bone porosity and pore size distribution using low field pulsed NMR approach. *J. Orthopaedic Research*, **21**, pp. 312–319.
45. Whittal, K. P., and MacKay, A. L. (1989). Quantitative interpretation of NMR relaxation data. *J. Magnetic Resonance*, **84**, pp. 134–152.
46. Zhang, M., Efremov, M. Y., Schiettekatte, F., Olson, E. A., Kwan, A. T., Lai, S. L., Wisleder, T., Greene, J. E., and Allen, L. H. (2000). Size-dependet melting point depression of nanostructures: nanocalorimetric measuremenst. *Phys. Rev. B*, **62**, pp. 10548–10557.

Chapter 5

Dynamic Light Scattering

Walther Burchard
Institute of Macromolecular Chemistry, University of Freiburg, Stefan-Meier-Str. 31, D-79104 Freiburg, Germany
walther.burchard@makro.uni-freiburg.de

5.1 Introduction

Dynamic light scattering is at present the most frequently used characterization technique to receive information on the size of macromolecules or supramolecular structures and compact colloidal particles. With the instruments available on the market, measurements are quickly performed with agreeable reproducibility. The instruments are equipped with powerful softwares such that only a few manipulations are needed to obtain the desired results. Easy handling of the instruments gives a safe feeling of reliability. However, the easiness and high reproducibility of measurements is deceptive and can become very misleading if one relies too much

Polysaccharide Hydrogels: Characterization and Biomedical Applications
Edited by Pietro Matricardi, Franco Alhaique, and Tommasina Coviello

ISBN 978-981-4613-61-3 (Hardcover), 978-981-4613-62-0 (eBook)
www.panstanford.com

on the computer and is not keeping a skeptical mind in relation to physical expectations. Mostly the software of the instruments is based on two significant conditions:

(i) The particles or molecules have to be small compared to the wavelength used, and
(ii) The solution or the suspension has to be diluted such that interparticle interactions can be disregarded.

Within these limits a quick result for the diffusion coefficient D is obtained and with the Stokes–Einstein relationship the hydrodynamic radius R_h that gives a valuable impression on the size of the particle. It does not mean that the particle is a hard sphere or even has a spherical shape, and therefore the comparison with molecular radii obtained from other techniques gives additional and more detailed information on the shape and conformational properties of the sample under investigation. There are available a few rather expensive instruments which allows for a check for these conditions. If the dynamic light scattering is recorded in a narrow range of scattering angles or even measured at only one scattering angle a misleading conclusion may be obtained. In addition to the diffusion coefficient these results may be of minor interest in application but they are of great academic value since they give information on the segmental mobility. The knowledge of the internal mobility may gather interest also to biochemists and pharmacists, because it may be essential if the loading and release of drugs should be optimized.

Static light scattering was introduced to physical chemistry about 70 years ago [18, 58, 59], but the corresponding method of dynamic light scattering started much later, more than 20 years, with the pioneering work of R. Pecora [38–40]. One may wonder why this apparently easy technique of *dynamic* light scattering has become applicable so late with the wonderful electronic development in the last three decades. This question may arouse suspect that a much higher complexity is hidden underneath the highly developed technique for dynamic light scattering. This suspect is affirmed when studying the long history before a suitable instrument could be developed. A short review on this history may be helpful to understand the hidden complexity.

5.2 Short Outline of the History in the Development of Optics

On a clear night we all are fascinated by the beauty of the firmament with the many stars and the graph representation by the ancient Greeks, filled with the myth of gods. Scientists never gave up to find an understanding of the universe, and the development of dynamic light scattering is one branch in this effort: it probably started in 1802 with the detection of dark lines in the continuous spectra of stars by Joseph Fraunhofer [23, 24]. He noticed that the same lines are also found in laboratory, for instance, if the yellow sodium light is placed in front of the bright arc of light from a Mercury lamp. This detection immediately induced a closer investigation of the spectral lines. Some years later Gustav Kirchhoff and his colleague Robert Bunsen found the explanation [4] and assigned the dark lines as the element specific absorption of a cooler gas before a hot light source.

James Clark Maxwell [34], who derived the electromagnetic theory of light, which in 1888 was experimentally proven by Heinrich Herz [28, 29], made the next important step in the understanding of light.

This theory permitted the prediction of natural spectral line width, and in parallel to this development in theory also the instrumental design of high quality spectrometers was developed that permitted measurements of the spectral line profile, the so-called power spectrum. The calculated natural line width turned out far too narrow and appeared not being measurable even with a Fabry–Perot spectrometer [41–43]. In fact a much broader line width was measured than expected which aroused the question what may be the reason. A hint was given by the observation that the line width increased with temperature that gave indications to molecular velocities in a hot gas. At the same time (1842) Doppler [20] published an explanation of the well-known increase of the sound pip if the source of noise is moving towards and a decrease if it is moving away. He derived the relationships

$$\omega = \omega_0(1 \pm v/c) \qquad \text{and} \qquad \lambda = \lambda_0(1 \pm v/c) \qquad (5.1\text{a,b})$$

For a frequency ω and wavelength λ, the velocity of the light source may be v and c the light velocity. In a gaseous atmosphere the molecules have a high velocity moving erratically in all directions in space with a Maxwell velocity distribution, i.e., the motion of the molecules causes a fairly broadening of the spectral lines with half line-width that is given by the Doppler shift

$$\frac{\Delta\omega_{\text{Doppler}}}{2} = \frac{\omega_0}{c}\sqrt{\frac{RT}{M}} \tag{5.2}$$

M is the molar mass of the molecules, R the gas constant, and T the temperature in Kelvin. The shape of the spectral line is not fully symmetric due to loss of radiation energy that leads to a damping of the oscillation. Taking this fact into account, the spectral line has the shape (the power spectrum) described by the Eq. 5.3.

$$I = I_0 \frac{(\gamma/2)^2}{(\omega-\omega_0)^2 + (\gamma/2)^2} \tag{5.3}$$

where γ is the damping factor [20]. Figure 5.1 shows the shape of such a spectral line. The equation is applicable also to small particles in a viscous medium. Here the damping arises from the friction in the viscous medium. This fact caused much interest because, due to Brownian motion, the random motion of the molecules is related to the diffusion coefficient.

$$\frac{I(\Delta\omega)}{I_0} = A\frac{1}{(1-x)^2 + \Gamma^2} \tag{5.3a}$$

Via the Stokes–Einstein relationship, the diffusion coefficient is related to the hydrodynamic radius R_h of the molecule or particle

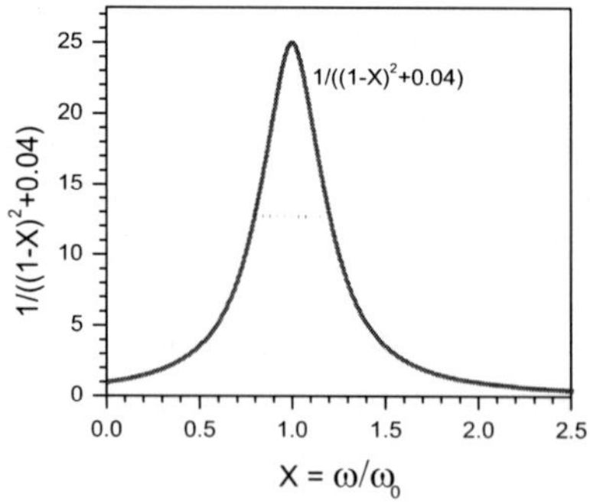

Figure 5.1 Shape of a spectral line, which can be schematically expressed with $x = \omega/\omega_0$, $\Gamma = (\gamma/2\omega_0)$ and a normalization parameter A by Eq. 5.3a with Γ^2 =0.25

in a viscous medium.

$$D = \frac{kT}{6\pi \eta_0 R_{\mathrm{h}}} \tag{5.4}$$

This hydrodynamic radius R_{h} corresponds to a hydrodynamic equivalent sphere but it does not mean that the molecule or an immersed particle has a spherical shape. The high interest in measurements of spectral line widths originated from the fact that even in the first decade of the 19th century the existence of atoms or molecules of discrete size was considered by many scientists as a doubtful philosophical conception. The desired measurements were not successful because of the high friction coefficient in a viscous medium. The corresponding slow motion of the particles causes a much too low broadening of the spectral line. In addition the spectral line-width of the light was insufficiently monochromatic even when interference filters were used.

A new aspect came up with the realization of gas ion-lasers whose light is considered to be essentially monochromatic with a very narrow line width. Even with these light sources the diffusion coefficient could be measured only for fairly low molar mass molecules that have a sufficiently high diffusion coefficient. About the same time another very efficient technique was developed and permitted the desired experiments, which was the design of short time-resolving photomultipliers. These two developments lead to very successful measurements of time correlation functions that turned out to be the Fourier transform of the frequency line width into the corresponding time domain. This correlation requires some words on the Brownian motion before the technique of dynamic light scattering can be understood.

The Brownian motion belongs to one of the most demanding issues in theory. The beauty of these theories cannot be made fully apparent in this overview. It will be of great value if the following outline will give at least some feeling of the complexity of dynamic light scattering which, by modern online evaluating programs, remains hidden. To avoid serious misleading conclusions it is advisable to contact an expert whenever doubts arise. Readers who are mostly interested in experimental results may find the following simplified discussion too complex and may omit some of the next sections and can go directly to the last sections.

5.3 Brownian Motion and Relation to Stochastic Processes

In 1827 the botanist Robert Brown watched in his microscope pollen dusts immersed in water. He noticed with surprise a continuous irregular motion of the particles that never ceased to come to a standstill. This irregular motion stimulated Albert Einstein to the following considerations: due to an accelerating force, the particle gains a certain initial velocity but this velocity will be slowed down by friction in a viscous medium. After a certain time, the motion of the particle must come to a standstill. This can be expressed by the differential equation for the particle motion

$$\frac{dv}{dt} + \frac{f}{m} = 0 \text{ with the solution } v = v_0 \exp\left(-\frac{f}{m}t\right) \tag{5.5}$$

which indeed causes an exponential slowing down of the velocity. However, this is not observed. Therefore, another force must exist which still keeps the Brownian particle in motion and Eq. 5.5 should be completed by an additional term

$$\frac{dv}{dt} = -\frac{f}{m} + A(t) \tag{5.6}$$

$A(t)$ is a fluctuating force which must arise from the thermal energy of the system. Einstein made now the following consideration: if some particles in the system get a push they will receive a certain directed velocity. These motions cause a certain increase of the local concentration above the equilibrium concentration. This higher concentration is responded by a local increase of the osmotic pressure, and the corresponding force tries to push the concentration back to equilibrium. This osmotic response must be the force to compensate the frictional force that initiates the diffusion back to equilibrium concentration of the moving particle. This immediately led Einstein to the conclusion that $A(t) = f/m$ and $A(t)$ must arise from the force of the increased osmotic pressure. For the frictional coefficient f Einstein used the Stokes equation $f = 6\pi\eta R_{\mathrm{h}}$ for hard spheres in a viscous medium and for the osmotic force that stimulates back diffusion toward equilibrium one has kT/mD. The mass of the particle is neglected in this relationship and finally the famous Stokes–Einstein equation is obtained.

$$R_{\mathrm{h}} = \frac{kT}{6\pi\eta D} \tag{5.7}$$

It should be mentioned that this derivation holds only if initial velocity almost immediately loses its original direction and comes to random fluctuation and thus the acceleration force can be neglected. Also this derivation gives us detailed information on the time dependent stochastic force. At this point it should be mentioned that, independently of Einstein, Mario Smoluchowsky [48, 49], from the Krakow University, derived a detailed description of the Brownian stochastic process, which complements Einstein's considerations.

5.4 Time Correlation Functions

Indeed, it is difficult to elucidate the special features of such random processes. Only two facts are known for sure: (i) on average over a sufficiently long time, the particles will have not moved from their original position in space, i.e., $\langle \Delta r \rangle = 0$, but (ii) the mean square displacement $\langle \Delta r^2 \rangle$ is a positive quantity and a characteristic parameter of the type of fluctuations. Already Einstein had the idea to cut a longer time into N intermediate time differences, for instance

$$t = \langle t_1 - t_0 \rangle + \langle t_2 - t_1 \rangle + \langle t_3 - t_2 \rangle + \cdots \langle t_N - t_{N-1} \rangle \tag{5.8}$$

and to combine this with the length the particle moved within the time intervals. Assuming that these movements in such time intervals occur at random, the statistics results in an average path length of

$$\langle \Delta r \rangle = \langle \Delta r(t_1) \rangle + \langle \Delta r(t_2) \rangle + \langle \Delta r(t_3) \rangle + \cdots \langle \Delta r(t_N) \rangle \tag{5.9}$$

It is sensible to assume that for all times the $\langle \Delta \mathrm{r}(\mathrm{t_j}) \rangle$ are of an average length l, and furthermore, as a crude approximation, that all the N processes occur independent from each other. We then can ask for a probability to find a particle at the position r after N time intervals. This question corresponds to a random walk and leads to the Gaussian probability distribution

$$W(r) = \frac{1}{\sqrt{2\pi N l^2}} \exp\left(-\frac{\Delta r^N}{2Nl^2}\right) dr \tag{5.10}$$

with a mean square displacement

$$\left\langle \Delta r^2 \right\rangle = 2Nl^2 \tag{5.11}$$

Equation 5.10 resembles the solution of the differential equation for the diffusion of a particle

$$\frac{\partial c}{\partial t} = D\frac{\partial^2 c}{\partial r^2} \tag{5.12}$$

with the solution

$$W(c, t) = \sqrt{\frac{1}{2\pi Dt}} \exp\left(-\frac{\Delta r^2}{4Dt}\right) \tag{5.13}$$

and mean square displacement

$$\left\langle \Delta r^2 \right\rangle = 4Dt \tag{5.14}$$

Comparing Eq. 5.11 with Eq. 5.14 one obtains

$$Nl^2 = 2Dt \tag{5.15}$$

which discloses the translational diffusion as a random walk.

Statistical independence of each step in a fluctuating system can be expressed by the product of $\langle w(\Delta r_j)w(\Delta r_k)\rangle = 0$ when $j \neq k$. Such independence is the simplest assumption one can make. More general, the two intervals can occur with different probabilities, and this leads to a time correlation function (TCF) [3, 61, 62], as reported in Eq. 5.16

$$C(t) = \langle \Delta r(t_j)\Delta r(t_k)\rangle > 0 \tag{5.16}$$

The steps of length l in the three-dimensional space can occur toward all directions and therefore $\Delta\mathbf{r}$ is a vector. Furthermore, for each step a velocity is defined by the ratio

$$\Delta r(t_j)/(t_j - t_{j-1}) = v_j$$

and $C(t)$ in Eq. 5.16 becomes a time correlation function of velocities.

5.5 Comment on Averages and the Effect of External Forces

The sum of all N values of $\Delta r(t)$ for one selected Brownian particle in the course of time is normalized by the N steps. Since N is proportional to the elapsed time t the average is a time average. On the other hand one can take for each Δr_j the ensemble average. In an ergodic system the ensemble average and the time average are the same. A system becomes non-ergodic when a motion is frozen in, e.g., in the glass formation. A simple generalization of a random process is a correlation with only the preceding time interval. Such a correlation is a Markov process, but even this simple modification causes troubles in the mathematical treatment and the generalized Markov process remains an unsolved problem. Still time correlation functions can be measured in any case, and realistic or intuitive interpretations are possible on the basis of physical and chemical properties of the investigated sample. Three special examples may shortly be discussed.

(i) One is the effect of the initial velocity in a Brownian process. Now the effect of the inertia force, shown in Eq. 5.6, has to be taken in account. Equation 5.6 can be formally solved and gives with the frictional coefficient $\zeta = f/m$ [3, 12, 35]

$$u(t) \equiv (v(t) - v_0) = \exp(-\zeta\, t) \int_0^t \exp(\zeta\, t')A(t')dt' \qquad (5.17)$$

with

$$\langle u \rangle = 0 \text{ and } \left\langle u^2(t) \right\rangle = \frac{3kT}{m}(1 - \exp(-2\zeta t)) \qquad (5.18)$$

If we insert the two averages of Eq. 5.18 into the Maxwell velocity distribution one obtains the probability distribution

$$W(v, t, v_0) = \left[\frac{m}{2\pi kT(1 - \exp(-2\zeta t))}\right]^{3/2} \times \exp\left(-\left[\frac{m(v - v_0\exp(-\zeta t))}{2kT(1 - \exp(-\zeta t))}\right]^2\right) \qquad (5.19)$$

This shows that the velocity distribution process corresponds to a time dependent Gaussian process. If $\zeta t >> 1$ the already known Gaussian distribution is obtained. Mostly the friction is very large such that the initial velocity can be scarcely measured.

(ii) Another example is the motion of particles that are not hard spheres but are long flexible chains or branched structures. With such structures the viscosity of the medium is significantly influenced by what is called the draining effect [31, 32, 57]. It is clear that the Brownian motion will have influence on the mobility of segments, and since segments are shorter than the diameter of such soft samples, the relaxation modes are faster than those of the diffusion mass centers.
The two models differ significantly in their relaxation spectra. Further details are given when specific measurements will be discussed.

(iii) The third example is of particular interest in pharmacy and biochemistry. Actually, proteins, nucleic acids, and drugs are the mostly ionically charged samples. In these examples the electrophoretic mobility is of essential interest [45, 54, 55]. Here the applied electric field is the external force that induces a steady state of the motion of the particles that is superimposed by the irregular Brownian motion. For these measurements special instruments are available which permit a fast determination of the electrophoretic mobility.
These three issues will be commented at some detail. The treatment of example (i) in combination with (ii) is highly involved and will be only superficially commented.

5.6 Flexible Particles: Linear and Branched Macromolecules in Solution

Often an interpretation of experimental data is made under the assumption of hard spheres or compact particles. For such particles the time correlation function is uniquely determined by the differential equation of diffusion as given by Eq. 5.12. For soft materials the behavior is far more complex. Such soft materials can consist of flexible linear or branched macromolecules in solution, but it also could be a supramolecular sample formed by cluster formation of nanoparticles. In the first case the thermal motion of solvent molecules causes not only the irregular diffusion of the center of mass but also acts onto the flexibly linked repeat units in

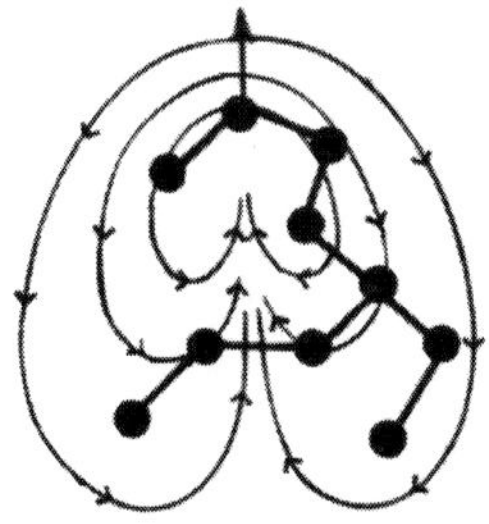

Figure 5.2 Behavior of a chain-like macromolecule in a flow field of an incompressible solvent. The arrow on the top may indicate the force acting on the macromolecule due to the Brownian motion. This motion drives away the solvent and because of the incompressibility a backflow arise as indicated by the small arrows in the streaming lines. In the end the different beads are pushed somewhat together which looks like an attraction. The solvent becomes inhibited to move through the coiled chains and this is called the inhibited draining.

a chain and induces oscillations according to their eigenfrequencies. A further effect is the hydrodynamic interaction which means that the motions of the individual beads, which denote the repeat units, are related to each other due to the forces in a flow field: the spherical repeat units of the particle start to attract each other if they are within such flow field. The special streaming field is caused by moving particle in the incompressible solvent. Figure 5.2 may schematically elucidate the effect of this hydrodynamic interaction.

In such a system the Langevin differential equation is modified by this additional force:

$$\frac{dv}{dt} = -\zeta v + \mathbf{F} + A(t) \tag{5.20}$$

In this example the external force **F** is represented by the Oseen tensor [37, 50] of hydrodynamic interaction, and $A(t)$ the fluctuating force of the Brownian motion. Equation 5.20 is a so-called Fokker–Planck differential equation. The hydrodynamic interaction has influence mainly on the local structure and for this reason the special architecture of the particle has to be taken into account as a marginal condition. A complete solution has been possible only for some special cases. One is the viscosity and diffusive behavior of uniform linear chains in a dilute solution. Without going into detail it is intuitively clear that linear chains can be described as a spring-bead model, where the bead represents the monomer repeating unit, and the segment length between two such repeating units can be approximately described by an elastic spring. Due to the Brownian motion, and even more pronounced, due to a shearing force, the elastic segments are excited to eigenvalue oscillations. These eigenvalues could be obtained by diagonalization of the interaction matrix of the segment pairs [9, 60]. These normal modes are much faster than the diffusive mobility of the center of mass. Therefore the effect of these internal modes of motion requires instruments with which the fast vibrations can be measured. Rouse [46] neglected the hydrodynamic interactions but B. H. Zimm [60] succeeded in performing the matrix diagonalization also for linear chains with hydrodynamic interactions.

5.7 The Time Correlation Function in Dynamic Light Scattering

The simplest behavior is observed with compact particles much smaller than the wavelength of the light ($r << \lambda$). Then the particles or molecules can be considered as point-like scatterers, and only the distance the particles move at random is important. In an experiment a scattering intensity can be recorded resulting in highly fluctuating scattering intensities. At a first sight it appears hopeless to extract from such irregular fluctuations any sensible information (see Fig. 5.3).

However, a time correlation function can be constructed. The height of the peaks corresponds to the number of photons collected within a time interval of 100 ms. The attached numbers denote the nth *interval*; the *height of the peaks* indicate the *number* of photons which can be read-off from the ordinate. Then the time correlation function (TCF) can be constructed as follows. The scattering intensity at time t_j is multiplied by that recorded at time t_{j+k}, and this proceeding is repeated about 10^5 times for increasing j but keeping the interval $(k - j)$ constant. A fairly smooth average curve is obtained after summing the products of the intensities from the two enters k and j over all j. Equation 5.21 shows the maneuver

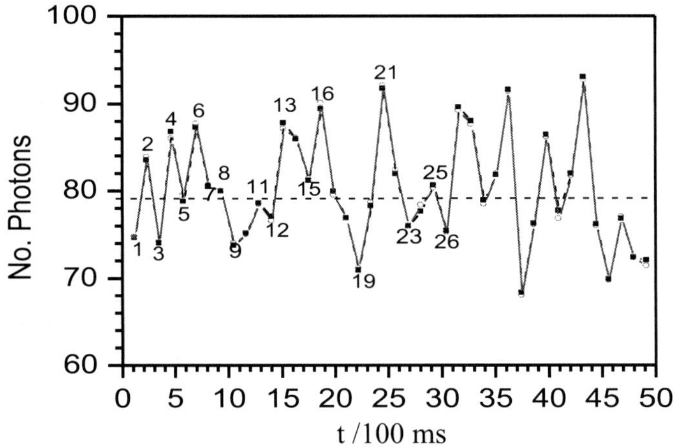

Figure 5.3 Fluctuation of the scattering intensity around the average value (dashed line). The fluctuating data are presented by the number of photons collected in the short time intervals of 100 ms.

with a few examples

$$\langle i(0)i(1)\rangle = C\sum_{j=0}^{N}(i_j i_{j+1});$$
$$\langle i(0)i(2)\rangle = C\sum_{j=0}^{N}(i_j i_{j+2});$$
$$\langle i(0)i(k)\rangle = C\sum_{j=0}^{N}(i_j i_{j+k}) \tag{5.21}$$

where $t = k\Delta t$ is the delay time. The function

$$< i(t)i(t+k\Delta t) > \; = G_2(t) \tag{5.22}$$

is the scattering intensity time correlation function and is called homodyne DLS. The intensity TCF has to be distinguished from the field TCF

$$g_1(t) = \langle E(t=0)E^*(t)\rangle / \langle E(t)E^*(t)\rangle \tag{5.23}$$

that is the quantity needed in theory for the data interpretation. $E(t)$ denotes the amplitude and $E(t)^*$ the conjugated complex amplitude of an electromagnetic wave, and $<E(t)E^*(t)>= i(t)$ is the scattering intensity. Under homodyne conditions the intensity correlation function is related to the field TCF $g_1(t)$ by the Siegert relationship where $g_2(t) = G_2(t)/i^2$ is the normalized intensity

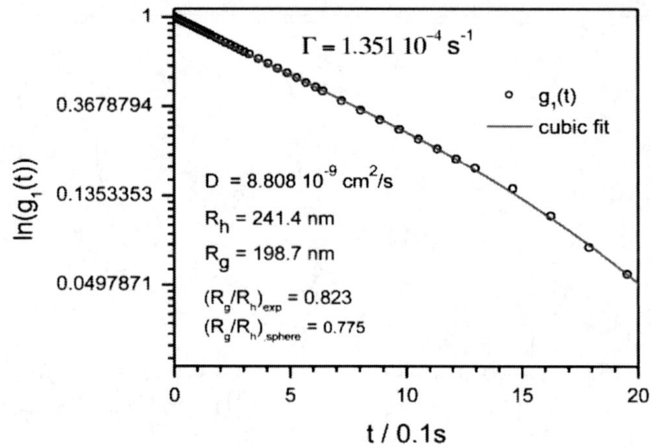

Figure 5.4 The $g_1(t)$ correlation function from spherical latex particles with radius $R = 246$ nm. Data from J. Schneider, Dept. Phys. Chem., University of Freiburg.

correlation function. Note that $G_2(t)$ decays toward a plateau of $<i>^2 = G_2(t \to \infty)$. The field correlation function is given by the equation [3]

$$g_1(t) = \sqrt{g_2(t) - 1} \tag{5.24}$$

An example of $g_1(t)$ as a function of the delay time t is given in Fig. 5.4 for a sample of monodisperse colloidal spheres in water.

Initially the function decreases like a single exponential but at larger delay times the curves in a plot of $\ln(g_1(t))$ flattens or decays stronger. The initial slope in a plot of $\ln(g_1(t))$ as a function of t is called the first cumulant Γ_1 and has the dimension of 1/time. It can be taken as a frequency term. The question is which information such a correlation function can provide us. In other words, to which physical parameter the first cumulant is related? Evidently the strong decay in the scattering light means that the correlation becomes weaker and weaker as the time proceeds and eventually decays to zero, i.e., no correlation at all. We know that a time correlation function is closely related to the Brownian motion and is determined by the translational diffusion. To check this statement it is useful to consider a Fourier transform of a function $G(r, t)$

$$F(q, t) = \int_0^\infty e^{iqr} G(r, t) 4\pi r^2 dr \tag{5.25}$$

It is possible to transform a function $G(r, t)$ in space into a reciprocal space, here denoted by the parameter q of dimension $1/r$. If $G(r, t)$ is the time correlation function in space, caused by Brownian motion, the Fourier transform becomes 1/3 of the mean square displacement. This pair of Fourier transforms is given by the following equations:

$$G(r, t) = \frac{1}{N} \left\langle \sum_{j=1}^{N} r_j(t) - r_j(0) \right\rangle \tag{5.26}$$

$$F(q, t) = \frac{1}{N} \left\langle \sum_{j=1}^{N} e^{iq(r_j(t) - r_j(0))} \right\rangle \tag{5.27}$$

$F(q, t)$ is the time correlation function of the scattered light, if $q = (4\pi/\lambda)\sin(\theta/2)$ is chosen, θ is the scattering angle and λ the wavelength of the light in the solvent. In a Brownian motion $G(r, t)$ is a probability distribution and therefore

$$\frac{\partial^2 F(q)}{\partial q^2} = -\frac{1}{3} \int_0^\infty 4\pi r^2 (G(r, t)) dr = \frac{1}{3} \left\langle r^2(t) \right\rangle \tag{5.28}$$

is 1/3 of the mean square displacement as outlined before. This mean square displacement can be expressed by the translational diffusion coefficient. In fact, if the differential equation of diffusion

$$\frac{\partial c(x, t)}{\partial t} = D \frac{\partial^2}{\partial x^2} c(x, t) \tag{5.29}$$

is Fourier transformed with Eq. 5.29, the common differential equation is obtained

$$\frac{dF(q, t)}{dt} = -Dq^2 F(q, t) \tag{5.30}$$

which can be easily integrated to give

$$F(q, t) = \exp(-Dq^2 t) \tag{5.31}$$

With the meaning of the q parameter in scattering experiments this function is the dynamic light scattering TCF.

$$F(q,t) = g_1(q,t) \tag{5.32}$$

Thus the first cumulant is

$$\Gamma_1 = q^2 D \tag{5.33}$$

The relationships of Eqs. 5.26 to 5.33 hold for even more complex systems, whenever a differential equation similar to the diffusion type is present. Then, of course, the prefactor no longer refers to a simple diffusion process. In addition to the motion of the center of mass the first cumulant contains further parameters that are related to other time dependent structural properties.

Typical additional parameters arise from the mobility of segments, or from rotational diffusion observed with long rod-like samples. Another influence on Γ arises from the hydrodynamic interaction among the monomer units in a polymeric sample, as was demonstrated in Fig. 5.2. The theoretical treatment needed for the interpretation of experimental results became a demanding issue to theoreticians. A solution has been possible for a few special structures [8, 46, 60].

5.8 Intermediate Summary

Dynamic light scattering (DLS) is based on a correlation of two scattering intensities that are separated by a certain delay time. The handling of the experimental instruments for DLS is easy and does not require a deeper understanding of the involved theory behind this method. However, before starting an interpretation some facts have to be kept in mind and are summarized below without mathematical arguments.

(i) A scheme has been presented to make clear how a time correlation functions (TCF) is constructed.
(ii) Applied to light scattering an exponentially decaying scattering intensity time correlation function TCF is obtained.

$$g_2(t) = \langle i(0)i(t)\rangle / \langle i\rangle^2$$

(iii) This normalized *intensity* correlation function $g_2(t)$ is related to the *field* correlation function $g_1(t)$ by the equation

$$g_1(t) = (g_2(t) - 1)^{1/2} \text{ where } g_1(t) = \langle E(0)E^*(t)\rangle / \langle i \rangle.$$

$E(0)$ is the electric field at a time zero and $E^*(t)$ is the conjugated complex field at a delay time t later.

(iv) The relaxation parameter Γ_1 (first cumulant) in

$$g_1(t) = \exp(-\Gamma_1 t)$$

is related to the translational diffusion coefficient $\Gamma(q) = Dq^2$, where

$$q = (4\pi/\lambda)\sin(\theta/2)$$

with λ the wavelength of the light in the solvent, and θ is the scattering angle.

(v) In complex systems the TCF is not a single exponential. In this case a multi cumulant fit of $\ln(g_1(t))$ has to be applied (see below Eq. 5.34). From the coefficient Γ_1 the diffusion coefficient $D = \Gamma_1/q^2$ is obtained, i.e., in the limit of $q \to 0$. For uniform hard spheres a single exponential decay is obtained and permits the determination of the diffusion coefficient at any scattering angle. Problems arise if at the probing angle the static particle scattering intensity has a deep minimum which is characteristic of uniform colloid particles.

(vi) Time correlation functions are based on irregular Brownian motions. These motions can affect also the internal structure of a large particle. For instance, mobile segments are excited to oscillations. In such cases the ratio $\Gamma/q^2 = D_{\text{app}}(q)$ is q-dependent and requires a careful recording of $g_2(t)$ at different scattering angles. A fit by two or three cumulants is afforded

$$-\ln g_1(t) = \Gamma_1 t + (\Gamma_2/2!)t^2 + (\Gamma_3/3!)t^3 + \ldots \quad (5.34)$$

in which Γ_1, Γ_2, and Γ_3 are the first, second, and third cumulant [33]. For instance the curve in Fig. 5.4 was fitted with three cumulants.

5.9 Relation Between Static LS and Dynamic DLS

It remains to define what is meant with the parameter q. To this end some thoughts have to be given on static light scattering (SLS). Evidently no time dependence exists if the intensities are collected over a sufficiently long time. Therefore, the dynamic (DLS) function $S(q, t)$ becomes $S(q)$ which is the static LS function. Let us consider two scattering centers j and k which are linked by a bond length $r_{jk} = nl$, where n is the static LS number of bonds between the two ends, j and k, of the segment. If a primary beam hits a unity at j and k, it excites the unit to emit a weak stray-light into all directions, i.e., two interfering spherical waves are emitted. It now depends on the scattering angle at which the scattering phenomena are recorded. The situation is illustrated in Fig. 5.5.

If the scattering intensity is measured at a scattering angle θ a phase difference occurs in the light from the two centers of $\exp(iqr_{jk})$, which turned out to be qr_{jk}. The primary light has a direction designed by $\mathbf{s_0}$ and the scattered light may be denoted by the vector $\mathbf{s}$, then $\mathbf{q}$ is proportional to the difference between the vectors $\mathbf{s}$–$\mathbf{s_0}$, the prefactor $4\pi/\lambda$ arises from the light of the wavelength λ. The $\mathbf{q}$ vector is called scattering vector, in neutron scattering impetus transfer. The magnitude of the $\mathbf{q}$ vector has the dimension of 1/length, and therefore qr_{jk} is dimensionless.

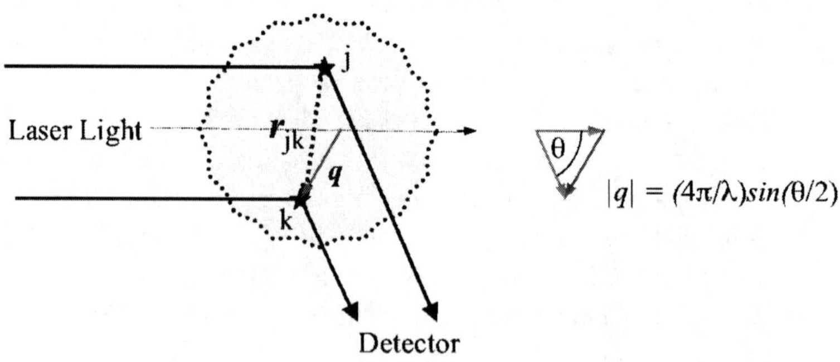

Figure 5.5 Schematic illustration of the scattering from two points in the particle. The black lines indicate the direction of the light from the primary beam and the scattered light at an angle θ. The red line indicates the direction of the scattering vector $\mathbf{q}$ that is the difference between the two unit vectors in direction of the primary beam and the scattering ray.

A macromolecule may have N scattering centers corresponding to a degree of polymerization DP $= N$. This means that the interference effects arise from N^2 pairs of scattering centers. Thus the static scattering function $S(q)$ is given by the double sum

$$S(q) = \sum_{j=1}^{N}\sum_{k=1}^{N} \exp(iqr_{jk}) \tag{5.35}$$

In *dynamic* light scattering the corresponding time correlation function (TCF), from the intensities at times t_0 and a certain time t later, is given in the normalized form by the equation

$$g_2(t) = \frac{\langle i(q)i(q,t)\rangle}{< i(q) >^2} = \frac{A(q)^2 + S(q,t)^2}{S(q)^2} \tag{5.36}$$

or

$$g_2(t) - 1 = \left(\frac{S(q,t)^2}{S(q)}\right)^2 \tag{5.37}$$

and

$$g_1(t) = \frac{S(q,t)}{S(q)} \tag{5.38}$$

Both, the *static* $S(q)$ and *dynamic* structure factors $S(q,t)$ are functions of q^2, and therefore $g_1(t)$ is also a function of q^2. The q^2 dependence becomes clear when exp*(iqr_{jk})* is expanded in series of q^2 and inserted in Eq. 5.35

$$S(q) = N\left(1 - \frac{1}{3} < R_g^2 > q^2 + \text{terms of } (q^2)^n\right) \tag{5.39}$$

and

$$S(q,t) = N\left[1 - \Gamma_1(q^2)t + \Gamma_2(q^2)t^2/2 - \Gamma_3(q^2)t^3/3! + \cdots\right] \tag{5.40}$$

with $\Gamma_1 = Dq^2$ or, according to Stokes–Einstein,

$$D = (kT/(6\pi\eta_0)) < 1/R_\text{h} > \tag{5.41}$$

Thus, in *static* light scattering the *radius of gyration* R_g is a leading structure and scaling parameter, whereas in *dynamic* light scattering it is *the first cumulant* Γ_1. The two radii, R_g and R_h, are related to each other. It is sensible to introduce a new parameter ρ defined by the ratio of both radii [8]

$$\rho \equiv \left\langle R_\text{g}^2\right\rangle^{1/2}\left\langle\frac{1}{R_\text{h}}\right\rangle \equiv \frac{R_\text{g}}{R_\text{h}} \tag{5.42}$$

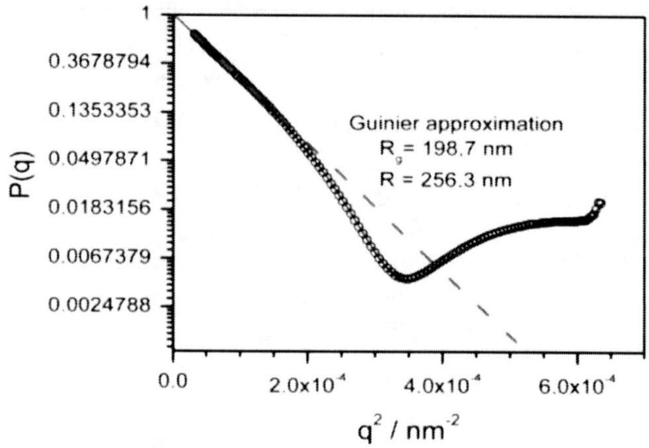

Figure 5.6 Plot of $\ln(P(q))$ as a function of q^2 from the same spherical colloid sample shown in Fig. 5.4 with $P(q) = I(q)/I(q=0)$ the normalized particle scattering factor. The broken line in red indicates the Guinier approximation $P(q) \cong \exp(-q^2 R_g^2/3)$. Data from J. Schneider, Dept. Phys. Chem., University of Freiburg.

which deviates from unity and is a characteristic branching parameter or a measure of the average segment density [8]. For hard spheres the ratio is $\rho_{\text{sphere}} = 0.774$ but for flexible chains the ρ value is larger and lies between 1.5 (theta solvent) and 1.8 (good solvent). If branching occurred the ratio is less than 1.5 and decreases with increasing branching. Actually the short hand writing R_g/R_h encloses a rather complex average formation, mainly due the -1 moment for R_h and the 2nd moment for R_g. Figure 5.6 shows the q^2 dependence of $S(q)$ from the same colloid sample shown in Fig. 5.4 for $S(q, t)$. The radius of a hard sphere is by a factor 1.29 larger than the radius of gyration, and has a ratio of $R_g/R_{HS} = 0.774$ where the index HS denotes hard sphere.

5.10 Summary

Three features are observed. (i) the static light scattering follows only over a limited q^2 region a linear decay, this is in agreement with the theory of Lord Rayleigh [44] for hard and uniform spheres.

(ii) The dynamic scattering curve does not decay as a single exponential over the whole delay time regime but displays a slowed-down decay. (iii) The $\rho = R_g/R_h$ ratio for the dynamic cumulant is for the sample in Figs. 5.4 and 5.6 about (6%) larger than predicted by theory. Still, within experimental errors of about $\pm 5\%$, the agreement of the result from the colloid sample with theoretical prediction for hard spheres is good.

The phenomenon of deviation from naive expectation can be far more pronounced for other structures. Such deviations are rather the rule and are not much considered in theories. To explore the reason for this behavior further model considerations and modifications of current theories are needed.

5.10.1 *Polydispersity*

The deviation of the time dependent $g_1(t)$ from a single exponential can arise from two facts which so far has not been taken into account. (i) The ensemble of particles may be of identical topology but the particles in the molar mass distribution are of different size, a feature that originates from the kinetics of polymerization [22, 30]. Due to this polydispersity the *initial* slope (first cumulant) in a plot of $\ln(g_1(t))$ against the delay time t will still follow a q^2 dependence but the first cumulant now represents the ensemble average over all particle sizes, i.e.,

$$< \Gamma_1 >= q^2 < D >_z= (kT/6\pi\eta_0)\left\langle \frac{1}{R_h} \right\rangle_z q^2 \qquad (5.43)$$

$$D = (kT/(6\pi\eta_0)) < 1/R_h >$$

is a z average. It means the average have not to be taken over the hydrodynamic radii $<R_h>$ but over the inverse radius $<1/R_h>$. This is a characteristic feature of the hydrodynamics of particles.

To understand the mentioned deviations from the single exponent it is helpful going back to the definition of the dynamic structure factor, which for the ensemble is given by the equation

$$\langle S(q,t)\rangle_z = \frac{\sum\limits_{\text{all}\, j} S_j(q) g_{1,j}(t) w_j}{\text{DP}_w} \qquad (5.44)$$

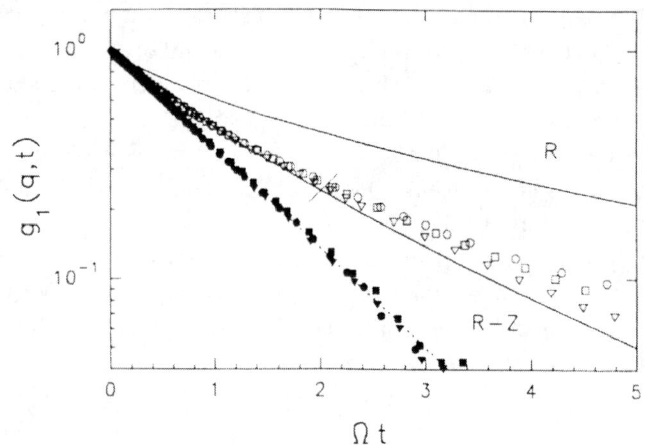

Figure 5.7 Time correlation functions from a linear polyester chain of $M_w = 1.03 \times 10^5$ (g/mol)(filled symbols) and a randomly branched polyester $M_w = 1.7 \times 10^6$ (g/mol) (open symbols) in dimethylformamide (DMF) [51–53]. In the plot of $\log(g_1(t))$ versus Ωt, $\Omega = \Gamma$ is the first cumulant of the TCF. For the fairly low molecular weight of the liner chain a straight line over the whole time regime was obtained corresponding to a single exponential. For the about six times larger branched samples deviation occurred already at very short delay times. *R* means Rouse behavior (no hydrodynamic interactions) [46] and *R–Z* behavior [60] with hydrodynamic interactions. Both samples were measured at angles of 30°, 70°, and 110°. Reprinted with permission from Ref. 51. Copyright © 1997, American Chemical Society.

in which w_j is the weight fraction of the sample distribution with a degree of polymerizarion DP_j. The small particles in the ensemble have a high diffusion coefficient and cause a fast decay, whereas, at the same time, the large particles with the much slower diffusion will show only a weak decay. On the other hand, at a same q value but at long delay times, the TCF from the small particles has already much decayed but the contributions of the large particles in the ensemble now become apparent with their weaker decay constant. Figure 5.7 shows an example of a linear polymer chain.

The effect of polydispersity can become very pronounced if the system consists of an ensemble of single macromolecules and cluster of several macromolecules. Such systems are obtained if the actual concentration c is considerably larger than the overlap concentration c^*. An example with the same linear polyester shown in Fig. 5.7 is reported in Fig. 5.8. Two relaxation times $\tau_1 = 1/\Gamma_1$ and $\tau_{\text{cluster}} = 1/\Gamma_{\text{cluster}}$ are recognized which are separated by factors of 10^3–10^4 from τ_1. This made necessary to use a logarithmic time scale. In the example of Fig. 5.8 the time correlation function (TCF) starts with a fast motion of single molecules that quickly decays to

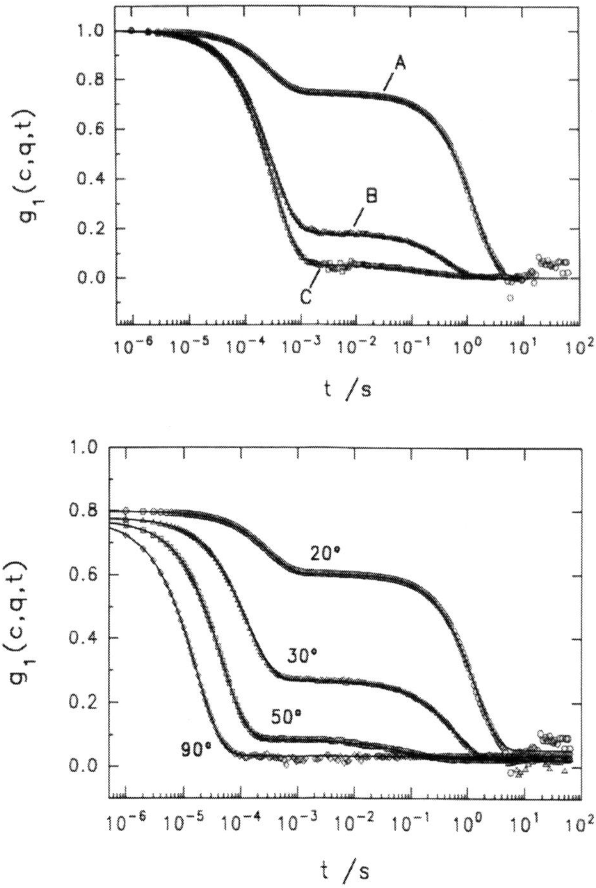

Figure 5.8 TCF of a semidilute solution ($c > c^*$) of the same linear polyester shown in Fig. 5.7 in DMF but at concentrations larger than the overlap concentration c^*; (up) data recorded at angle of 20° and at three overlap concentrations: A: $c/c^* = 7.76$; B: $c/c^* = 6.19$; C: $c/c^* = 5.69$; (down): data from the sample with concentration A at 4 different angles [52].

a constant plateau before the slow motion of the clusters becomes detectable and decay to zero. Both, an angular and a pronounced concentration dependence, are observed.

5.10.2 *Internal Modes of Motion*

We mentioned already that the molecules or clusters are not rigid particles as are realized with colloidal hard spheres or long cylinders. Such rigidity cannot be expected with the chain-like structure of large macromolecules. If such chains are affected by Brownian motion the chains start to oscillate [46, 60] which reminds of the overtones from a violin string but the chain modes differ from the harmonic overtones of a string. The oscillating modes are much faster than the translation motion of the center of mass and have a relaxation distribution of τ_1/j^2 for a Rouse chain [46] without hydrodynamic interaction and τ_1/j^3 for the Zimm model [60] of chains with hydrodynamic interactions where τ_1 is the longest relaxation time of the oscillating linear chain. The effect of these motions could be made visible in experiments at large qR_g values where only the vibration modes contribute to the TCF. But the effect is noticeable also at small q values and can be expressed by an apparent diffusion coefficient [1, 2, 5–7, 9, 10]

$$D_{\mathrm{app}}(q) = \frac{\Gamma_1}{q^2} = D_{\mathrm{trans}}\left[1 + C(q R_{\mathrm{g}})^2 - \cdots\right] \tag{5.45}$$

The coefficient C reflects the influence of the longest relaxation mode of the particle which depends on the topology of the particle (linear, branched or crosslinked chains, etc.). It also depends on a geometrically anisotropic structure, as found for long cylinders. Such anisotropy may be observed with double or triple helixes. In this case the C factor depends largely on the rotational diffusion coefficient, which also may be considered as the longest internal mode. An example was realized in a work by Brown and Schillén [25, 47]. A list of data for different structures is given in [6]. For chain-like flexible macromolecules, i.e., when the repeat units can rotate more or less freely around their bond length, a characteristic asymptotic q dependence was predicted [1, 2, 33]. If hydrodynamic interactions are neglected, Rouse modes of motion should be observed with an asymptotic q^{-4} dependence,

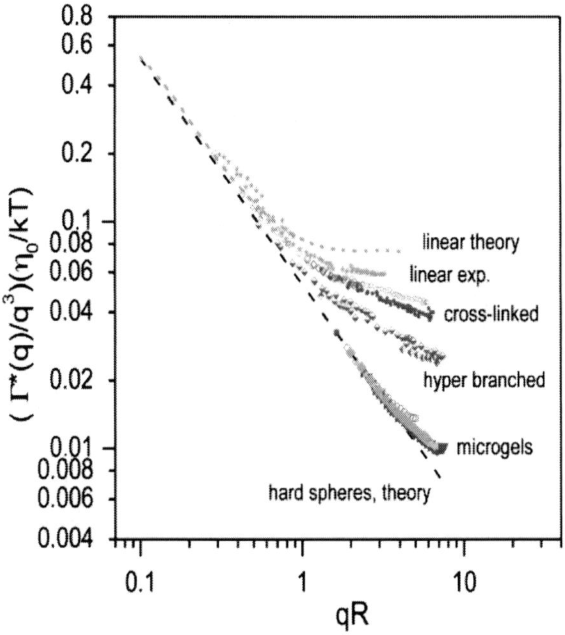

Figure 5.9 Dynamic scattering behavior of linear chains, randomly and hyper-branched chains and nano-sized microgels. In this plot the first cumulant Γ_1 was divided by q^3 and normalized by η_0/kT. In this plot a plateau of height 0.078 was predicted for linear chains with hydrodynamic interactions in a good solvent. The straight line corresponds to the q^2 dependence of hard spheres.

but with hydrodynamic interactions the Zimm–Rouse relaxations dominate resulting in a q^{-3} asymptote. For flexible linear chains all measurements prove the q^{-3} dependence of the Zimm theory [2, 52, 53]. The behavior has been studied by a long list of authors, which are quoted in the paper by Trappe et al. [52, 53]. The predicted Rouse modes could not be realized with polymers in solution. The hydrodynamic behavior of more complex molecules, like branched chains or stiff anisotropic objects, is even more complex. Some examples are shown in Fig. 5.9 [5, 6, 7, 9, 46].

The crosslinked and hyper-branched chains seem to approach a power law asymptote with an exponent of $q^{2.43}$ [28, 29, 41, 42, 43]. Microgels are well-swollen spherical particles and are not as rigid as spherical polystyrene colloids. A flexibility became visible at $qR_g > 5$ [27], which corresponds to segment lengths 5 time shorter than the radius of gyration. For the other structures the onset of deviation from the q^2 behavior started much earlier at $qR_g > 0.8$.

The Zimm behavior is obtained if in a plot of

$$\Gamma^* = (\Gamma(q)/q^3)(\eta_0/kT)$$

a constant plateau is reached at large delay times. This indeed was found with random coil linear chains but with slightly lower value than predicted. For the randomly and hyper-branched samples such asymptote was not found. From these preliminary results it can be stated that the segmental motion becomes increasingly be suppressed when the branching density is increased. For stiff chains recent theory predicted a $q^{-2.43}$ asymptote, which again was found experimentally. Figure 5.10 demonstrates this observation.

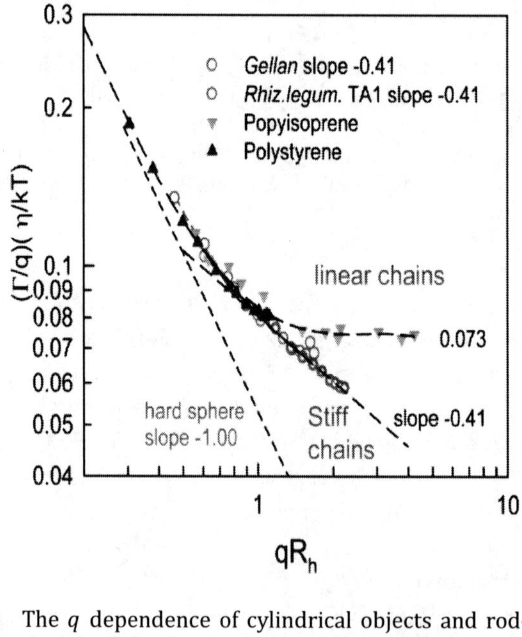

Figure 5.10 The q dependence of cylindrical objects and rod-like chains compared to that from randomly crosslinked chains in a plot of Γ/q^3 as a function of qR_h [51, 52].

Table 5.1 Molar mass M, diffusion coefficient D, the radii R_h and R_g from some proteins, and the ratio of both radii which is a valuable quantity that gives a hint whether a globular particle is present or a more elongated one [27]

Protein	M ($g\,mol^{-1}$)	$D \times 10^7$ ($cm^2\,s^{-1}$)	R_h (nm)	R_g (nm)	$\rho = R_g/R_h$
Lysozyme	14 400	11.80	1.814	1.52	0.838
Ribonuclease	17 000	10.2	2.098	1.48	0.705
β-Casein	24 100	6.05	3.587	7.50	2.120
Chymotrypsinogen α	38 000	7.9	2.709	1.81	0.668
Ovoalbumin	45 000	7.3	2.931	2.40	0.819
Bovine serum albumin	66 000	5.93	3.609	2.98	0.826
Hexokinase (yeast)	99 000	6.0	3.562	2.47	0.693
Lysine-tRNA ligase	138 000	4.3	4.976	3.75	0,754
γ-Globulin (human blood serum)	162 000	3.7	5.783	7.00	1.210
Catalase (bovine liver)	225 000	4.1	5.220	3.98	0.763
Fibrinogen	390 000	1.86	11.505	14.20	1.234
Myosin (Rabbit muscle)	493 000	1.16	18.448	46.80	2.537
α_2-Myoglobin	820 000	2.41	8.879	6.37	0.717
Satellite tobacco necrosis virus	1 700 000	2.04	10.491	6.80	0.648
Pyruvate-dehydrogenase	3 780 000	1.2	12.833	15.65	0.876
Turnip yellow mosaic virus	4 970 000	1.55	13.806	10.80	0.782
Alfalfa mosaic virus (bottom)	6 920 000	1.05	20.381	21.60	1.060
Hemocyanin helix pomatia	8 905 000	1.07	20.000	16.40	0.820
Bacteriophage λ (enlarged prehead)	17 000 000	0.69	31.014	28.50	0.9189
Tobacco mosaic virus	39 000 000	0.38	56.316	92.40	1.641

5.10.3 *Proteins*

Proteins are fairly compact and small compared to the wavelength used. Table 5.1 gives a list of data from well-known proteins. In most of these examples the normalized static structure factor $P(q) = S(q)/S(q = 0)$ shows no detectable angular dependence. This behavior is always observed when the particle is smaller than the light wavelength in the solvent (i.e., $R_g < \lambda/30$). For example, in water the wavelength of the HeNe laser is $\lambda = 633/1.333$ and therefore the radius of gyration should be $R_g > 15$ nm in order to detect the angular dependence. In DLS experiments somewhat smaller radii can be measured but below 2 nm problems come up which only in rare cases can overcome.

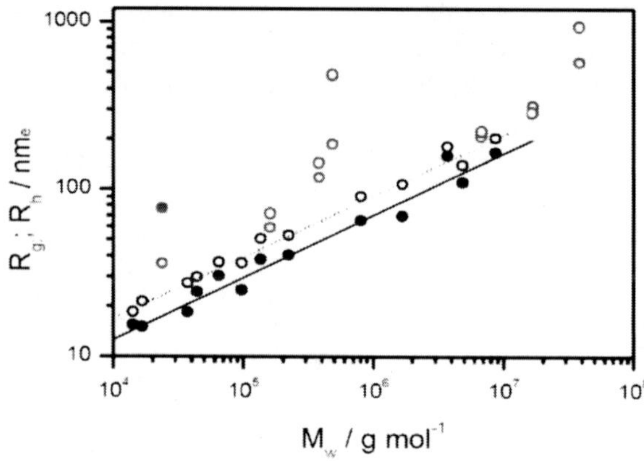

Figure 5.11 Molar mass dependence of the radii of gyration (filled symbols) and of the hydrodynamic radii (open symbols). Most of the examples follow a fairly well defined power law with an exponent of 0.369 ± 0.003. A number of proteins display marked deviations (red symbols) whose origin is made clear in Fig. 5.12.

The data of Table 5.1 were modified from a paper by He and Niemeyer [27].

Up to about 50 nm the first cumulant $\Gamma_1 = Dq^2$ shows no further q dependence. In other words, the particle can be considered as small and close to hard sphere behavior such that exclusively the translational diffusion coefficient counts for the interference of the particle at $x(t = 0)$ and $x(\Delta(t))$ some time later. The data of Table 5.1 are plotted in Figs. 5.11 and 5.12 for the radii R_g and R_h and the ratio $\rho = R_g/R_h$.

5.11 Polydispersity and Effects of Cluster Formation

Proteins often have a strong tendency to aggregate to form a dimer or even larger bundles. This fact has been observed with serum albumin, insulin and thrombin activated fibrinogen to mention only a few examples. Often the aggregates consist of a defined number of unimers. The time correlation function displays noticeable deviations from a single exponential decay but develops, after a fairly fast decay, a single weaker but still a single exponential

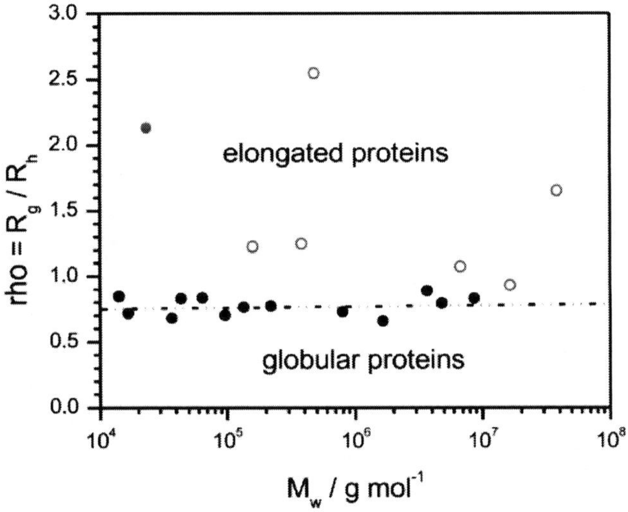

Figure 5.12 Plot of the ρ parameter of proteins in a wide rage of molecular weights. The black symbols are positioned around a constant value of 0.774 that is the predicted value for compact spheres, and therefore give indication for a globular structure of the proteins. Also the data of the red symbols can be interpreted by theory and the high values arise from an elongated structure but this requires additional measurements, at different scattering angles.

asymptote. In such cases the TCF can satisfactorily be described by a two exponential fit with two different relaxation times or two different diffusion coefficients which give information on the unimer and aggregated sizes. Frequently the two relaxation times differ by 3 to 4 decimals. In these cases the fast process decays to a plateau which extends up to large correlation times but then a second decay takes place. The origin of such slow motions in most cases cannot be disclosed by DLS alone and further experiments must be carried out with a different method. It is advisable to perform static LS with the same sample and same concentration because the slow motion often indicates a large object that in static LS will display a pronounced angular dependence; Figure 5.8 shows such an example.

The second now mostly applied method to receive quantitative data on various components in the system is the application of the CONTIN inversion program. In general the DLS time correlation function can be expressed by the integral

$$g_1(q, t) = \int_0^\infty H(\tau)\exp(-t/\tau)d\tau$$

$$\frac{1}{\tau} = Dq^2 \qquad (5.46)$$

The integral in Eq. 5.46 presents a Laplace transform of the relaxation time distribution. In contrast to Fourier transform a Laplace transform cannot be analytically transformed. Thus an attempt of inversion has to be made by sensible conjectures established by the information theory. The elaborated algorithm works fine for continuous single-peaked distributions of $H(\tau)$, but if the distribution consists of two or three well separated peaks, additional constrains have to be inserted in the conversion program like that of CONTIN. The development of this program is based on the Information Theory and the algorithm has been continuously improved and gave the experimentalist a safe feeling. Still one has to be aware that the derived prediction is not fully determined and a critical judgment of the obtained results remains important.

5.12 Electrophoretic/Dynamic Light Scattering

One possibility for such a check is possible for proteins by the further developed electrophoretic dynamic light scattering (ELS). The method was already outlined by Ware and Flygare [54], based on the method of heterodyne scattering (an excellent description was given by Wilcoxon and Schurr [56]). A short description is given below. The idea consists in the study of what can be observed if, vertically to the plane of common scattering, an electric field is applied to a protein. It turns out that nothing special can be observed in the homodyne SLS, but a very different and informative result is obtained if the heterodyne technique is applied.

Heterodyne means that a small portion of constant not-scattering light is added to the scattering field. This setup causes a mixing of the local light with the homodyne scattering field by which the static light (i.e., the base line) is much enhanced and only the time-dependent field correlation function is obtained as scattering source. The situation is described by the equation

$$\langle i(0)i(t)\rangle \cong (I_{\text{local}}^2 + 2I_{\text{local}}\ \text{Re}(I_1(t)) \tag{5.47}$$

where

$$I_1(t) = \left\langle A(0)^* A(t)\right\rangle \tag{5.48}$$

is the *electric field* TCF, and only the real part has a physical meaning. This heterodyne correlation function has the advantage that the effect of an external force can be taken into account. Due to this force the scattering particle undergoes not only the Brownian motion but in addition a constant flow occurs. It has been already outlined that the field correlation function can easily be obtained by the differential equation of a diffusing particle in space. The diffusion equation can now easily be obtained by adding the gradient of the external flow to the differential equation. In addition, after performing the Fourier transform, this partial differential equation becomes a common differential equation with the solution of

$$\text{Re}[g_1(q,t)] = \text{Re}[\exp(\mathbf{q}\cdot\mathbf{v}t)]\exp(-q^2Dt) = \langle N\rangle\cos(\mathbf{q}\cdot\mathbf{V}t)\exp(-q^2Dt) \tag{5.49}$$

This equation makes apparent that, besides the relaxation due to diffusion, now a cosine wave–like time dependence is superimposed. The use of vector notation (bold letters) is required since the direction of scattering vector **q** will in general form an angle with the direction of the velocity **V**. The scalar product $\mathbf{q}\cdot\mathbf{V}$ can be written in detail as

$$qV\cos(\phi) = \frac{4\pi}{\lambda}\sin(\theta/2)V\cos(\phi) \tag{5.50}$$

On the other hand the product

$$\frac{2\pi}{\lambda}V = \Delta\omega$$

is a frequency that in the present experimental setup can be considered as the Doppler shift caused by the moving charged particle due to the external electric force. This makes clear the meaning of the cosine of the angle ϕ between the direction of scattering vector **q** and the direction of flow velocity **V**. Of course if the electric field causes a flow vertically to the scattering vector no Doppler effect can be observed, the angle ϕ permits via the choice of cos(ϕ) to observe the cosine component of the moving velocity (the two angles θ and ϕ may be somewhat confusing, so again, the definition: θ is the scattering angle in the plane of the primary beam and the detector, whereas ϕ is defined as the angle between the flow direction and the direction of the **q** vector. Inspection of Fig. 5.5 may help to clear the meaning of the two angles). The equivalence of

$$\mathbf{q} \cdot \mathbf{V} = \Delta\omega_{\text{Doppler shift}} \tag{5.51}$$

permits a twofold detection of the flow (i) by heterodyne auto correlation function of the light scattering, i.e., the observation of how fast the time dependence of the cosine waves of light move across the detector or (ii) the corresponding power spectrum can be recorded and the Doppler shift is detected from the line width. The first type of experiments was realized by Ware and Flygare and is shown in Fig. 5.13, which is a slightly modified copy from the original paper.

At zero electric field the common exponential decay to the diffusion is obtained. With increasing field strength (V/cm) a wave is observed within the same time interval as recorded at zero field strength with increasing number of waves in the same time interval of 2–3 minutes. From these curves the electrophoretic mobility μ_{E} and the electric charges of the particle can be obtained. Wilcoxon and Schurr checked the equivalence with the Doppler shift, by carrying out online a fast Fourier transformation and recording the obtained power spectrum.

The obtained line width in their experiments was fairly noisy but still a complete agreement in the electrophoretic mobility was found.

At present three types of ELS spectrometers are on the market. Not many details of the instruments are given in the advertisement, but it seems that mostly the TCF is converted by fast Fourier transformation into a power spectrum resulting in a spectral line and the Doppler shift is obtained from the measured line width broadening.

In the meantime the recording of the Doppler shift could be much improved, such that quick measurements of high quality can be made. The measurements are made at only one fixed angle of ϕ and θ. This means that only small particles of $R_{\text{h}} < 30$ nm can reliably be measured, which is well in the regime of interest with proteins and other globular shape.

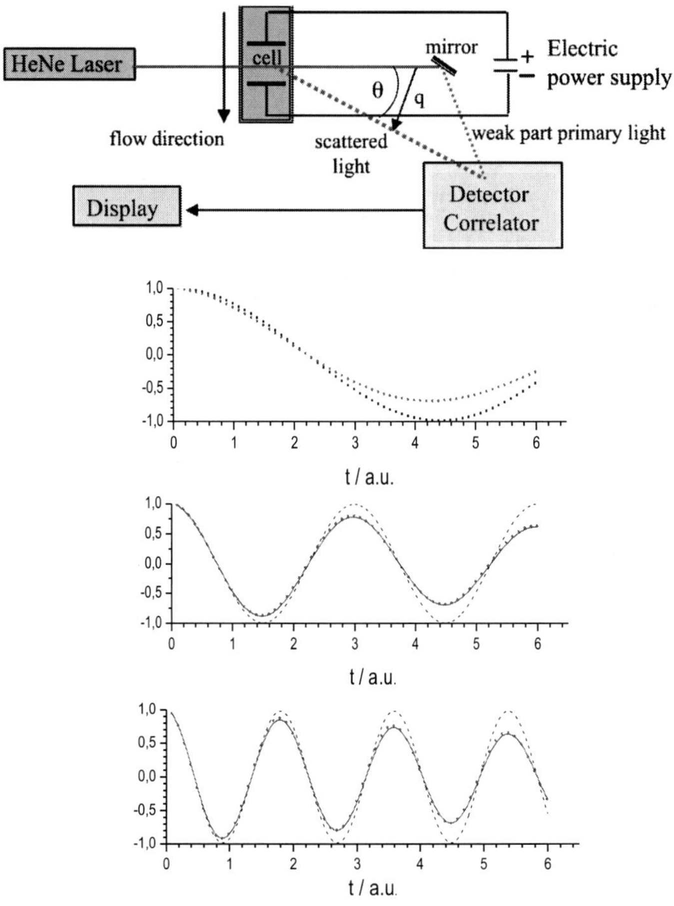

Figure 5.13 Draft of the optical arrangement for measuring electrophoretic mobility via dynamic light scattering. θ is the scattering angle and q the corresponding scattering vector. The mixing of a local constant intensity with the fluctuating scattering intensity is arranged in this example by a mirror after attenuation before the two intensities enter the detector and correlator. The applied electric voltage causes a flow of the charged sample. The response is a weakly attenuated cosine wave. At higher voltages a faster flow occurs indicated by more wave cycles within the same time interval. For the three examples, observed within the same time interval, the voltage was increased three and five times. The damping of the amplitudes due to the Brownian fluctuations is mostly fairly weak and can be measured, from the first cumulant and the diffusion coefficient, when the electric voltage is switched off.

The variation of the angles is restricted by the construction of the instruments that becomes clear from the graph given by Wikipedia, which is reproduced here with minor modifications.

5.13 Summary of ELS/DLS

Mixing of the time dependent scattering intensity with the light of a local light, e.g., a weak part of the primary beam, leads to a heterodyne correlation function. This correlation function consists of the squared intensity of the local light source plus the product of the local intensity and the field correlation function. This design of experiments has the advantage that the effect of external forces can be analyzed. If the external force is caused by an electric field and the particles are somewhat charged the heterodyne field TCF is modified by a damped cosine wave modulation from which the electrophoretic mobility can be obtained. The damping arises from diffusion of the Brownian particles. The diffusion constant and the related hydrodynamic radius of the charged particle can be measured at zero electric field. Commercially available instruments permit quick measurements of the electrophoretic mobility and estimation of the charges of the particle.

5.14 Concentration Dependence

So far the diffusion coefficient at one concentration has been considered. Often no significant concentration dependence is obtained, in particular, if small particles are investigated. In such cases the hydrodynamic radius can directly obtained from the Einstein relationship. Actually, two counteracting interactions are effective at finite concentration. One arises from the concentration dependence of the osmotic compressibility and the other from the increasing friction as the concentration of particles is increased. Diffusion is a process of irreversible thermodynamics and always is effective if the solution is not at thermodynamic equilibrium and the particles move to re-establish equilibrium. The driving force is the osmotic compressibility

$$(M/kT)(\partial\Pi/\partial c) = M/M_{\text{app}}(c) = 1 + 2A_2Mc + 3A_3Mc^2 + \cdots \quad (5.52)$$

It can easily be obtained from static light scattering at zero scattering angle

$$\frac{kc}{R_{\theta=0}} = \frac{1}{M_{\text{app}}(c)}$$
$$\frac{1}{M_{\text{app}}(c)} = \frac{1}{M} + 2A_2c + 3A_3c^2 + \ldots \quad (5.53)$$

The concentration dependence of the mutual diffusion coefficient then is

$$D_{\text{m}}(c) = \frac{kT}{f(c)}\frac{M}{kT}\frac{\partial\Pi}{\partial c} = \frac{kT}{f(c)}(1 + 2A_2M_c + 3A_3Mc^2 + \cdots) \quad (5.54)$$

The concentration dependence of the frictional coefficient can be expressed by

$$f(c) = f_0(1 + k_f c + \cdots) \quad (5.55)$$

Combining Eqs. 5.53 and 5.54 one has a concentration dependence which in a first approximation is given by

$$D_{\text{m}}(c) = D_0(1 + k_D c) = \frac{kT}{6\pi\eta_0 R_h}(1 + 2A_2c - k_f c) \quad (5.56)$$

with

$$k_{\text{D}} = 2A_2 - k_{\text{f}} \quad (5.57)$$

A vanishing concentration dependence of the diffusion coefficient is the result of two different interactions of opposite behavior. Mostly the thermodynamic effect is stronger than the increasing friction and a positive k_{D} is observed. For charged proteins the effect can be very strong such that in any case a dependence from the concentration should be measured. Otherwise a too small hydrodynamic radius is obtained. It has to be kept in mind: the Stokes–Einstein relationship holds valid only in the limit of zero concentration, or if no concentration dependence is observed. The behavior of Eq. 5.56 can be a reason for somewhat confusing effect, which is well known in polymer science. Polymers of different molar mass are often self-similar, i.e., a large polymer develops essentially the same behavior as smaller ones of the same type and differs only by a scaling parameter. Such behavior is frequently observed also for the diffusion coefficient D_0 but not when data at finite concentration are used. Such effect is observed for marginally good solvents when the virial coefficient is fairly low but the friction remained effective. In such systems one observes a large positive k_D for high molecular weights but the value decreases continuously for smaller molecular weights, and k_D becomes negative when $M < 5000$ g/mol. The friction increase is stronger than the thermodynamic repulsion occurring among different particles.

A special case is obtained with polymers in a so-called theta solvent, in which the second virial coefficient $A_2 = 0$. The particles appear to be in an ideal solution. A similar situation can be found with proteins by adding salt. The particle seems to be unperturbed by interparticle interactions and one may also be inclined to expect no concentration dependence of the diffusion coefficient. On the contrary a very pronounced decrease of D_m is observed since the self-diffusion

$$D_{self} = kT/(6\pi \eta_0 R_h(1 + k_f c)) \tag{5.58}$$

remains effective. A quantitative interpretation of the concentration dependence on the self diffusion is a not yet fully understood issue, because with increasing concentration also the solvent viscosity can be increased. Certainly also the Brownian motion through an ensemble of obstacles has a marked effect on the increased friction which is extremely more difficult to predict than the non-correlated Brownian motion.

5.15 Summary

The easy measurable dynamic light scattering leads to reliable results if one is sure of probing *small particles*, i.e., $R_h << \lambda$, the wavelength of the light used. For strong thermodynamic interactions, measurements at finite concentration give smaller apparent hydrodynamic radii and this effect is even more evident with increasing concentration. The effect is especially pronounced with highly charged samples. Data obtained from measurements at different concentrations have to be extrapolated to $c = 0$. Only then the hydrodynamic radius can be determined by the Stokes–Einstein relationship. The concentration dependence arises from thermodynamic interactions between the particles and the friction caused by the viscosity of the solvent. In dilute solutions the former can be well approximated by the second virial coefficient, e.g., measured by static light scattering, but the friction should be measured separately by other techniques, e.g., from sedimentation measurements or by applying a special NMR technique.

For *large particles* with $R_g \approx \lambda$ the first cumulant often develops deviations from the q^2 dependence. This effect is caused by internal motions of the segments and, to a lower part, by a broad molar mass distribution. Measurements have to be made in a wide angular range. The application of the Stokes–Einstein relationships is, in fact, applicable only after extrapolation towards scattering angle $\theta \rightarrow 0$. For weak deviations the apparent diffusion coefficient can be represented by a linear approximation in terms of qR_h. The coefficient in this linear approximation gives indications on the elastic motions. At large qR_h values an asymptotic behavior is obtained which is characteristic of special segmental motions. For linear and star-branched chains the local dynamics could be solved in theory. Experimental data are largely missing mainly since the required spin-echo measurements by SANS require an exorbitant long recording time.

At present the local motions are considered as less important. However, in order to understand the mechanisms of drug loading and release, the local dynamics will become of more interest.

References

1. Akcasu, A. Z., Gurol, J. (1976). Quasielastic scattering by dilute polymer solutions. *J. Polym. Sci. Polym. Phys. Ed.* (**14**), pp. 1–10.
2. Akcasu, A. Z., Benmouna, M., Han, C. (1980). Interpretation of dynamic scattering from polymer solutions. *Polymer* (**21**), pp. 866–890.
3. Berne, B. C., Pecora, R. (1976). *Dynamic Light Scattering,* Wiley & Son, New York.
4. Bunsen, R., Kirchhoff, G. (1861). Untersuchungen über das Sonnenspektrum und die Spektren der chemischen Elemente. *Abh. kgl. Akad. Wiss,* Berlin.

5. Burchard, W. (1978). Application of cascade theory to calculation of quasielastic scattering functions. 1. Polydisperse, ideal, linear chains in dilute solutions. *Macromolecules* (**11**), pp. 455–459.
6. Burchard, W. (1983). Static and dynamic light scattering from branched polymers and biopolymers. *Adv. Polym. Sci.* (**48**), pp. 1–124.
7. Burchard, W. (1985). New aspects of polymer characterization by dynamic light-scattering. *Chimia* (**39**), pp. 10–18.
8. Burchard, W. (1999). Solution properties of branched macromolecules. *Adv. Polym. Sci.* (**143**), pp. 113–194.
9. Burchard, W., Kajiwara, K. (1984). Rotational isomeric state calculations of the dynamic structure factor and related properties of some linear chains. 2. First cumulant of the dynamic structure factor. *Macromolecules* (**17**), pp. 2674–2679.
10. Burchard, W., Schmidt, M., Stockmayer, W. H. (1980). Information on polydispersity and branching from combined quasi-elastic and integrated scattering. *Macromolecules* (**13**), pp. 2365–2372.
11. Burne, P. M., Sellen, D. B. (1994). A laser light scattering study of gellan gels. *Biopolymers* (**34**), pp. 371–382.
12. Chandrasekhar, S. (1943). Stochastic problems in physics and astronomy. *Rev. Mod. Phys.* (**15**), p. 1.
13. Coviello, T., Burchard, W., Dentini, M., Crescenzi, V. Morris, V. J. (1995). Solution properties of exocellular polysaccharides of rhizobium leguminosarum. Static and dynamic light-scattering: Characterization in dilute solution. *J. Polym. Sci. Part B* (**33**), pp. 1833–1846.
14. Coviello, T., Kajiwara, K., Burchard, W., Dentini, M., Crescenzi, V. (1986). Solution properties of xanthan. 1. Dynamic and static light scattering from native and modified xanthans in dilute solutions. *Macromolecules* (**19**), pp. 2826–2831.
15. Coviello, T., Burchard, W. (1992). Criteria for the point of gelation in reversibly gelling systems according to dynamic light scattering and oscillatory rheology. *Macromolecules* (**25**), pp. 1011–1012.
16. De Gennes, P.-G. (1967). Quasi-elastic scattering of neutrons by dilute polymer solutions. I : Dree draining limit. *Physics* (**3**), pp. 37–45.
17. De Gennes, P.-G., Dubois-Violette, E. (1967). Quasi-elastic scattering by dilute, ideal, polymer solutions. II: effects of hydrodynamic interaction. *Physics* (**3**), pp. 181–198.
18. Debye, P. (1945). Technical Report CR-637. Private communication to Reconstruction Finance Corporation, Office of Rubber Reserve.

19. Dentini, M., Coviello, T., Burchard, W., Crescenzi, V. (1988). Solution properties of exocellular microbial polysaccharides. 3. Light scattering from gellan and from the exocellular polysaccharide of Rhizobium trifolii (strain TA-1) in the ordered state. *Macromolecules* (**21**), pp. 3312–3320.
20. Doppler, C. (1842). *Abh. kgl. Böhmischer Ges. d. Wiss.* (**2**), p. 465.
21. Einstein, A. (1905). *Ann. Phys.* (**17**), pp. 549–560.
22. Flory, P. J. (1952). *Principles of Polymer Chemistry*, Cornell University Press, Ithaca, N.Y.
23. Wollaston, W. H. (1802). A method of examining refractive and dispersive powers, by prismatic reflection. *Phil. Trans. Roy. Soc. Lond.*, pp. 365–380.
24. Fraunhofer, J. (1817). Bestimmung des Brechungs- und des Farbenzerstreungs-Vermögens verschiedener Glasarten, in Bezug auf die Vervollkommnung achromatischer Fernröhre. *Annalen der Physik* (**56**), pp. 264–313.
25. Fytas, G., Patkowski, A. (1993). Dynamic light scattering from polymers in solution and in bulk, in *Dynamic Light Scattering*, Brown, W. (ed), Oxford Science Publications, Clarendon Press, Oxford.
26. Harnau, L., Winkler, R., Reinecker, P. (1997). Remarks on the interpretation of dynamic light scattering from gellan in dilute solution. *Macromolecules* (**30**), pp. 6974–6976.
27. He, L., Niemeyer, B. (2003). A novel correlation for protein diffusion coefficients based on molecular weight and radius of gyration. *Biotechnol. Prog.* (**19**), pp. 544–548.
28. Herz, H. (1888). *Progr. Ann. Phys. Chem.* (**24**), p. 551.
29. Herz, H. (1888). *Berlin. Ber.*, p. 197.
30. Hiemenz, P. C., Lodge, T. P. (2007). *Polymer Chemistry*, CRC Press, Taylor & Francis, New York.
31. Kirkwood, J. G., Riseman, J. (1948). The intrinsic viscosities and diffusion constants of flexible macromolecules in solution. *J. Chem. Phys.* (**16**), pp. 565–572.
32. Kirkwood, J. G. (1967). Macromolecules, in *Documents of Modern Physics*, Auer, P. L., Kirkwood, J. G. (eds.), Gordon Breach, New York.
33. Koppel, D. E. (1972). Analysis of macromolecular polydispersity in intensity correlation spectroscopy: the method of cumulants. *J. Chem. Phys.* (**57**), pp. 4814–4820.
34. Maxwell, J. C. (1873). *Electricity and Magnetism*, Oxford.

35. McQuarrie, D. A. (1979). *Statistical Mechanics*, Harper & Row, New York.
36. Murphy, R. M. (1997). Static and dynamic light scattering of biological macromolecules: what can we learn? *Curr. Opin. Biotechnol.* (**8**), pp. 25–30.
37. Oseen, C. W. (1927). *Hydrodynamik,* Akad. Verlag Ges., Leipzig.
38. Pecora, R. (1964). Doppler shifts in light scattering from pure liquids and polymer solutions. *J. Chem. Phys.* (**40**), pp. 1604–1613.
39. Pecora, R. (1968). Spectral distribution of light scattered by monodisperse rigid rods. *J. Chem. Phys.* (**48**), pp. 4126–4127.
40. Pecora, R. (1968). Spectral distribution of light scattered from flexible-coil macromolecules. *J. Chem. Phys.* (**49**), pp. 1032–1035.
41. Perot, A., Fabry, C. (1897). *Ann. Phys. Chim.* (**12**), p. 459.
42. Perot, A., Fabry, C. (1899). *Ann. Phys. Chim.* (**16**), p. 115.
43. Perot, A., Fabry, C. (1901). *Ann. Phys. Chim.* (**22**), p. 564.
44. Rayleigh, J. W. (1914). On the diffraction of light by spheres of small relative index. *Proc. Roy. Soc. London A* (**90**), pp. 219–225.
45. Raj, R., Flygare, W. H. (1974). Diffusion studies of bovine serum albumin by quasielastic light scattering. *Biochem.* (**13**), pp. 3336–3340.
46. Rouse, P. E. (1953). A theory of the linear viscoelastic properties of dilute solutions of coiling polymers. *J. Chem. Phys.* (**21**), pp. 1272–1280.
47. Russo, P. S. (1993). Dynamic light scattering from rigid and nearly rigid rods, in *Dynamic Light Scattering*, Brown, W. (ed.), Oxford Science Publications, Clarendon Press, Oxford.
48. Smoluchowski, M. (1906). *Ann. Phys.* (**25**), p. 205.
49. Smoluchowski, M. (1916). *Phys. Zeitschr.* (**17**), p. 557.
50. Stockmayer, W. H. (1973). Dynamics of chain molecules, in *Molecular Fluids,* Balian, R., Weill, G. (eds.), Gordon and Breach, London, New York, Paris.
51. Trappe, V., Bauer, J., Weissmuller, M., Burchard, W. (1997). Angular-dependence in static and dynamic light-scattering from randomly branched systems. *Macromolecules.* (**30**), pp. 2365–2372.
52. Trappe, V., Burchard, W. (1997). Local dynamics in branched polymers, in *Light Scattering and Photon Correlation Spectroscopy*, Pike, E. R., Abbiss, J. A. (eds.), Kluver Academic Press, Dordrecht, London.
53. Trappe, V. (1994). Ph.D. thesis, University of Freiburg.
54. Ware, B. R., Flygare, W. H. (1971). The simultaneous measurement of the electrophoretic mobility and diffusion coefficient in bovine serum

albumin solutions by light scattering. *Chem. Phys. Letters* (**12**), pp. 81–85.

55. Ware, B. R., Flygare, W. H. (1972). Light scattering in mixtures of BSA, BSA dimers, and fibrinogen under the influence of electric fields. *J. Coll. Int. Sci.* (**39**), pp. 670–675.
56. Wilcoxon, J. P. and Schurr, J. M. (1983). Electrophoretic light scattering studies of poly(L-lysine) in the ordinary and extraordinary phase. Effects of salt, molecular weight, and polyion concentration. *J. Chem. Phys.* (**78**), pp. 3354–3364.
57. Yamakawa, H. (1971). *Modern Theory of Polymer Solutions,* Harper & Row, New York.
58. Zimm, B. H. (1948). The scattering of light and the radial distribution function of high polymer solutions. *J. Chem Phys.* (**16**), pp. 1093–1098.
59. Zimm, B. H. (1948). Apparatus and methods for measurement and interpretation of the angular variation of light scattering; preliminary results on polystyrene solutions. *J. Chem Phys.* (**16**), pp. 1099–1116.
60. Zimm, B. H. (1956). Dynamics of polymer molecules in dilute solution: viscoelasticity, flow birefringence and dielectric loss. *J. Chem. Phys.* (**24**), pp. 269–278.
61. Zwanzig, R. (1965). Time-correlation functions and transport coefficients in statistical mechanics. *Annu. Rev. Phys. Chem.* (**16**), pp. 67–102.
62. Zwanzig, R. (1973). Linear response theory, in *Molecular Fluids,* Balian, R., Weill, G. (eds.), Gordon and Breach, London/New York/Paris.

Chapter 6

NMR Methodologies in the Study of Polysaccharides

Chiara Di Meo,[a] **Noemi Proietti,**[b] **Luisa Mannina,**[a,b] **and Donatella Capitani**[b]

[a]*Department of Drug Chemistry and Technologies, University of Rome "La Sapienza", Piazzale Aldo Moro 5, I-00185 Rome, Italy*
[b]*Magnetic Resonance Laboratory "Annalaura Segre," Institute of Chemical Methodologies, CNR Research Area of Rome, Via Salaria km 29,300, 00015 Monterotondo (Rome), Italy*
donatella.capitani@cnr.it

6.1 NMR in Solution

Polysaccharides offer the possibility to obtain physical, physico-chemical or chemical hydrogels due to the intrinsic properties of such systems, such as the high molecular weight, the peculiar viscoelastic properties in water solution, the presence of functional groups such as hydroxyls, carboxylates, amines, sulphated moieties.

The existence of networks due to intramolecular and intermolecular hydrogen bonds often allows one to obtain physical hydrogels in aqueous solutions both in absence and in presence of suitable counterions. The functional groups on the chains can also be used

Polysaccharide Hydrogels: Characterization and Biomedical Applications
Edited by Pietro Matricardi, Franco Alhaique, and Tommasina Coviello

ISBN 978-981-4613-61-3 (Hardcover), 978-981-4613-62-0 (eBook)
www.panstanford.com

to chemically crosslink the polysaccharide using omo- or etero-bifunctional crosslinkers that directly react with hydroxyl, amine, or carboxylic groups on the polymer chain.

In many cases polysaccharide chains are properly derivatized in order to provide the system with further chemical functionalities able to react in a rapid and selective manner with a wide range of crosslinkers giving new hydrogel systems with improved performances.

The chemical derivatization of polysaccharide chains with a linear arm often needs a large excess of reactants and other chemicals useful as catalysts. As a consequence, final products need an efficient purification. In this frame NMR spectroscopy is a suitable, accurate method to assess both the derivatization degree (dd) of obtained products and their purity.

The assignment of NMR spectra of polysaccharide derivatives in solution can be obtained using various NMR experiments such as ^{1}H–^{1}H COSY, ^{1}H–^{1}H TOCSY, ^{1}H–^{1}H NOESY, ^{1}H–^{1}H ROESY, and ^{1}H–^{13}C HMQC experiments with a z gradient coherence selection [8]. In particular ^{1}H–^{1}H COSY [3] and ^{1}H–^{1}H TOCSY [5] provide ^{1}H–^{1}H spin–spin coupling connectivities for hydrogen atoms within a spin system that forms an unbroken chain of couplings, whereas NOESY [33] allows the detection of spatial proximities between ^{1}H spins by taking advantage of the dipolar coupling interaction. In the case of large molecules having a short correlation time NOESY is used to correlate nuclei through space; however, the NOE cross peak and the exchange cross peak cannot be distinguished. In this case, ROESY (NOE in the rotating frame) pulse sequence should be used [4].

In addition to ^{1}H, other nuclei, in particular ^{13}C are also of interest. Although heteronuclei can be observed directly it may be desirable to observe them indirectly, especially those exhibiting low sensitivity such as ^{13}C through the use of inverse detection 2D experiments [8]. The most common heteronuclear correlation NMR experiments used in the analysis of polysaccharides are ^{1}H–^{13}C HMQC [44] that shows correlations between hydrogen atoms and the carbon atom to which they are attached, and ^{1}H–^{13}C HMBC [6] that provides correlations between hydrogen and heteronuclei that are generally two or three bonds away.

The complete and correct signal assignment allows a quantitative evaluation of dd to be obtained [23] by integrating a suitable signal of the grafted molecule with respect to a typical polysaccharide signal (usually anomeric protons).

As reported above, another important parameter to be controlled is the purity of the product. It can be assessed by means of DOSY NMR experiment which allows several compounds having different molecular weights (i.e., molecules not linked to the polysaccharide chains after the reaction) to be discriminated. DOSY experiment introduced by Morris and Johnson [38] is based on recording a series of 1D NMR spectra with increasing diffusional attenuation, i.e., increasing the magnetic field gradient strength [47]. The resulting dataset is subsequently inverted to yield a pseudo 2D NMR spectrum showing NMR chemical shifts on the horizontal axis and self-diffusion coefficients on the vertical one. Therefore a diffusion spectrum at each chemical shift can be obtained and the self-diffusion coefficient of the investigated compound can be measured. As a consequence, DOSY experiment allows a selection of compounds according to differences in their effective size provided that the corresponding resonances are well resolved in the chemical shift dimension. DOSY experiments have been used for the analysis of a variety of mixtures and aggregates [35] and for the study of intermolecular interactions [36]. Moreover, DOSY has been proposed as a versatile tool for achieving a simple estimation of the molecular weight of uncharged mono-, oligo-, and polysaccharides [53]. DOSY experiment has also been proposed for controlling the purification process (dialysis) of derivatized polysaccharides [19]. Specifically, hyaluronic acid (HA) has been investigated after a derivatization process with suitable moieties to introduce amino functionalities on the polysaccharide chains. Because of the presence of variable amounts of unreacted reagents and/or the formation of low molecular weight compounds after the synthesis, HA derivatives were extensively dialyzed against distilled water. However, the choice of a suitable time of dialysis was rather difficult to establish. In fact, even after days, the presence of low molecular weight unwanted compounds impaired the determination of the true dd percentage due to the overlapping of signals of these compounds to those of the derivatized polysaccharide. In this case

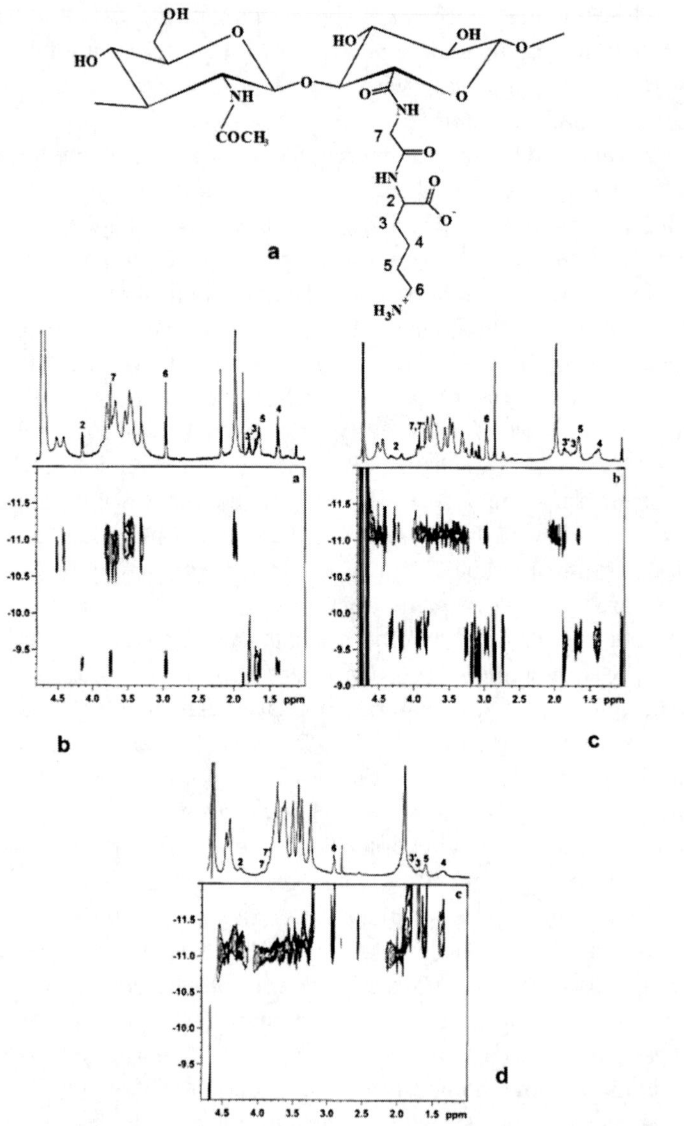

Figure 6.1 Sketch of glycine–lysine HA derivative (a), ^{1}H detected 2D DOSY of (b) a mixture of HA and unreacted glycine-lysine (GK), (c) HA-GK sample dialyzed for 20 days against distilled water, and (d) HA-GK sample dialyzed first against NaCl saturated solution for 3 days and then dialyzed for 20 days against distilled water. In all samples, the polysaccharide concentration was 5 mg/0.7 mL D_2O. Reprinted with permission from Ref. 19. Copyright © 2003, American Chemical Society.

DOSY allowed the choice of the correct time of dialysis. By way of example, we report results obtained by applying the DOSY technique to HA derivatized with glycine-lysine (Fig. 6.1). In Fig. 6.1a the sketch of derivatized HA is reported. Because of the marked difference in the self-diffusion coefficient, DOSY easily distinguished between glycine–lysine free to move and glycine–lysine bound to HA chains. In fact, after the binding, the self-diffusion coefficient of the substituent must be the same as that of HA. The method was tuned by adding a mixture of glycine–lysine to a HA/D_2O solution, in this case, no reaction between the two compounds took place (Fig. 6.1b).

The DOSY map showed that the self-diffusion coefficient of HA was definitely lower than that of free glycine–lysine. In Fig. 6.1c the DOSY map of the HA–glycine–lysine derivative dialyzed against distilled water for 20 days is reported. The map showed the presence of both free and bound glycine–lysine. Finally, Fig. 6.1d shows the DOSY map of an HA–glycine–lysine solution dialyzed against NaCl saturated solution for 3 days. In the map all resonances showed the self-diffusion coefficient of HA. With this method, the presence of low molecular weight molecules not chemically linked to the polysaccharide can be detected, and the progress of the purification process can be monitored and optimized until a pure product is obtained. Thereafter dd can be calculated applying a suitable integration procedure.

6.2 Solid State NMR

When polysaccharide chains or derivatized polysaccharide chains are crosslinked by using a suitable bifunctional linker, a tridimensional network able of swelling in water is obtained.

The chemical structure of these unsoluble amorphous networks can be investigated by solid state NMR [19]. By way of example Fig. 6.2 shows the ^{13}C CP-MAS spectrum of a network obtained from the reaction between partially acetylated chitosan and aqueous glutaraldehyde [19]. Resonances of anomeric carbons (1) of the polysaccharide as well as the carboxylate (7) and methyl carbons (8) are clearly observed along with carbon resonances of the crosslinker (X, Y, and Z) confirming the success of the crosslinking reaction.

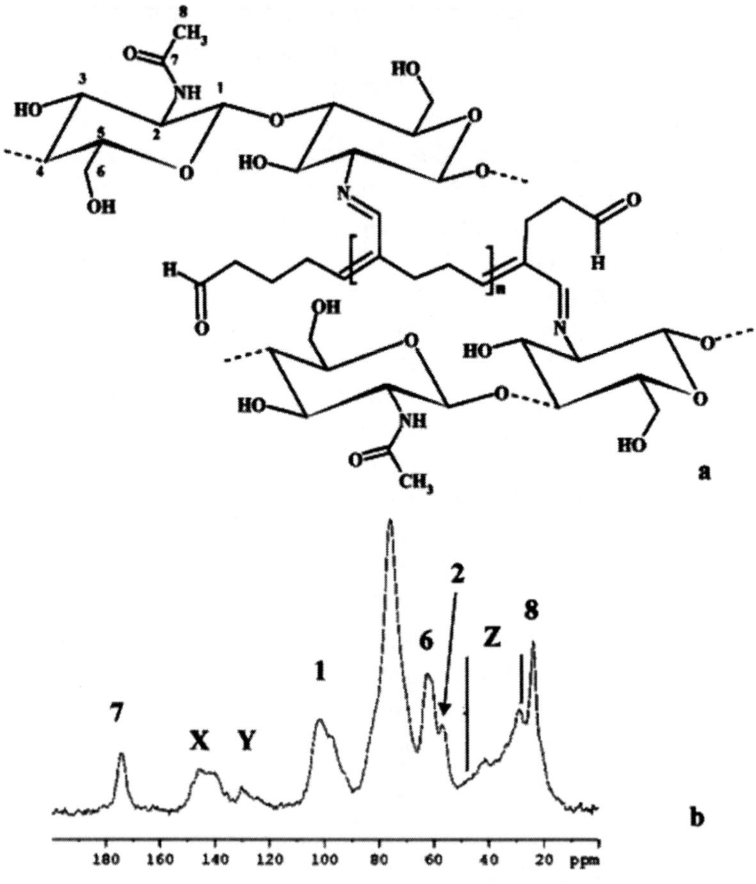

Figure 6.2 (a) Schematic representation of the proposed structure of the network formed from the reaction between partially acetylated chitosan and aqueous glutaraldehyde. (b) ^{13}C CPMAS NMR spectrum of a chitosan-based network. Beside the resonance of chitosan, other carbon resonances are observed: Z = methylene carbons of chemical bridge, X = quaternary carbons of chemical bridges, and Y = methine carbons of chemical bridges. Reprinted with permission from Ref. 19. Copyright © 2003, American Chemical Society.

The degree of crosslink is a structural parameter of primary importance to evaluate the swelling properties of the polysaccharide network in an aqueous medium. Generally the determination of the effective degree of crosslink can be obtained by using equilibrium swelling measurements or by performing uniaxial compression tests on swollen networks [7]. However, both techniques allow a straightforward determination of the crosslinking degree if the chemical structure is "simple" and structural complexity is absent. Interpretation of experimental data is further complicated by the possible presence of microporosity [7]. The availability of an alternative experimental technique able to supply a value for the chemically effective degree of crosslinking, obtained independently of the knowledge of other physico-chemical and structural parameters is an important aid for the characterization of crosslinked systems.

Solid state ^{13}C CP-MAS (cross-polarization magic angle spinning) NMR spectroscopy [43] applied on dried or freeze-dried samples allows both the investigation of the chemical structure of crosslinked systems and the determination of the degree of crosslink [14, 15]. In solid state NMR dipolar interactions cause a marked enlargement of peaks. Furthermore, the position of signals depends on the orientation of molecules with respect to the magnetic field. As a consequence each orientation gives rise to a distribution of lines (powder spectrum) which causes a further enlargement of the line width (chemical shift anisotropy, CSA). To overcome these difficulties, techniques such as high power dipolar decoupling (DD), and magic angle spinning (MAS) have been developed [25]. DD allows the removal of heteronuclear dipolar couplings which in solid samples may be of the order of 80–100 kHz. Many interactions which cause line broadening in the solid phase involve the term $(3\cos^2\theta - 1)$. By spinning the sample at the magic angle ($\theta = 54.7^\circ$) with respect to the direction of the magnetic field, the term $(3\cos^2\theta - 1)$ is averaged out to zero and interactions may be either removed or reduced. MAS at a sufficient spin rate is therefore used to remove the effect of CSA and to assist in the removal of dipolar couplings.

Cross-polarization (CP) is applied to dilute nuclei with a low natural abundance and a low gyromagnetic ratio γ that give rise to a very weak observable net magnetization [48] such as ^{13}C and ^{29}Si. With CP the magnetization is transferred from abundant to dilute nuclei. The overall effect is to enhance the signal-to-noise ratio in spectra of dilute nuclei. It is noteworthy that in spectra acquired with CP T_1 relaxation time of protons and not that of heteronuclei, dictates the suitable recycle delay for signal averaging.

Because in solid samples T_1 of proton nuclei is definitely shorter than that of dilute nuclei, the time of acquisition of CP spectra of dilute nuclei is markedly shortened with respect to that of spectra acquired without CP. The magnetization transfer from abundant to dilute nuclei is obtained by varying the rf pulses ($\mathbf{B}_1$) of the two nuclear species to obtain the matching of their energy levels (Hartman–Hahn condition), i.e., $\gamma_I \mathbf{B}_1(I) = \gamma_S \mathbf{B}_1(S)$, where I is the abundant spin, S is the rare spin, and γ_I and γ_S are the corresponding magnetogyric ratios. Provided that the Hartman–Hahn condition is satisfied, $\omega_I = \omega_S$, the magnetization is transferred from I to S and the intensity of S magnetization is increased of the value γ_I/γ_S, which in the case of $I = {}^1$H and $S = {}^{13}$C is equal to 4 in a single scan. The Hartman–Hahn condition must be kept for a time τ (contact time) that allows for maximization of the NMR signal of dilute nuclei. Nevertheless τ must not exceed a duration which depends on the analyzed sample above which, due to relaxation processes, a loss of signal intensity occurs [30].

6.2.1 *Cross-Polarization (CP) Dynamic*

Because the efficiency of the cross-polarization process depends on the number of abundant spin I and on their distance from dilute spin S, the intensity of resonances is affected by the kinetics of the transfer of magnetization. The question rises when we wish to obtain a quantitative analysis of the intensity of resonances of carbon of different types. To obtain a quantitative analysis the full cross-polarization process must be investigated [30, 55]. At this aim spectra must be collected as a function of the contact time τ. To obtain the true intensity of peaks, the intensity of selected resonances vs. the contact time must be fit to a suitable equation. As an example data may be fit using the two-phase polarization transfer model [29, 57]:

$$\frac{S}{S_0} = \exp\left(-\frac{\tau}{T_{1\rho}^{1\mathrm{H}}}\right)\left[1 - s\exp\left(-\frac{\tau}{T_{\mathrm{CHR}}}\right) - (1-s)\exp\left(-\frac{3\tau}{2T_{\mathrm{CHR}}}\right)\exp\left(-\frac{\tau^2}{2T_{\mathrm{CHD}}^2}\right)\right] \quad (6.1)$$

where

S_0 is the intensity of the resonance at the time $\tau = 0$
$T_{1\rho}^{1H}$ is the proton spin-lattice relaxation time in the rotating frame
T_{CHD} is the cross-relaxation time between carbons and covalently bonded protons
T_{CHR} is the cross-relaxation time between carbons and more remote protons
$(1 - s)/s$ is the ratio between the magnitude of fast and slow polarization transfer phases [32].

Note that S_0 value obtained fitting the experimental data to Eq. 6.1 is the "true" intensity of the resonance and can be used for quantitative analysis.

An example of quantitative analysis to obtain the degree of crosslink of HA-based hydrogels is reported in Ref. 23. In this case, by means of UGI reaction using the spermidine as diamine and formaldehyde and cyclohexylisocyanide, an instantaneaous crosslinking of HA chains was obtained (scheme 1). The theoretical crosslinking degree (TCD) defined as the stoichiometric ratio between moles of diamine and moles of the carboxylic groups of HA was modulated by dosing the suitable amount of spermidine.

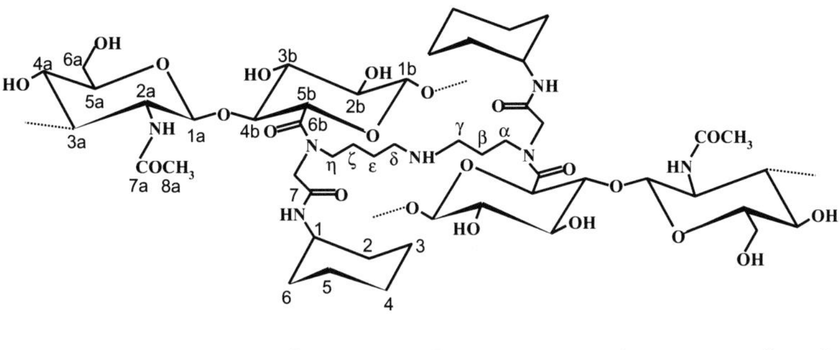

Scheme 6.1 Structure of HA-spermidine UGI networks. Reprinted with permission from [23]. Copyright © 2006, American Chemical Society.

Spectral editing helped to assign all carbon resonances. Specifically the spectral editing of the solid state spectrum of these networks was obtained by applying ^{13}C CP SPI pulse sequence (cross-polarization with a simultaneous phase inversion) that allows different types of carbon to be discriminated. In fact a spectrum was obtained where methine carbon resonances were zeroed, methylene

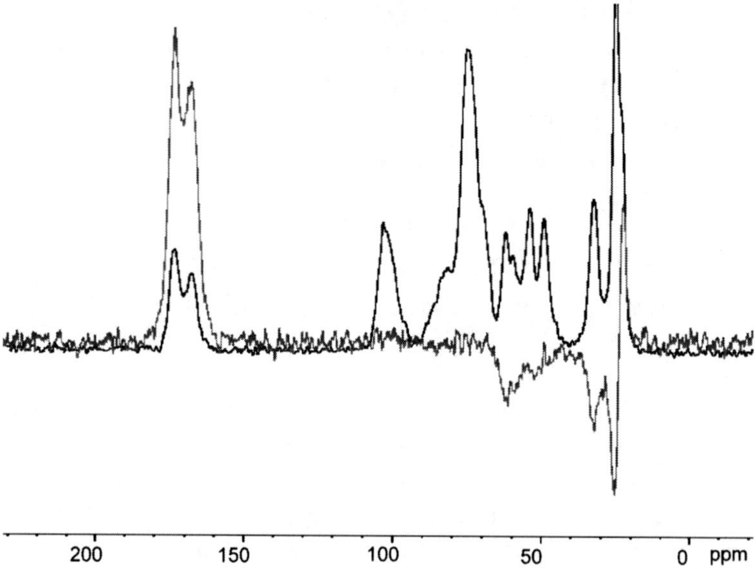

Figure 6.3 ^{13}C CP-MAS NMR spectrum of HA-spermidine network with TCD = 25% (black trace). ^{13}C CP-SPI spectrum of the same sample (blue trace). Reprinted with permission from Ref. 23. Copyright © 2006, American Chemical Society.

carbon resonances were inverted, and resonances due to quaternary and methyl carbon atoms were intense and positive, see Fig. 6.3.

Besides resonances of HA other resonances due to the crosslink are observed. It is noteworthy that the intensity of some resonances increased with increasing TCD, see Fig. 6.4. The resonance at 168 ppm is due to the amidic carbon 7, the resonance centered at 47 ppm is due to methine carbon 1 of cyclohexyl rings and to methylene carbons γ and δ of spermidine The resonance at 33 ppm is due to methylene carbons 2 and 6 of cyclohexyl rings and to methylene carbons α and η of spermidine. Methylene carbons 3, 4, and 5 of cyclohexyl rings and methylene carbons ε, β, and ζ of spermidine are observed at about 24 ppm and partly overlap the intense resonance of the acetyl carbon of HA (see Scheme 6.1).

By plotting the intensity of two selected resonances vs the contact time τ it was possible to obtain the real crosslinking degree (RCD) of five UGI networks with different TCD. The intensity of

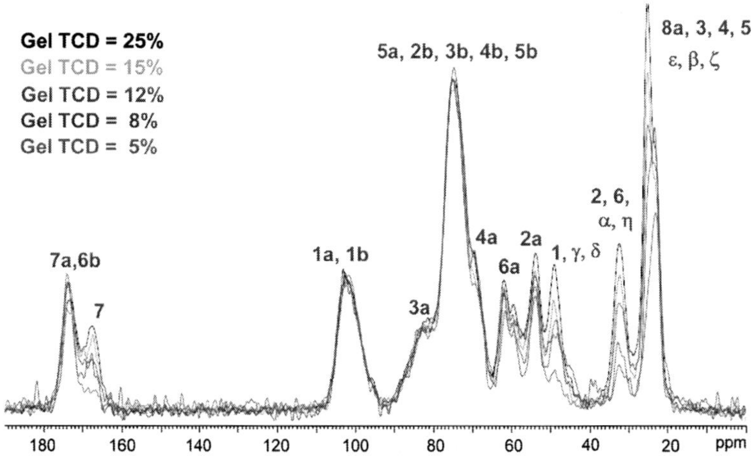

Figure 6.4 ^{13}C CP-MAS NMR spectra of HA-spermidine UGI networks along with the complete resonances assignment referred to Scheme 6.1. Reprinted with permission from Ref. 23. Copyright © 2006, American Chemical Society.

anomeric carbons signal at 105 ppm and that of signals at 33 ppm of carbons belonging to the crosslinker were plotted vs the contact time and fit to equation (6.1). As an example Fig. 6.5 shows the correlation between the intensity of these resonances and the contact time for two spermidine-based networks with TDC = 12% (top) and 25% (bottom) respectively.

In the case of the network with TCD = 12% the "true" intensity of the anomeric carbons and that of signal of crosslinker were found to be 1.7 ± 0.1 and 0.60 ± 0.01 respectively. These values were used to quantitatively evaluate RCD:

$$\mathrm{RCD} = \frac{\frac{S_0(33)}{n}}{\frac{S_0(105)}{m}} \tag{6.2}$$

where $n = 6$ is the number of carbon atoms which contribute to the integral of the resonance at 33 ppm and $m = 2$ is the number of anomeric carbon atoms which contribute to the area of the resonance at 105 ppm.

Table 6.1 shows RCD values obtained from NMR data and compared with the corresponding TCD values.

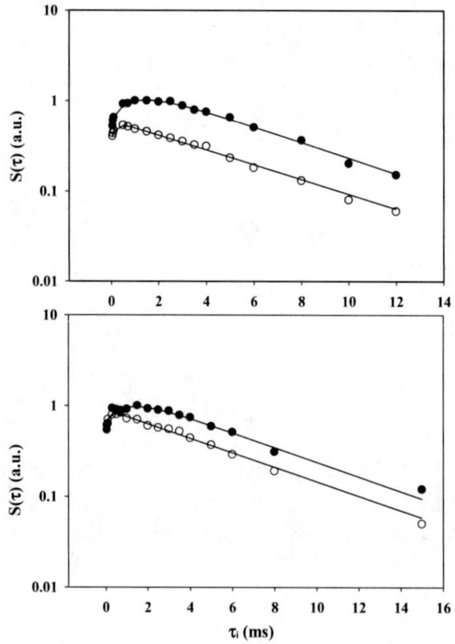

Figure 6.5 Intensity of resonances at 105 ppm (•) and 33 ppm (○) vs. contact time of spermidine-based networks with TDC = 12% (top) and 25% (bottom). Reprinted with permission from Ref. 23. Copyright © 2006, American Chemical Society.

Table 6.1 TCD values of spermidine-based networks compared with RCD values obtained by solid state NMR

TCD (%)	RCD (%)
25	20.0 ± 0.5
15	13.0 ± 0.4
12	11.8 ± 0.4
8	8.0 + 0.5
5	4.9 + 0.4

Source: Reprinted with permission from [23]. Copyright © 2006, American Chemical Society.

6.3 ^{1}H HR-MAS NMR

High-resolution magic angle spinning (HR-MAS) NMR is a relatively young technique developed in the 1990s to use MAS technology for nonsolid or semisolid materials. In fact by spinning the sample at the magic angle, line broadening effects due to dipolar interactions and susceptibility differences within the sample are removed resulting in high-resolution quality spectra. In HR-MAS probes a *Z* gradient aligned along the magic angle enables to access a wide variety of NMR experimental techniques including gradient enhanced solvent suppression and artifact free 2D homonuclear and heteronuclear experiments. A scheme of an HR-MAS stator is pictured in Fig. 6.6.

In the last few years ^{1}H HR-MAS NMR technique has found application in many fields of research [2, 21, 22, 34, 37, 52]. Hydrogels are very suitable materials for ^{1}H HR-MAS analysis [31] that enables the structural characterization of the network in its hydrate state [41, 45].

For example "click-gels" based on HA derivatives have been characterized by means of ^{1}H HR-MAS NMR [20]. Gels were obtained

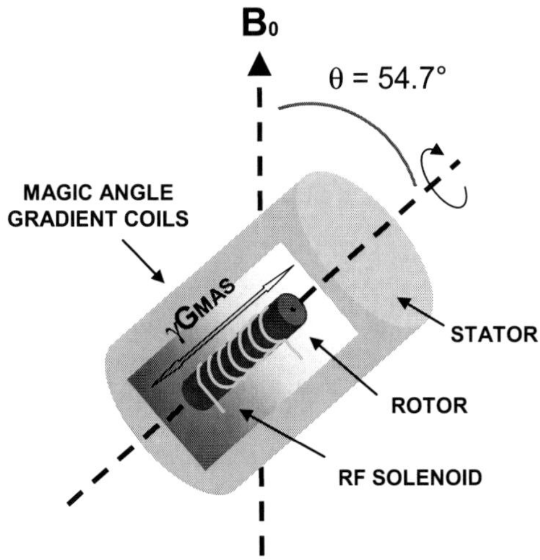

Figure 6.6 Schematic representation of an HR-MAS stator with magic angle gradient along the rotor spinning axis.

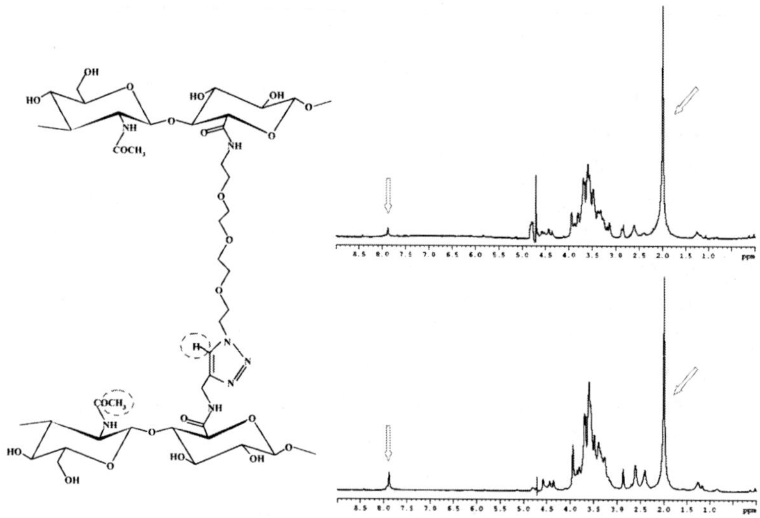

Figure 6.7 Left: structure of HA-based click-gels, protons used for the quantitative evaluation of RCD are evidenced. Right: ^{1}H HR-MAS NMR spectra at 600.13 MHz of two HA-based click-gels with a different TCD. The resonances used for quantitative analysis, i.e., those of triazole ring proton and acetyl protons, are indicated by arrows. Reprinted with permission from Ref. 20. Copyright © 2007, American Chemical Society.

by a "click reaction" between an azide derivative and an alkyne derivative of HA, Cu(I) catalyzed, leading to the formation of a triazolic ring whose proton was well observable at 7.9 ppm in the ^{1}H HR-MAS NMR spectrum (see Fig. 6.7).

Two click-gel networks with a different TCD, hydrated with phosphate-buffered D_2O solution 100 mM at pH = 7, were investigated. Well-resolved ^{1}H HR-MAS spectra were obtained allowing the integration of representative signals, namely the signal of acetyl protons at 2.0 ppm and the signal of the triazolic proton at 7.9 ppm. The obtained integrals allowed a quantitative evaluation of RCD for both systems, that was found to be 8% and 21% mol (linker)/mol (HA) respectively.

^{1}H HR-MAS NMR spectroscopy was also applied to study the HA azide derivative crosslinked by using different dialkynes as linkers [50]. Dialkynes were different in the chemical structure (aromatic or aliphatic) and in the chain length (see Scheme 6.2).

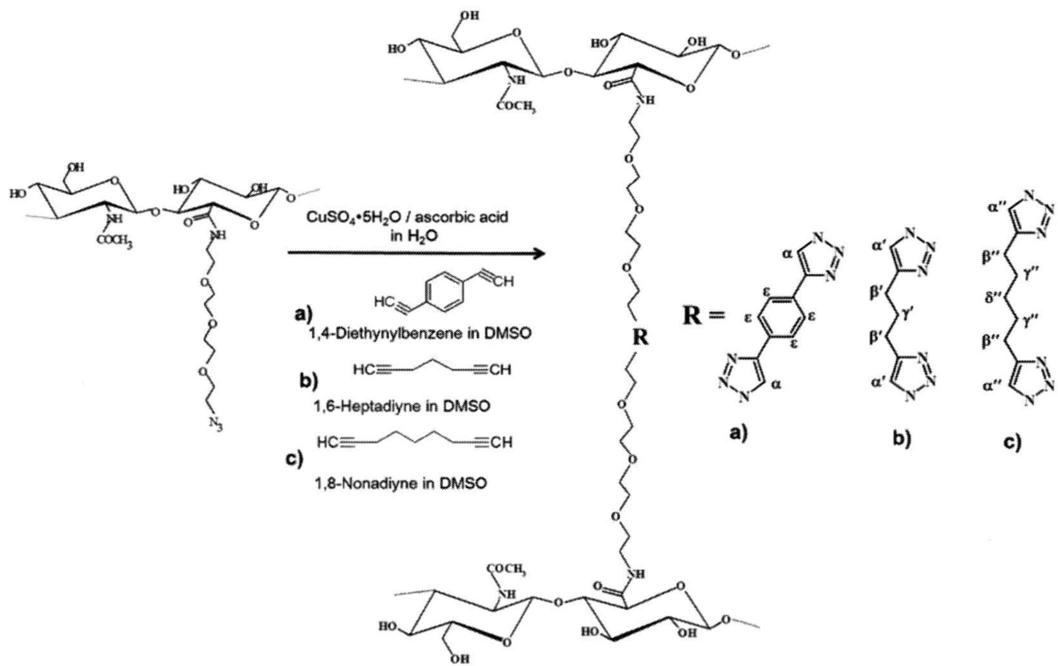

Scheme 6.2 Schematic representation of HA-based click-gels formation starting from the azide derivative of HA and three different dialkynes used as crosslinkers. Reprinted from [50] with permission from Elsevier.

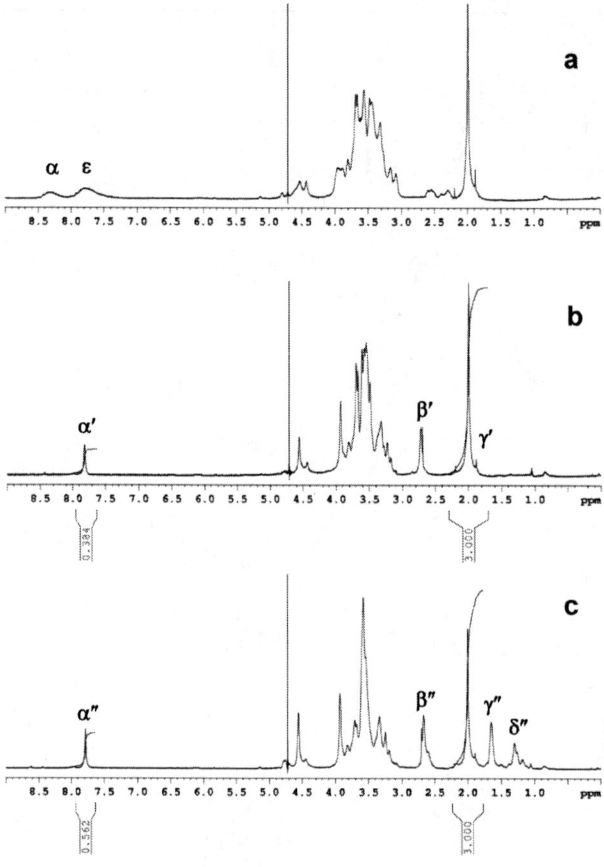

Figure 6.8 ^{1}H HR-MAS NMR spectra at 600.13 MHz of three HA-based click-gels formed using different dialkynes as crosslinkers (see Scheme 6.2). Resonances used for the quantitative analysis are indicated. Reprinted with permission from Ref. 50. Copyright © 2009, Elsevier.

In ^{1}H HR-MAS NMR spectra, besides signals of HA moiety, protons of the two triazolic rings were clearly observable (see Fig. 6.8). A quantitative evaluation of RCD was obtained by integrating the resonances of triazolic ring protons with respect to the resonance of OCH_3 group of HA. RCD was found to be 22% and 28% in samples b and c respectively, whereas, in the case of sample a the stiffness of the system of the three rings caused a broadening of the aromatic signals that prevented the possibility of integrating them. In the case of aliphatic linkers results obtained indicated that the longer and flexible was the linker, the higher the crosslinking efficiency.

6.3.1 ^1H HR-MAS Diffusion

It has been shown that translational diffusion can be conveniently studied using a combination of pulsed field gradient (PFG) and MAS technique [28, 40, 54].

The diffusion behavior of water in scleroglucan (Sclg) gels has recently been investigated [24].

Sclg is able of forming a thermoreversible gel by a weak crosslinking interaction mechanism which depends on the mobility of side chains mediated by water molecules of the first hydration shell. In the presence of borax, Sclg forms a self-sustaining gel. In Sclg tablets produced by unidirectional compression the swelling occurs almost exclusively in the axial direction. Sclg/borax gel appears as an entangled super structure of soft nanochannels made by Sclg triple helix aggregates. Triple helices tend to arrange in domains with an intrinsic ordered structure that, enhanced upon compression, leads to an anisotropic arrangement of chains along the compression direction. Water molecules diffuse inside this network that modifies their diffusion properties. Water diffusion in Sclg and Sclg/borax systems has been studied by pulsed gradient spin-echo (PGSE) NMR technique [24] (see Fig. 6.9). Measurements were carried under magic angle spinning condition at a spinning rate of 4 kHz. PGSE was

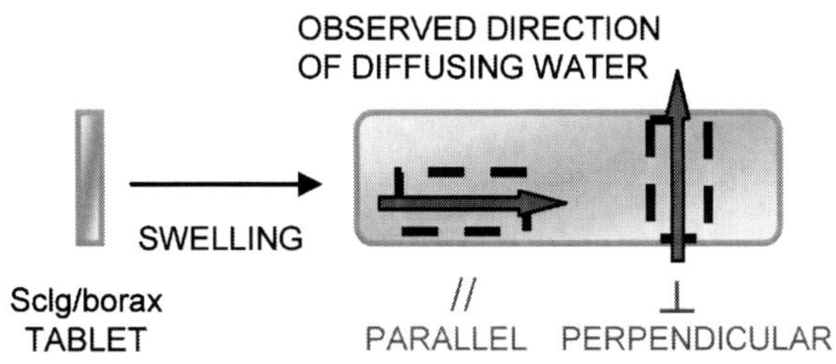

Figure 6.9 Direction of diffusion measurements in relation to principal swelling elongation in scleroglucan/borax tablets. Reprinted from Ref. 24 with permission of the Royal Society of Chemistry.

carried out by keeping the gradient pulse duration at 1.0 ms to avoid relaxation contributions to echo decay, whereas the diffusion time t_d in different experiments was varied from 20 to 800 ms. The gradient strength **g** was logarithmically incremented in 32 steps from 2% up to 95% of the maximum gradient strength.

In PGSE the nuclear spin magnetization is dephased and refocused by two gradient pulses during a spin-echo experiment [49, 56]. The first gradient encodes nuclei position in the spin phase, after a time Δ the second gradient pulse refocuses spin decoding their phase. The resulting echo intensity of the NMR signal is proportional to the number of nuclei that during the time Δ have not significantly changed their position. In particular the echo attenuation $E(\mathbf{q}, t_d)$ is the Fourier transform of the average propagator $\overline{\mathrm{P}}(\mathbf{r}, t_d)$:

$$E(\mathbf{q}, \Delta) = \int \overline{\mathrm{P}}(\mathbf{r}, \Delta)\, e^{(i\mathbf{q}\cdot\mathbf{r})} d\mathbf{r} \tag{6.3}$$

where $\mathbf{q} = \gamma\delta\mathbf{g}$, γ is the magnetogyric ratio of the observed nucleus, and **r** is the dynamical displacement of molecules during the time Δ.

If molecular displacements occurring during gradient pulses δ are negligible with respect to the size of possible barriers, $\delta << \Delta$, and the motion is along one dimension z, the echo attenuation is given by [12]

$$E(q, t_d) = \int \overline{\mathrm{P}}(z, t_d)\, e^{(iq\cdot z)} dz \tag{6.4}$$

where $t_d = \Delta$.

If the average propagator is gaussian the echo attenuation is

$$E(q, t_d) = e^{\left(-\frac{1}{2}q^2\overline{z^2}(t_d)\right)} \tag{6.5}$$

and

$$\overline{z^2}(t_d) = 2Dt_d \tag{6.6}$$

where D is the self-diffusion coefficient.

The combination of Eqs. 6.5 and 6.6 gives the well-known Stejskal–Tanner equation [47].

In general for anomalous diffusion the relationship between MSD (mean squared displacement) $\overline{z^2}$ and the diffusion time t_d is complex as it depends on the statistical properties of the diffusion process. Nevertheless, in several cases a power–law relationship is found:

$$\overline{z^2} = Kt_d^{\gamma} \tag{6.7}$$

Depending on the value of the exponent γ the diffusion process may be classified as subdiffusive if $0 \leq \gamma < 1$, gaussian if $\gamma = 1$, and superdiffusive if $\gamma > 1$. In the case of a gaussian process Eq. 6.7 reduces to Eq. 6.6, and K has the dimension of a self-diffusion coefficient.

Figure 6.10a shows the log–log plot of MSD of water in Sclg sample as a function of the diffusion time. Each point of the plot was calculated from the corresponding NMR echo decay by regression to Eq. 6.5. By fitting MSD data to Eq. 6.7, a self-diffusion coefficient $D = (2.51 \pm 0.03) \times 10^{-5}$ cm^2 s^{-1} was obtained and $\gamma = 1.00 \pm 0.01$. The D value was slightly smaller than that found in bulk water measured with the same method ($D_{H2O} = (2.63 \pm 0.01) \times 10^{-5}$ cm^{2}s^{-1}).

By adding borax to Sclg, a gel was formed and the water diffusion behavior definitely changed, in fact the log–log plot of MSD of water vs the diffusion time showed the existence of two or more water diffusing phases in a slow exchange. Figure 6.10b shows MSD for each water phase vs the diffusion time in Sclg/borax in a relative ratio $R = 0.5$. MSD trend of the normal diffusing water phase (solid circles) indicated a gaussian process with $\gamma = 1.00 \pm 0.02$ and $D = (2.90 \pm 0.04) \times 10^{-5}$ cm^2 s^{-1}. MSD of the anomalous water phase displayed a double diffusive regime (triangles) with a crossover at $\Delta_c = 500$ ms. At short diffusion times ($t_d < \Delta_c$) a superdiffusive regime with $\gamma = 1.4 \pm 0.2$ and $K = (46 \pm 14) \times 10^{-5}$ cm^2 s$^{-1.4}$ was observed, whereas at long diffusion times ($t_d > \Delta_c$) a Gaussian regime with $\gamma = 1.0 \pm 0.1$ and $D = (15 \pm 1) \times 10^{-5}$ cm^2 s^{-1} was observed.

By compressing in one direction the Sclg/borax system, anisotropic swelling properties were obtained. To verify if this anisotropy was also transferred to water transport properties, self-diffusion was measured along and perpendicularly to the compression direction in swollen tablets. Figure 6.10c shows MSD of water in Sclg/borax tablets with $R = 0.5$ vs. the diffusion time, circles and up triangles represent the slow and fast diffusing components along the compression direction. Down triangles represent the diffusion along the direction perpendicular to the compression. Data were fit to Eq. 6.7 with $\gamma = 1.33 \pm 0.0\,3$ and $K = (40\pm 16) \times 10^{-5}$ cm^2 s$^{-1.4}$ for the fast diffusing component along the compression direction, with $\gamma = 1.012 \pm 0.008$ and $D = (3.00 \pm 0.03) \times 10^{-5}$ cm^2 s^{-1} for the slow diffusing component along the compression direction, and with $\gamma = 1.006 \pm 0.003$ and $D = (2.139 \pm 0.006) \times 10^{-5}$ cm^2 s^{-1} in the case of diffusion along the direction perpendicular to the compression. In conclusion, the diffusion behaviour along the compression direction (up triangles and circles) was found to be similar to that observed in the uncompressed Sclg/borax system (Fig. 6.10b) whereas in the perpendicular direction (down triangles), any anomalous behavior disappeared and a single phase Gaussian diffusion was found.

To summarize, the analysis of water diffusion properties in the Sclg/borax gel showed that, with the formation of the gel network, an anomalous enhanced diffusion behavior appeared. This enhanced water diffusion can be related to the formation of polysaccharide nanochannels that induce the system to behave as a strong adsorber. On the nanochannel network water molecules perform bulk mediated Levy diffusion. Upon Sclg/borax tablet compression, anomalous diffusion disappeared in the perpendicular direction, where, due to a mechanical alignment of the polysaccharide nanochannels, the Sclg network behaves as a weak adsorber [1, 11, 18, 58]. Obviously these findings are important in the view of a deeper interpretation of the drug release kinetics from these matrices.

6.3.2 ^{1}H *Pulsed Low-Resolution NMR*

NMR is a sensitive and informative method to study physicochemical processes such as wetting, swelling, changes in the macromolecular status, interactions between water and macromolecules. This technique provides an insight into the dynamics of water molecules and their environment.

The simplest NMR experiment consists of applying an rf pulse to the spin system, as the rf pulse is switched off nuclei relax back to equilibrium generating an interferogram (amplitude vs. time) namely the free induction decay (FID). One can choose to

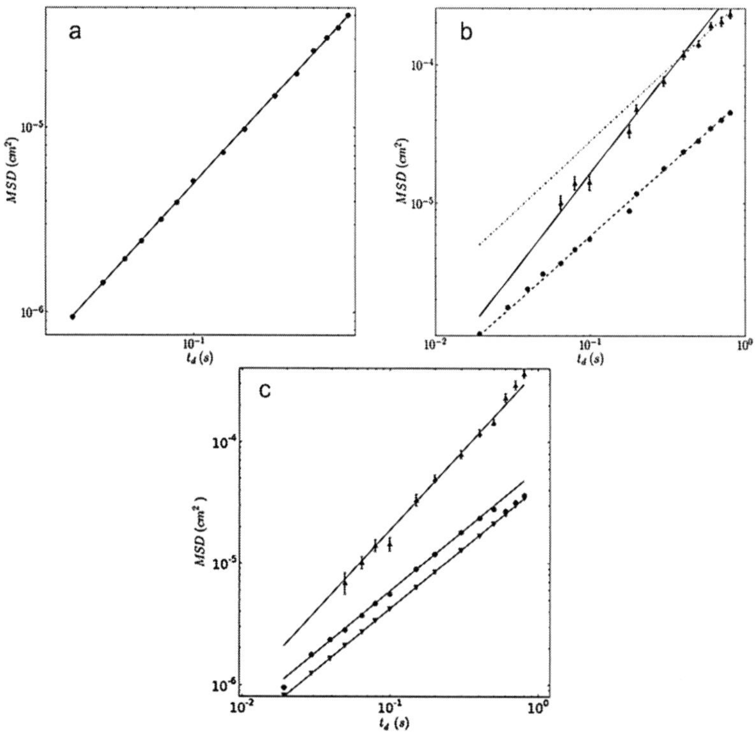

Figure 6.10 MSD of water vs. the diffusion time: (a) Sclg sample, (b) Sclg/borax sample ($R = 0.5$), and (c) Sclg/borax tablets ($R = 0.5$). Lines represent the fit of data to Eq. 6.7. Reprinted from Ref. 24 with permission of the Royal Society of Chemistry.

operate in the time domain, and this is the form we know as pulsed low resolution NMR [26]. Measurement of FID is the basic way in which the magnitude and other characteristics of the macroscopic magnetization of the spin system are determined. Different parts of the FID contain different kinds of information. The FID contains signals from different magnetization components precessing at different rates. In the case of a system consisting of narrow and broad lines, the nuclei having the narrowest line contribute more to the tail end of the FID than those having broader lines which contribute mostly to the beginning of the FID. Figure 6.11 shows FIDs of a solid material (a), a liquid sample (b), and a material

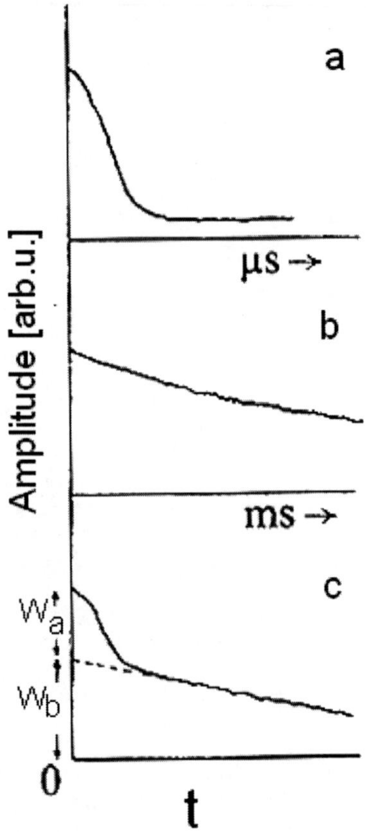

Figure 6.11 Free induction decay of (a) a solid sample, (b) a liquid sample, and (c) a sample with both rigid and mobile components.

having both rigid and mobile components (c). The time scale for a FID of a solid material is of the order of microseconds, whereas that of a liquid material is of the order of milliseconds. Intensities of a composite FID extrapolated back to $t = 0$, such those reported in Fig. 6.1c, denote the number of nuclei contributing to each phase.

In pulsed low-resolution NMR, FID obtained after applying two or more pulses is used in the determination of relaxation times. In fact, after perturbing a spin system with a proper rf pulse sequence, the system will return back to equilibrium through a process called "relaxation" characterized by a decay time constant known as relaxation time [27]. The longitudinal (or spin-lattice) relaxation time T_1 is the decay time constant for the recovery of the z component M_z of the nuclear spin magnetization toward its thermal equilibrium value. Longitudinal relaxation is due to energy exchange between nuclear spins and the surrounding lattice re-establishing thermal equilibrium. As spins go from a high-energy state back to a low energy state, energy is released back into the surrounding lattice.

The transverse (or spin–spin) relaxation time T_2 is the decay time constant for the component M_{xy} of the nuclear spin magnetization in the xy plane. Transverse relaxation results from spins getting out of phase. As spins move together, their magnetic fields interact (spin-spin interaction), slightly modifying their precession rate. These interactions are temporary and random causing a cumulative loss in phase resulting in transverse magnetization decay.

6.4 ^1H Pulsed Low-Resolution NMR in Polysaccharide-Based Hydrogels

Certain materials when placed in excess of water are able of swelling and retain large volumes of water in the swollen structure. These materials (hydrogels) do not dissolve in water and maintain three dimensional networks [42, 46]. Hydrogels are usually made of hydrophilic polymer molecules which are crosslinked either by chemical bonds or other cohesion forces. Hydrogels can be purposely tuned to respond to changes in environmental conditions and the extent of the response can be controlled. The environmental conditions include pH, temperature, electric field, ionic strength, salt type, solvent, external stress, light, and combination of these. It is because of these unique properties that hydrogels find numerous applications in pharmaceutical, agricultural, biomedical, and consumer-oriented fields.

Hydrogels can only fulfill their intended purpose when hydrated, consequently the need for a full understanding of the complex manner in which water behaves in host matrices is exceedingly important. Actually water behaves anomalously when adsorbed into matrices in the sense that the sharp first order transitions which characterize bulk water are modified. The anomalous behavior of water is ascribed to several physical and chemical sources. Many factors further complicate the interpretation of data in fact water may act as plasticizer or antiplasticizer depending on concentration and temperature, the organization of adsorbed water is affected by polymer mobility, hydration may induce conformational changes, additives may also play an important role.

In aqueous media the degree of crosslink is a structural important parameter in determining the swelling properties of a polysaccharide network. To obtain a gel with an amount of water as high as possible the amount of crosslinked units must be kept as low as possible and yet able to give an insoluble macromolecule, i.e., a superabsorbent network. The macroscopic behavior of an hydrogel is obtained by measuring the swelling properties as a function of the ionic strength and pH [9, 10].

The different types of water that coexist in a polymer have been described in literature as bound and free, freezable and nonfreezable, and associated water. Hydration typically proceeds from tightly bound water at a low level of hydration to increasing amounts of bulk water at saturation levels [39, 51]. Specifically NMR is able of distinguishing between water still mobile at low temperature and bulk water [13] that freezes at 273 K.

The NMR investigation of polymer–water interactions cannot be done using fully swollen hydrogels, as the amount of free water in these systems is so high as to hide completely the NMR signals both of the polymer and water either tightly bound to the polymer or located in a solvation shell around the chains. However, it is possible to perform the hydration of the dry polymeric network with known amounts of water and use pulsed low resolution NMR to study the water–polymer interaction in the network [16, 42, 46].

The state of water in a superabsorbing network based on hydroxyethylcellulose (HEC) and carboxymethylcellulose sodium salt (CMCNa) crosslinked with divinylsulfone (DVS) was investigated [17] in the 200 $\div$ 300 K temperature range by pulsed low resolution NMR. Known amounts of water (W_0 = [mol of added water]/[mol of the polysaccharide repeating unit]) were added to dry networks with a different degree of crosslink. The obtained results were compared with those obtained in the case of an uncrosslinked CMCNa/HEC mixture.

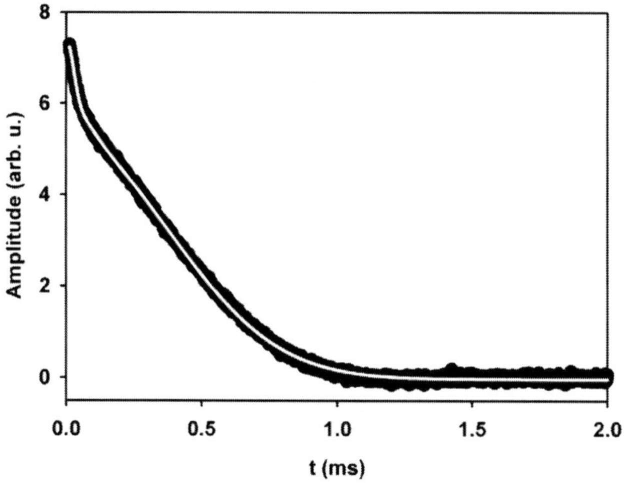

Figure 6.12 FID of a CMC/HEC/DVS network with added water (W_0 = 5.5). The solid line through the experimental points was obtained by fitting data to Eq. 6.8. Reprinted with permission from Ref. 17. Copyright © 2014, Elsevier.

In the full investigated range of temperature, when no water is added to the dry network ($W_0 = 0$), the FID showed just one Gaussian component, whereas when $W_0 > 0$ the deconvolution of the FID showed the presence of at least two Gaussian components, a fast decaying one ("a" component), and a slow decaying one ("b" component).

Figure 6.12 reports the FID of a CMC/HEC/DVS network with $W_0 = 5.5$. The solid line through the experimental points was obtained by fitting data to the sum of two Gaussian components:

$$F(t) = C_0 + \left[W_a \exp\left(-\frac{t^2}{G_a^2}\right)\right] + \left[W_b \exp\left(-\frac{t^2}{G_b^2}\right)\right] \qquad (6.8)$$

where W_a and W_b are the spin densities of the two components, C_0 is the average value of the noise, and G_a and G_b are the times needed for the two components to decay to $1/e$ of their starting value at $t = 0$.

Figure 6.13 shows the trend of the spin densities W_a and W_b vs. temperature measured on a CMC/HEC/DVS network with known amounts of added water (W_0 = 0.8, 5.5, 10.9, 20.2, 30.0).

W_a is the spin density of the fast decaying component that was ascribed to the polysaccharide matrix with its exchangeable hydroxyl

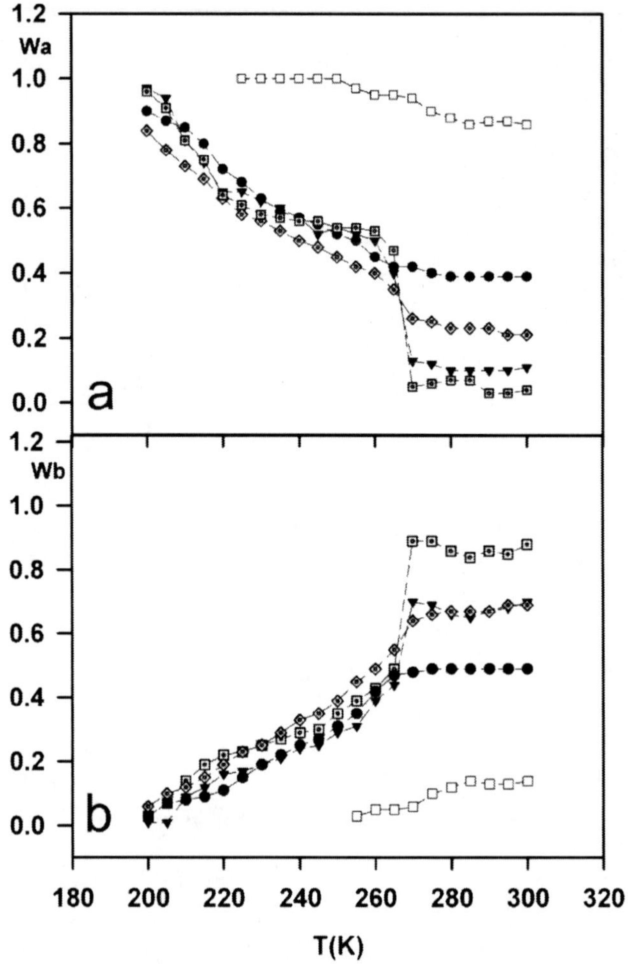

Figure 6.13 Normalized spin density of the fast (a) and slow (b) decaying components of the FID vs. the temperature. (□) $W_0 = 0.8$, (•) $W_0 = 5.5$, (◈) $W_0 = 10.9$, (▼) $W_0 = 20.2$, (⊡) $W_0 = 30.0$. Reprinted with permission from Ref. 17. Copyright © 2014, Elsevier.

groups and to the water tightly bound to the polysaccharide. W_b is the spin density of the slow decaying component due to the water in other shells of solvation. A net increase of W_a was observed when a fraction of gel freezes. This effect was observed near the melting point of a confined solvent in the presence of its free state. At about

265 K a net decrease of W_b together with a net increase of the corresponding W_a was observed in samples with $W_0 \geq 10.9$. At about 200 K $W_a \cong 1$ and $W_b \cong 0$, suggesting that all bound water was frozen. In samples with $W_0 = 5.5$, W_a and W_b showed a rather continuous trend vs. temperature. These data indicated that the polysaccharide was able of tightly coordinating 6–9 water molecules for repeating unit [17].

Figure 6.14 reports the spin–lattice relaxation times vs. 1000/T [K] of samples without added water, and with different amounts of added water. Samples with $W_0 \geq 5.5$ showed a flat and well-defined minimum at $1000/T \cong 4$ ($T = 250$ K) followed by a linear increase of log T_1 at low temperatures. From the slope of this linear trend, the activation energy (about 27 kJ/mol) for the overall motions of the polysaccharide–water–ice system was obtained. On the contrary, samples with $W_0 < 5.5$ did not show any minimum.

The experimental echo decays obtained applying the CPMG sequence to networks with different amounts of added water were fit to a mono-exponential function:

$$Y(t) = C_0 + \left[W* \exp\left(-\frac{t}{T_2}\right)\right] \tag{6.9}$$

Figure 6.15 shows the spin–spin relaxation times vs. $1000/T$ [K] of samples without added water, and with different amounts of added water. In samples with $W_0 \geq 5.5$, T_2 undergoes a transition at about 265 K, the transition was found to be steeper with increasing the water content. These data indicated that about 7–8 water molecules were tightly bound to the crosslinked network. Furthermore, as the water content increased, an increase in the amount of free or nearly free water was observed.

The presence of a single T_2 value in all samples and in the full investigated range of temperature was explained with the presence of a fast exchange. All solvation shells but the first one including the very tightly bound water were in a very fast exchange with water molecules continuously jumping from one solvation shell to another one.

An easy representation of spin–spin relaxation times may be obtained applying an inverse Laplace transformation to CPMG echo decays. In this representation a distribution of T_2 relaxation times is

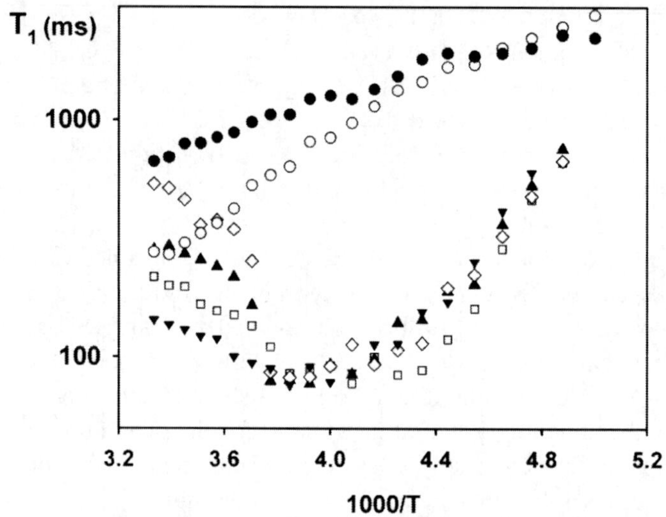

Figure 6.14 Spin–lattice relaxation time (T_1) vs. 1000/T. (•) W_0 = 0, (○) W_0 = 0.8, (▼) W_0 = 5.5, (□) W_0 = 10.9, (▲) W_0 = 20.2, (◊) W_0 = 30.0. Reprinted with permission from Ref. 17. Copyright © 2014, Elsevier.

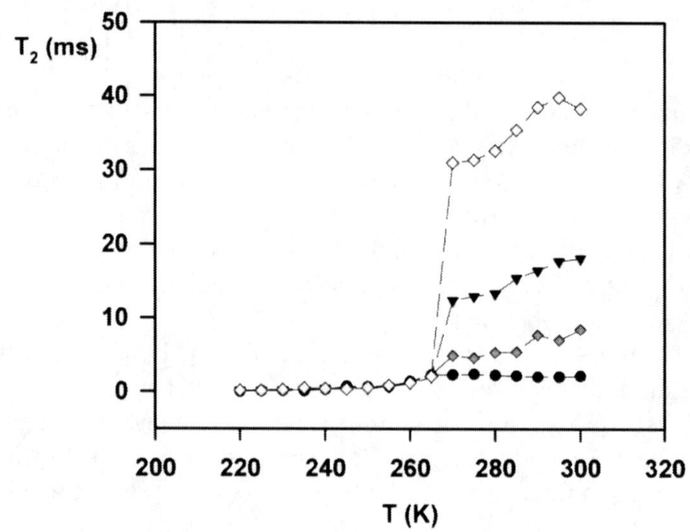

Figure 6.15 Spin–spin relaxation time T_2 vs. temperature obtained for CMC/HEC/DVS network (•) W_0 = 5.5, (♦) W_0 = 10.9, (▼) W_0 = 20.2, (◊) W_0 = 30.0. Reprinted with permission from Ref. 17. Copyright © 2014, Elsevier.

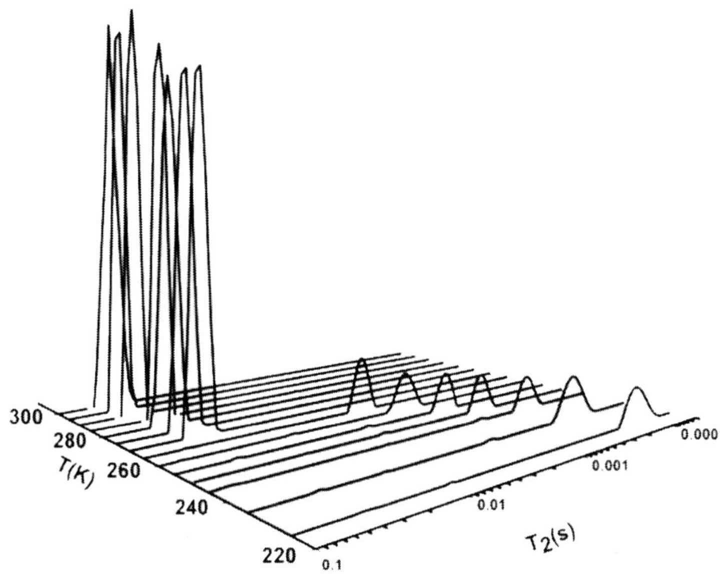

Figure 6.16 Spin–spin relaxation times distribution vs. temperature obtained for a CMC/HEC/DVS network with $W_0 = 30$. Reprinted with permission from Ref. 17. Copyright © 2014, Elsevier.

obtained where the peaks of the distribution correspond to the most probable T_2 values and the area of each peak corresponds to the normalized spin density. Figure 6.16 shows the distribution of spin–spin relaxation times vs. temperature obtained for a CMC/HEC/DVS network with $W_0 = 30$.

In the distribution, the peak of unfrozen water is observed even at temperatures as low as 220 K.

To gather information on polysaccharide–water interaction, T_2 distributions obtained for crosslinked sample with known amounts of added water were compared with those obtained for the uncrosslinked mixture with same amounts of added water. Note that both crosslinked and uncrosslinked CMC/HEC/DVS systems were able of absorbing water, nevertheless the uncrosslinked system did not give rise to a superabsorbing system. This observation was also suggested by the analysis of T_2 distributions vs. the amount of added water obtained at 300 K for the uncrosslinked mixture (a) and the crosslinked polymer (b) reported in Fig. 6.17. In the case of the

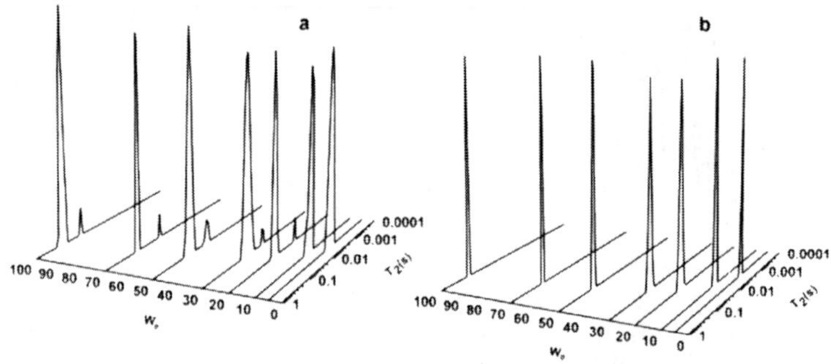

Figure 6.17 Distributions of T_2 relaxation times at 300 K. (a) Uncrosslinked mixture and (b) crosslinked sample with known amounts of added water: $W_0 = 5.5$, 10.9, 20.2, 30.0, 50.0, 70.0, and 100. Reprinted with permission from Ref. 17. Copyright © 2014, Elsevier.

crosslinked sample, even at very high amount of added water, up to $W_0 = 100$, a single T_2 component was observed with the T_2 value increasing by increasing the water content. On the contrary, in the case of the uncrosslinked mixture, from $W_0 = 20$ up to 100, two T_2 components were observed. The second component accounts for about 10% of the total amount of water. This result indicated the existence of at least two phases in the uncrosslinked mixture. As a matter of fact, the uncrosslinked mixture starts exhibiting microheterogeneity at $W_0 \geq 20$.

6.5 Conclusion

Application of NMR methodologies to investigate polysaccharide-based gels was reported. Mono- and bidimensional NMR techniques applied to liquid samples allows both the structural characterization of the native and chemically modified polysaccharide, the degree of derivatization of the polysaccharide can also be obtained. Solid state NMR permits the structural characterization of unsoluble and amorphous networks constituting the guest matrix of hydrogel and the assessing of the real degree of crosslink of these networks. ^{1}H HR MAS spectroscopy allows the characterization of networks in their hydrate state, and the use of a gradient along the rotor spinning axis allows translational diffusion measurements in hydrogels. ^{1}H pulsed low resolution NMR provides an insight into the dynamics of water molecules and their interaction with the polysaccharide network.

References

1. Ajdari, A. (1995). Transport by active filaments, *Europhys. Lett.*, **31**, pp. 69–74.
2. Alam, T. M., and Jenkins, J. E. (2012). *Advanced Aspect of Spectroscopy*, ed. INTECH,. Farruch, M. A., Chapter 10 "HR-MAS NMR spectroscopy in material science," pp. 279–306.
3. Bax, A., and Freeman, R. J. (1981). Investigation of complex networks of spin-spin coupling by two-dimensional NMR, *J. Magn. Reson.*, **44**, pp. 542–561.
4. Bax, A., and Davis, D. G. (1985). Practical aspects of two-dimensional transverse NOE spectroscopy, *J. Magn. Reson.*, **63**, pp. 207–213.
5. Bax, A., and Davis, D. G. (1985). MLEV-17 based two-dimensional homonuclear magnetization transfer spectroscopy, *J. Magn Reson.*, **65**, pp. 355–360.
6. Bax, A., and Summers, M. F. (1986). ^{1}H and ^{13}C assignments from sensitivity enhanced detection of heteronuclear multiple-bond connectivity by two-dimensional multiple quantum NMR, *J. Am. Chem. Soc.*, **108**, pp. 2093–2094.
7. Brannon-Peppas, L., and Peppas, N. A. (1990). *Absorbent Polymer Technology*, ed. Brannon-Peppas, L., and Harland, R. S., "The equilibrium swelling behavior of porous and nonporous hydrogels" (Elsevier, Amsterdam), pp. 67–102.
8. Braun, S., Kalinowski, H. O., and Berger, S. (1998). *150 and More Basic NMR Experiments* (Wiley-VCH, Weinheim, Germany).
9. Brondsted, H., and Kopacek, J. (1992). Polyelectrolite gels, *ACS Symposium* Series 480, American Chemical Society, Washington, DC, pp. 285–304.
10. Buchholz, F. L., and Peppas, N. A. (1994). Superabsorbent polymers: science and technology, *ACS Symposium* Series 573, American Chemical Society, Washington, DC.
11. Bychuk, O. V., and O'Shaughnessy, B. (1995). Anomalous diffusion at liquid surfaces, *Phys. Rev. Lett.*, **74**, pp. 1795–1798.
12. Callaghan, P. T. (1991). *Principles of Nuclear Magnetic Resonance Microscopy* (Oxford University Press).

13. Capitani, D., Segre, A. L., Attanasio, D., Blicharska, B., and Focher, B. (1996). ^{1}H NMR relaxation study of paper as a system of cellulose and water, *TAPPI J.*, **79**, pp. 113–122.
14. Capitani, D., de Nooy, A. E. J., Masci, G., and Crescenzi, V. (2000). Ionic polysaccharide hydrogels via the passerini and ugi multicomponent condensations: synthesis, behavior and solid-state NMR characterization, *Biomacromolecules*, **1**, pp. 259–267.
15. Capitani, D., Del Nobile, M. A., Mensitieri, G., Sannino, A., and Segre, A. L. (2000). ^{13}C Solid State NMR determination of degree of crosslinking in super-absorbing cellulose based networks, *Macromolecules*, **33**, pp. 430–437.
16. Capitani, D., Crescenzi, V., De Angelis, A. A., and Segre, A. L. (2001). Water in hydrogels. An NMR study of water/polymer interactions in weakly cross-linked chitosan networks, *Macromolecules* **34**, pp. 4136–4144.
17. Capitani, D., Mensitieri, G., Porro, F., Proietti, N., and Segre, A. L. (2003). NMR and calorimetric investigation of water in a superabsorbing crosslinked network based on cellulose derivatives, *Polymer*, **44**, pp. 6589–6598.
18. Chechkin, A. V., Zaid, I. M., Lomholt, M. A., Sokolov, I. M., and Metzler, R. (2009). Bulk-mediated surface diffusion along a cylinder: Propagators and crossovers, *Phys. Rev. E*, **79**, 040105.
19. Crescenzi, V., Francescangeli, A., Taglienti, A., Capitani, D., and Mannina, L. (2003). Synthesis and partial characterization of hydrogels obtained via glutaraldehyde crosslinking of acetylated chitosan and of hyaluronan derivatives, *Biomacromolecules*, **4**, pp. 1045–1054.
20. Crescenzi, V., Cornelio, L., Di Meo, C., Nardecchia, S., and Lamanna, R. (2007). Novel hydrogels via click chemistry: Synthesis and potential biomedical applications, *Biomacromolecules*, **8**, pp. 1844–1850.
21. Cruciani, O., Mannina, L., Sobolev, A. P., Segre, A. L., and Luisi, P. (2004). Multilamellar liposomes formed by phosphatidyl nucleosides: an NMR-HR-MAS characterization, *Langmuir*, **20**, pp. 1144–1151.
22. Cruciani, O., Mannina, L., Sobolev, A. P., Cametti, C., and Segre, A. L. (2006). An improved NMR study of liposomes using 1-palmitoyl-2-oleoyl-sn-glycero-3-phospatidylcholine as model, *Molecules*, **11**, pp. 334–344.
23. Di Meo, C., Capitani, D., Mannina, L., Brancaleoni, E., De Luca, G., and Crescenzi, V. (2006). Synthesis and NMR characterization of new hyaluronan-based NO donors, *Biomacromolecules*, **7**, pp. 1253–1260.

24. Di Meo, C., Coviello, T., Matricardi, P., Alhaique, F., Capitani, D., and Lamanna, R. (2011). Anisotropic enhanced water diffusion in scleroglucan gel tablets, *Soft Matter*, **7**, pp. 6068–6075.

25. Duer, M. J. (2002). *Solid State NMR Spectroscopy: Principles and Applications* (Blackwell Science).

26. Farrar, T. C., and Becker, E. D. (1982). *Pulse and Fourier Transform NMR Introduction to Theory and Methods* (Academic Press, New York).

27. Fukushima, E., and Roeder, S. B. W. (1981). *Experimental pulse NMR A Nuts and Bolts Approach* (Addison-Wesley).

28. Gaede, H. C., and Gawrisch, K. (2004). Multi-dimensional pulsed field gradient magic angle spinning NMR experiments on membranes, *Magn. Reson. Chem.*, **42**, pp. 115–122.

29. Ha, M. A., Evans, B. W., Jarvis, M. C., Apperley, D. C., and Kenwright, A. M. (1996). CP-MAS NMR of highly mobile hydrated biopolymers: polysaccharides of Allium cell walls, *Carbohydr. Res.*, **288**, pp. 15–23.

30. Harris, R. K. (1988). *Multinuclear Magnetic Resonance in Liquids and Solids-Chemical Application*, ed. Granger, P., and Harris, R. K., NATO ASI Series Vol. 322, Chapter XVI (Kluwer Academic, Dordrecht, the Netherlands).

31. Iqbal, S., Rodriguez-LLansola, F., Escuder, B., Miravet, J. F., Verbruggen, I., and Willem, R. (2010). HRMAS ^{1}H NMR as a tool for the study of supramolecular gels, *Soft Matter*, **6**, pp. 1875–1878.

32. Jarvis, M. C., Fenwick, K. M., and Apperley, D. C. (1996). Cross-polarisation kinetics and proton NMR relaxation in polymers of Citrus cell walls, *Carbohydr. Res.*, **288**, pp. 1–14.

33. Jeener, J., Meier, B. H., Bachmann, P., and Ernst, R. R. (1979). Investigation of exchange processes by two-dimensional NMR spectroscopy, *J. ChemPhys.*, **71**, pp. 4546–4553.

34. Jiménez, B., Mirnezami, R., Kincross, J., Cloarec, O., Keun, H. C., Holmes, E., Goldin, R. D., Ziprin, P., Darzi, A., and Nicholson, J. K. (2013). ^{1}H HR-MAS NMR spectroscopy of tumor-induced local metabolic "field-effects" enables colorectal cancer staging and prognostication, *J. Proteome Res.*, **12**, pp. 959–968.

35. Johnson, C. S. (1999). Diffusion ordered nuclear magnetic resonance spectroscopy: principles and applications, *Prog. Nucl. Magn. Reson. Spect.*, **34**, pp. 203–256.

36. Kapur, G. S., Cabrita, E. J., and Berger, S. (2000). The qualitative probing of hydrogen bond strength by diffusion-ordered NMR spectroscopy, *Tetrahedron Lett.*, **41**, pp. 7181–7185.

37. Lucas, L. H., Cerny, M. A., Koen, Y. M., Hanzlik, R. P., and Larive, C. K. (2004). ^{1}H High-resolution magic angle spinning (HR-MAS) analysis of ligand density on resins using a resin internal standard, *Anal. Bioanal. Chem.*, **380**, pp. 627–631.
38. Morris, K. F., and Johnson Jr., C. S. (1992). Diffusion-ordered two-dimensional nuclear magnetic resonance spectroscopy, *J. Am. Chem. Soc.*, **114**, pp. 3139–3141.
39. Overloop, K., and Van Gerven, L. (1993). Exchange and cross-relaxation in adsorbed water, *J. Magn. Reson.*, **A 101**, 147–156.
40. Pampel, A., Reszka, R., and Michel, D. (2002). Pulsed field gradient MAS NMR studies of the mobility of carboplatin in cubic liquid crystalline phases), *Chem. Phys. Lett.*, **357**, pp. 131–136.
41. Power, W. P. (2011). High-resolution magic angle spinning-enabling applications of NMR spectroscopy to semi-solid phases, *Annu. Rep. NMR Spectrosc.*, **72**, pp. 111–156.
42. Quinn, F. X., Kampff, E., Smyth, G., and McBrierty, V. (1988). Water in hydrogels. 1. A study of water in poly(*N*-vinyl-2-pyrrolidone/methyl methacrylate) copolymer, *Macromolecules*, **21**, pp. 3191–3198.
43. Schmidt-Rohr, K., and Spiess, H. W. (1994). *Multidimensional Solid-State NMR and Polymers* (Academic Press, London).
44. Shaka, A. J., Barker, P. B., and Freeman, R. (1985). Computer-optimized decoupling scheme for wideband applications and low-level operation, *J. Magn. Reson.*, **64**, pp. 547–552.
45. Shapiro, Y. E. (2011). Structure and dynamics of hydrogels and. organogels: an NMR spectroscopy approach, *Prog. Polym. Sci.*, **36**, pp. 1184–1253.
46. Smyth, G., Quinn, F. X., and McBrierty, V. (1988). Water in hydrogels. 2. A study of water in poly(hydroxyethylmethacrylate), *Macromolecules*, **21**, pp. 3198–3204.
47. Stejskal, E. O., and Tanner, J. E. (1965). Spin diffusion measurements: spin echoes in the presence of a time-dependent field gradient, *J. Chem. Phys.*, **42**, pp. 288–292.
48. Stejskal, E. O., and Memory, J. D. (1994). *High-Resolution NMR in the Solid State Fundamentals of CP-MAS* (Oxford University Press, New York).
49. Stilbs, P. (1987). Fourier transform pulsed-gradient spin-echo studies of molecular diffusion, *Prog. Nucl. Magn. Reson. Spectrosc.*, **19**, pp. 1–45.
50. Testa, G., Di Meo, C., Nardecchia, S., Capitani, D., Mannina, L., Lamanna, R., Barbetta, A., and Dentini, M. (2009). Influence of dialkyne structure on

the properties of new click-gels based on hyaluronic acid, *Int. J. Pharm.*, **378**, pp. 86–92.

51. Vackier, M. C., Hills, B. P., and Rutledge, D. N. (1999). An NMR relaxation study of the state of water in gelatin gels, *J. Magn. Reson.*, **138**, pp. 36–42.
52. Vermathen, M., Marzorati, M., and Vermathen, P. (2012). Exploring high resolution magic angle spinning (HR-MAS) NMR spectroscopy for metabonomic analysis of apples, *Chimia*, **66**, pp. 747–751.
53. Viel, S., Capitani, D., Mannina, L., and Segre, A. L. (2003). Diffusion-Ordered NMR Spectroscopy: a versatile tool for the molecular weight determination of uncharged polysaccharides, *Biomacromolecules* **4**, pp. 1843–1847.
54. Viel, S., Ziarelli, F., and Calderelli, S. (2003). Enhanced diffusion-edited NMR spectroscopy of mixtures using chromatographic stationary phases, *Proc. Natl. Acad. Sci. USA*, **100**, pp. 9696–9698.
55. Voelkel, R. (1988). High-resolution solid-state ^{13}C-NMR spectroscopy of polymers, *Angew. Chem. Int. Ed. Engl.*, **27**, pp. 1468–1483.
56. Walderhaug, H., Söderman, O., and Topgaard, D. (2010). Self-diffusion in polymer systems studied by magnetic field-gradient spin-echo NMR methods, *Progr. Nucl. Magn. Reson. Spectrosc.*, **56**, pp. 406–425.
57. Wang, X., Zhang, X. S., and Wu, X. (1988). Two-stage feature of Harimann-Hahn cross relaxation in magic-angle sample spinning, *Phys. Rev. B*, **37**, pp. 9827–9829.
58. Zumofen, G., Klafter, J., and Blumen, A. (1990). Enhanced diffusion in random velocity fields, *Phys. Rev. A*, **42**, pp. 4601–4608.

Chapter 7

Small-Angle Neutron Scattering of Polysaccharide Hydrogels

Mitsuhiro Shibayama
Institute for Solid State Physics, The University of Tokyo, 5-1-5 Kashiwanoha, Kashiwa, Chiba 277-8581, Japan
sibayama@issp.u-tokyo.ac.jp

Small-angle neutron scattering (SANS) is a powerful tool for structural investigation of polysaccharide hydrogels because the scattering contrast between gel network chains and the solvent can be significantly enhanced by using D_2O. SANS functions are usually obtained in absolute intensity scale, allowing one to evaluate the molecular and structure parameters, such as molecular weight and number of aggregates. Furthermore, contrast-variation SANS allows extraction of targeting part in the gel networks. In this chapter, the characteristic features and advantages of SANS are described, followed by recent progress of SANS studies on hydrogels with the emphasis of polysaccharides.

Polysaccharide Hydrogels: Characterization and Biomedical Applications
Edited by Pietro Matricardi, Franco Alhaique, and Tommasina Coviello

ISBN 978-981-4613-61-3 (Hardcover), 978-981-4613-62-0 (eBook)
www.panstanford.com

7.1 Introduction

SANS is a powerful technique for structural investigation of polymeric systems in a few to 10 nm range. SANS is characteristic of (i) capability of isotope labeling and/or contrast variation, (ii) strong penetration power, (iii) low level of parasitic scattering from aperture edge, and (iv) absolute intensity compatible to thermodynamic quantities. Hence, SANS allows one to analyze scattering functions quantitatively and as well as to compare experimental scattering functions with those of theoretical prediction. SANS was first used in polymer science in early 1970s to study polymer chain conformation and molecular weight (M_w) dependence of its size. A mixture of amorphous hydrogeneous and deutereous polystyrenes (PSs) were investigated with SANS and revealed that PS chains take a random Gaussian conformation and its size scales with $M_w^{1/2}$ [16, 26]. Since then, SANS has been widely used in structural investigations of polymers, namely miscibility of polymer blends [16, 18], structure and/or order–disorder transition of block copolymers [3], polymer conformation in external fields, and polymer gels [2, 9, 23].

Polysaccharides are carbohydrate polymers consisting of two or more monosaccharides linked together covalently by glycosidic linkages in a condensation reaction. Since polysaccharides are extremely important in organisms for the purposes of energy storage and structural integrity, the structure and properties have been extensively investigated. In this chapter, basic understanding on SANS is reviewed, which is followed by SANS applications to structural investigations of polymer gels.

7.2 Basic Features of Neutron Scattering

7.2.1 *Scattering Contrast*

The most important parameter in neutron scattering is the scattering lengths and cross sections. The scattering lengths and cross sections for typical isotopes (or elements) consisting of hydrogels, b and σ, respectively, are listed in Table 7.1.

Table 7.1 Neutron scattering lengths and cross sections for typical elements and isotopes

Isotope	c	b^{coh}	b^{inc}	σ^{coh}	σ^{inc}	σ^{scat}	σ^{abs}
Unit	%	fm	fm	bahn	bahn	bahn	Bahn
H	—	**−3.739**	—	1.7568	**80.26**	**82.09**	0.3326
^{1}H	99.985	**−3.7406**	25.274	1.7583	**80.27**	**82.03**	0.3326
^{2}H (D)	0.15	6.671	4.04	5.592	2.05	7.64	0.000519
C	—	6.646	—	5.551	0.01	5.551	0.0035
N	—	9.36	—	11.01	0.5	11.51	1.9
O	—	5.803	—	4.232	0.0008	4.232	0.00019
P	100	5.13	0.2	3.307	0.005	3.312	0.172

Source: http://www.ncnr.nist.gov/resources/n-lengths/elements/li.html

The symbols c, b^{coh}, b^{inc}, σ^{coh}, σ^{inc}, σ^{scat} $(= \sigma^{coh} + \sigma^{inc})$ and σ^{abs} denote the concentration (natural abundance), coherent and incoherent scattering lengths, coherent and incoherent scattering cross sections, and absorption cross sections, respectively. The units of b and σ are fm ($= 10^{-13}$ cm) and bahn ($= 10^{-24}$ cm^2), respectively. Note that the unique features of ^{1}H (or simply H) are (i) negative coherent scattering length and (ii) large incoherent cross section as marked in bold font. This is why (i) a large scattering contrast is obtained by D/H substitution or by using deuterated solvent, e.g., D_2O, while (ii) the amount of H in the sample should be minimized to reduce incoherent scattering from H as low as possible. Another important feature is that the coherent scattering length is of the order of 10^{-13} cm and is much smaller than the wavelength of neutron (of the order of 10^{-8} cm), neutron scattering from a single atom is scattering angle independent. This means that the atomic scattering factor is simply given by b.

The scattering contrast of neutron contrast is obtained from the difference in the coherent scattering length densities of the solute (polymer; P) and the solvent (S), i.e., $\Delta\rho = (\rho_P - \rho_S)$, where ρ_i is obtained by dividing the coherent scattering length for the molecule or molecular unit by that of volume:

$$\rho_i = \frac{\sum_j^{(molecule\_i)} b_j}{v_i} \tag{7.1}$$

Typical examples are H_2O, D_2O, and glucose; $\rho_{H2O} = -0.560 \times 10^{10}$ cm^{-2}, $\rho_{D2O} = 6.339 \times 10^{10}$ cm^{-2}, and $\rho_{glu} = 1.534 \times 10^{10}$ cm^{-2}.

7.2.2 *Transmission*

Transmission of a sample can be easily obtained by measuring transmitted intensities with and without the sample. In the case of neutron scattering, attenuation of incident neutron beam is mainly due not to absorption (see small values of σ^{abs} in Table 7.1) but to multiple scattering from hydrogen atom because of the large incoherent scattering cross section, σ^{inc}, of H (Table 7.1). Since the scattering intensity is larger for a thicker sample and incoherent scattering increases with increasing sample thickness, the sample thickness should be optimized to $t \approx 1/\Sigma_{tot}$ in order to succeed in a SANS experiment. Here, Σ_{tot} is the total cross section per unit volume of the sample.

7.2.3 *Incoherent Scattering*

As shown in Table 7.1, hydrogen atom has a large incoherent scattering cross section and is a major source of incoherent scattering from the sample. The incoherent scattering intensity, $(d\Sigma/d\Omega)_{inc}$, can be determined by a transmission measurement or can be estimated by knowing the total cross section per unit volume (cm^{-1}) (from Table 7.1; $\Sigma_{tot} = \Sigma_j(\sigma_j^{scat} + \sigma_j^{abs})/v_i$), Σ_{tot}, in terms of the following equation [21, 22],

$$\left(\frac{d\Sigma}{d\Omega}\right)_{inc} \cong \frac{\exp(\Sigma_{tot}t) - 1}{4\pi t} = \frac{1 - T_r}{4\pi t T_r}, \Sigma_{tot} \equiv -\frac{\ln T_r}{t} \tag{7.2}$$

Here, $(d\Sigma/d\Omega)$ is the differential cross section with the cross section per unit volume (Σ, (cm^{-1})) and the solid angle (Ω). T_r is the transmission, and t is the sample thickness (cm). Figure 7.1 shows sample thickness dependence of the transmission for a mixture of light (H_2O) and heavy waters (D_2O). As shown in Fig. 7.1a, the transmission dramatically changes as a function of the volume

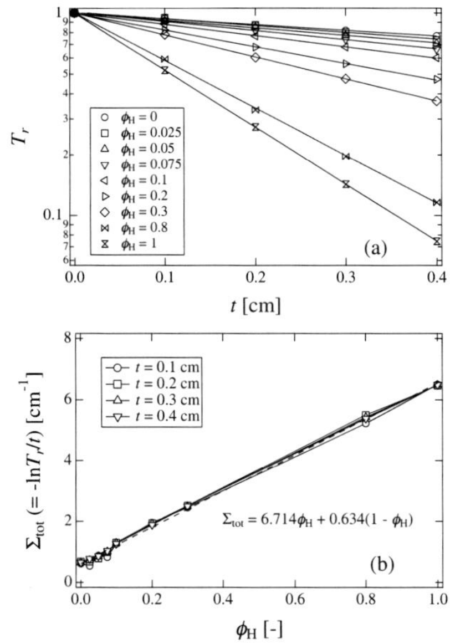

Figure 7.1 (a) Sample thickness dependence of neutron transmission for mixtures of H_2O/D_2O water, where ϕ_H is the fraction of H_2O, and (b) the total cross section per unit volume, Σ_{tot}. Reprinted with permission from Ref. 22, copyright IUCr.

fraction of H_2O, ϕ_H. Note that Lambert–Beer law is satisfied even in the high ϕ_H region having a very strong absorption (multiple scattering). Figure 7.1b shows a reduced plot of the total cross section as a function of ϕ_H. Σ_{tot} is given by a linear function of ϕ_H, i.e.,

$$\Sigma_{tot} = 6.714\phi_H + 0.634\,(1 - \phi_H) \tag{7.3}$$

This enables one to evaluate Σ_{tot} for any composition of aqueous systems having a mixed solvent of H_2O/D_2O. Figure 7.2 shows the variation of the incoherent scattering $(d\Sigma/d\Omega)_{tot}$ of a mixture of H_2O/D_2O at zero-thickness limit. Since the scattering intensities of typical gels are in the order of a few tenths of cm^{-1} to tens of cm^{-1}, one has to keep in mind to reduce the incoherent scattering from the solvent as low as possible.

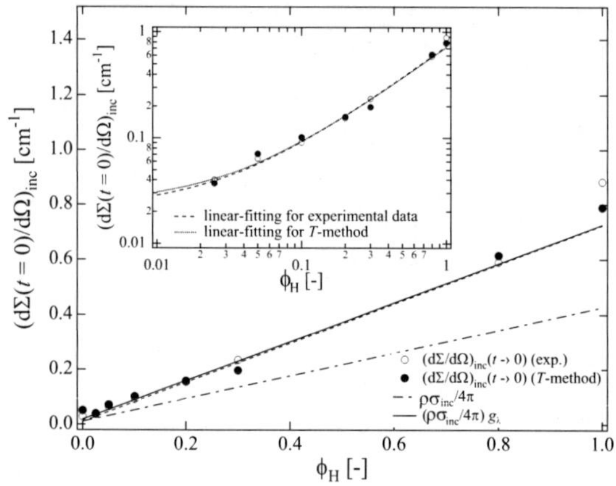

Figure 7.2 Incoherent scattering intensity from H_2O/D_2O mixtures having different H_2O fractions at zero-thickness limit. Reprinted with permission from Ref. 22, copyright IUCr.

7.3 Theoretical Background

7.3.1 *Scattering Function of Semi-dilute Polymer Solutions*

Typical polymer concentrations of polymer gels are in semi-dilute regime. Hence, the corresponding system of a gel is a semi-dilute polymer solution in which no crosslinking is present in the system. The scattering intensity from a semi-dilute polymer solution is commonly given by an Ornstein-Zernike (OZ) function,

$$I(q) = \frac{I(0)}{1 + \xi^2 q^2} \tag{7.4}$$

Here, q is the magnitude of scattering vector and ξ is the correlation length, i.e., a measure of a mesh size of the gel or the screening length shielded by neighboring polymer chains. If the system is far from phase separation boundary, ξ represents a mesh size of the gel, while it diverges at the spinodal no matter how small the mesh is [19]. The thermodynamic limit of the scattering intensity, $I(0)$, is related to the osmotic modulus K_{os} of the solution,

$$I(0) = \left(\frac{d\Sigma}{d\Omega}\right)_{q=0} = (\Delta\rho)^2 \frac{k_B T \phi^2}{K_{os}} \tag{7.5}$$

and

$$K_{os} = \phi\,(\partial\Pi/\partial\phi)_T \tag{7.6}$$

where k_B is the Boltzmann constant, T the absolute temperature, and ϕ the polymer volume fraction, Π the osmotic pressure. Since Π and ξ are scaled to ϕ as $\Pi \approx \xi^{-3}$ and $\xi \approx \phi^{-3/4}$ [4], the scattering intensity for a polymer solution in semi-dilute regime is given by [4]

$$I\left(q\right) = I_{\text{sol}}\left(q\right) \sim \left(\Delta\rho\right)^2 \frac{k_B T \phi \xi^{5/3}}{1+\xi^2 q^2} \tag{7.7}$$

This means that the scattering intensities for semi-dilute polymer solutions can be reduced by ϕ and ξ.

7.3.2 *Scattering Functions of Polymer Gels and Inhomogeneities*

When crosslinks are introduced to a polymer solution by chemically (covalent bond) or physically (hydrogen, crystallization, complexation, etc.), additional concentration fluctuations appear in the system. Particularly, when crosslinking is introduced during polymerization process, polymerization/crosslinking occurs inhomogeneously in the reaction bath, resulting in formation of frozen-in concentration fluctuations [1]. Figure 7.3 illustrates (a) thermal concentration fluctuations in a polymer solution, $\rho^{\text{th}}(r)$, (b) frozen-concentration fluctuations introduced by crosslinking, $\rho^{\text{cross}}(r)$, and (c) concentration fluctuations in a gel, $\rho^{\text{gel}}(r)$, i.e. [20],

$$\rho^{\text{gel}}\left(r\right) = \rho^{\text{th}}\left(r\right) + \rho^{\text{cross}}\left(r\right) \tag{7.8}$$

Since $\rho^{\text{gel}}(r)$ is a superposition of $\rho^{\text{th}}(r)$ and $\rho^{\text{cross}}(r)$ with very different wavenumbers, it is better to treat in Fourier space, in another word, scattering. Phenomenologically, the scattering intensity function for polymer gels is given by

$$I(q) \approx I_{\text{sol}}(q) + I_{\text{ex}}(q) = \frac{I_{\text{sol}}(0)}{1+\xi^2 q^2} + I_{\text{ex}}(q) \tag{7.9}$$

and

$$I_{\text{ex}}\left(q\right) = \frac{I_{\text{DB}}\left(q\right)}{\left(1+\Xi^2 q^2\right)^2}, \quad I_{\text{ex}}\left(q\right) = I_{\text{G}}\left(q\right)\exp\left(-\frac{R^2 q^2}{3}\right) \tag{7.10}$$

The left and right equations in 7.10 denote a Debye–Bueche (DB) equation for two-phase system with chord length of Ξ [27] and the Guinier function (G) for a dilute system with the radius of gyration of R_g [11, 12].

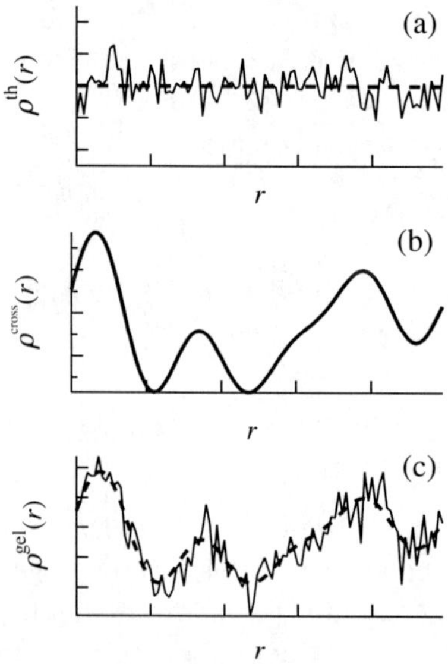

Figure 7.3 Schematic representation of (a) thermal fluctuations in a polymer solution, ρ^{th}, (b) frozen-concentration fluctuations introduced by crosslinking, ρ^{cross}, and (c) concentration fluctuations in a gel, ρ^{gel}.

There are other types of functions describing additional concentration fluctuations introduced by crosslinking, namely a power-law function [10],

$$I\ (q) \sim q^{-\alpha} \tag{7.11}$$

where α is an scattering exponent. This value is dependent on the structure of gels. As a matter of fact, there is no unified agreement in description of the additional scattering.

7.3.3 *Inhomogeneities*

Inhomogeneities are characteristic of polymer gels, and extensive studies on inhomogeneites have been carried out by many researchers. As shown in Fig. 7.4, inhomogeneities are classified to spatial, topological, connectivity, and mobility inhomogneities [20]. These inhomogeneites are introduced during gelation process due to inhomogeneous chemical reaction, polymerization, or mismatch of the concentrations of crosslinker and polymer reacting cites, etc.

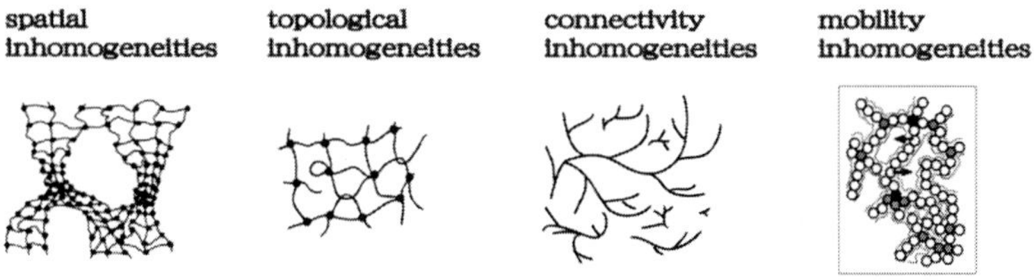

Figure 7.4 Various types of inhomogenities: spatial, topological, connectivity, and mobility inhomogeneities.

Among these inhomogeneities, SANS are sensitive to spatial inhomogeneities, while the other inhomogeneities are well characterized by swelling/mechanical measurements, dynamic light scattering (DLS), rheological measurements and others. Since inhomogeneities degrade the mechanical properties and increase haze and/or opacity, it is desired to obtain inhomogeneity-free gels.

7.4 Progress in Structural Investigations of Polymer Gels by Neutron Scattering

7.4.1 *Synthetic Hydrogels*

Neutron scattering studies on synthetic gels have been extensively carried out on hydrogels consisting of acrylamide [12], thermosensitive poly(*N*-ispropylacrylamide) [19], and others [2]. In most cases, strong forward scattering originated from frozen inhomogeneities described by Eq. 7.10 are observed. Figure 7.5 shows SANS function of end-crosslinked poly(tetrahydrofuran) (PTHF) gel [24].

This type of gel is called a "model network" since the gel is prepared by end-coupling of monodisperse telechelic polymer chains with tetrafunctional crosslinker. However, as shown in the figure, there are nontrivial forward scattering (the dotted line in Fig. 7.5), indicating presence of inhomogeneities.

Recently, Sakai and Shibayama developed nearly ideal polymer gels free from crosslink inhomogeneitis called tetra-PEG gels [15]. The tetra-PEG gels can be obtained by cross-end-coupling of symmetric tetra-arm PEG polymers having a complementary end

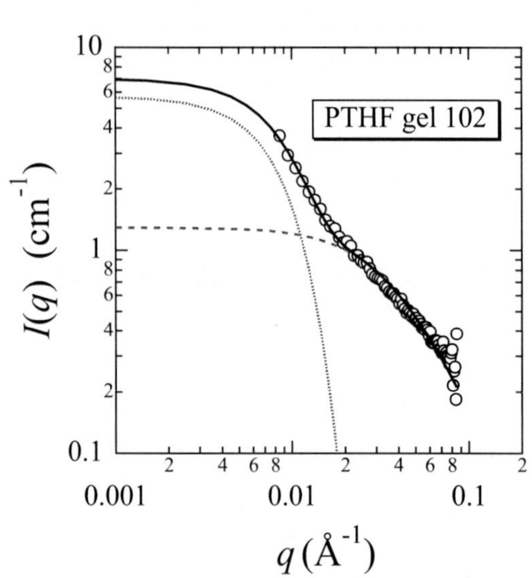

Figure 7.5 SANS intensity function of end-coupled PTHF model network. The function is decomposed with Eq. 7.9. Dashed line: Ornstein–Zernike function; dotted line: Guinier function; and solid line: reproduced by adding the two components. Reprinted with permission from Ref. 24. Copyright © 1995, American Chemical Society.

groups, such as amino and active ester groups. Figure 7.6 shows SANS functions of tetra-PEG gels with various molecular weights and concentrations [13]. Note that all of SANS functions can be represented by Ornstein–Zernike functions (solid lines) (without an inhomogeneity term in Eq. 7.10). In any case, synthetic polymer gels are relatively well characterized in comparison with natural gels.

7.4.2 *Polysaccharides*

There are a variety of polysaccharide gels, such as agarose, alginate, carrageenan, galactomannan, gellan gum, locust bean gum, pectin, xanthan gels, and their mixtures. Their gelation mechanisms and gel structures also vary one to another. Typically, gelation occurs either by changing temperature, solvent quality, ionic environment, or mixing two or more different polysaccharides. Since the structures and physical properties of these gels are described elsewhere in this book, the details are not discussed here and focus on recent

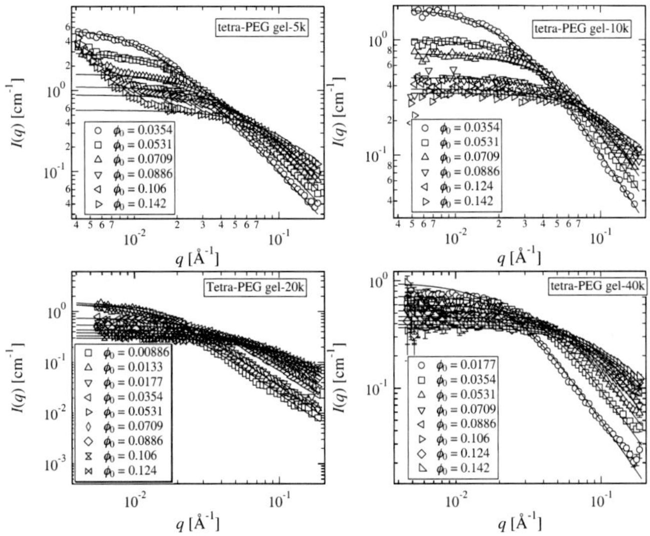

Figure 7.6 SANS functions of tetra-PEG hydrogels with various molecular weights (5k, 10k, 20k, and 40k) and polymer volume fractions, ϕ_0. Reprinted with permission from Ref. 13. Copyright © 2009, American Chemical Society.

topics from the viewpoint of gel structure. Since monosaccharide units forming polysaccharide are rather stiff in comparison with those of most synthetic polymers, polysaccharide chain behaves as a wormlike chain or stiff chain with a finite cross section. Because of these characteristic features in addition to relatively low water solubility and strong association via hydrogen bonding, etc., structural characterization of polysaccharide gels by scattering methods is not as simple as in the case of synthetic polymer gels.

For a neutral polysaccharide solution in semi-dilute regime, scattering function is the same as that for synthetic polymers and, in principle, is given by Eq. 7.4. Typical scattering function from a neutral polysaccharide, galactomannan, solution is shown in Fig. 7.7. The obtained correlation length was obtained to be 265 Å with Eq. 7.4.

Due to strong hydrogen-bond capability, however, SANS or SAXS curves for polysaccharide solutions sometimes show anomalous strong forward scattering due to associated or aggregated structures, which leads to overestimation of the molecular weight and

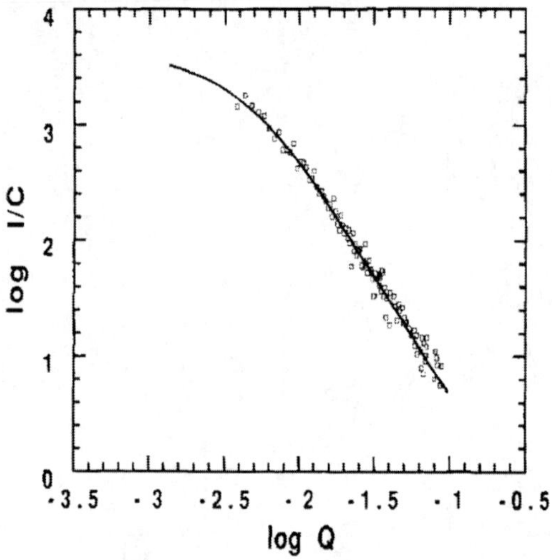

Figure 7.7 Log–log plot of SAXS intensity I/C vs. q for galactomannnan solution at 0.2 w/w polymer concentration. The solid line is the fit with Eq. 1.4 with $\xi = 265$ Å [25]. Reprinted from Ref. 25 with the permission of John Wiley and Sons.

or structural parameters. Hence, one should be careful in analyzing scattering data. In the case of highly charged polyelectrolytes, on the other hand, they exhibit a broad peak in the scattering curve, which is suppressed by added salt or is amplified by gelation with metal ions. Figure 7.8 shows SAXS curves for κ-carrageenan gel in cesium form (1.2 and 1.8 w/w polymer concentrations) [25]. The scattering maximum corresponds to packing of rodlike structure.

In some cases, it is possible to extract the information of the thickness (radius) of the polymer chain in the larger q-region via so-called cross-section Guinier plot,

$$\ln\left[qI\left(q\right)\right] = (\text{const.}) - \frac{R_c^2 q^2}{2} \tag{7.12}$$

By applying Eq. 7.12, the cross-section radius, R_c, was obtained for the above κ-carrageenan gel (Fig. 7.8) to be 8 Å.

Ramzi et al. investigated structure-property relationship for agarose gels in binary solvent by SAXS as well as SANS. They interpreted the scattering curves by considering a random assembly

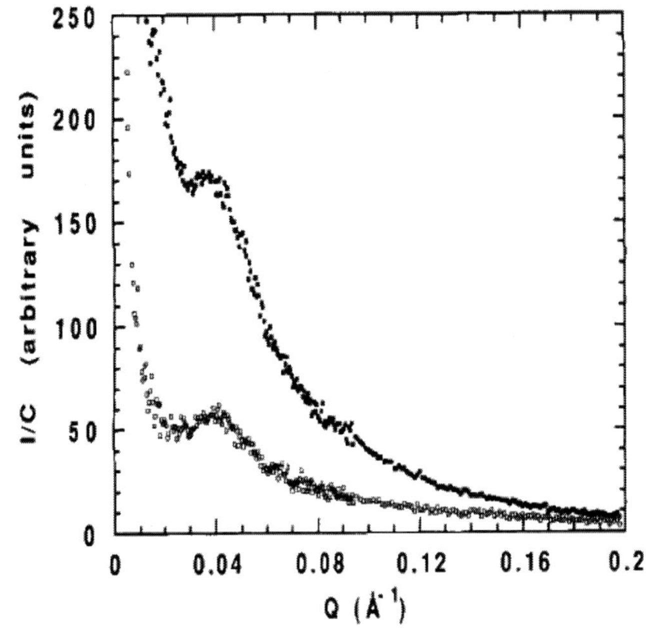

Figure 7.8 SAXS curves for κ-carrageenan gel in cesium form (1.2 and 1.8 w/w polymer concentrations). Reprinted from Ref. 25 with the permission of John Wiley and Sons.

of straight fibers having cross-section polydispersity with Guenet model for fibrillar gels) [7].

$$I(q) = \frac{4\pi \mu_L}{q} \frac{J_1^2(qr)}{q^2 r^2} \tag{7.13}$$

where J_1 is the Bessel function of the first kind, μ_L is the mass per unit mass per unit length and r is the cross-section radius [14].

Here, the Guenet model is a network model with cross-section-polydispersed cylinders in which the mesh size is far larger than the largest cross-sectional radius is calculated by considering a weight distribution function of the type, $w(r) \sim r^{\lambda}$ with two cut-off radii, r_{min} and r_{max} [7]. Figure 7.9 shows a departure from Eq. 7.13. They concluded that departure from the Porod regime at wide scattering angles suggested the existence of free and/or dangling chains that did not participate in the network elasticity.

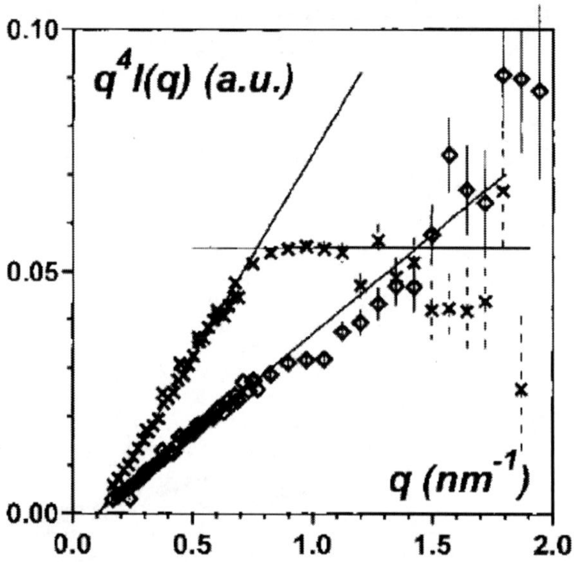

Figure 7.9 SANS curves for agarose gels prepared from D2O/DMSO-d6 binary mixtures. Agarose concentrations were 30 g/L (crosses) and 70 g/L (diamonds). Reprinted with permission from Ref. 14. Copyright © 1998, American Chemical Society.

Esquenet and coworkers studied structure of hydrophobically modified chitosanes (HMCs) associative networks by rheological measurements, fluorescence spectroscopy, SANS, light scattering, and DLS [6]. The scattering experiments were carried out in the associative gel phase ($c > c^*$; the chain overlap concentration). They observed a q^{-3} (or q^{-4}) regime in low-q (light scattering) regime. They assigned this regime to be flower aggregates as shown in Fig. 7.10. At high SANS q-regime, on the other hand, q^{-1} behavior was observed, indicating a rodlike structure. The flower aggregates are destroyed to low connected network microgels.

They concluded that in the semi-dilute regime, SANS and DLS showed presence of ca. 50 nm flower aggregates. As polymer concentration was increased, a connected network is formed from the percolation of bridges as shown in Fig. 7.10, leading to a large increase in solution viscosity.

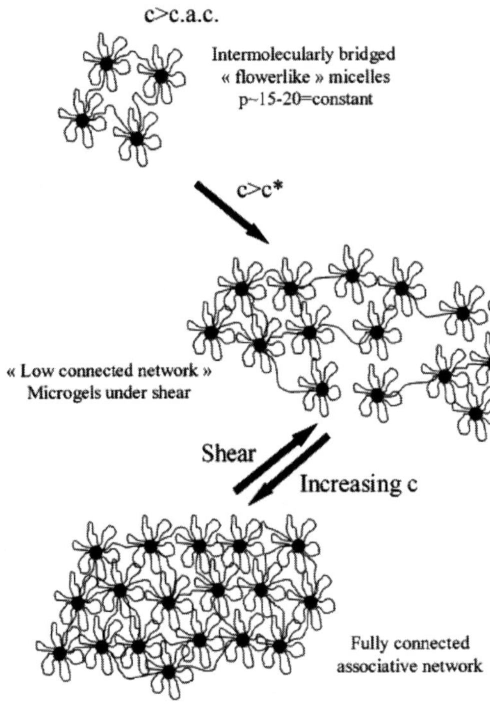

Figure 7.10 Schematic diagram depicting the concentration and the shear of the associative network. Reprinted with permission from Ref. 6. Copyright © 2004, American Chemical Society.

Dror et al. investigated the structure of gum Arabic (GA), a commonly used material for food hydrocolloid, in aqueous solutions. They reported that scattering measurement revealed an intricate shape composed of many spheroidal aggregates. A distinct scattering maximum was obtained in a SANS curve as shown in Fig. 7.11. This peak was obtained at moderate-to-high concentrations, while it disappeared by addition of salt, indicating electrostatic nature. The q^{-2} behavior at low q-region suggested Gaussian chain like behavior of GA.

Schmidt et al. investigated the structure of polysaccharide–protein complexes by SANS [17]. The system was lysozyme–pectine complexes. Lysozyme is a positively charged globular protein and pectin is a natural anionic semiflexible polysaccharide. They

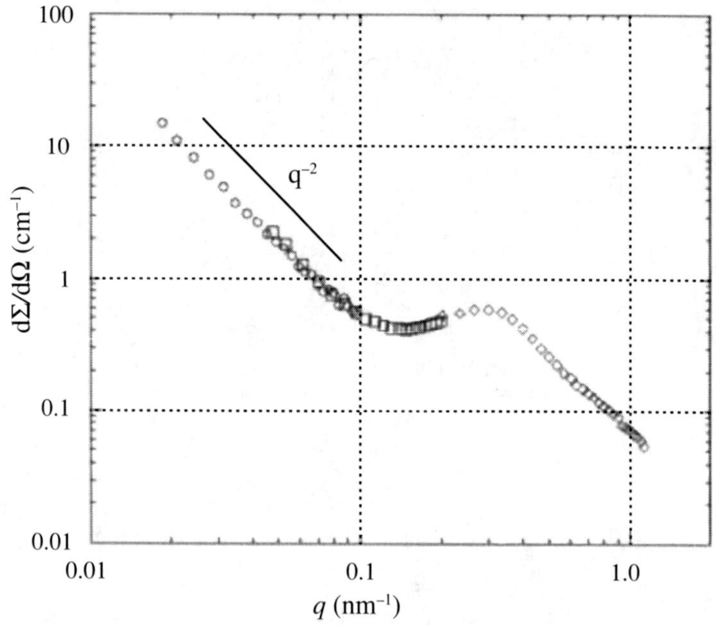

Figure 7.11 SANS pattern of 5% (w/w) gum Arabic (GA) obtained at three sample-detector distances shown by different symbols. Reprinted from Ref. 5 with the permission of John Wiley and Sons.

observed a correlation peak at 0.2 Å^{-1}, corresponding to complexes where proteins are clustering in. On large scale, on the other hand, spherical globules were observed with a well-defined radius of 10–50 nm. On the basis of these observation, they proposed a model as shown in Fig. 7.12 Globular lysozyme (shown by circles) are bound together via pectin (curved lines) to form a globular complexes. These complexes are further connected with a fractal dimension of D_{app}.

Hyland et al. studied mutually reinforced multicomponent polysaccharide networks by mechanical testing, scanning electron microscopy, and SANS. They emphasized that networks made from chitosan and alginate could be utilized as prospective tissue engineering scaffolds due to material biocompatibility and degradability and Ca^{2+} could be added as a modifier for mechanical strength enhancement [8].

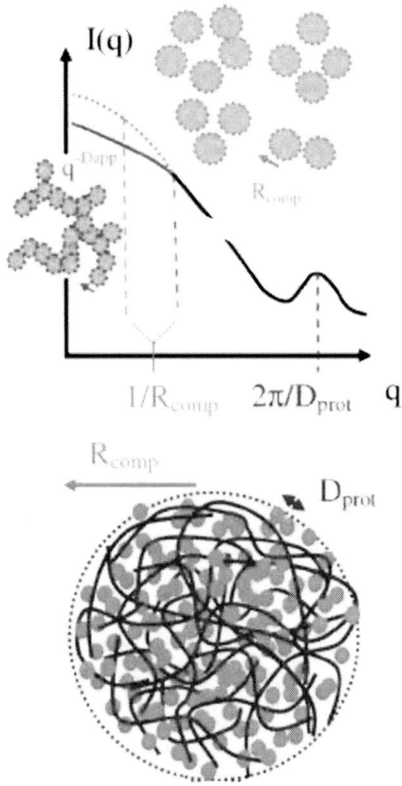

Figure 7.12 Schematic illustration of the scattering in log–log plot for globular pectin-lysozyme complexes with their model in low- and high-q regions. R_{comp} and D_{prot} denotes the radius of spherical globules and the inter globular protein distance, respectively, and D_{app} is the scattering exponent. Reprinted with permission from Ref. 17. Copyright © 2009, American Chemical Society.

7.5 Concluding Remarks

Recent SANS studies on polymer hydrogels have been reviewed. Advantages and characteristic features of neutron scattering are summarized with the emphasis of the treatment of incoherent scattering. In the case of synthetic polymer gels, novel gels with extremely small inhomogeneities, called tetra-PEG gels, are introduced. In the case of polysaccharide gels, in spite of their importance in science as well as industries, the structure–property relationship has not been well elucidated. This is partially due to their variety in structure, gelation mechanism, complexity, etc. Recent investigations on polysaccharides with SANS together with other means have unveiled many unsolved problems, some of which are introduced in this chapter. As discussed in this chapter, neutron scattering is one of the best means for structural investigation of polysaccharide with high quality of data. Although access to neutron facilities is still limited because of insufficient neutron beam time at any facility, neutron scattering will be a standard method for structural investigations of polysaccharide gels after upgrade of ILL and ISIS and completion of European Spallation Source (ESS).

Acknowledgments

The author is grateful to Profs. Tommasina Coviello and Franco Alhaique for giving a chance to write this chapter. This work was partially supported by Grant-in-Aid of Ministry of Education, Science, Sports, and Culture, Japan (Nos. 12450388, 14350493, 16350120, 18205025, 22245018).

References

1. Bastide, J., and Leibler, L. (1988). Large-scale heterogeneities in randomly cross-linked networks. *Macromolecules* **21**, pp. 2647–2649.
2. Bastide, J., and Candau, S. J. (1996). Structure of gels as investigated by means of static scattering techniques, in *The Physical Properties of Polymer Gels*, J. P. Cohen Addad, editor. John Wiley: NY, p. 143.
3. Bates, F. S., and Wignall, G. D. (1986). Isotope-induced quantum-phase transitions in the liquid state. *Phys. Rev. Lett.* **57**, pp. 1429–1432.
4. de Gennes, P. G. (1979). *Scaling Concepts in Polymer Physics*. Ithaca: Cornell University.
5. Dror, Y., Cohen, Y., and Yerushaulmi-Rosen, R. (2006). Structure of gum arabic in aqueous solution. *J. Polym. Sci. Part B: Polym. Phys.* **44**, pp. 3265–3271.

6. Esquenet, C., Terech, P., Boue, F., and Buhler, E. (2004). Structural and rheological properties of hydrophobically modified polysaccharide associative networks. *Langmuir* **20**, pp. 3583–3592.

7. Guenet, J. M. (1994). Scattering by a network of prolate, cross-section-polydispersed cylinders applicable to fibrillar thermoreversible gels. *J. Phys. II France* **4**, pp. 1077–1082.

8. Hyland, L. L., Taraban, M. B., Hammouda, B., and Yu, Y. B. (2011). Mutually reinforced multicomponent polysaccharide networks. *Biopolymers* **95**, pp. 840–851.

9. Higgins, J. S., and Benoit, H. C. (1994). *Polymers and Neutron Scattering*. Oxford: Clarendon Press.

10. Horkay, F., Basser, P. J., Hecht, A. M., and Geissler, E. (2000). Osmotic and SANS observations on sodium polyacrylamide hydrogels in physiological salt solutions. *Macromolecules* **33**, pp. 8329–8333.

11. Horkay, F., Hecht, A. M., Mallam, S., Geissler, E., and Rennie, A. R. (1991). Macroscopic and microscopic thermodynamic observations in swollen poly(vinyl acetate) networks. *Macromolecules* **24**, pp. 2896–2902.

12. Mallam, S. Horkay, F., Hecht, A. M., Rennie, A. R., and Geissler, E. (1991). Microscopic and macroscopic thermodynamics observations in swollen poly(dimethylsiloxane) networks. *Macromolecules* **24**, pp. 543–548.

13. Matsunaga, T., Sakai, T., Akagi, Y., Chung, U., and Shibayama, M. (2009). Structure characterization of tetra-PEG gel by small-angle neutron scattering. *Macromolecules* **42**, pp. 1344–1351.

14. Ramzi, M., Rochas, C., and Guenet, J. M. (1998). Structure-properties relation for agarose thermoreversible gels in binary solvents. *Macromolecules* **31**, pp. 6106–6111.

15. Sakai, T., Matsunaga, T., Yamamoto, Y., Ito, C., Yoshida, R., Suzuki, S. et al. (2008). Design and fabrication of a high-strength hydrogel with ideally homogeneous network structure from tetrahedron-like macromonomers. *Macromolecules* **41**, pp. 5379–5384.

16. Schelten, J., Wignall, G. D., and Ballard, D. G. H. (1974). Chain conformation in molten polyethylene by low angle neutron scattering. *Polymer* **15**, pp. 682–685.

17. Schmidt, I., Cousin, F., Huchon, C., Boué, F., and Axelos, M. A. V. (2009). Spatial structure and composition of polysaccharide-protein complexes from small angle neutron scattering. *Biomacromolecules* **10**, pp. 1346–1357.

18. Shibayama, M., Yang, H., Stein, R. S., and Han, C. C. (1985). Study of miscibility and critical phenomena of deuterated polystyrene

and hydrogenated poly(vinyl methyl ether) by small-angle neutron scattering. *Macromolecules* **18**, pp. 2179–2187.

19. Shibayama, M., Tanaka, T., and Han, C. C. (1992). Small angle neutron scattering study on poly(*N*-isopropyl acrylamide) gels near their volume-phase transition temperature. *J. Chem. Phys.* **97**, pp. 6829–6841.
20. Shibayama, M. (1998). Spatial inhomogeneity and dynamic fluctuations of polymer gels. *Macromol. Chem. Phys.* **199**, pp. 1–30.
21. Shibayama, M., Nagao, M., Okabe, S., and Karino, T. (2005). Evaluation of incoherent neutron scattering from softmatter. *J. Phy. Soc. Jpn.* **74**, pp. 2728–2736.
22. Shibayama, M., Matsunaga, T., and Nagao, M. (2009). Evaluation of incoherent scattering intensity by transmission and sample thickness. *J. Appl. Crystallogr.* **42**, pp. 621–628.
23. Shibayama, M. (2011). Small-angle neutron scattering on polymer gels: phase behavior, inhomogeneities, and deformation mechanisms. *Polym. J.* **43**, pp. 18–34.
24. Takahashi, H., Shibayama, M., Fujisawa, H., and Nomura, S. (1995). Equilibrium swelling and small-angle neutron-scattering study on end-linked poly(tetrahydrofuran) networks. *Macromolecules* **28**, pp. 8824–8828.
25. Turquois, T., Rochas, C., Taravel, F. R., Doublier, J. L., and Axelos, M. A. V. (1995). Small-angle X-ray scattering of κ-carrageenan based systems: sols, gels, and blends with carob galactomannan. *Biopolymers* **36**, pp. 559–567.
26. Wignall, G. D., Ballard, D. G. H., and Schelten, J. (1974). Measurement of persistence length and temperature dependence of the radius of gyration in bulk atactic polystyrene. *Eur. Polym. J.* **10**, pp. 861–865.
27. Wu, W., Shibayama, M., Roy, S., Kurokawa, H., Coyen, L. D., Nomura, S. et al. (1990). Physical gels of aqueous poly(vinyl alcohol) solutions: a small-angle neutron-scattering study. *Macromolecules* **23**, pp. 2245–2251.

Chapter 8

The Method of Small-Angle X-ray Scattering and Its Application to the Structural Analysis of Oligo- and Polysaccharides in Solution

Kanji Kajiwara[a] **and Isao Wataoka**[b]

[a]*Faculty of Textile Science and Technology, Shinshu University, Ueda-City, Tokota, 386-8567 Japan*
[b]*Faculty of Fiber Science and Engineering, Kyoto Institute of Technology, Kyoto-City, Matsugasaki, 606-8585 Japan*
kajiwara@shinshu-u.ac.jp, wataoka@kit.ac.jp

General principles of electromagnetic wave scattering are summarized, with emphasis laid on small-angle X-ray scattering. The conventional analysis of scattering data is based on the various ways of plotting the data according to the simple model approximation. In order to avoid the artifact due to the simple model approximation, the analysis of the scattering data is performed by either adapting a triaxial model of homogeneous density distribution, a molecular model, or a density distribution function.

The application is shown mainly for $(1 \rightarrow 4)$-α-D-glucan (basic amylose sequence) and $(1 \rightarrow 4)$-β-D-glucan (basic cellulose sequence). The conformation of oligo- and polysaccharides are analyzed by combining the results of small-angle X-ray scattering

Polysaccharide Hydrogels: Characterization and Biomedical Applications
Edited by Pietro Matricardi, Franco Alhaique, and Tommasina Coviello

ISBN 978-981-4613-61-3 (Hardcover), 978-981-4613-62-0 (eBook)
www.panstanford.com

from their solutions with the computer simulation. The molecular model for a supramolecular structure is considered for a gelling system of xyloglucan aqueous solutions.

8.1 Introduction

The synchrotron radiation provides a powerful X-ray source, which is almost 10^6 times stronger than a conventional X-ray generator. According to this instrumental development, the method of small-angle X-ray scattering (SAXS) has become a common tool for the investigation of the state of materials in solution in nano order. For example, the structural change was observed in real time by synchrotron SAXS for many biopolymers in aqueous solution [1]. The principle of the scattering method is common to all light sources including light, X-ray and neutron. The scattering process is characterized by the reciprocity law, i.e., an inverse relationship between particle size and scattering angle. Here X-ray is scattered by electrons, so that X-ray scattering is observed when electron density inhomogeneity exists in the order of colloidal dimensions (between a few and several hundred nanometers).

Table 8.1 summarizes the characteristics of X-ray (light), electron and neutron scattering. Despite the difference in the wavelength and scatterer, the scattering process can be treated in a similar way following the reciprocity law. The difference in wavelength and scatterer can be applied for the combined measurements of light, X-ray and/or neutron to yield supplemental information.

8.2 Method of Small-Angle X-ray Scattering

8.2.1 *General Description of Scattering*

Small-angle X-ray scattering (SAXS) is described in terms of a reciprocal law. Imagine that a particle of diameter D is immersed in X-ray beam (see Fig. 8.1). X-rays are scattered by all the electrons of the particle and interfere with each other, and result a scattered intensity decreasing with increasing scattered angle. The scattered

Table 8.1 Characteristics of X-ray (light), electron and neutron scattering

Beam source	Wave length (nm)	Wave nature	Interaction	Obtained information	Scattering characteristics ($\times 10^{12}$ cm)	Scattering amplitude*
X-ray (light)	~ 0.1 ($1 \sim 10^5$)	Electro magnetic wave	electrons	Electron density distribution	H(0.14) D(0.14) C(1.01) Pd(19.30)	Increases with atomic number
Electron	~ 0.005	Particle wave	Electro static potential	Potential distribution composed of atomic nucleus and electrons	H(-3100) D(-3100) C(-14400) Pd(-80600)	Very large
Neutron	0.1~1	Particle wave	Atomic nucleus	Position and motion of atoms	H(-0.38) D(0.67) C(0.66) Pd(0.95)	Irregular, positive, or negative values

*The scattering amplitude is specified at the position 1 cm from the atom scattering the incident plane wave of the absolute amplitude 1 cm.

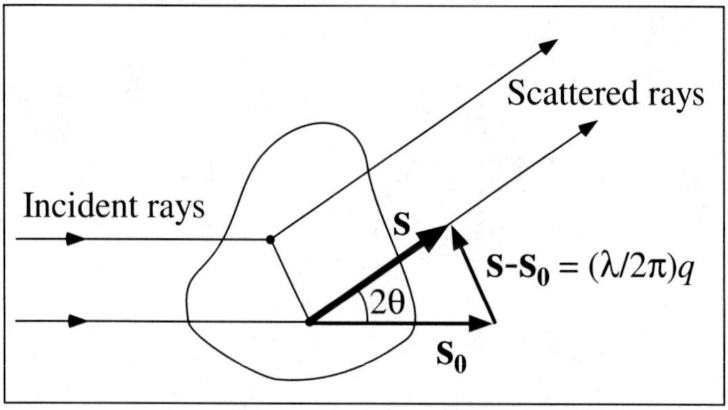

Figure 8.1 Diffraction by a single particle.

intensity has its maximum in the direction of the incident beam (i.e., at zero scattering angle), where the scattered rays are all in phase. This intensity at the zero scattering angle is proportional to the number of electrons in the particle. The phase difference between scattered rays of wavelength λ becomes more pronounced as the scattering angle increases, and eventually diminishes at a scattering angle of the order of λ/D. Figure 8.2 illustrates this situation, where the total scattered intensity is composed of scattering from three pairs of points. For example, the limiting scattering angle to be observed is equal to 0.45° when $D = 10$ nm, or to 0.045° when $D =$ 100 nm, for the X-ray wavelength of 0.154 nm from the CuK_α line. That is, a nanoparticle of the size 100 nm yields a scattering profile at the scattering angle smaller than 0.045°, and the profile (the scattered intensity as a function of scattering angle) is determined the shape of the particle.

The scattered X-rays are coherent, and the scattered intensity is given by the square of the amplitude resulting from the summation of all the amplitudes of scattered X-rays. The amplitudes are of the same magnitude but differ in their phase, depending on the position of electrons in space. Accordingly, the total amplitude $A(\mathbf{q})$ of the scattered rays at a point $\mathbf{q}$ is given by [2]

$$A(q) = A(\mathbf{q}) = \int_0^\infty \rho(\mathbf{r}) \cdot \exp(-i\mathbf{q} \cdot \mathbf{r}) d\mathbf{r} \tag{8.1}$$

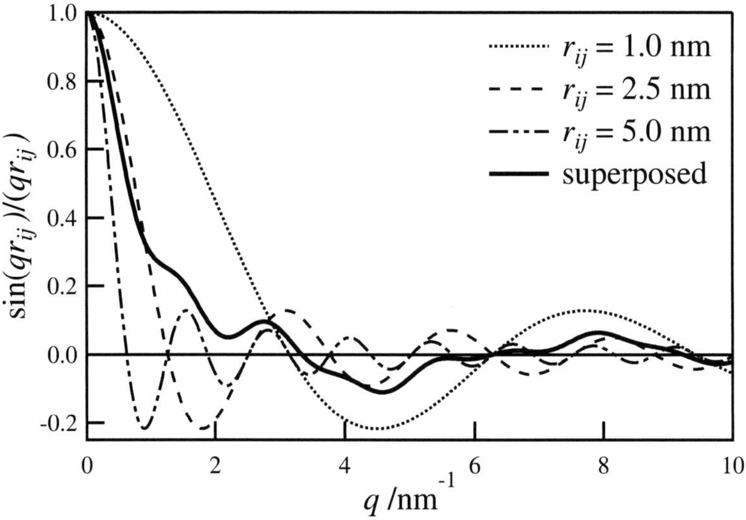

Figure 8.2 The profile of a scattering function $\sin qr_{ij}/(qr_{ij})$ for three pairs of scattering centers separated by the distance r_{ij}; $q = (4\pi/\lambda)\sin(\theta/2)$. A solid line indicates an integrated profile observed from a hypothetical particle composed of three pairs of scattering centers separated 1.0 nm, 2.5 nm and 5.0 nm, respectively.

where $\rho(\mathbf{r})$ is the electron density of the particle at a point $\mathbf{r}$, and the scattering vector $\mathbf{q}$ is defined as

$$\mathbf{q} = (2\pi/\lambda)(\mathbf{s} - \mathbf{s}_0) \tag{8.2}$$

Here $\mathbf{s}$ and $\mathbf{s}_0$ denote the unit vectors in the direction of the scattered and incident X-rays, respectively. The scattering vector $\mathbf{q}$ has the same direction as $(\mathbf{s} - \mathbf{s}_0)$ as seen from Fig. 8.1, and its magnitude is given by

$$q = (4\pi/\lambda)\sin(\theta/2) \tag{8.3}$$

with θ being the scattering angle specified by the vectors of the scattered and incident X-ray directions.

Equation 8.1 implies that the amplitude $A(\mathbf{q})$ of scattering in the direction of $\mathbf{q}$ is the Fourier transform of the electron density distribution $\rho(\mathbf{r})$. The scattered intensity is then given by the product of the amplitude and its complex conjugate A^* as

$$I(\mathbf{q}) = \int_0^\infty \int_0^\infty \rho(\mathbf{r}_1)\rho(\mathbf{r}_2) \cdot \exp[-i\mathbf{q} \cdot (\mathbf{r}_1 - \mathbf{r}_2)]\mathrm{d}\mathbf{r}_1\mathrm{d}\mathbf{r}_2 \tag{8.4}$$

which involves the separation ($\mathbf{r}_1 - \mathbf{r}_2$) for every pair of scattering points Eq. 8.4 can be modified as

$$I(\mathbf{q}) = \int_0^\infty \tilde{\rho}^2(\mathbf{r}) \cdot \exp(-i\mathbf{q} \cdot \mathbf{r})\mathrm{d}\mathbf{r} \tag{8.5}$$

where $\tilde{\rho}^2(\mathbf{r})$ is regarded as the density of electron pairs with separation $\mathbf{r}$. Eq. 8.5 also implies that $I(0)$ denotes the total density of electron pairs in the particle which corresponds to the mass of the particle. We discuss more details on the scattered intensity at the angle 0 and the scattered intensity extrapolated to the scattering angle 0 in Section 8.2.6. $\tilde{\rho}^2(\mathbf{r})$ is determined by the structure of the particle, and is given by the inverse Fourier transform of the scattered intensity as

$$\tilde{\rho}^2(\mathbf{r}) = (1/2\pi)^3 \int_0^\infty I(\mathbf{q}) \cdot \exp(i\mathbf{q} \cdot \mathbf{r})\mathrm{d}\mathbf{r} \tag{8.6}$$

Assuming the system is statistically isotropic, the phase factor $\exp(-q \cdot r)$ is replaced by its space average according to Debye [3]:

$$\langle \exp(-i\mathbf{q} \cdot \mathbf{r}) \rangle = \frac{\sin qr}{qr} \tag{8.7}$$

and Eq. 8.5 reduces to

$$I(q) = \int_0^\infty 4\pi^2 \tilde{\rho}^2(r) \frac{\sin qr}{qr} \mathrm{d}r = V \int_0^\infty 4\pi r^2 \gamma(r) \frac{\sin qr}{qr} \mathrm{d}r \tag{8.8}$$

in terms of the correlation function $\gamma(\mathrm{r})$ introduced by Debye and Bueche [4]. The correlation function $\gamma(\mathrm{r})$ is given by the average of the product of electron density fluctuations at two positions separated at a distance r, defined by the electron distribution density fluctuation $\eta(\equiv \rho - \bar{\rho})$ as $\langle \eta(x)\eta(x + r) \rangle$.

The Fourier transform of $I(q)$ yields a distance distribution function $p(r)$, which is related to the correlation function $\gamma(\mathrm{r})$ as [5]

$$p(r) = \frac{1}{2\pi^2} \int_0^\infty I(q) \cdot qr \sin(qr)\mathrm{d}q = Vr^2 \cdot \gamma(r) \tag{8.9}$$

Here the correlation function $\gamma(r)$ represents the correlation of the electron density fluctuation at a distance r. In the special case of $r = 0$, the mean square fluctuation of electron density $\gamma(0)$ is obtained from Eq. 8.9 as

$$Q \equiv V\gamma(0) = \frac{1}{2\pi^2} \int_0^\infty q^2 I(q)\mathrm{d}q \tag{8.10}$$

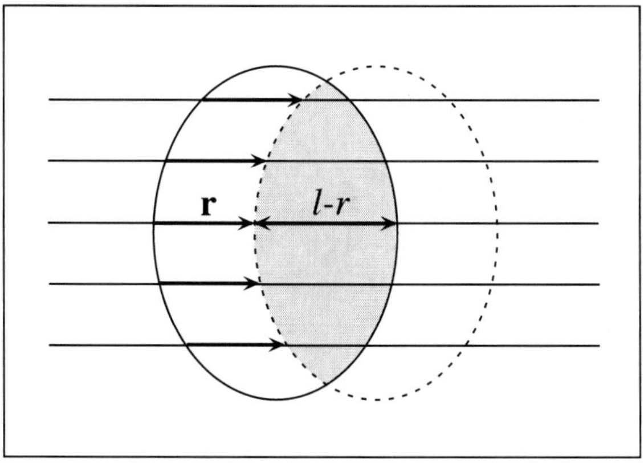

Figure 8.3 A particle and its ghost obtained by shifting the particle by a vector r.

which is invariant with the particle structure. The integral of the scattered intensity in Eq. 8.10 is taken over the whole reciprocal space, so that the invariant Q is related to the mean-square electron density fluctuation. In practice, the electron density difference $(\Delta\rho)$ is assumed to be constant, and the correlation function is then rewritten as

$$\gamma(r) = (\Delta\rho)^2 \cdot \gamma_0(r); \quad \gamma_0(0) = 1; \quad \gamma_0(r \geq D) = 0 \tag{8.11}$$

where $\gamma_0(r)$ is determined by the geometry of the particle and will vanish when $r \geqq D$ (the largest diameter of the particle). The physical definition of $\gamma_0(r)$ is given by Porod [6] as the average of the volume shared by the particle and its ghost shifted by a vector **r** (see Fig. 8.3). Suppose the particle composed of a group of rods of varying the length l, which is termed *chord*. Define a distribution function $G(l)\mathrm{d}l$ as the probability that a chord chosen at random is of a length between l and $l+\mathrm{d}l$. A piece of $(l-r)$ from any chord of the length $l<r$ is contained in the volume shared with the particle and its ghost. The average volume $\bar{V}_{\text{shared}}$ shared with the particle and its ghost is then given by extending an integral of $(l-r)$ over the chord distribution $G(l)$ as

$$\bar{V}_{\text{shared}} \propto \int_r^D (l-r)G(l)\mathrm{d}l \tag{8.12}$$

Accordingly, the normalized correlation function $\gamma_0(r)$ is calculated as

$$\gamma_0(r) = \frac{1}{\bar{l}} \int_r^D (l - r)G(l)\mathrm{d}l \text{ with } \bar{l} = \int_0^D lG(l)\mathrm{d}l \tag{8.13}$$

In some cases, it is easier to specify $G(l)$ for a particle and then calculate $\gamma_0(r)$. For example, a simple geometry yields

$$G(l) = \frac{2}{D^2} \tag{8.14}$$

for the sphere of a diameter D, so that

$$\gamma_0(r) = 1 - \frac{3}{2}\left(\frac{r}{D}\right) + \frac{1}{2}\left(\frac{r}{D}\right)^3; \quad \bar{l} = \frac{2}{3}D \tag{8.15}$$

The scattered intensity Eq. 8.8 can be rewritten by introducing the largest diameter D as

$$I(q) = (\Delta\rho)^2 \cdot V \cdot \int_0^D 4\pi r^2 \gamma_0(r) \cdot \frac{\sin qr}{qr} \cdot \mathrm{d}r \tag{8.16}$$

Since the correlation length l_c is defined as the mean width of the correlation function, it is calculated from Eq. 8.13 as

$$l_c \equiv 2\int_0^D \gamma_0(r)dr = \overline{l^2}/\bar{l} \tag{8.17}$$

which indicates l_c is the weight average of l. By integrating Eq. 8.16 with respect to q, one obtains

$$\int_0^\infty qI(q)dq = 4\pi(\Delta\rho)^2 V \int_0^D \gamma_0(r)dr = 2\pi(\Delta\rho)^2 V l_c \tag{8.18}$$

and the correlation length is given as

$$l_c = \pi \int_0^\infty qI(q)\mathrm{d}q/Q \tag{8.19}$$

where the variant Q takes a form

$$Q = \int_0^\infty q^2 I(q)dq = 2\pi^2(\Delta\rho)^2 V \tag{8.20}$$

and the volume V can be evaluated from the scattered intensity at the zero scattering angle and the invariant as

$$V = 2\pi^2 I(0)/Q \tag{8.21}$$

A series expansion of the Debye factor in Eq. 8.16 yields

$$\frac{\sin qr}{qr} = 1 - \frac{q^2\overline{r^2}}{3!} + \frac{q^4\overline{r^4}}{5!} - \cdots \tag{8.22}$$

and the scattered intensity is expressed in terms of the sum of the moments with respect to the distribution function $\gamma_0(r)$ as

$$I(q) = (\Delta\rho)^2 V^2 \left\{1 - \frac{q^2\overline{r^2}}{3!} + \frac{q^4\overline{r^4}}{5!} - \cdots\right\} \tag{8.23}$$

with

$$\overline{r^n} = \frac{1}{V}\int 4\pi r^2 \cdot r^n \cdot \gamma_0(r)\mathrm{d}r \tag{8.24}$$

where $\overline{r^2}$ corresponds to the radius of gyration R_G^2 by definition.

The parameters such as V, Q, l_c, and $\overline{r^2}$ represent the characteristics of the overall dimensions of the particle, and thus are termed as integral parameters [5]. At larger scattering angles, the scattered intensity reflects the finer structure of the particle, which is related to the behavior of $\gamma_0(r)$ at small r. By expanding $\gamma_0(r)$ into a power series of r, one obtains

$$\gamma_0(r) = 1 - ar + br^2 + cr^3 + \cdots \tag{8.25}$$

where the coefficients a, b, c, ... can be calculated by the successive differentiation of $\gamma_0(r)$ with respect to r at $r = 0$, and especially

$$a = \frac{1}{\bar{l}} \tag{8.26}$$

When a ghost is shifted only slightly in Fig. 8.3, the volume $\overline{V}_{\text{shared}}$ shared with the particle and its ghost is different from the particle volume by the shift of surface. Accordingly the volume $\overline{V}_{\text{shared}}$ is given approximately as

$$\overline{V}_{\text{shared}} \cong V - \frac{1}{4}Sr \tag{8.27}$$

with S being the total surface area. Here the contribution of a surface element dS to the shell is d$S \cdot r\cos\theta$, where θ is defined as the angle between the vector $\mathbf{r}$ and the surface normal. Taking into account all directions of $\mathbf{r}$ ($\overline{|\cos\theta|} = 1/2$) and the contribution of only the part of $\mathbf{r}$ directed inwards (1/2), the contribution of each surface element is given by $(1/4)\mathrm{d}S \cdot r$ in average, and $(1/4)Sr$ for the total surface. Equation 8.27 implies

$$\gamma_0(r) \cong 1 - \frac{1}{4}Sr; \quad a = \frac{S}{4V} = \frac{1}{\bar{l}} \tag{8.28}$$

By substituting Eq. 8.28 into Eq. 8.16, one obtains

$$I(q) \cong (\Delta\rho)^2 V \int_0^D 4\pi r^2 \left(1 - \frac{r}{\bar{l}} + \cdots\right) \cdot \frac{\sin qr}{qr}\mathrm{d}r \tag{8.29}$$

which yields Porod's fourth law at larger q as [6]

$$I(q) \to (\Delta\rho)^2 V \cdot \frac{8\pi}{\bar{l}q^4} = (\Delta\rho)^2 \cdot \frac{2\pi}{q^4} \cdot S \tag{8.30}$$

or alternatively the specific surface is determined as

$$\frac{S}{V} = \pi \lim_{q\to\infty} q^4 I(q)/Q \tag{8.31}$$

8.2.2 *Structure Parameters*

The scattering factors derived by assuming a single particle in solvent can be applied in practice to a dilute solution system consisting of N scattering particles, when they are sufficiently isolated so as not to interact with each other. At very small angles, the sine term in Eq. 8.8 is expanded in the series of q^2 and yields

$$P(q) = \frac{I(q)}{I(0)} = \frac{1 - \frac{1}{3}q^2 \int_V \rho(r) r^2 \mathrm{d}V}{\int_V \rho(r)\mathrm{d}V + O(q^4)} \tag{8.32}$$

Here the homogeneous electron density distribution is assumed in the particle and

$$4\pi r^2 \tilde{\rho}^2(r)\mathrm{d}r$$

in Eq. 8.8 is replaced with $\rho(r)\mathrm{d}V$. The second term in the right side of Eq. 8.32 corresponds to the radius of gyration R_G of the particle. The particle scattering factor $P(q)$ corresponds to a normalized scattered intensity with respect to the scattered intensity at the zero angle, and represents the scattering from a single particle. When the sine expansion of Eq. 8.22 is replaced approximately with the exponential function, the particle scattering factor reduces to the Guinier approximation [5]:

$$P(q) \approx \exp(-q^2 R_\mathrm{G}^2/3) \tag{8.33}$$

Table 8.2 Radius of gyration R_G of a simple triaxial body

Triaxial body	Mean-square radius of gyration R_G^2
Sphere of radius R	$3R^2/5$
Hollow sphere of radii R_1 (inner) and R_2 (outer)	$3(R_2^5 - R_1^5)/5(R_2^3 - R_1^3)$
Ellipsoid of semi-axes a, b, c	$(a^2 + b^2 + c^2)/5$
Prism with edge lengths A, B, C	$(A^2 + B^2 + C^2)/12$
Elliptic cylinder of semi-axes a, b and height h	$[(a^2 + b^2)/4] + (h^2/12)$
Hollow cylinder of radii R_1, R_2 and height h	$[(R_1^2 + R_2^2)/2] + (h^2/12)$

Equation 8.33 indicates that the radius of gyration is evaluated from the initial slope of the Guinier plot where $\ln P(q)$ is plotted against q^2. The radius of gyration gives a measure of the size of the particle, and is related to the dimension of the particle of various shapes as summarized in Table 8.2.

The Guinier approximation is also applied to evaluate the radius of gyration corresponding to the cross section of a rod-like particle or the thickness of a flat particle. The scattering function from a cylinder is composed of two factors corresponding to the height and the cross section, and is approximately given as

$$P_{\text{cylinder}}(q) \approx \frac{\pi}{2Hq} \cdot P_{\text{c}}(q) \tag{8.34}$$

where $P_{\text{c}}(q)$ and $2H$ denote the scattering function from the cross section and the height of the cylinder, respectively. The factor $1/q$ is characteristic of a rod-like particle, and Eq. 8.34 is valid only when the condition $Hq < 1$ is satisfied. The scattering function from the cross section is given by the Guinier approximation in terms of the cross-sectional radius of gyration R_{c} as

$$P_{\text{c}}(q) = \exp(-q^2 R_{\text{c}}^2/2) \tag{8.35}$$

Equations 8.34 and 8.35 imply that the initial slope of $\ln qP_{\text{cylinder}}(q)$ plotting against q^2 (the Guinier plot for cross section) yields the quantity equivalent to $R_{\text{c}}^2/2$. The cross-sectional radius of gyration depends on the cross-sectional shape as summarized in Table 8.3. Similarly, the scattering function from a flat particle is given by the product of two factors corresponding to the flat surface and the thickness as

$$P_{\text{disk}}(q) \approx \frac{2\pi}{Aq^2} \cdot P_t(q) \tag{8.36}$$

Table 8.3 Cross-sectional radius of gyration R_c

Cross-sectional shape	Mean-square cross-sectional radius of gyration R_c^2
Circle of radius R	$R^2/2$
Ring of inner radius R_1 and outer radius R_2	$(R_1^2 + R_2^2)/2$
Ellipse of semi-axes a and b	$(a^2 + b^2)/4$
Rectangle with edge length A and B	$(A^2 + B^2)/12$

where $P_t(q)$ and A denote the scattering function from the thickness and the cross-sectional area, respectively. The scattering function from the thickness $P_t(q)$ is given by the Guinier approximation in terms of the thickness radius of gyration R_t as

$$P_t(q) = \exp(-q^2 R_t^2) \tag{8.37}$$

Equations 8.36 and 8.37 imply that the initial slope of $\ln q^2 P_{\text{disk}}(q)$ plotting against q^2 (the Guinier plot for thickness) yields the quantity equivalent to R_t^2. Here the thickness radius of gyration R_t is given in terms of the thickness T as

$$R_t = \frac{T}{\sqrt{12}} \tag{8.38}$$

Following Eqs. 8.34 or 8.36, the cross-sectional radius of gyration or the thickness radius of gyration is evaluated from the initial slope of the cross-sectional Guinier plot

$$(\ln q\, P_{\text{cylinder}}(q))$$

plotted against q^2, or the thickness Guinier plot

$$(\ln q^2\, P_{\text{disk}}(q))$$

plotted against q^2, respectively.

Equations 8.34 and 8.36 also imply that the mass corresponding to the cross section (the mass per unit length) or the thickness (the mass per unit area) could be evaluated from the respective scattered intensity at the zero scattering angle as $(2q/\pi)\cdot I(0)$ or $(q^2/2\pi)\cdot I(0)$ by extrapolating the respective Guinier plots to the zero angle. More details will be found in Section 8.2.6.

In some special cases of homogeneous triaxial bodies, the scattering factor is explicitly calculated in a closed form [6] as follows:

(a) A sphere of the radius R:

$$P(q) = \Phi^2(qR)$$
$$\Phi(qR) = \frac{3(\sin qR - qR\cos qR)}{(qR)^3} \tag{8.39}$$

where the function $\Phi(qR)$ corresponds to the scattering amplitude of a rigid sphere of radius R.

(b) A hollow sphere of the outer radius R_1 and the inner radius R_2:

$$P(q) = \left[\frac{R_1^3\Phi(qR_1) - R_2^3\Phi(qR_2)}{R_1^3 - R_2^3}\right] \tag{8.40}$$

(c) An ellipsoid of revolution with semi-axes a, a and αa:

$$P(q) = \int_0^{\pi/2} \Phi^2(qa\sqrt{\cos^2\theta + \alpha^2\sin^2\theta)}\cos\theta d\theta \tag{8.41}$$

(d) A cylinder of the radius R and the height $2H$:

$$P(q) = \int \frac{\sin^2(qH\cos\theta)}{q^2H^2} \times \frac{4J_1^2(qR\sin\theta)}{q^2R^2\sin^2\theta}\sin\theta d\theta \tag{8.42}$$

(e) A rod of length $2H$ (equivalent to the cylinder of the infinitesimal radius and the height $2H$):

$$P(q) = \frac{\mathrm{Si}(2qH)}{\mathrm{qH}} - \frac{\sin^2(qH)}{q^2H^2}$$
$$Si(x) = \int_0^x \frac{\sin t}{t}dt \tag{8.43}$$

(f) A disk of the radius R (equivalent to the cylinder of the radius R and infinitesimal thickness):

$$P(q) = \frac{2}{q^2R^2}\left[1 - \frac{1}{qR}J_1(2qR)\right] \tag{8.44}$$

(g) A linear chain consisting of N units feely jointed by a link of length b, which is equivalent to a Gaussian chain [7]:

$$P(q) = \frac{2N^2}{x^2}[\exp(-x) - 1 + x]$$
$$x = q^2R_G^2 \tag{8.45}$$

where the mean-square radius of gyration R_G^2 is given by $R_G^2 = Nb^2/6$.

(h) A ring-shaped polymer consisting of N units freely jointed by a link of length b [8]

$$P(q) = (2/x^{1/2})\exp(-x/4)D(x^{1/2}/2)$$
$$D(X) = \int_0^X \exp(t^2)dt \tag{8.46}$$

where $D(X)$ denotes the Dawson integral, and x is given by the second equation in Eq. 8.45. The mean-square radius of gyration of the Gaussian ring differs by a factor 2 from that of a linear Gaussian chain, and is given as $R_G^2 = Nb^2/12$.

(i) A star-shaped polymer of total N units freely jointed by a link of length b with f branches of equal length radiating from the center

$$P(q) = \frac{2}{x^2}[f(f-1)\exp(-2x/f)/2 - f(f-2)\exp(-x/f) + x + f(f-3)/2] \tag{8.47}$$

where x is given by the second equation in Eq. 8.45. The mean-square radius of gyration differs by a factor $(3f-2)/f^2$ taking into account the branching effect from that of a linear Gaussian chain, and is given by

$$R_G^2 = \frac{(3f-2)}{f^2}\cdot\frac{Nb^2}{6}.$$

The chain stiffness is another factor to be considered in a coiled molecule. The scattering behavior of such a molecule is expected to be described by a hybrid function of Eq. 8.43 for a rigid rod and Eq. 8.45 for a completely flexible chain.

Figure 8.4 illustrates the scattering profiles from model triaxial bodies and a Gaussian chain. Here the radius of gyration is fixed to $R_G = 5$ nm, and the lines are drawn with the slopes expected from Eqs. 8.33, 8.35, and 8.37 with the corresponding radii of gyration summarized in Tables 8.2 and 8.3.

8.2.3 *Scattering from Non-particulate Systems*

The scattering from a well-defined object can be treated explicitly by the models shown above. However, in reality we encounter more often the scattering from not-so-well-defined objects. Here

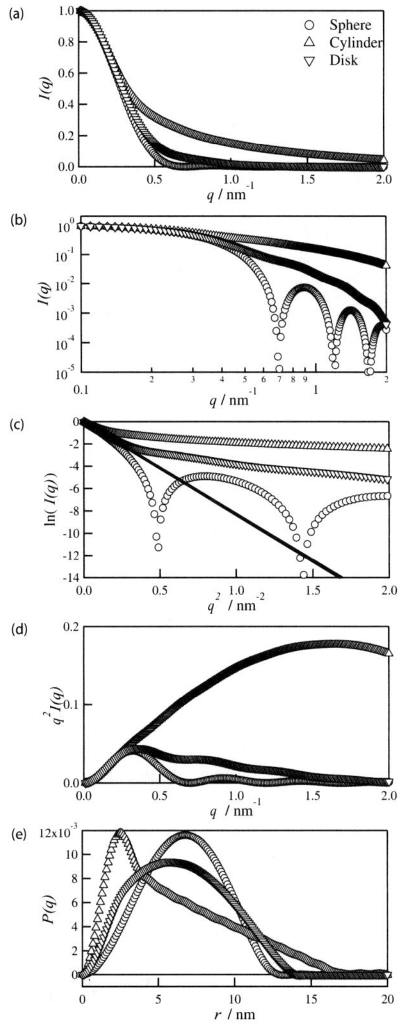

Figure 8.4 Scattering profile from a sphere of radius 6.46 nm, a cylinder of cross-sectional radius 1.68 nm and height 16.82 nm, and a disk of radius 6.98 nm and thickness 2.79 nm (where R_G is fixed to 5 nm), calculated according to Eqs. 3.89, 8.42 and 8.44, respectively. The profiles are shown by various ways of plotting including the normal plots (a), the double logarithmic plots (b), the Guinier plots (c) where the solid line represents the initial slope specified by the value $R_G^2/3$ (Eq. 8.33), the Kratky plots ($q^2 I(q)$ vs. q) (e), and the distance distribution function defined by Eq. 8.9 (e).

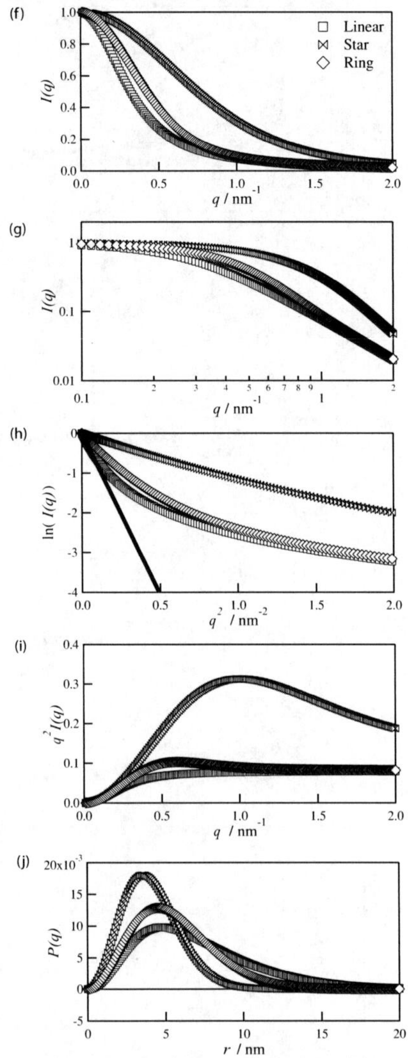

Figure 8.4 (*continued*): Scattering profile from Gaussian chains of various architecture (linear, star-shaped and ring) composed of 600 (N) links of length $b = 0.5$ nm ($R_{\mathrm{G}} = 5$ nm). The profiles are shown by various ways of plotting including the normal plots (f), the double logarithmic plots (g), the Guinier plots (h) where the solid line represents the initial slope specified by the value $R_{\mathrm{G}}^2/3$ (Eq. 8.33), the Kratky plots ($q^2 I(q)$ vs. q) (i), and the distance distribution function defined by Eq. 8.9 (j).

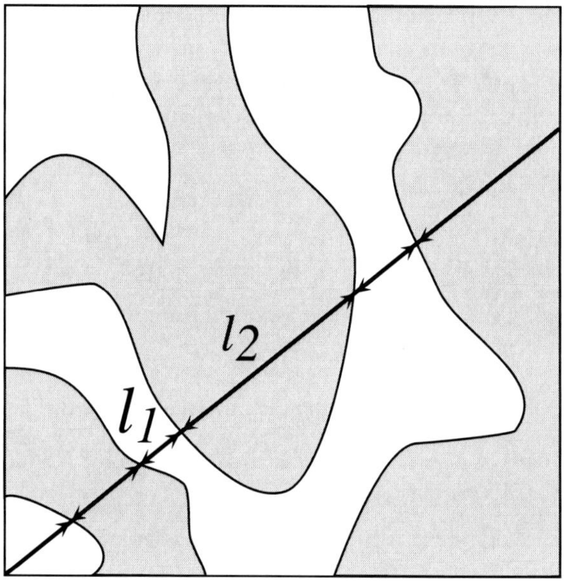

Figure 8.5 Schematic view of non-particulate two-phase system.

the scattering object cannot be defined with a tri-axial body or a coiled chain, where the particle scattering function is given in an explicit form. There are several factors which cause various degrees of inhomogeneity. In general, we may define some domain in which the correlation function can be described (at least approximately) by a simple function to specify the electron density distribution. Let suppose the scattering system composed of two phases which can be distinguished by respective constant electron density [6] (see Fig. 8.5). Two phases fill the fractions φ_1 and φ_2 of the total volume of V. The average electron density $\overline{\rho}$ and a mean-square electron density fluctuation $\overline{\eta^2} \equiv \langle(\rho - \overline{\rho})^2\rangle$ are given as

$$\overline{\rho} = \varphi_1 \rho_1 + \varphi_2 \rho_2 \tag{8.48}$$

$$\overline{\eta^2} \equiv \langle(\rho - \overline{\rho})^2\rangle = (\rho_1 - \rho_2)^2 \cdot \varphi_1 \varphi_2 \tag{8.49}$$

where the mean-square electron density fluctuation is directly related to the invariant Eq. 8.10 as

$$Q \equiv \int_0^\infty q^2 I(q) \mathrm{d}q = 2\pi^2 V \cdot \langle(\rho - \overline{\rho})^2\rangle = 2\pi^2 V \cdot (\rho_1 - \rho_2)^2 \cdot \varphi_1 \varphi_2 \tag{8.50}$$

Now the correlation function (defined by Eq. 8.9) should be extended to a two-phase system in order to understand the scattering profile from such a system. Now we ask the probability $P_{ij}(r)$ to find a point belonging to the phase j (1 or 2), which is separated by a distance r from the point belonging to the phase i (1 or 2). Obviously

$$\begin{aligned} &P_{11}(0) = P_{22}(0) = 1; \qquad P_{12}(0) = P_{21}(0) = 0 \\ &P_{11}(\infty) = P_{21}(\infty) = \varphi_1;\ P_{22}(\infty) = P_{12}(\infty) = \varphi_2 \end{aligned} \tag{8.51}$$

Thus defining the normalized correlation function $\gamma_0(r)$ as

$$\gamma_0(0) = 1; \quad \gamma_0(\infty) = 0 \tag{8.52}$$

and considering the mathematical symmetry, we obtain

$$\begin{aligned} &P_{11}(r) = \varphi_1 + \varphi_2\gamma_0(r);\ P_{12}(r) = \varphi_2 - \varphi_2\gamma_0(r) \\ &P_{21}(r) = \varphi_1 - \varphi_1\gamma_0(r);\ P_{22}(r) = \varphi_2 + \varphi_1\gamma_0(r) \end{aligned} \tag{8.53}$$

Since only two values of the fluctuation (assigned by $\eta_1 = \varphi_2(\Delta\rho)$ and $\eta_2 = -\varphi_1(\Delta\rho)$ where $\Delta\rho = \rho_1 - \rho_2$) are possible in the two-phase system and four combinations corresponding to the probabilities given by Eq. 8.53 are available, the electron density correlation function is calculated as

$$\gamma(r) \equiv \langle\eta(x)\eta(x+r)\rangle = P_{11}\eta_1^2 + 2P_{12}\eta_1\eta_2 + P_{22}\eta_2^2 = \varphi_1\varphi_2(\Delta\rho)^2\gamma_0(r) \tag{8.54}$$

where the front factor of $\gamma_0(r)$ is the mean-square fluctuation of electron density given by Eq. 8.52.

The scattered intensity from the whole system is given by the Fourier transformation of Eq. 8.54 as

$$I(q) = V \cdot \varphi_1\varphi_2(\Delta\rho)^2 \int_0^\infty 4\pi r^2\gamma_0(r) \cdot \frac{\sin qr}{qr} \cdot \mathrm{d}r \tag{8.55}$$

which has an extra front factor due to two phases I order to take into account an inhomogeneous electron density in the volume (see the particle scattering Eq. 8.8 for a homogeneous particle). The correlation function will be given by the Fourier transformation of Eq. 8.55 as in Eq. 8.9, but the physical meaning of the correlation function $\gamma(r)$ (or $\gamma_0(r)$) is less intuitive than in the case of the particle scattering. However, the format of the scattered intensity is similar to the homogeneous particle scattering factor, so that the concept of size and shape may still be applied by considering that the correlation function $\gamma(r)$ represents an averaged surrounding electron densities. The scattered intensity at the zero angle yields the correlation volume v_c which can be defined in analogy with the case of the particle scattering:

$$I(0) = V\varphi_1\varphi_2(\Delta\rho)^2 \cdot v_c;\ v_c \equiv \int_0^\infty 4\pi r^2\gamma_0(r)\mathrm{d}r \tag{8.56}$$

or in terms of the invariant Q;

$$v_c = \frac{2\pi^2}{Q} \cdot I(0) \tag{8.57}$$

The correlation volume v_c can be regarded as a measure of the size of the heterogeneity. However, in the case of a densely packed system, v_c becomes extremely small and even reduces to zero as seen from the liquid type system with interparticular interactions, since $\gamma_0(r)$ can take a negative value.

The concept of shape in the non-particulate system refers to globular, fibrillar and lamellar systems. The criterion for the particle scattering is still valid as the factor $\approx q^{-1}$ for fibrils and $\approx q^{-2}$ for lamellae. Since the mean lengths of l_1 and l_2 (see Fig. 8.5) should be in proportion to the respective volume fraction,

$$\bar{l}_1 : \bar{l}_2 = \varphi_1 : \varphi_2 \tag{8.58}$$

and Eq. 8.28 can be applied to respective phases. The volume of each phase is now given by $V\varphi_1$ and $V\varphi_2$, respectively, we will obtain

$$\bar{l}_1 = \frac{4V}{S} \cdot \varphi_1;\quad \bar{l}_2 = \frac{4V}{S} \cdot \varphi_2 \tag{8.59}$$

Here we introduce the average $\bar{l}$ of the mean lengths $\bar{l}_1$ and $\bar{l}_2$ to express $\gamma_0(r)$ in terms of the power series of r with the coefficient of r being equal to $1/\bar{l}$, that is

$$\gamma_0(r) = 1 - r/\bar{l} + \cdots \tag{8.60}$$

Comparing Eqs. 8.56 and 8.57 with Eq. 8.63, the average $\bar{l}$ is given by

$$\bar{l} = \bar{l}_1\varphi_2 = \bar{l}_2\varphi_1 = 4\varphi_1\varphi_2\frac{V}{S};\quad \text{or}\quad \frac{1}{\bar{l}} = \frac{1}{\bar{l}_1} + \frac{1}{\bar{l}_2} \tag{8.61}$$

In the limiting case of an infinitely dilute system, the two-phase system reduces to the case of particle scattering, where $\varphi_1 \to 0$ and $\varphi_2 \to 1$, so that $\bar{l} \to \bar{l}_1$ and $\bar{l}_2 \to \infty$.

The correlation function in Eq. 8.60 is specified by a single parameter $\bar{l}$ as in Eq. 8.28, and a similar argument leads to Porod's fourth power law at larger q as

$$I(q) \to (\Delta\rho)^2 \varphi_1 \varphi_2 V \cdot \frac{8\pi}{\bar{l} q^4} = (\Delta\rho)^2 \cdot \frac{2\pi}{q^4} \cdot S \tag{8.62}$$

By normalizing the scattered intensity with respect to q^4 and the invariant Q, we shall be able to evaluate the specific surface of the system as

$$\pi \lim_{q\to\infty} q^4 I(q)/Q = \frac{1}{\varphi_1 \varphi_2} \cdot \frac{S}{V} \tag{8.63}$$

as shown for the case of particle scattering Eq. 8.31. Porod's fourth power law is thus valid for any system which is characterized by a sharp boundary. Equation 8.60 indicates the initial slope of $\gamma_0(r)$ vs. r will give the correlation length $\bar{l}$, so that

$$\left(\frac{d\gamma_0}{dr}\right)_{r=0} = \frac{1}{4\varphi_1 \varphi_2} \cdot \frac{S}{V} \tag{8.64}$$

When the boundary is smeared by a fine structure such as small pores or cleavages, $I(q)$ may depend on the lower power of $1/q$. That is, Eq. 8.64 yields no asymptotic (constant) value but increases with increasing q (scattering angle). Porod's fourth power law is found to describe the scattering behavior well at higher q for solid polymers with a high crystallinity (linear polyethylene, polyoxymethylene, polyethylene oxide, etc.) and globular proteins in solution. When the electron density at the phase boundary varies over the distance E by the presence of transition layers, the actual electron density distribution is considered to be given by the convolution of the electron densities of corresponding layers, which can be replaced with a smoothing function $h(x)$ [8]. If the electron density in the phase boundary varies linearly, the smoothing function $h(x)$ is given by a top-hat function of width E, and the scattering intensity is modified as

$$I(q) = I^0(q) \cdot \frac{\sin^2(\pi q E)}{(\pi q E)^2} \tag{8.65}$$

where $I^0(q)$ denotes the ideal scattering from a two-phase system obeying Porod's fourth power law. In the Porod region, Eq. 8.65 reduces approximately to

$$I(q) = \frac{Q}{\pi \varphi_1 \varphi_2 q^4} \cdot \frac{S}{V} \left(1 - \frac{\pi^2 q^2 E}{3}\right) \tag{8.66}$$

which is used to estimate the specific volume and the width of transition layer E from the intercept and slope, respectively, by plotting, for example, $q^4 I(q)$ against $1/q^2$. If E is relatively large, no linear relation will be observed by the above plotting, and more general data treatment is required [9].

8.2.4 *Scattering by Density Fluctuation [10]*

Scattering is observed by density fluctuation which is closely related with the thermodynamic state of the system. So far we have observed scattering from well-defined objects, and thus the scattering is a deterministic phenomenon. Fluctuation on the other hand can be described in terms of a stochastic process where the system is composed of an ensemble undergoing constant change of state over time and space. If the density correlation is given, the scattering function can be calculated by its Fourier transform. Instead of relying on Debye's formula of particle scattering, we consider the fluctuations of thermodynamic quantities and the form of the density correlation in the system relevant in the sol and gel state of macromolecules in solution.

8.2.4.1 Typical distributions

Fluctuation is specified by a distribution function $P_N(r)$ (or the probability density) for a distance r where $P_N(r)\mathrm{d}r$ gives the probability of finding a point in the interval between r and $r+\mathrm{d}r$ after N steps. First we consider a simple example of the event occurrence A or B which takes place according to the probability p (the event A) and q (the event B). If there is no other event except for A and B, $p + q = 1$ and the probability of obtaining a particular sequence composed of $x\,A$ events and $(n-x)\,B$ events after n trials is given by $p^x q^{n-x}$. Since there are

$$\frac{n!}{x!(n-x)!}$$

ways to compose a sequence of $x\,A$ events and $(n-x)\,B$ events, the probability of finding such a sequence after n trials is given by

$$W_n(x) = \frac{n!}{x!(n-x)!} \cdot p^x q^{n-x} = {}_nC_x\, p^x q^{n-x} \qquad (8.67)$$

where ${}_nC_x$ is the binomial coefficients, and the distribution of x expressed by Eq. 8.67 is called the Bernoulli distribution. The mean of x, x^2 and the mean-square of the fluctuation Δx $(\equiv x - \langle x\rangle)$ are calculated by the successive differentiation of the generating function $(pu+q)^n$ with respect to the dummy variable u at $u = 1$ as

$$\rangle x \langle \equiv \sum_{x=0}^{n} x W_n(x) = \left[\frac{\partial(pu+q)^n}{\partial u}\right]_{u=1} = np \tag{8.68}$$

$$\langle x^2\rangle \equiv \sum_{x=0}^{n} x^2 W_n(x) = \left\{\frac{\partial}{\partial u}\left[u\cdot\frac{\partial(pu+q)^n}{\partial u}\right]\right\}_{u=1} = n^2p^2 + npq \tag{8.69}$$

$$\langle(\Delta x)^2\rangle = \langle(x-\langle x\rangle)^2\rangle = \langle x^2\rangle - \langle x\rangle^2 = npq \tag{8.70}$$

Two limiting cases at $n \to \infty$ of Eq. 8.67 are of practical interest. In the first case, we consider that p is finite and the mean value $\langle x\rangle(= np)$ (see Eq. 8.68) increases indefinitely as $n \to \infty$. By applying Stirling's formula and $\Delta x << \langle x\rangle$, Eq. 8.67 reduces to

$$W_n(x) = \frac{1}{(2\pi npq)^{1/2}}\cdot\exp\left[-\frac{(\Delta x)^2}{2npq}\right] \tag{8.71}$$

where $x = \langle x\rangle + \Delta x$. The distribution function of the form given by Eq. 8.71 is called the Gaussian distribution. Here the mean square of the fluctuation $\langle(\Delta x)^2\rangle$ is identical to that in the Bernoulli distribution Eq. 8.70.

If np remains finite as $n \to \infty$, $n >> x$ and $x \approx \langle x\rangle \approx np$. By replacing p with $\langle x\rangle/n$ and using the condition $\langle x\rangle/n << 1$, Eq. 8.67 reduces to

$$W_n(x) = \frac{\langle x\rangle^x}{x!}\lim_{n\to\infty}\left(1-\frac{\langle x\rangle}{n}\right)^n = \frac{\langle x\rangle^x}{x!}\cdot\exp(-\langle x\rangle) \tag{8.72}$$

which is called the Poisson distribution and yields

$$\langle x^2\rangle = np + n^2p^2 \tag{8.73}$$

$$\langle(\Delta x)^2\rangle = \langle x^2\rangle - \langle x\rangle^2 = np \tag{8.74}$$

Equations 8.73 and 8.74 are understood as a special case when $q \to 1$ in Eqs. 8.69 and 8.70.

A one-dimensional random walk is a good example to be represented by the Gaussian distribution even when p is large and n is relatively small. Consider a point which moves either left or right by a distance b. After N moves, the point will reach to the position apart from the origin with the distance either of $-Nb$, $(-N+2)b, \ldots (N-2)b$, or Nb. The probability is designated as $W_N(m)$ to reach the position specified by the distance mb from the origin. Here m denotes the difference in the moves between to right and to left, so that $n = N$ and $x = (N+m)/2$ in Eq. 8.67. If the move will take place with an equal probability to right or to left, $p = q = 1/2$. Thus

$$W_N(m) = \frac{N!}{\{(N+m)/2\}!\{(N-m)/2\}!} \cdot \left(\frac{1}{2}\right)^N \tag{8.75}$$

which reduces to the Gaussian distribution when $N >> 1$

$$W_N(m) = \left(\frac{2}{\pi N}\right)^{1/2} \exp\left(-\frac{m^2}{2N}\right) \tag{8.76}$$

Equation 8.76 can be rewritten for the continuous distribution of the coordinate x by defining the probability density function (the distribution function) $P_N(x)$ as $P_N(x)\mathrm{d}x$ being the probability to find the point between x and $x + \mathrm{d}x$ after N steps as

$$P_N(x) = \frac{1}{(2\pi Nb^2)^{1/2}} \cdot \exp\left(-\frac{x^2}{2Nb^2}\right) \tag{8.77}$$

since there are $\mathrm{d}x/2b$ positions available for the point to reach within the distance $\mathrm{d}x$ and thus $P_N(x)\mathrm{d}x = W_N(m) \cdot (\mathrm{d}x/2b)$.

If a point moves on the three-dimensional cubic lattice (the lattice distance b) for $N/3$ steps in each x, y and z, then the distribution function $P_N(\mathbf{R})$ to reach $\mathbf{R} = (x, y, z)$ is given by the product of the distribution functions in respective directions as

$$P_N(\mathbf{R}) = P_{N/3}(x)P_{N/3}(y)P_{N/3}(z) = \frac{1}{\pi^{3/2}\tau^3} \cdot \exp(-R^2/\tau^2)$$
$$\tau^2 \equiv \frac{2Nb^2}{3} \tag{8.78}$$

The normalization condition of this distribution function is given as

$$\int P_N(\mathbf{R})\mathrm{d}\mathbf{R} = \int_0^\infty 4\pi R^2 P_N(\mathbf{R})\mathrm{d}R = 1 \tag{8.79}$$

which yields the mean-square distance $\langle R^2 \rangle$ as

$$\langle R^2 \rangle = \int R^2 P_N(\mathbf{R})\mathrm{d}\mathbf{R} = 4\pi \int R^4 P_N(\mathbf{R})\mathrm{d}R = \frac{3\tau^2}{2} = Nb^2 \quad (8.80)$$

Each step of a point on the cubic lattice corresponds to a single link of a polymer chain, so that **R** is equivalent to the end-to-end distance of a polymer chain composed of freely jointed N links. Equation 8.80 confirms the mean-square end-to-end distance of a polymer chain being in proportion to the number of links N so as the mean-square radius of gyration Eq. 8.48.

8.2.4.2 Pair correlation function

A pair correlation function $\gamma(\mathbf{r})$ for a polymer chain is defined as the average of the number density at the distance **r** from an arbitrary monomer overall monomers in the polymer chain. The Fourier transform of $\gamma(\mathbf{r})$ yields Eq. 8.8 which can be observed directly by scattering experiments. A pair correlation function is defined as

$$\gamma(\mathbf{r}) = \frac{1}{c}\left[\langle c(0)c(\mathbf{r})\rangle - c^2\right] \quad (8.81)$$

and a simple Ornstein-Zernike type is believed to represent $\gamma(\mathbf{r})$ in the region where $r \geq \xi$, that is, $\gamma(\mathbf{r})$ is given in terms of the correlation length ξ specifying the range of density fluctuations effectively correlated as

$$\gamma(\mathbf{r}) \cong c\frac{\xi}{r}\exp(-r/\xi) \quad (8.82)$$

which yields the structure factor $S(q)$ as

$$S(q) \cong \frac{c\xi^3}{1+\xi^2 q^2} \quad (8.83)$$

The denominator in Eq. 8.83 is given by the product of density and volume, so that it corresponds to the number of monomers in one blob defining the unit of correlated density fluctuation. From the thermodynamic point of view [10–12], the structure factor at the zero scattering angle is related to the osmotic pressure π and Eq. 8.83 reduces in the limit of $q \to 0$ to

$$S(0) = c\xi^3 = \frac{KRT}{(\partial\pi/\partial c)_{T,\mu_0}} \quad (8.84)$$

where K denotes an experimental constant which depends on the apparatus and radiation source. Equation 8.83 indicates that the quantity ξ determines how fast $S(q)$ decays with q in the region of small q. The correlation length ξ is thus taken as a measure of the range of r in which density fluctuations are effectively correlated, and Eq. 8.83 holds only for small q. That is, the Ornstein–Zernike form Eq. 8.82 is valid only for $r > \xi$, and characterizes the decay rate of $\gamma(\mathbf{r})$.

For an isolated Gaussian spring-bead chain, the intrachain bead-pair radial distribution function is given by Eq. 8.78, where the step N should be reread as the number of links $|i - j|$ between a pair of beads (e.g., i and j), and the particle scattering factor is calculated by taking the sum of $P_{ij}(\mathbf{R})$ overall pairs of i and j, and Fourier-transforming the sum as (see Eq. 8.46)

$$P(q) = \int \sum_{i=1}^{N} \sum_{j=1}^{N} P_{ij}(\mathbf{R}) \cdot \exp(i\mathbf{q} \cdot \mathbf{R})d\mathbf{R} = \frac{2N^2}{x^2}\left[\exp(-x) - 1 + x\right] \tag{8.85}$$

where

$$x = \frac{Nb^2q^2}{6} \tag{8.86}$$

Since Eq. 8.85 reduces to

$$P(q) \approx \frac{1}{1 + (x/3)} \tag{8.87}$$

for small q (when $x << 1$), the correlation length ξ is found to correspond in this case to

$$\xi = \frac{R_g}{\sqrt{3}} \tag{8.88}$$

by comparing Eq. 8.87 with Eq. 8.83. Here $R_G = Nb^2/6$ is the radius of gyration of an isolated Gaussian chain, and Eq. 8.88 indicates that the density fluctuation is correlated effectively in an isolated Gaussian chain up to the distance comaparable to the chain dimension.

The density fluctuation specified by Eq. 8.82 is not explicitly identical to that specified for the Gaussian chain. In any system, we can define some region where the density fluctuation is effectively correlated but not beyond that region. This region is

Table 8.4 Density correlation function and corresponding scattering function

Specification	Density correlation function $\gamma(r)$	Scattering function $P(q)$
Sphere*	$1-\frac{3}{4}\left(\frac{r}{R}\right)+\frac{1}{16}\left(\frac{r}{R}\right)^3$	$\left[\frac{3(\sin qR - qR\cos qR)}{(qR)^3}\right]^2$
Ornstein–Zernicke	$\frac{\xi}{r}\exp\left(-\frac{r}{\xi}\right)$	$\frac{1}{1+\xi^2q^2}$
Generalized Ornstein–Zernicke**	$\left(\frac{\xi}{r}\right)^{3-D}\exp\left(-\frac{r}{\xi}\right)$	$\frac{1}{\left[1+(D+1)\xi^2q^2/3\right]^{D/2}}$
Debye–Bueche	$\exp\left(-\frac{r}{\xi}\right)$	$\frac{1}{(1+\xi^2q^2)^2}$
Guinier	$\exp\left(-\frac{r^2}{\xi^2}\right)$	$\exp(-\xi^2q^2/4)$
Combined	$\frac{1}{2}\left(1+\frac{\xi}{r}\right)\exp\left(-\frac{r}{\xi}\right)$	$\frac{1}{(1+\xi^2q^2)^3}$

* R denotes the diameter of the sphere.
** D denotes the fractal dimension.

defined as a blob. The density fluctuation within the blob is described by various types of the correlation functions which afford unique scattering profiles respectively. Table 8.4 summarizes the correlation function and corresponding scattering function [13], and Fig. 8.6 demonstrates the scattering profiles specified by each corelation function listed in Table 8.4. The density correlation function can be generalized in terms of the product of two terms in the form of

$$\gamma(r) = \left(\frac{\xi}{r}\right)^{\nu} \exp\left\{-\left(\frac{r}{\xi}\right)^{a}\right\} \tag{8.89}$$

where ν and a are the parameters specifying the mode of density correlation in terms of the correlation length ξ. The first term in the right side of Eq. 8.89 is characteristic of the cut-off effect and the second exponential term specifies the density decay profile. $\nu < 2$ since the distance distribution function should converge to zero at $r \to 0$ (see Eq. 8.9). Some of the correlation functions and the corresponding scattering profiles are shown in Fig. 8.7 for 5 sets of ν and a.

Here the scale factor ν determines the density correlation in the short distance ($r \leq \xi$) and the scale factor a describes the

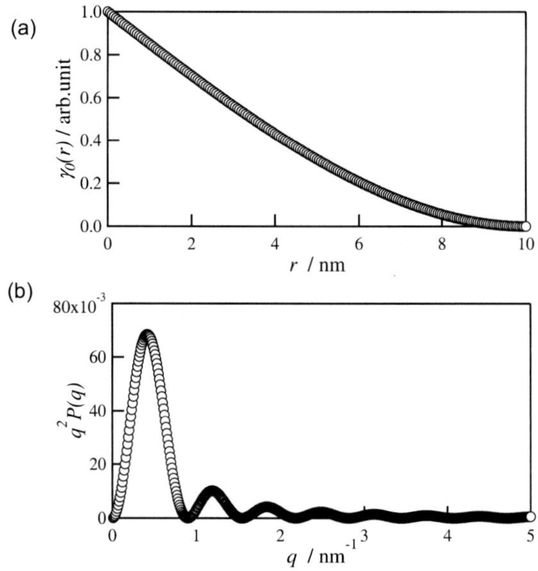

Figure 8.6A Correlation function (a), and a corresponding scattering profile in terms of Kratky plots (b). A case of a sphere.

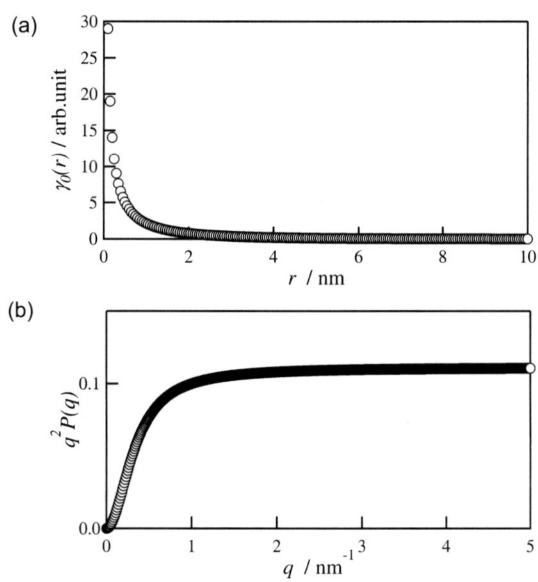

Figure 8.6B Correlation function (a), and a corresponding scattering profile in terms of Kratky plots (b). A case of an Ornstein–Zernicke type.

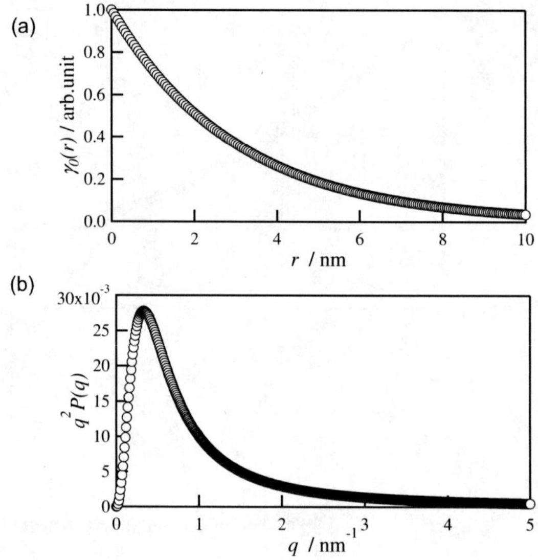

Figure 8.6C Correlation function (a), and a corresponding scattering profile in terms of Kratky plots (b). A case of a Debye–Bueche type.

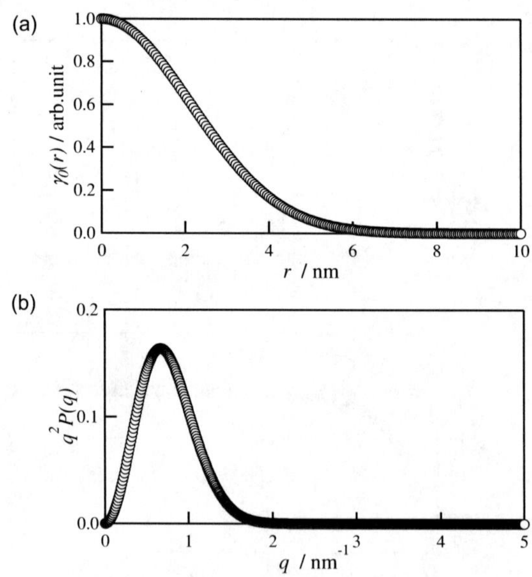

Figure 8.6D Correlation function (a), and a corresponding scattering profile in terms of Kratky plots (b). A case of a Guinier type.

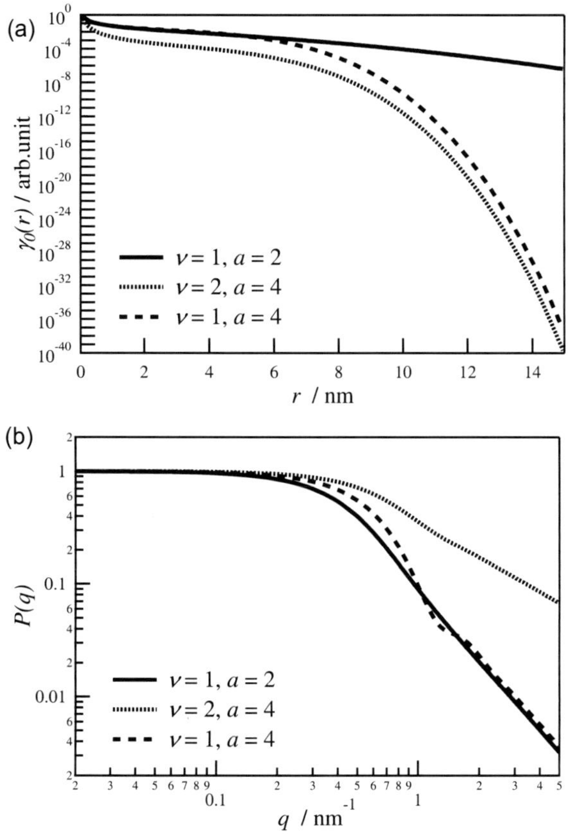

Figure 8.7 Generalized correlation function (the parameters are listed in the figure) (a), and a corresponding scattering profile in terms of the double logarithmic plots (b).

correlation at $r > \xi$. Here we have considered a blob which is defined as a unit of correlated density fluctuation. The system is composed of multipule domains which behave independently of each other. No thermodynamic interaction will be exerted between two domains, and we can treat the system as a random assembly of a number of domains (blobs). If the system is inhomogeneous and is composed of several types of domains (blobs) specified by different correlation lengths, the total scattering profile can be expressed by the linear combination of the scattering profiles given by the different correlation lengths and/or the different types of correlation functions. The examples will be shown in the later section where the inhomogeneous gel structure is analysed in terms of the linear combination of two correlation functions.

The generalized density correlation function Eq. 8.89 takes into account the fractal structure as well as the cut-off effect due to a specific density distribution by two parameters ν and a. As seen from Table 8.4, Eq. 8.89 covers the correlation functions listed in Table 8.4 by varying the parameters ν and a. See Fig. 8.7, which demonstrates the correlation function and its corresponding scattering profile by varying two parameters.

8.2.5 *Interference Effect*

So far the discussion is restricted to the scattering profile from the individual domains (particles) where no interdomain (interparticular) interaction is appreciable. If the interdomain (interparticular) interaction becomes stronger, the correlation between domains (particles) appears on the scattering profile in the form of a sharp peak at the q range corresponding to the interaction distance. The change of the scattering profile due to interparticular interaction is ascribed to that of the state from gas to liquid (see Fig. 8.8).

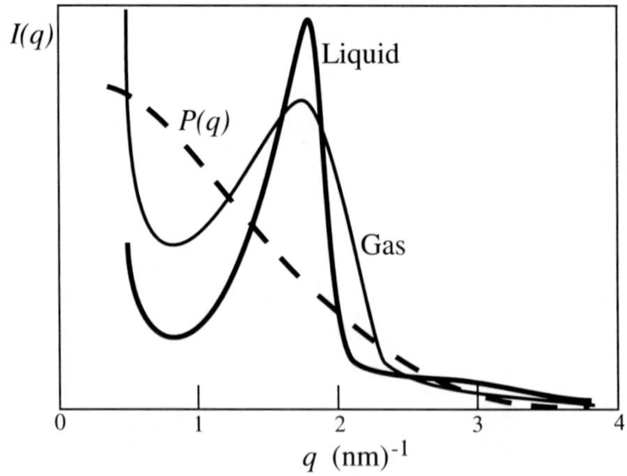

Figure 8.8 Gas-type and liquid-type scattering profiles. A broken line represents a scattering profile free from interparticular interaction.

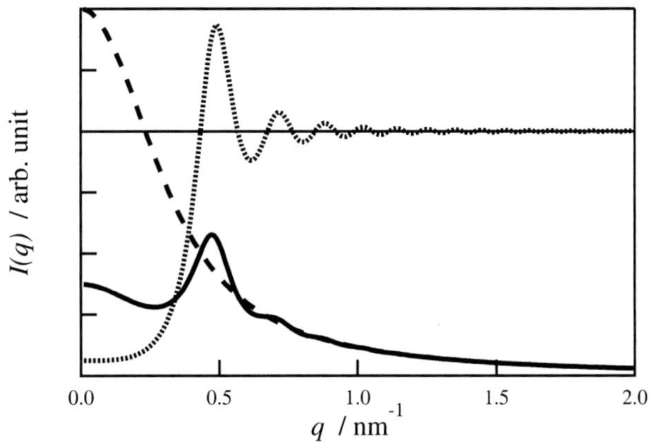

Figure 8.9 Interference effect on scattering profile. A broken line represents a scattering profile from an isolated particle, a dotted line a rigid-sphere-like interference term, and a solid line a scattering profile from the interacting particles where a hard-sphere repulsion is assumed.

When the interparticular (or interdomain) interaction is isotropic and approximately modeled by the hard-sphere repulsion, the scattering profile is decomposed into the product of two terms $P(q)$ and $S_I(q)$ due to the particle scattering factor and the interference (Fig. 8.9), respectively, as [5]

$$I(q) \approx P(q) \cdot S_I(q) \tag{8.90}$$

where the interference term is written as

$$S_I(q) \approx \frac{1}{1 - (2\pi)^{3/2}(\varepsilon/v_1)\beta(q)} \tag{8.91}$$

with ε and v_1 being a constant close to a unity and the average volume allocated to each solute particle in space, respectively. $\beta(q)$ represents the interaction potential in the Fourier space:

$$q\beta(q) = \frac{2}{\sqrt{2\pi}} \int_0^\infty r\alpha(r)\sin(qr) \cdot \mathrm{d}r \tag{8.92}$$

$$\alpha(r) = \exp(-\Phi(r)/kT) - 1 \tag{8.93}$$

The potential function $\Phi(r)$ is given for the hard-sphere repulsion by the scattering amplitude of a rigid sphere of the radius R

$$\Phi(r) = \frac{\sin qr - qr\cos qr}{3\,(qR)^3} \tag{8.94}$$

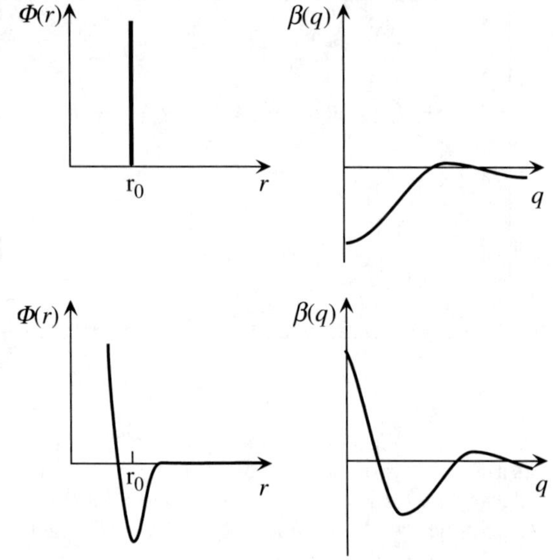

Figure 8.10 Schematic relation of the potential function in the real space $\Phi(r)$ and the potential function in the Fourier space $\beta(q)$. A hard-sphere potentials in the real and a Fourier space are shown in top, and a Lennard–Jones potential in bottom.

and then Eq. 8.94 reduces to [5]

$$S_I(q) \approx \frac{1}{1 + 8(v_0/v_1)\varepsilon\Phi(2qR)} \tag{8.95}$$

where v_0 denotes the volume of the sphere. Equation 8.95 indicates that the hard-sphere interaction is represented by the sphere of a uniform radius of $2R$, but it is often found that the radius of the interacting particles (or domains) is not necessarily equal to R. The potential function $\Phi(r)$ is schematically related to the $\beta(q)$ function as shown in Fig. 8.10, where two types of potential functions (hard-sphere repulsion and Lennard–Jones) yield characteristic $\beta(q)$ functions, respectively.

The scattering profile cannot be expressed in terms of the product of two independent contributions of the particle scattering and interference in the case that the interparticular interaction is non-isotropic. In general the interference depends on the particle scattering factor. However, in solutions and even in gels, we can assume that particles or domains move freely with almost no restriction. That is, the molecular motion in this state is fast enough to cancel the anisotropic effect of interaction. Considering the size distribution and the average effect of particular motion, the interparticular interaction cannot be represented by a single hard sphere, but an average over size distribution and mode of motion should be taken into account in the particular interaction. Let assume the average effect is taken into account by a random process. Then the function $\beta(q)$ is approximately given by the Gaussian function as

$$\beta(q) \approx -\exp(-\xi^2 q^2) \tag{8.96}$$

and Eq. 8.91 is rewritten as

$$S_{\mathrm{I}}(q) = \frac{1}{1 + 2A_2 M_{\mathrm{w}} c \exp(-\xi^2 q^2)} \tag{8.97}$$

where the Gaussian type interaction potential is characterized by the correlation length ξ of interaction. The volume factor $8(v/v_1)\varepsilon$ in Eq. 8.95 is replaced with the second virial coefficient A_2 by adjusting the second term in the virial expansion of Eq. 8.90.

In more general treatment of interaction, the attractive force should be taken into account in addition to the repulsive force. A typical example is a Lennard–Jones potential where the interaction potential is expressed in terms of the inverse power series of the distance r as [14]

$$\Phi(r) = 4\varepsilon\left[(\sigma/r)^{12} - (\sigma/r)^6\right] \tag{8.98}$$

The first term $(\sigma/r)^{12}$ and the second term $(\sigma/r)^6$ in Eq. 8.98 represent the repulsion and attraction, respectively, and the parameter ε specifies the depth of the minimum at $r_{\mathrm{e}} = 2^{1/6}\sigma$ in the potential function. Applying a similar argument that the ineraction potential is smeared by the size distribution and thermal fluctuation, we may approximately represent the attraction term by a stretched exponential (Kohlrausch–Williams–Watts function) $\exp(-\xi^a q^a)$ [15] where the parameter a specifies the mean correlation length of attraction ξ_{att} as

$$\xi_{\mathrm{att}} = \xi a/\Gamma(1/a) \tag{8.99}$$

with Γ being the gamma function. Thus Eq. 8.99 is extended to include the attractive force as

$$\beta(q) \approx -\left[\exp(-\xi^2 q^2) - C \cdot \exp(-\xi^a q^a)\right] \tag{8.100}$$

which yields the interference

$$S_{\mathrm{I}}(q) = \frac{1}{1 + 2A_2 M_{\mathrm{w}} c \left[\exp(-\xi^2 q^2) - C \cdot \exp(-\xi^a q^a)\right]} \tag{8.101}$$

where the parameter C and a are determined by the mode of the attractive force. The coefficient C and the factor a characterize the shape of $\beta(q)$ as demonstrated in Fig. 8.11. As seen from Fig. 8.11, the potential function $\beta(q)$ reproduces qualitatively a correct interparticular interference.

8.2.6 *Scattering at Zero Angle*

The scattered intensity at zero angle is constituted of the amplitudes of the sum of the waves scattered by all electrons in the particle and should be equal to [5, 6]

$$I(0) = I_{\mathrm{e}}(0) n^2 \overline{N^2} \tag{8.102}$$

where $I_{\mathrm{e}}(0)$, n and $\overline{N^2}$ denote the scattered intensity at zero angle by one electron, the total number of electrons in the particle and the mean-square average of the number of particles, respectively. The intensity at the angle θ scattered by one electron is given by the Thomson formula as

$$I_{\mathrm{e}}(q) = 7.90 \times 10^{-26} \cdot \frac{I_0}{a^2} \cdot \frac{1 + \cos^2\theta}{2} \tag{8.103}$$

where I_0 is the intensity of the incident beam, a the distance from the object to the point of detection. The numerical factor (the Thomson factor in cm^2) is equal to the square of electron radius (e^2/mc^2).

When extrapolated to zero angle from finite angles, the scattered intensity at zero angle would not reduce to Eq. 8.102 but to

$$I(q \to 0) = I_{\mathrm{e}}(0) n^2 (\overline{N^2} - \overline{N}^2) \tag{8.104}$$

Comparing Eq. 8.102 with Eq. 8.103, the intensity $I(0)$ at $q = 0$ is of the order of $\overline{N^2}$ while the limiting value $I(q \to 0)$ is of the order of $\overline{N^2} - \overline{N}^2$. A real scattered intensity thus has a singularity (a sharp peak) in the neighborhood of $q = 0$, but the width of

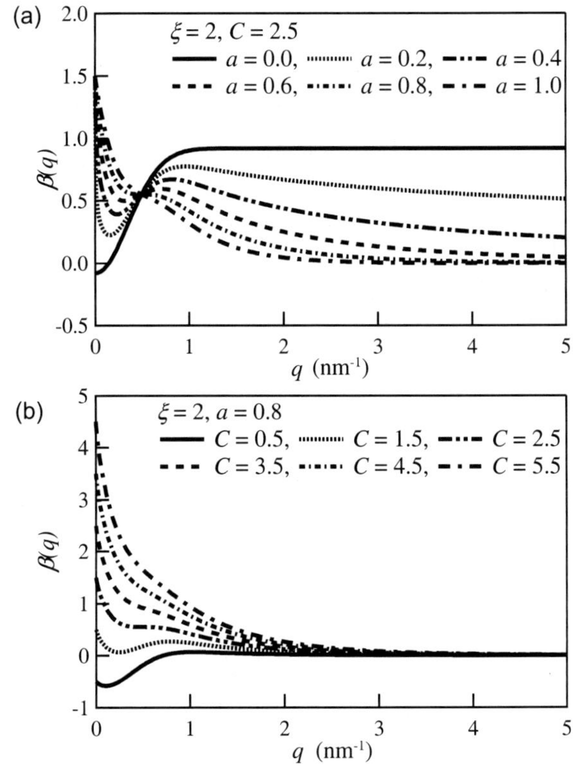

Figure 8.11 The potential function in the Fourier space with varying the parameters C and a (see Eq. 101). The case of $C = 0$ corresponds to the state of no attraction represented by Eq. 95.

this singularity is by several orders of magnitude smaller than the minimum observable angle. In practice, the extrapolation of the scattered intensity to zero angle yields Eq. 8.104, and we never observe the scattered intensity at zero angle (Eq. 8.102) even with the most perfect experimental system.

We consider the simplest case where the particles possessing an isopotential specific volume v_2 and z_2 mole electrons per gram are dispersed in a homogeneous solution and no interparticular interaction affects the scattering. That is, one particle contains $z_2 \cdot M_w$ electrons if the molecular weight of the particle is given by M_w. Accordingly the number of effective mole electrons per gram is given by

$$\Delta z = (z_2 - v_2 \cdot \rho_0) \tag{8.105}$$

where ρ is the mean electron density of the solvent. In the limiting case where the particles are ideally separated and dispersed in a complete random fashion,

$$\overline{N^2} - \overline{N}^2 = \overline{N} \tag{8.106}$$

and the average number of particles is given in terms of the concentration c (g/cm^3) and the molecular weight M_w as

$$\overline{N} = N_A \cdot M_w \cdot c \cdot d \tag{8.107}$$

for the sample of the thickness d. (N_A denotes Avogadro number.) Introducing Eqs. 8.106 and 8.107 into Eq. 8.104, the molecular weight can be determined from the scattered intensity extrapolated to zero angle as [16]

$$M_w = \frac{I(q \to 0)}{I_0} \cdot \frac{21.0 \cdot a^2}{\Delta z^2 \cdot d \cdot c} \tag{8.108}$$

In this system, the scattered intensity at an arbitrary q is expressed in the absolute scale as

$$I(q) = \frac{7.90 \times 10^{-26} \cdot \Delta z^2 \cdot M_w \cdot d \cdot N_A \cdot c \cdot I_0}{a^2} \cdot P(q) \tag{8.109}$$

since $\theta << 1$ and $I_e(q)$ is regarded as a constant value. Here $P(q)$ denotes the particle scattering factor defined as a normalized scattered intensity with respect to the scattered intensity at the zero angle.

By applying a similar argument to rod-like particles and flat particles, the mass per unit length M_c and the mass of unit area M_t can be evaluated from the scattered intensity extrapolated to zero angle. Using Eqs. 8.34 for a cylinder and 8.36 for a disk, the scattered intensity Eq. 8.109 can be rewritten, respectively, as

$$q \cdot I(q) = \frac{7.90 \times 10^{-26} \pi \cdot \Delta z^2 \cdot (M_w/2H) \cdot d \cdot N_A \cdot c \cdot I_0}{a^2} \cdot P_c(q) \tag{8.110}$$

$$q^2 \cdot I(q) = \frac{7.90 \times 10^{-26} (2\pi) \cdot \Delta z^2 \cdot (M_w/A) \cdot d \cdot N_A \cdot c \cdot I_0}{a^2} \cdot P_t(q) \tag{8.111}$$

where the particle scattering factors $P_c(q)$ for the cross section and $P_t(q)$ for the thickness are given by the Guinier approximation Eqs. 8.35 and 8.37 in terms of the cross-sectional radius of gyration R_c and the thickness radius of gyration R_t, respectively. In the limit of $q \to 0$, Eqs. 8.110 and 8.111 yield the mass per unit length and the mass per unit area as

$$M_c \equiv \frac{M_w}{2H} = \frac{\lim_{q\to 0} q \cdot I(q)}{I_0} \cdot \frac{6.68a^2}{\Delta z^2 \cdot d \cdot c} \tag{8.112}$$

$$M_t \equiv \frac{M_w}{A} = \frac{\lim_{q\to 0} q^2 \cdot I(q)}{I_0} \cdot \frac{3.34a^2}{\Delta z^2 \cdot d \cdot c} \tag{8.113}$$

That is, the extrapolation of the Guinier plots to $q \to 0$ yields the quantity proportional to the corresponding mass.

8.3 Application to the Conformational Analysis of Polysaccharides

Oligo- and polysaccharides are biopolymers found commonly in living organism, and are known to reveal the physiological functions by forming a specific conformation. Although a monosaccharide unit is common to many polysaccharides, its linkage mode varies and characteristic functions/properties will appear accordingly. A good example is demonstrated by water-soluble digestible amylose and water-non-soluble non-digestible cellulose, both composed of D-glucose units but by a different mode of linkages, i.e., $(1 \to 4)-\alpha-$ and $(1 \to 4)-\beta$-linkages, respectively. When cellulose is chemically modified by introducing alkyl groups, resulted cellulose derivative becomes soluble in water and forms gel upon heating.

The conformational analysis of polysaccharide chains involves two aspects: (i) the characterization of a single chain conformation and (ii) the analysis of the chain assembly (suprastructure) of polysaccharides. A single chain conformation of polysaccharides is determined primarily by the chemical structure specified by the types of sugar residues, sugar linkages and side groups. For example, the $(1 \to 4)-\alpha-$linkage (amylose) and the $(1 \to 4)-\beta-$linkage (cellulose) of D-glucosidic residues yield a wobbled helix and a

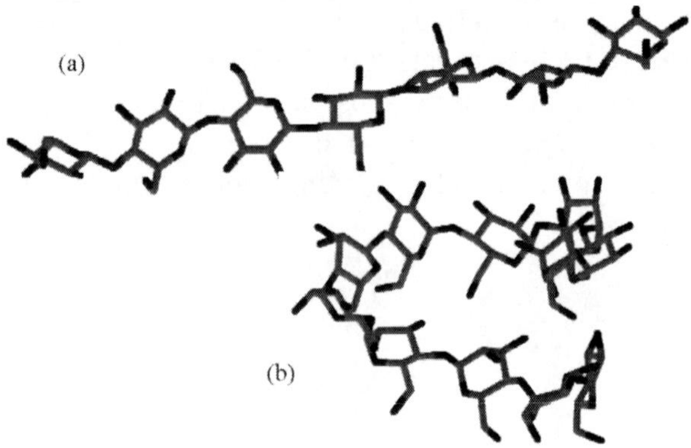

Figure 8.12 Stretched zigzag conformation (a) and wobbled helical conformation (b), representing the basic conformations of cellulose and amylose, respectively.

stretched zigzag chain, respectively, by jointing the D-glucosidic residues in a simple manner so as to place the chain on a plane [17] (see Fig. 8.12). Those single chain conformations account in some extent for the formation of suprastructures such as the complexing capability of amylose and the fringed micelle formation of cellulose.

Unlike cellulose and amylose, most of polysaccharides have no regular homopolymeric structure, where the regularity is interrupted by the random intrusion of different types of linkage and/or sugar units. The introduction of such an irregularity hampers crystallization and promotes the formation of suprastructure which is characteristic of the polysaccharide species. The interchain interaction of polysaccharides seems to be specific as exemplified by the suprastructure depending on the chemical structure and counterions (in the case of the polysaccharides possessing carboxyl or sulfate groups). The formation of the suprastructure results often gelation.

This chapter introduces the application of the SAXS method in the structural analysis of oligo- and polysaccharide chains in solution. The complexity in characterizing polysaccharide chain conformation is due to the fact that the interchain interaction of polysaccharides is so specific that polysaccharide chains are seldom dispersed in solvent as a single chain. Thus a first task to understand the structure and function relationship of polysaccharides is to evaluate the intrinsic chain (single chain) characteristics free from interchain interaction. Once the intrinsic chain conformation is specified, the interchain interaction can be analyzed in terms of the mode of suprastructure composed of several polysaccharide chains.

8.3.1 *General Strategy*

The structure of polysaccharides in aqueous solution is neither completely ordered nor disordered. Polysaccharide is generally regarded as a semi-flexible polymer chain, and its conformation has been analyzed in terms of a semi-flexible chain model such as a helical wormlike chain [18] or a broken-rod chain [19]. Although the semi-flexible polymer chain model presents a consistent view over the dilute solution behavior of polysaccharide chians, the model is unable to provide information on precise local structure, which is considered to be a key factor for understanding the mechanism of the physiological function of polysaccharide chains. Many polysaccharide chins are constituted of a series of locally ordered structures linked probably by disordered chain, and assume wormlike chian conformation as a whole. The distance order of the localy ordered region is around 3 to 20 nm, and is covered by the observation range of small-angle X-ray scattering (SAXS). Light scattering (LS) is often employed for the investigation of the chain conformation of polymer chains in solution. Since the wavelength of visible light is from 350 to 700 nm, LS is suitable to observe the distance order from 50 to 200 nm. Thus a combination of LS and SAXS will yield the the structural information of polysaccharide in solution over the range from the quasi-atomic level to a whole chain conformation, although there is a missing gap between 20 and 50 nm. The scattering profile contains in principle all structural information, but the conventional analysis of the scattering profile allows only to evaluate the statistical averages of the chain size specified by the radius of gyration and/or the cross-sectional/thickness radius of gyratrion (see Eqs. 8.33, 8.35, and 8.37). In order to explore the information borne in the scattering profile fully, molecular models are generated by the Monte Carlo method or molecular mechanics/molecular dynamics, and the scattering profile is directly calculated from the ensemble of the generated molecular models to compare with the observed scattering profile. If the calculated and observed scattering profiles are identical, we may expect the generated chains reflecting the structure of a real chain, and visualize the chain structure in solution from the computer-generated chains.

The particle scattering from a single molecule is in principle calculated from the coordinates of the constituent atoms according to the Debye formula (see Eq. 8.7)

$$I(q) = \sum_{i=1}^{n} g_i \phi_i^2(q) + 2\sum_{i=1}^{n-1}\sum_{j=i+1}^{n} g_i g_j \phi_i(q)\phi_j(q) \cdot \frac{\sin d_{ij}q}{d_{ij}q} \quad (8.114)$$

where q denotes the magnitude of the scattering vector given by $(4\pi/\lambda)\sin(\theta/2)$ with λ and θ being the wavelength of the incident beam and scattering angle, respectively, and g_i is an atomic scattering factor. d_{ij} is the distance between the ith and jth atoms, and the form factor for a single atom $\phi_i(q)$ is assumed to be given by the form factor for a sphere with a van der Waals radius of the ith atom

$$\phi_i = \frac{3\left[\sin(R_i q) - (R_i q)\cos(R_i q)\right]}{(R_i q)^3} \quad (8.115)$$

with R_i being the van der Waals radius of the ith atom (see also Eq. 8.40).

If a molecule is rigid, the distance d_{ij} is fixed and Eq. 8.114 affords directly the particle scattering factor of such a molecule which moves freely in space. If a molecule (for example, a flexible polymer molecule) has a large internal freedom, the distance d_{ij} fluctuates with time due to the internal motion of such a molecule. In this case, the particle scattering factor should be calculated as an average over a statistical ensemble generated by the Monte Carlo procedure [20, 21] according to the conditional bond conformation probability [22].

The computer simulation by molecular dynamics [23] is an alternative way to generate molecular models. Molecular dynamics allow to simulate multiple chains with interchain interaction in solution, but the simulation of a single chain needs a considerable computation time. Thus in practice the particle scattering factor is calculated from a single molecule simulated by the molecular dynamics. If the system is supposed to be ergodic, the space average should be equivalent to the time average. Since the radius of gyration of a model molecule simulated by the molecular dynamics hardly changes for the last 1000 steps, the conformation of a model molecule is regarded as almost identical to the time-averaged conformation, and in turn is regarded as being equivalent to that averaged over a statistical ensemble. The consistency of this procedure has been confirmed for several systems [24, 25]. The observed scattering profile should be modified by taking into account the concentration-dependent factor and/or the interference effect.

Since the scattering profile is decomposed into the product of two terms of the particle scattering factor $P(q)$ and the interference S_I (q) (see Eq. 8.90), the interference term can be evaluated from the observed scattering profile dividing by the particle scattering factor which can be calculated from the molecular model or be reduced by extrapolating the observed scattering profile to the zero concentration. The correlation length of interaction is then evaluated according to Eq. 8.97 or 8.101.

8.3.2 *Conformation of Single Poly-*$(1 \rightarrow 4)$*-*α*-*D*-glucan and Poly-*$(1 \rightarrow 4)$*-*β*-*D*-glucan*

Poly-$(1 \rightarrow 4)$-α-D-glucan and poly-$(1 \rightarrow 4)$-β-D-glucan are most abundantly available natural polymers, as represented by amylose and cellulose, respectively. Both polyglucans are constituted of simple $(1 \rightarrow 4)$ linked glucopyranose units, but differ in the mode of glucosidic linkage. This results in the different potential energy map, which shows the 2-D countor potential energy as a function of two dihedral angles ϕ and ψ defined as the rotational angles of two successive glucopyranose unit planes with respect to the glycosidic bond (C1-O-C4 in the case of amylose or cellulose) in a disaccharide. When the glucose residue is assumed to be rigid and the bond angle τ formed by two virtual bonds connecting the neighboring oxygene atoms of the glycosidic linkage is fixed (110° in the case of amylose and cellulose), the polymer chains are generated by the Monte Carlo method according to the occurrence probability $P(\phi, \psi)$

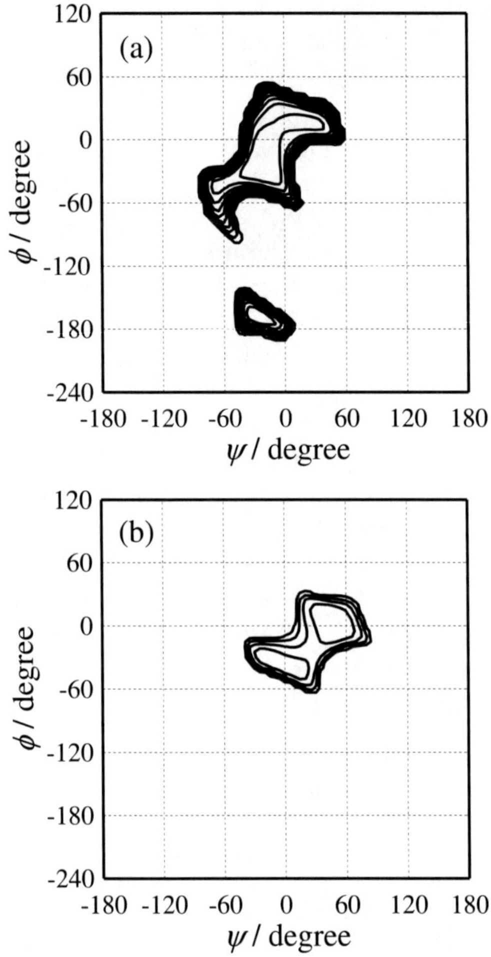

Figure 8.13 2-D contour energy map (the potential energy as a function of two dihedral angles ϕ and ψ) for (a) maltose (Glcpα1$\rightarrow$4Glc) and (b) cellobiose (Glcpβ1 $\rightarrow$ 4Glc).

(the Boltzmann factor) given by the potential energy $E(\phi, \psi)$ of a disaccharide with a set of ϕ and ψ as

$$P(\phi, \psi) = C \exp[-E(\phi, \psi)/k_B T] \tag{8.116}$$

Here k_B is the Boltzmann constant, T an ablsolute temperature, and C the normalization constant.

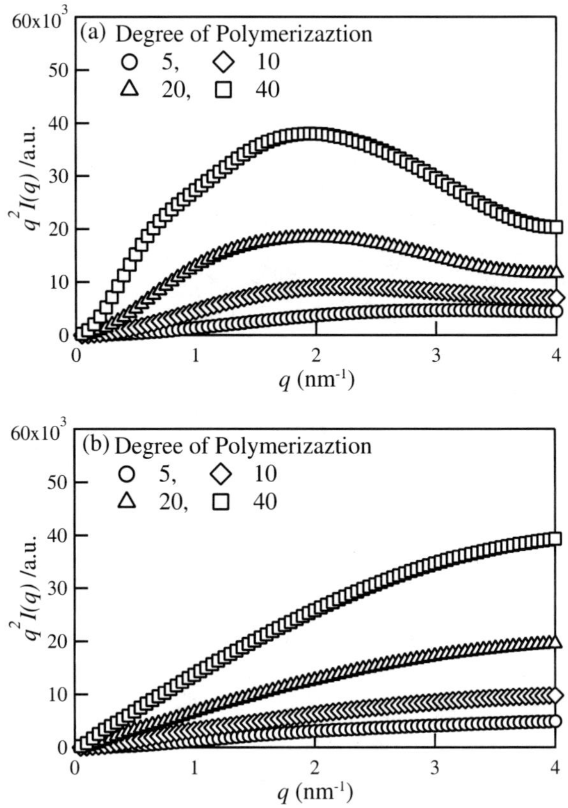

Figure 8.14 Simulated scattering profiles of (a) 1,4-α-D-glucan (amylose) and (b) 1,4-β-D-glucan (cellulose). The number of glucosidic residues (DP) is indicated in the figure.

A chain is generated step by step, and its geometry is determined by a set of the dihedral angles ϕ and ψ and the virtual bond angle τ. A set of the dihedral angles ϕ and ψ is determined by the Boltzmann factor in each step. Scattering profiles calculated by Eq. 8.114 from the atomic coordinates of the Monte Carlo generated chains are shown in Fig. 8.14 for $(1 \rightarrow 4)$-α-D-glucan and $(1 \rightarrow 4)$-β-D-glucan as a function of the number of glucosidic residues, where the scattering profiles are averaged over 500 independently generated chains. The plots in Fig. 8.14 represent the Kratky plots plotted $q^2 I(q)$ against q. The simulated $(1 \rightarrow 4)$-α-D-glucan chain (amylose)

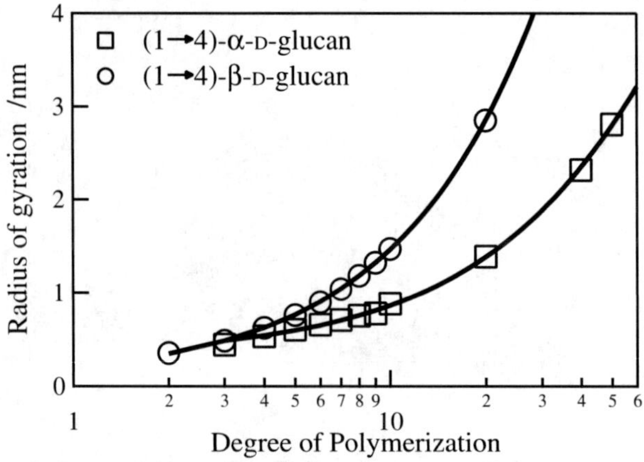

Figure 8.15 Radius of gyration of 1,4-α-D-glucan and 1,4-β-D-glucan oligomers.

reveals the wobbled helical conformation with localized highly ordered helical region and its scattering profile has a characteristic peak due to a helical structure at $q = 2$ nm, whereas the $(1 \rightarrow 4)$-α-D-glucan chain (cellulose) is rather extended and its scattering profile exhibits typical rigid rod-like characteristics. The cross-sectional radius of gyration is evaluated from respective cross-sectional Guinier plots (Eq. 8.34) as 0.5 nm for $(1 \rightarrow 4)$-α-D-glucan chain of over 20 glucosidic residues, and 0.21 nm for $(1 \rightarrow 4)$-β-D-glucan chains. The larger cross-sectional radius of gyration in $(1 \rightarrow 4)$-α-D-glucan chain is due to the quasi-helical conformation stabilized by intramolecular hydrogen bonding. The radius of gyration increases with increasing the number of residues in both polyglucans as shown in Fig. 8.15, but its number dependence reflects the conformational difference.

The $(1 \rightarrow 4)$-α-D-glucan and $(1 \rightarrow 4)$-β-D-glucan oligomers dissolve in water, and the small-angle X-ray scattering (SAXS) was observed from their aqueous solutions. As seen from Figs. 8.16 and 8.17, no effect of association was observed from the SAXS profiles of maltohexaose and maltooctaose ($(1 \rightarrow 4)$-α-D-glucan oligomers), whereas cellopentaose ($(1 \rightarrow 4)$-β-D-glucan oligomer) forms large aggregates in its aqueous solution and its SAXS profile exhibits a sharp upturn toward lower q in the Kratky plots due to large

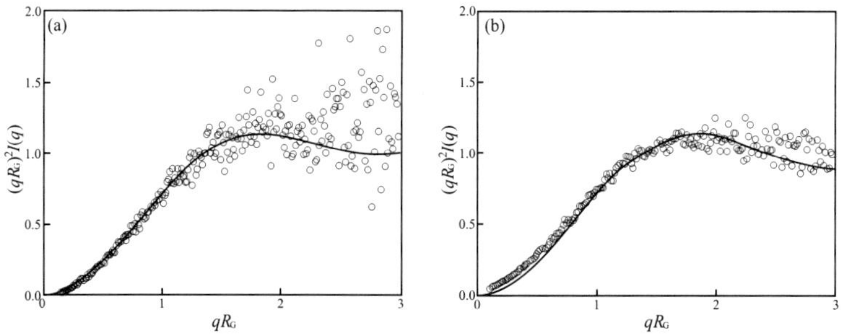

Figure 8.16 Simulated (solid lines) and observed (circles) SAXS profiles from maltohexaose (a) and maltooctaose (b). The oligomer concentration is 30 mg/mL in both solutions.

aggregates. The SAXS profile of (1 → 4)-α-D-glucan oligomers shows a characteristic peak for quasi-helical conformation in the Kratky plots, and the peak becomes more pronounced in maltooctaose than in maltohexaose (Fig. 8.16). Cellopentaose ((1 → 4)-β-D-glucan oligomer) is not dissolved in water completely because of a strong intermolecular interaction by hydrogen bonding through OH groups on C6. The SAXS profile from the aqueous solution of cellopentaose is characterized by a sharp upturn toward lower q in the Kratky plots (Fig. 8.17a), and deviates significantly from the simulated scattering profile. When added 1 M urea, the intermolecular hydrogen bonds are broken and a good agreement was seen between the observed and simulated SAXS profiles (Fig. 8.17b). The cross-sectional radius of gyration is evaluated as 0.21 nm from the cross-sectional Guinier plots (see Fig. 8.17b in the Kratky plots), which corresponds to that of a single (1 → 4)-β-D-glucan chain. The SAXS profile is typical of a stiff chain. The cross-sectional radius of gyration of cellopentaose is estimated as 0.35 nm at the intermediate q range, and as over 7 nm at the smaller q range (see Fig. 8.17a). Comparing with the cross-sectional radius of gyration 0.21 nm for a single chain, two cellopentaose chains seem to form stable aggregates by stacking side by side, and some aggregate further into larger clusters.

The molecular dynamics simulation was performed on maltopentaose with currently available force fields [24]. The results are shown in Fig. 8.18, together with the Monte Carlo simulation (see above) and a regular helix model [24–28]. The scattering profiles are

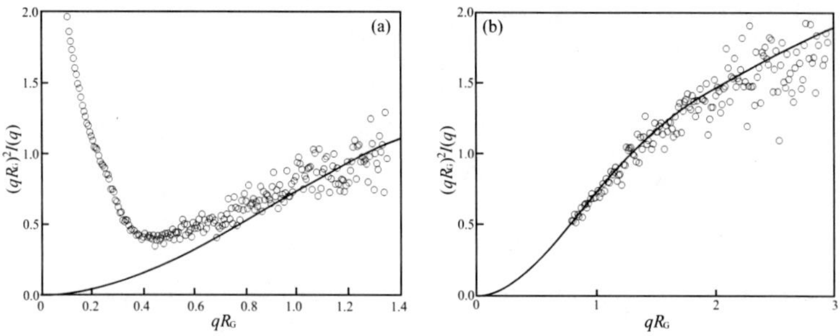

Figure 8.17 Simulated (solid lines) and observed (circles) SAXS profiles from cellopentaose in water (a) and in 1 M urea aqueous solution (b). The cellopentaose concentration is 30 mg/mL in both cases.

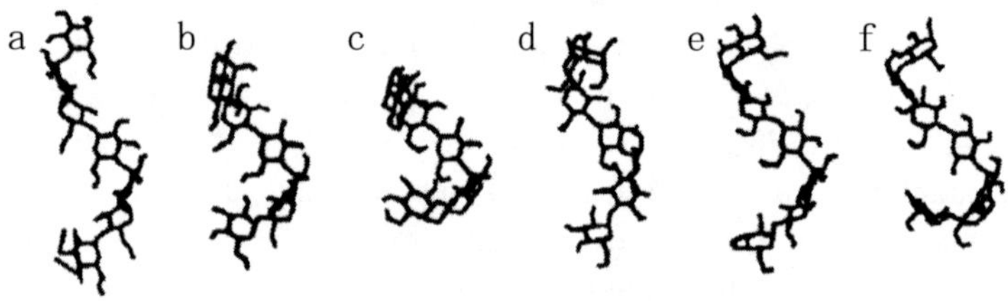

Figure 8.18 Snapshot conformations of maltopentaose. (a) Regular helix 0.83 nm), (b) regular helix (0.74 nm), (c) regular helix (0.59 nm), (d) Monte Carlo/MM3 (0.757 nm), (e) Glycam99 (0.808 nm), (f) modified Glycam93 (0.772 nm). The value in each bracket denotes the radius of gyration R_G calculated from respective models, and R_G was evaluated as 0.74 ± 0.002 nm from the observed SAXS profile.

calculated directly from the atomic coordinates of those models, and are compared with the SAXS profiles observed from maltopentaose in aqueous solution (Fig. 8.16). Figure 8.18 shows the scattering profiles calculated from the atomic coordinates of the molecular models proposed as regular helices or simulated by the Monte Carlo method. The molecular models are also simulated with available force field. Here the Monte Carlo simulations were performed with two probability maps (the probability map Fig. 8.13a and that calculated by MM3), and the MD (molecular dynamics) with various force fields including Glycam93, Glycam99, modified Glycam93, Cff91, Cff, and Gromos. Both Monte Carlo results are consistent with observed SAXS profiles, where a minor difference due to the glucose geometry was observed at higher q. The results of MD simulation vary with the force fields, and the Cff91 seems to yield the best fit to an observed profile. The helix model by Goldsmith et al. (Fig. 8.18b) [28] exhibits a scattering profile fitting satisfactorily to the observed SAXS profile. In fact the molecular model simulated by Cff91 is almost identical to the helix model by Goldsmith et. al. That is, amylose in aqueous solution may assume at least locally a quasi-helical conformation specified by a radius of 0.538 nm, a rise of 0.244 nm, a pitch of 1.760 nm, a repeat of 0.72 nm, $\phi = 105^\circ$ and $\psi = -135^\circ$.

8.3.3 *Conformation of Poly-*$(1 \to 4)$*-*β*-*D*-glucan with Side Chains*

Poly-$(1 \to 4)$-β-D-glucan (cellulose) is not water-soluble because of numerous intermolecular hydrogen bond. As discussed above, a rather stretched chain conformation promotes this strong intermolecular hydrogen bonding. Cellulose can be dissolved in water when cellulose is modified chemically by acetylation or carboxylation of OH groups. Introduction of bulky side chains breaks intermolecular hydrogen bonds, and modified cellulose dissolves in water. Most of those cellulose derivatives form lyotropic liquid crystal due to the semi-rigid nature of $(1 \to 4)$-β-D-glucan backbone.

Branched$(1 \to 4)$-β-D-glucan is also abundantly found in nature. Xyloglucan, for example, is normally contained in plant seed, and its flour has been traditionally used as a food additive in everyday life. Xyloglucan is a general term applied to non-starch plant polysaccharides composed of a $(1 \to 4)$-β-D-glucan spine (cellulose spine) with $(1{\to}6)$-α-branched xylose, which is partially substituted by $(1{\to}2)$-β-galacto-xylose [29]. Xyloglucan consists of four types of monomer units represented as XXXG (composed of total 7 sugar residues), XXLG (composed of 8 sugar residues), XLXG (composed of 8 sugar residues) and XLLG (composed of 9 sugar residues) (Fig. 8.19). Monomer units are made of a sequence of four $(1 \to 4)$-β-D-glucans but differ in the number of galactose side chains [29]. Tamarind seed xyloglucan (TPS; *Tamarindus indica*)

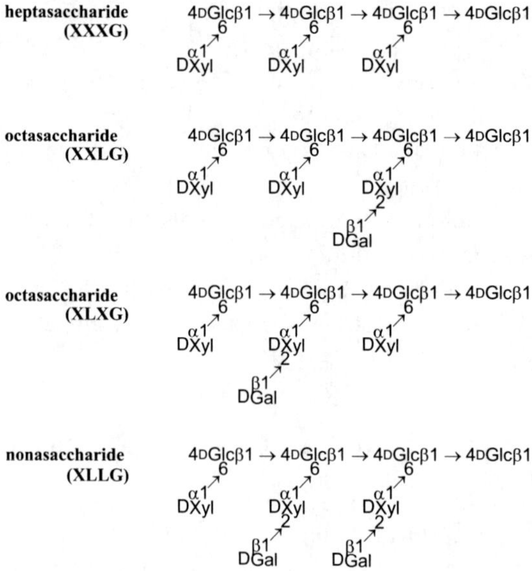

Figure 8.19 Four types of monomer units of xyloglucan.

and detarium seed xyloglucan (*Detarium senegalense* Gmelin) are typical xyloglucans, extracted from respective plant seeds, but the composition of four monomer units is slightly different in each species. Xyloglucan dissolves in water, and yields high viscous solution. However, its gelation is prevented by the steric hindrance and hydrophilicity of (1→2)-β-galacto-xylose branches.

The conformations of four monomer units of xyloglucan are simulated by the molecular dynamics, and the scattering profiles are calculated according to Eq. 8.114 from their atomic coordinates. The small-angle X-ray scattering was observed from the aqueous solutions of those monomer units, where xyloglucan octasaccharide unit contains two isomers of XXLG and XLXG. The observed and calculated scattering profiles are compared in Fig. 8.20, where a reasonable agreement was observed. The snapshots of simulated xyloglucan monomers show rather flat zigzag conformation of (1 → 4)-β-D-glucan spine with xylose and galacto-xylose branches extending and folding upright on the (1 → 4)-β-D-glucan flat surface [30]. A similar conformation of galacto-xylose side chains

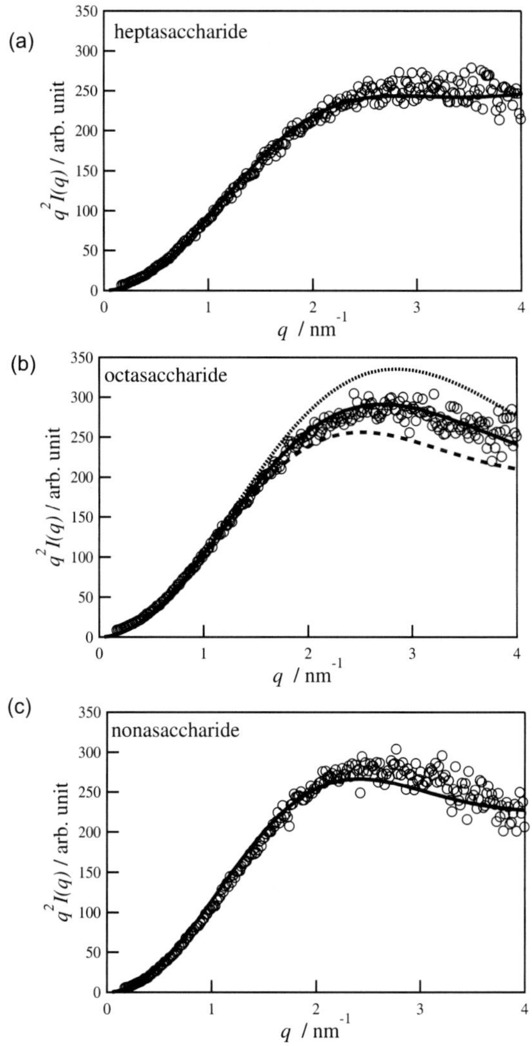

Figure 8.20 Small-angle X-ray scattering observed (circles) and calculated (solid lines) from four types of TSP monomer units. Since octasaccharide has two isomers (XLXG and XLXG), the scattering profiles from XXLG (dotted line) and XLXG (broken line) are combined linearly to the scattering profile from octasaccharide (solid line) [30].

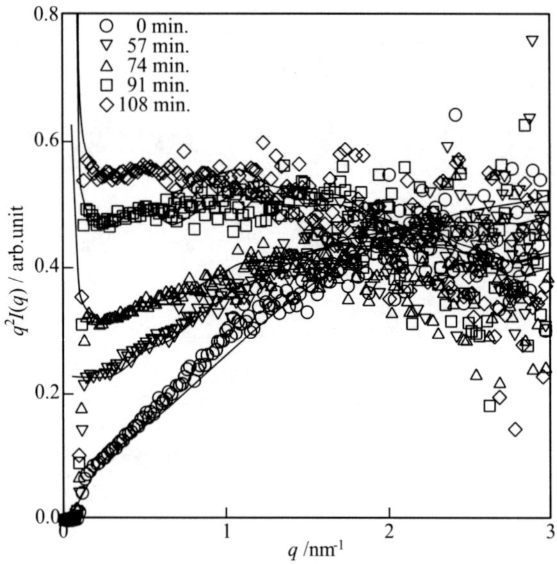

Figure 8.21 Observed and calculated scattering profiles at various reaction times. Symbols represent the observed SAXS intensities and solid lines are calculated by Eq. 8.117 where the scattering amplitude $A^2(q)$ is assumed to be given by the scattering factor of the domain composed of 14 aligned xyloglucan chains composed of 30 $(1 \rightarrow 4)$-β-D-glucans with 30 $(1 \rightarrow 6)$-α-xylose branches.

is observed by Levy at al. [31] from the simulation by assuming a fixed flat (cellulose-like) or twisted (cellobiose-like) backbone conformation. The shape of those monomers are approximately represented by flat ellipsoids, where the shortest (0.22 nm) and longest (1.45 nm) semi-axes do not vary but the width of the cross-section increases from 0.62 nm (heptasaccharide) to 0.71 nm (octasaccharide) and 0.75 nm (nonasaccharide) as the number of residues increases. This result also confirms the side chains protruding horizontally from the flat $(1 \rightarrow 4)$-β-D-glucan spine surface.

The removal of galactose side chains results in the increase of hydrophobicity and the reduction of steric hindrance which prevent further association of xyloglucan into a large cluster. In fact, the enzymatic degradation by β-galactosidase results in gelation of xyloglucan aqueous solution [32]. At room temperature, 1% TPS aqueous solution, for example, forms opaque gel at about 45% release of galactose residues. The release of galactose residues proceeds with reaction time during the enzymatic degradation, and the 45% release of galactose residues in 1% TPS aqueous solution corresponds the reaction time 57 minutes after the enzyme is added. When (1→2)-β-galactose is removed, hydrophobicity increases and steric hindrance reduces. When the part of xyloglucan chain missing the terminal galactose exceeds a certain length, those parts associate laterally and form a quasi-ordered domain composed of laterally arranged xyloglucan chains. The small-angle X-ray scattering was observed from 1% TPS aqueous solution during the course of enzymatic degradation as shown in Fig. 8.21. The conventional analysis of the observed scattering profiles (see Eq. 8.36) for the thickness Guinier plots) indicates the formation of flat objects with 1.1 nm thickness upon gelation [34], which is considered to correspond to the quasi-ordered domain composed of parallel-stacked cellulose-like chains. The molecular model of the quasi-ordered domain is thus made of 14 xyloglucan chains each composed of 40 (1 → 4)-β-D-glucans with 30 (1→6)-α-xylose branches in sequence of XXXG, and the scattering profile is calculated from the atomic coordinates of the model (Fig. 8.22).

The observed SAXS profiles during the course of enzymatic reaction exhibit the characteristics of a rod-like scattering at higher q regions, while the scattering curve at $q \to 0$ shows the tendency to diverge after 57 minutes. This symptom is a typical behavior of the scattering from gelling systems, and is simply explained by assuming the gelation taking place according to the classic Flory–Stockmayer polyfunctional condensation scheme (a tree-like model) as [34]

$$I(q) = A^2(q)(1 + \alpha\phi)/[1 - (f - 1)\alpha\phi]$$
$$\phi = \exp(-b^2q^2/6) \qquad (8.117)$$

Here f denotes the functionality of the cross-linking domain (the number of branches radiating from the domain), α the conversion (the probability that an arbitrary chosen unit is reacted), b^2 the mean-square average of the distance between the neighboring scattering units, and $A^2(q)$ the scattering amplitude of each scattering unit. If the scattering unit is represented by a point,

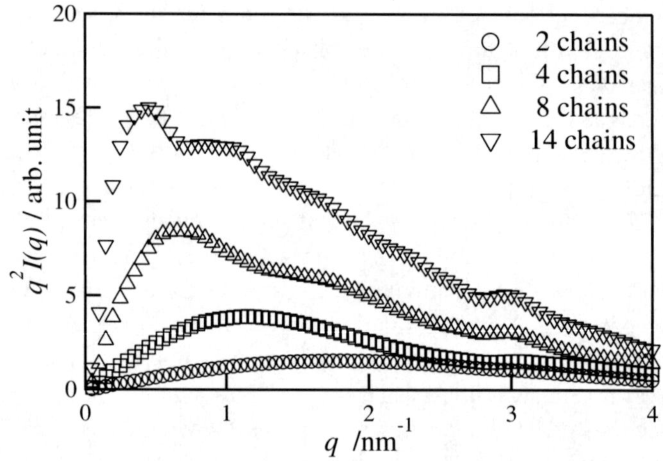

Figure 8.22 Molecular model for xyloglucan single chain and quasi-ordered domains. The scattering profiles are calculated from the model xyloglucan aggregates, which are formed by stretched xyloglucan chains aligned horizontally in a flat layer (Kratky plots).

$A^2(q) = 1$. In the present gelling system, we consider a quasi-ordered domain composed of 14 xyloglucan chains as a new scattering unit. By assuming that the system consists of two phases of shingle chains (a dilute phase) and quasi-ordered domains (a condensed phase) for simplicity, the observed SAXS profiles are analyzed according to Eq. 8.117. The results are summarized in Table 8.5 and Fig. 8.21. The parameter $(f - 1)\alpha$ in Eq. 8.117 and Table 8.5 specifies the average branching degree, and $(f - 1)\alpha = 1$ corresponds to a gel point. The analysis involves three parameters, b, $(f - 1)\alpha$ and the weight fraction of the quasi-ordered domain. In the present system, TSP aggregates into quasi-ordered domains represented by 14 xyloglucan chains horizontally aligned in parallel, which function as crosslinking units. Gel is formed when about 75% of TSP chains are incorporated in quasi-ordered domains (after 57 minutes reaction). The thickness of the domain is 1.1 nm, indicating the domain composed of a single layer of stacked xyloglucan chains. At the end of reaction, most of the chains are incorporated in the quasi-ordered domains, which constitute a network made of thin walls.

Table 8.5 Evaluated parameters according to Eq. 8.117 as a function of reaction time

			Weight fraction of	
Reaction time (min)	$(f-1)\alpha$	b (nm)	Single chain	14-chain aggregate
0	0.45	6.0	1.0	0.0
40	0.5	11.0	0.49	0.51
57	1.0	13.5	0.24	0.76
74	1.04	13.7	0.13	0.87
91	1.06	14.0	0.07	0.93
108	1.07	14.0	0.05	0.95

8.3.4 *Conformation of Oligo-(1 $\rightarrow$ 4)-α-D-glucan as Side Chains*

Most of carbohydrates function as ligands or recognition signals, and in this context a considerable attention was paid to the glycoconjugate polystyrene composed of hydrophobic polystyrene as a main chain bearing hydrophilic oligosaccharides [35]. Synthetic glycoconjugate polystyrene was found to function as highly sensitive ligands possessing concentrated multiantennary glyco signals along hydrophobic main chain. The enhanced interaction with carbohydrate-binding proteins is not only due to the high density of glyco signals but also due to the presence of phydrophobic phenyl component, which may disturb the conformation of oligosaccharide side chains. That is, the physiological activity depends on the chain conformation of a glycoconjugate polymer in water.

A series of malto-oligosaccharide-carrying polystyrenes were prepared by varying the number of maltose residues from 1 to 7. Those samples (coded as shown in Fig. 8.23) were disolved in 0.1 M aqueous urea (2.0 wt%), and the small-angle X-ray scattering was observed from those solutions. The conformation of glycoconjugate polystyrenes was simulated by molecular dynamics (CERIUS ver. 3), and the scattering profiles were calculated from the atomic coordinates of the simulated molecules accroding to Eq. 8.114. As shown in Fig. 8.24, the observed and calculated scattering profiles are consistent, confirming the adequacy of the computer simulation. The results indicate that the main chain assumes a pseudo-helical

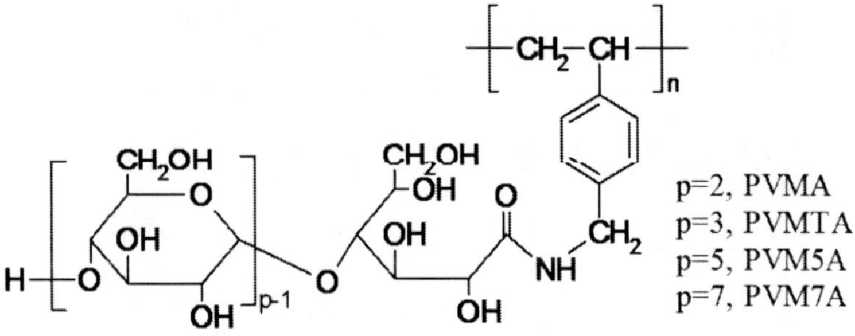

Figure 8.23 Chemical structure of malto-oligosaccharide carrying polystyrene.

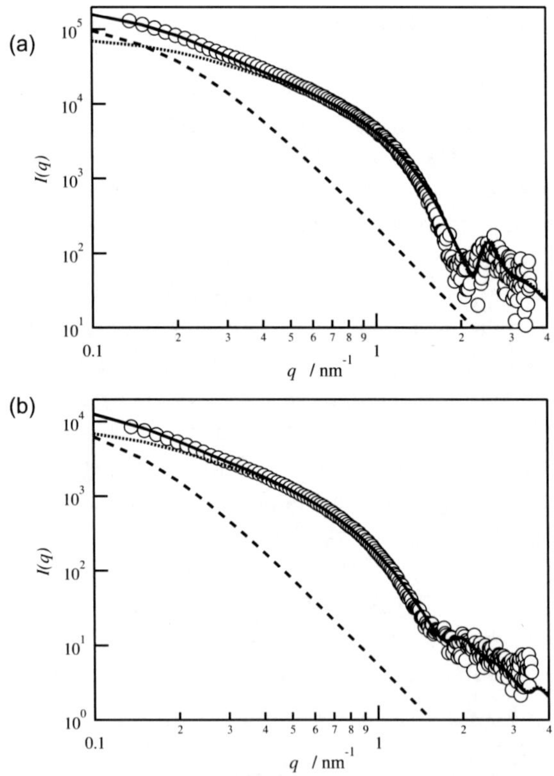

Figure 8.24 Calculated and observed scattering profiles for PVMTA (a) and PVM7A (b). The calculated profiles (represented by dotted lines) are corrected by taking into account aggregation (represented by broken lines). The molecular models are shown at the bottom.

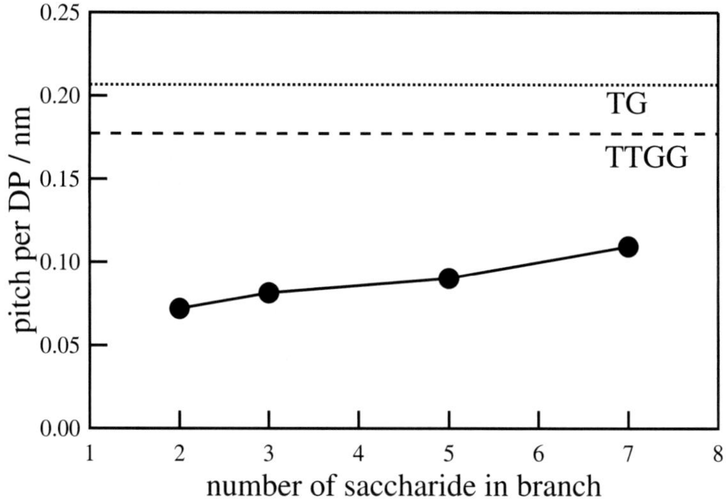

Figure 8.25 The side-chain length dependence of the pseudo-helical pitch. The helical pitches of two polystyrene crystals are also shown in the figure.

conformation where its pitch depends on the side-chain length (see Fig. 8.25) [36].

A pseudo-helical conformation of glycoconjugate polystyrene is basically composed of random-coil sequence, and is much more compressed than a regular polystyrene helix specified by the sequence of TG or TTGG. Here oligosaccharide side-chains seem to stick out from the polystyrene backbone, and the whole shape of glycoconjugate polystyrene resembles a bottle brush. The diameter of the pseudo-helix of the polystyrene backbone increases with decreasing length of oligosaccharide side-chains in order to cover the entire surface of hydrophobic polystyrene backbone. A specific physiological activity (such as cell recognition) and an optical activity of glycoconjugate polystyrene depend on the length of the oligosaccharide side-chains [37]. That is, the spatial arrangement of biofunctional oligosaccharide side-chains determines the physiological activity of glycoconjugate polystyrene. As seen from the molecular models in Fig. 8.24, the chain ends of oligosaccharide side-chains constitute the surface of cylindrical glycoconjugate polystyrene, so that the surface composed of saccharide tips are curved and its curvature depends on the number of saccharide units of the side chains.

The change of the curvature of a glycoconjugate polystyrene cylinder is observed in Fig. 8.25 as the DP dependence of the pitch, which is approximately equal to the distance between two neighboring saccharide tips. In this specific example, PVM5A was found to be the most efficient to recognize lever cells [37], so that the pitch of 0.085 nm is considered to correspond to the hepatocyte recognition site.

8.4 Concluding Remarks

Since an extremely strong X-ray source is available now owing to the development of synchrotron radiation technology, small-angle X-ray scattering (SAXS) provides an indispensable tool to evaluate nano-scale (from 0.5 to 200 nm) structural characteristics of materials including solutions, gels and solids. SAXS can be measured within milliseconds, and thus the structural change could be followed dynamically. Various molecular parameters are obtained from SAXS measurements as demonstrated above, providing information to understand the nano-scale process of the phenomena.

References

1. See, for example, Bianconi, A., Castellano, and A. C. (eds.) (1987). *Biophysics and Synchrotron Radiation*, Springer, Berlin; Saisho, H., Goshi, Y. (eds.). (1996). *Applications of Synchrotron Radiation to Materials Analysis*, Elsevier, Amsterdam.
2. Landau, L. D., and Lifshits, E. M. (1985). *Quantum Mechanics*, Pergamon, Oxford.
3. Debye, P. (1915). Zerstreuung von Röntgenstrahlen. *Ann. Phys.* (Leipzig), **351**, pp. 809–823.
4. Debye, P., and Bueche, A.M. (1949). Scattering by an inhomogeneous solid. *J. Appl. Phys.*, **20**, pp. 518–525.

5. See, for example, Guinier, A., and Fournet, G. (1955). *Small-Angle Scattering of X-rays*, Wiley, New York.
6. See, for example, Glatter, O., and Kratky O. (eds.). (1982). *Small Angle X-ray Scattering*, Academic Press, London.
7. Debye, P. (1947). Molecular-weight determination by light scattering. *J. Phys. Colloid Chem.*, (**51**), pp. 18–32.
8. Ruland, W. (1971). Small-angle scattering of two-phase systems: determination and significance of systematic deviations from Porod's law *J. Appl. Cryst.*, **4** pp. 70–73.
9. Vonk, C. G. (1973). Investigation of non-ideal two-phase polymer structures by small-angle X-ray scattering *J. Appl. Cryst.*, **6**, pp. 81–86
10. For a general description, see Kurata, M. (1982). *Thermodynamics of Polymer Solutions*, Harwood Acad., Chur.
11. de Gennes, P.-G. (1979). *Scaling Concepts in Polymer Physics*, Cornell University Press, Ithaca.
12. Fujita, H. (1990). *Polymer Solutions*, Elsevier, Amsterdam.
13. Ross, G. (1969). Light scattering in amorphous media *Opt. Act.*, **16**, pp. 95–109
14. See for example, Atkins, P. W. (1982). *Physical Chemistry*, Oxford University Press, Oxford.
15. Lindsey, C. P. and Patterson, G. D. (1980). Detailed comparison of the Williams–Watts and Cole–Davidson functions *J. Chem. Phys.*, **73**, pp. 3348–3357
16. Kratky, O., Porod G., and Kahovec, L. (1951). *Z. Electrochem.*, **55**, p. 53.
17. Burchard, W. (1993). *Plant Polymeric Carbohydrates* (F. Meuser, D. L. Manners, and W. Seibel, eds.), The Royal Society of Chemistry, Cambridge, p. 215.
18. Yamakawa, H. (1998). *Helical Wormlike Chains in Polymer Solutions*, Springer, Berlin.
19. Guenet, J.-M. (1992). *Thermoreversible Gelation of Polymers and Biopolymers*, Academic, London.
20. Zierenberg, B., Carpenter, D. K., and Hsieh, J. H. (1976). *J. Polymer Sci., Polym. Symp.*, **54**, p. 145.
21. Kajiwara, K. Burchard, W. (1984). Rotational isomeric state calculations of the dynamic structure factor and related properties of some linear chains. 1. The $\rho = < S^2 > 1/2 < R_H^{-1} >$ parameter. *Macromolecules*, (**17**), pp. 2669.
22. Flory, P. J. (1969). *Statistical Mechanics of Chain Molecules*, Interscience, New York.

23. Kollman, P. (1997). Free energy calculations: applications to chemical and biochemical phenomena. *Chem. Rev.*, **93**, pp. 2395–2417.
24. Shimada, J., Kaneko, H., Takaha, T., Kitamura, S., and Kajiwara, K. (2000). Conformation of amylose in aqueous solution: small-angle X-ray scattering measurements and simulations. *J. Phys. Chem. B*, **104**, pp. 2136–2147.
25. Kajiwara, K., and Miyamoto, T. (2000). *Polysaccharides: Structural Diversity and Functional Versatility*, 2nd ed. (S. Dumitriu, ed.), Marcel Dekker, New York, p. 1.
26. Imberty, A., and Perez, S. (1988). A revisit to the three-dimensional structure of B-type starch. *Biopolymers*, **27**, pp. 1205–1221.
27. Rappenecker, G., and Zugenmaier, P. (1981). Detailed refinement of the crystal structure of Vh-amylose. *Carbohydr. Res.*, (**89**), pp. 11–19.
28. Goldsmith, E., Sprang, S., and Fletterich, R. (1982). Structure of maltoheptaose by difference Fourier methods and a model for glycogen. *J. Mol. Biol.*, **156**, pp. 411–427.
29. York, W. S., van Halbeek, H., Darvill, A. G., and Albersheim, P. (1990). Structural analysis of xyloglucan oligosaccharides by 1H-n.m.r. spectroscopy and fast-atom-bombardment mass spectrometry. *Carbohydr. Res.*, **200**, pp. 9–31.
30. Yamanaka, S., Mimura, M., Urakawa, H., Kajiwara, K., and Shirakawa, M. (1999). Conformation of tamarind seed xyloglucan oligomers. *Sen'i Gakkaishi*, **55**, 590–596.
31. Levy, S., York, W. S., Stuike-Prill, R., Meyer, B., and Staehelin, A. L. (1991). Simulations of the static and dynamic molecular conformations of xyloglucan. The role of the fucosylated sidechain in surface-specific sidechain folding. *Plant J.*, **1**, pp. 195–215.
32. Shirakawa, M., Yamatoya, K., and Nishinari, K. (1998). Tailoring of xyloglucan properties using an enzyme. *Food Hydrocoll.*, **12**, pp. 25–28.
33. Yuguchi, Y., Mimura, M., Urakawa, H., Kajiwara, K., Shirakawa, M., Yamatoya, K., and Kitamura, S. (1996). *Proc. Intern. Workshop Green Polymer*, Bandung-Bogor, Indonesia, p. 306.
34. Kajiwara, K., Kohjiya, S., Shibayama, M., Urakawa, H. (1999). In *Polymer Gels: Fundamentals and Biomedical Applications* (R. De Rossi, K. Kajiwara, Y. Osada, and A. Yamauchi, eds.), Plenum, New York, p. 1.
35. Kobayashi, K., Sumitomo, H., and Ina, Y. (1985). Synthesis and functions of polystyrene derivatives having pendant oligosaccharides. *Polym. J.*, **17**, pp. 567–575.

36. Wataoka, I., Kobayashi, K., and Kajiwara, K. (2005). Effect of the carbohydrate side-chain on the conformation of a glycoconjugate polystyrene in aqueous solution *Carbohydr. Res.*, **340**, pp. 989–995.
37. Shimojo, S., Cho, C. S., Park, I. K., Kunou, M., Goto, M., and Akaike, T. (2003). Helical structure of sugar-carrying polystyrene in aqueous solution by circular dichroism. *Carbohydr. Res.*, **338**, pp. 2129–2133.

Chapter 9

Stimuli-Responsive Polysaccharide-Based Hydrogels

André R. Fajardo,[a,b] Antonio G. B. Pereira,[a,c] Adley F. Rubira,[a] Artur J. M. Valente,[d] and Edvani C. Muniz[a,e]

[a]*Department of Chemistry, State University of Maringa, Av. Colombo 5790, 87020-900 Maringá, Brazil*
[b]*Center for Chemical Sciences, Pharmaceutical and Food, Federal University of Pelotas, 96010-900, Pelotas, Brazil*
[c]*Federal University of Technology - Paraná (UTFPR), 85660-000 Dois Vizinhos, Brazil*
[d]*Department of Chemistry, University of Coimbra, 3004-535 Coimbra, Portugal*
[e]*The Graduate Program in Biotechnology Applied to Agriculture, Paranaense University (UNIPAR), 87502-210, Umuarama, Brazil*
ecmuniz@uem.br

9.1 Basic Concepts of Hydrogels

Hydrogels, which belong to a class of crosslinked polymeric materials, are good examples of a continuous technological development, which improves the society's benefits and, at same time, motivates and instigates researchers to go beyond the boundaries of knowledge. After the work of Wichterle published in 1960s [171], hydrogels developed from simply inert to complex stimuli-responsive (those that show noticeable change in their properties, such as volume, with environmental—pH, ionic strength,

Polysaccharide Hydrogels: Characterization and Biomedical Applications
Edited by Pietro Matricardi, Franco Alhaique, and Tommasina Coviello

ISBN 978-981-4613-61-3 (Hardcover), 978-981-4613-62-0 (eBook)
www.panstanford.com

temperature, magnetic field, specific molecules, light, etc.—stimulus variation) materials presenting great prospective in a wide range of applications such tissue engineering [2], soil conditioning[64], wound dressing[38], sensors [15], contact lenses [110], drug delivery systems [132], separation science [43], hygienic products [67], among others. More than 30,000 papers (published in the last 50 years, being ca. 80% in the last decade) can be found by searching at JCR® (*Journal of Citation Reports*) database under the topic "hydrogel." This demonstrates the importance, in both academic and technologic points of view, of such kind of material.

Hydrogels are formed by three-dimensional (3D) hydrophilic polymer network capable of absorbing and retaining large amounts of water or biologic fluids [6, 46, 73, 129, 187]. This ability is due to polymer chains that are chemically or physically crosslinked to each other, avoiding the hydrogel solubilization in aqueous environments and keeping its 3D network even in the swollen state [59, 69, 178]. Some hydrogels possess the interesting property to swell up to 1,500 times relative to its dry weight [57] and maintaining their 3D structure. In general, the hydrogel 3D network is constituted by porous structure (Fig. 9.1), in which the morphology, averaged porous size (ξ), porous-size distribution are dependent on several factors, as well as the molar mass between two neighbor crosslinks (M_c) are dependent on several factors some of them associated to the method used for hydrogel preparation [70] (concentration of polymer and/or monomer solution, crosslinking agent, etc.) and others.

In fact, the final properties of a hydrogel are dependent on factors associated to the preparation conditions and to the external stimuli (temperature, solvent, ionic strength, etc.) [2, 164, 187]. In swollen state, interactions among water molecules and hydrophilic groups (present in the main chains or pendant to polymer chains) prevail. On the other hand, in deswelled state such interactions are weakened. In specific conditions, a swollen stimuli-responsive hydrogel may collapse in response to environmental stimuli changes releasing the absorbed water or aqueous fluids. During the deswelling process, the porous size of a 3D network changes and any encapsulated solute can eventually diffuse outwards of 3D matrix [57, 151]. The swelling–deswelling–swelling cycle is often reversible

[57, 132, 133]. This is a key factor for application of stimuli-responsive hydrogels in several fields [63, 117]. The soft consistency [112], elasticity and similarity to body tissues [44], due mainly to the high water content [188], are important characteristics of hydrogels. Such similarity allows the hydrogels to be biocompatible and, for consequence, they are very attractive material to be applied as biomaterials [26, 33, 35, 137].

As mentioned, the polymer chains in a hydrogel can be physically and chemically crosslinked [15]. In the physically crosslinked (or physical hydrogels) the polymer chains are joined themselves just by physical interactions (mainly electrostatic interaction, dipole–dipole, H bonds, and hydrophobic forces) and by entanglements and junctions (see Fig. 9.1).

If the interactions weaken the physical crosslinked network disrupts and then the physical hydrogel matrix disintegrates. This behavior favors the releasing of an encapsulated drug from a physical hydrogel [116, 148, 182]. On the other hand, some internal

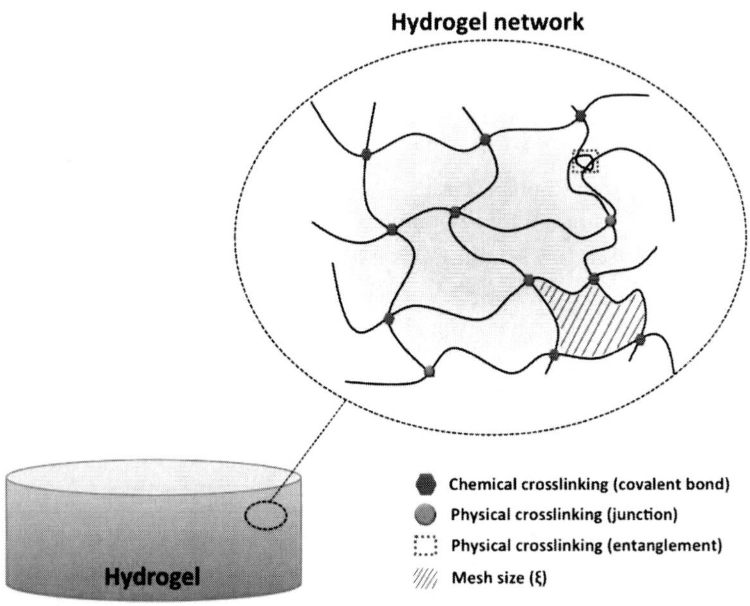

Figure 9.1 Schematic representation of different crosslinks in hydrogel networks.

or external stimuli allow rebuilding the 3D network of physical hydrogels [38, 39, 40, 41, 181]. Due to this feature such materials are called reversible hydrogels.

In chemical hydrogels the chains are crosslinked by covalent bonds and once destroyed (e.g., chemically or enzymatically) there is no possibility for rebuilding. This type of hydrogel has been used as biomaterial for more robust applications such as in situ site-specific application in humans and/or animals. Some hydrogels are classified as both physical and chemical [7, 31, 36, 147]. In this way, combination of procedures for preparing in one or more steps is needed [36, 174]. Obviously, the structure, morphology and final properties of such kind of material can be more interesting, as compared to only physical or chemical ones [101].

There are plenty of methods for preparing hydrogels [12, 37, 159]. Reversible hydrogels are widely prepared by mixing solutions from cationic and anionic polymers [8]. The final properties of the 3D network of physical hydrogels should depend on several factors such as the anionic/cationic charge ratio, the pH and the existence, or not, of stirring and the respective speed during the preparation, as well as several other parameters [40, 41]. The addition of di- or trivalent cations to sodium alginate aqueous solution is largely used, for instance, on preparation of alginate-Ca^{2+} hydrogels [122]. In this case, the addition of aqueous ethylenediaminetetraacetic (EDTA) sodium salt to a swelled alginate-Ca^{2+} 3D matrix will disrupt (partially or completely depending on the ratio of Ca^{2+} to EDTA) the matrix due to the competition to the Ca^{2+} between alginate and EDTA [13].

Chemical hydrogels can be prepared mainly by [37] (i) polymerization of monomers in presence of crosslinking agent, as co-monomers, (ii) crosslinking of polymer chains by use of crosslinking agent, or (iii) polymerization of monomers (or oligomers) with averaged functionality higher than 2. It is important to notice that the crosslinking agent is a reactant that should have functionality higher than 2.

Polysaccharide-based hydrogels can be physical, chemical or a combination of both [28]. For preparing chemical polysaccharide-based hydrogel, chemical modifications of raw polysaccharides are generally required to allow chemical crosslinking [162, 163].

Year by year, larger and crescent number of publications using polysaccharides to obtain 3D networks have been appeared in the specific literature [163]. This is mainly due to often biocompatibility of polysaccharides, to renewable sources and to new and "green" procedures for modifying polysaccharides [163]. The development of new methodologies for inserting specific chemical groups into the polysaccharides backbone rendering crosslinkable polymers has provided new possibilities for using different polysaccharides [54]. In certain cases, the inserted chemical groups allow the respective polysaccharide-based hydrogels be sensible to external stimuli beyond be chemically crosslinkable [111].

This chapter explores the methods for preparing stimuli-responsive polysaccharide-based hydrogels. Critical analysis about inherent characteristics and emphasis to some applications of such materials are given. Although the pharmacological application is the main focus of polysaccharide-based hydrogels, their application in other important fields is briefly discussed as well.

9.2 Main Characteristic of Stimuli-Responsive Hydrogels

In the last two decades significant changes occurred in concept and design of new soft materials. In this short period of time the society witnessed a pattern-shifting from the inert to smart (or "intelligent") materials that possess the ability to respond to different external or internal stimuli. In this way, responsive hydrogels constitute a new generation of soft materials to be applied in several fields, including biomedical devices, scaffolds for engineered tissues, biosensors and actuators [49]. These materials are sensitive to changes of temperature [51, 132], pH [105, 134], ionic strength [18, 87], magnetic field [120], and light [185]. The main characteristic properties of hydrogels sensitive to different external stimuli are described in this section.

9.2.1 *Temperature-Responsive Hydrogels*

Among different possible external and internal stimuli, changes in temperature have proven to be easy and safe for many applications, especially for biomedicals. In this sense, much of the research in stimuli sensitive hydrogels is related to temperature-responsive ones [49, 143]. For temperature-sensitive hydrogels, a small variation in temperature induces a phase transition and, therefore, discontinuous volume changes (shrinking or expansion) are observed [52, 141]. Some temperature-sensitive polymers undergo abrupt decrease in solubility in response to the increase of environmental temperature as the lower critical solution temperature (LCST) is approached. There is a consensus that the phase separation is governed by the balance of hydrophilic and hydrophobic moieties on the polymer chain and on the Gibbs free energy of mixing [49, 141, 158]. Enthalpy (H), entropy (S), and temperature (T) do influence the Gibbs free energy of association (G) according to the second law of thermodynamics [$\Delta G = \Delta H - T\Delta S$]. The value of $T\Delta S$ will increase as the temperature is raised and considering that the positive enthalpy term (ΔH) is often smaller than the entropy term (ΔS), an increase in T makes ΔG negative and favors polymer chain association. Also, the temperature dependence of certain molecular interactions, such as H bonds and hydrophobic forces, contributes by themselves to the phase separation. At the LCST, H bonds among polymer chains and water molecules are unfavorable compared to polymer–polymer and water–water interactions. As consequence an abrupt transition in the 3D matrix occurs: water-solvated macromolecule quickly dehydrates and changes to a more hydrophobic structure [141, 158].

The common characteristic of temperature-responsive polymers is the presence of hydrophobic pendant groups such as methylic, ethylic and propylic in a hydrophilic main chain [132, 133]. Poly(*N*-isopropylacrylamide) (PNIPAAm) has been extensively investigated and it is one of the most studied temperature-responsive polymer. A search at JCR® database using the keyword "poly(*N*-isopropylacrylamide)" provided more than 7,400 papers. There are many papers on the literature showing that the LCST of PNIPAAm, in aqueous solution or in swollen network, falls in the range from 32° to 34°C [49, 133, 141, 143]. Below the LCST, H bonds between hydrophilic segments of the PNIPAAm chains and water molecules dominate leading to enhance dissolution of linear PNIPAAm (or swelling of PNIPAAm hydrogel) in water. As the temperature increases and approaches to the LCST of PNIPAAm, hydrophobic interactions among hydrophobic segments (isopropyl) become strengthened, while H bonds become weaker [49, 58, 133]. When PNIPAAm is heated at temperatures higher than its LCST, the chains dramatically collapse due to the breakdown of H bonds (interactions among >N-H groups and water molecules) [58, 132, 141, 185]. As mentioned, the hydrophilic/hydrophobic phase transition may have different applications, but the poor mechanical resistance at a highly swollen state is a severe drawback for using PNIPAAm hydrogels [58, 132, 141, 185]. Copolymerization [119, 145], interpenetrated polymer network (IPN) formation [169, 170], and semi-IPN formation [102, 103] are methodologies used to improve the mechanical properties of PNIPAAm-based hydrogels. Takigawa et al. [157] showed that the elastic modulus of PNIPAAm hydrogels is higher in the collapsed state than in the swollen one. Zaroslov et al. [179] published a paper concerning the changes in the elastic modulus of strongly charged hydrogels near the collapsing transition, where a relation between the swelling ratio and the elastic modulus was described. By preparing either IPN or copolymerizing NIPAAm with more hydrophilic monomers, the LCST of PNIPAAm can be tailored to occur close to human body temperature. For instance, Wenceslau et al. [169, 170] prepared IPN of PNIPAAm and chemically modified PVA (PVA–Ma) using a two-step methodology. These authors showed that the LCST of PNIPAm shifts linearly to the PVA-Ma/NIPAAm ratio used in preparation of IPN. Semi-IPN based on polyacrylamide (PAAm) network having PNIPAAm entrapped presented higher compressive elastic modulus than the pure PAAm hydrogel, especially when the PNIPAAm chains were in the collapsed state [103]. Chetty et al. [25] investigated the influence of copolymer composition on poly(*N*-isopropylacrylamide-co-*N*,*N*′-methylene-bisacrylamide) (PNIPAAm-co-MBAAm) prepared in water by redox initiation. In this case, both storage (G') and loss (G'') moduli decreased with decreasing crosslinker content as revealed by dynamic rheometry. A very interesting temperature-responsive sandwich-like membrane composed by IPN hydrogels having crosslinked PNIPAAm interpenetrated into PAAm networks has been described [50]. On the basis of SEM images, it is possible to conclude that these hydrogel membranes are constituted by three different layers, being the internal layer, composed by crosslinked PNIPAAm, fully involved by the other two. Significant morphological differences between the internal and external layers were also observed as the temperature is changed due to the collapsing of internal layer. Temperature-responsive hydrogels in other polymers than PNIPAAm are also important. One example is the poly(*N*,*N*-diethylacrylamide) (PDEAAm) because its LCST falls in the range of 25–32°C, close to the body temperature [85, 86]. Temperature-responsive hydrogels based on different polymers have been often published in literature although number of works is much lower as compared to PNIPAAm [49, 86]. For instance, Lessard et al. [84] prepared a series of poly(*N*,*N*-diethylacrylamide-co-sodium acrylate) with a degree of crosslinking of 1 mol% for use as thermosensitive superabsorbents for water. The critical swelling temperatures or the volume phase transition temperature (VPTT) and the water absorption capacity of the polymers could be modulated by varying the amount of sodium acrylate (0–60 mol%) in the copolymers. Also, such work showed that the water absorption capacity of the copolymers depends on the media pH; for example, when the acrylate units are in dissociated form the poly(*N*,*N*-diethylacrylamide-co-sodium acrylate) shows higher water absorption capacity. Addition of urea into the media raised and narrowed the VPTT values of the sodium acrylate copolymers [84], probably due to competition of urea and water to the polar presents in the polymer.

9.2.2 *pH-Responsive Hydrogels*

The pH-sensitive polymers often contain side functional groups such as acid (e.g., carboxylic and sulfonic acids) or basic (e.g., ammonium salts), which either accept or release protons in response to changes of external pH. Poly(acrylic acid) (PAAc) becomes ionized at pHs higher than the pK_a of carboxylic acid functional groups (4.0–4.5), while the hydrogel obtained by use of chemically modified Arabic gum presented pH response in the range of 2 to 8 [134].

The acid- or basic-side functional groups in polyelectrolytes undergo dissociation in a similar way as the acidic or basic groups of monoacids or monobases. The ionization on polyelectrolytes, however, is more difficult due to electrostatic effects of other adjacent ionized groups. So, the apparent dissociation constant (K_a) may be different from that of the corresponding monoacid or monobase.

When macromolecules are constituted by a huge amount of ionizable groups such polymer is called polyelectrolyte. Hydrogels constituted by polyelectrolytes are, often, sensible to pH changes, which makes the hydrogel to swell (if the side functional groups are in ionized state) or to collapse (if the side functional groups are not ionized). The process of swelling/collapsing of pH-sensitive hydrogel is reversible and may have strong applications in pharmaceutical field because hydrogels can protect a given encapsulated drug from hostile environments, e.g., the presence of enzymes and low pH in the stomach region. A drug that can suffer degradation at acidic conditions can be transported through the gastric tract and be released in the colon if it is encapsulated in an appropriated pH-sensitive hydrogel. Hydrogels can also control drug release by changing the gel structure in response to environmental stimuli. Hydration under alkaline pH favors enzyme access to the hydrogel and its later enzymatic decomposition with the ensuing delivery of the drug entrapped in the hydrogel. In this way, Reis et al. [134] obtained pH-responsive hydrogel from Arabic gum (AG) chemically modified with glycidyl methacrylate (AG-MA). An appropriate mixture of water and DMSO was used to dissolve AG and GMA for obtaining AG-MA. The authors observed that AG-MA hydrogels showed significant water uptake pH-dependence. Swelling assays in wide pH range (2–10) showed that the transport mechanism of water into AG-MA hydrogel was governed by Fickian diffusion and polymer relaxation (anomalous transport). At high pH values, the water transport profile becomes more dependent to the polymer relaxation. This effect was attributed to the increase in the ionized groups of glucuronic acid segments which contributed to electrostatic repulsion among the groups and led the expansion of the polymer gel network. AG-MA hydrogels exhibited pH responsiveness, demonstrating that they are appropriate materials for further tests as drug carriers.

The sensibility of hydrogels to pH may be applied to health treatment using drugs with anti-cancer effect. For instance, Na et al.

[105] fabricated doxorubicin (a model anticancer drug)-loaded self-organized nanogels composed by hydrophobized pullulan (PUL)-*N*-alpha-Boc-L-histidine (bHis) conjugates. The bHis was grafted to PUL-deoxycholic acid (DO) conjugate (PUL-DO) via ester linkage. The authors synthesized PUL-DO/bHis conjugates with two different degrees of bHis substitutions (PUL-DO/bHis36 and PUL-DO/bHis78). The responses to tumor extracellular pH (pH 6–8) were determined, and they were also evaluated with regard to their anticancer efficacy against breast cancer cell lines (MCF-7). The results show that PUL-DO/bHis nanogels have potentially to be employed as anti-tumor drug carriers.

New pH-responsive polymers and hydrogel nanoparticles, which are constituted by either sulfapyridine or sulfamethoxypyridazine, were synthesized by Kang et al. [71]. The linear copolymers showed soluble/insoluble transition in water, while the respective hydrogel nanoparticles experienced, in aqueous solutions, association/dissociation transition in a narrow pH range (7–8). The authors confirmed the pH sensitivity by changes in turbidity or particle size as a function of pH. The ionization of $-SO_2NH$ group in sulfonamides is responsible for aggregation of polymers or hydrogel nanoparticles. The transition pH was determined by the amount of $-SO_2NH$ groups in the copolymers or on the hydrogel nanoparticle surface; at an optimum composition, the transitions occur near to physiological pH (7.4). The authors pointed out that such systems present potential for application on various biomedical and bioengineering fields, such as pulsatile drug delivery, targeting, embolization, sensors, and bioseparation.

Hydrogels may be, at same time, sensitive to both temperature and pH. For instance, Kang et al. [72] synthesized polymeric nanogel particles sensitive to temperature and pH based on NIPAAm, *N*-dimethylacrylamide and sulfamethoxypyridazine, by radical polymerization. The materials presented sharp change at 38–45°C (LCST) and at 7.3–76 (pH response). The loading/releasing efficiencies of doxorubicin (DOX) rates were influenced by four combinatorial release conditions of two pH levels (7.0 and 7.5) and two temperatures (37° and 42°C). Due to DOX release properties the materials presented high potential to be applied as biomaterial.

9.2.3 *Ionic Strength (IS)-Responsive Hydrogels*

The swelling-, loading-, and releasing-solute capabilities of hydrogels are, sometimes, dependent on ionic strength (IS) of external fluid. Due to the diffusion and convection, an osmotic pressure is produced by the difference in the ionic concentration between the exterior solution and interior hydrogel, which drives the swelling of the IS-smart hydrogel. A relevant paper giving details on the ionic response of IS-smart hydrogels was published by Li et al. [87] where a mathematical model based on osmotic pressure and on repulsive forces among the fixed ionic-charged moieties, another driving force that do influences on swelling, was also derived. The decrease in swelling due to the presence of salt may be significant [138].

According to Poujavadi et al. [130], the sensitivity of hydrogel to the presence of salts can be related by the dimensionless factor f, defined as

$$f = 1 - \left(\frac{W_\mathrm{s}}{W_\mathrm{w}}\right) \tag{9.1}$$

where W_s and W_w are, respectively, the swelling in saline solution and in deionized water. According to Eq. 9.1, f values close to 1.0 mean that hydrogels significantly shrinks inside a saline solution, whilst f values close to zero indicates that the swelling of the hydrogel is not influenced by the presence of salt. Candido et al. [18] investigated the influence of NaCl, $CaCl_2$, and $NaHCO_3$ on the swelling capability of superabsorbent hydrogels of poly(acrylamide-co-acrylic acid) (PAAm-co-PAAc) containing rice husk ash (RHA). Decreases in swelling degree were observed leading to changes in f from 0.43 (NaCl 0.001 M) to 0.91 (NaCl 0.1 M). A 0.1 M $CaCl_2$ aqueous solution has a stronger influence on the swelling degree, being the f value equal to 0.97. The presence of RHA in the matrix increased the swelling compared to neat PAAm-co-PAAc matrix but not minimized the influence of salt on swelling capability. The hydrogels of PAAm-co-PAAc containing rice husk ash (RHA) were pH-sensitive as well.

Molina et al. [100] investigated the swelling of chemically crosslinked poly(*N*-vinylimidazole) (PVI) immersed in salt aqueous solutions as a function of the IS, in the whole range of pH. In acid solutions at pH range from 0 to 4, imidazole moieties became protonated, and PVI behaved as a polyelectrolyte gel: the swelling decreased upon increasing the IS for both NaCl and for $CaCl_2$, with HCl as protonating acid. However, at pHs higher than 4 the hydrogel was practically neutral, and swelling ratio increased by increase of IS, showing a salting-in effect. The authors pointed out in details a correlation between the responses of hydrogel (swelling) at different IS and pH and the protonation of imidazole moieties. Valente et al. [160, 161] investigated the influence of different electrolytes on DNA releasing from PVA-DNA gel matrices. The release of DNA from gels to 1:1 sodium and nitrate salts showed that the transport properties are dependent on the ability of anions/cations solubilizing DNA in the aqueous phase, which with the exception of bromide, can be related to the Hofmeister series. It was pointed out that in the presence of multivalent electrolytes, or increasing the ionic strength, the condensation of DNA inside the gel is followed by a phase separation induced by the retention of DNA inside the polymer matrix, as seen by scanning electron microscopy (SEM). The authors also evaluated the effect of uni- and divalent cationic surfactants on the release properties of DNA. On the basis of their findings, the authors suggest that the kinetics of DNA release depends on a complex balance between different structural properties of the surfactants such as charge and the bulkiness of the head group, and alkyl chain length.

9.2.4 *Solvent-Responsive Hydrogels*

The response to solvent is also a very important issue on stimuli hydrogels. Several aspects are exploited in this way. The change from hydrophilic to hydrophobic solvent (or the reverse) is a common strategy to induce the shrinking or expansion of a 3D network. Beyond temperature response, PNIPPAm presents discontinuous volume-phase transitions (swollen and collapsed states) as a response to changes in solvent composition [60, 61]. Some authors used this system as model to study the phase transition and critical phenomena. Once collapsed such hydrogels underwent a reentrant phenomenon, and then swelled again when the ratio of water to mixing solvent, namely DMSO [74] and MeOH or EtOH [3], was systematically varied. The authors [3, 60, 61, 74] assumed that the free energy of contact among solvent molecules and polymer segments was not a monotonic function of the solvent composition to explain this phenomenon. Schild et al. [142] investigated the LCST of PNIPAAm in co-nonsolvent mixtures of water and methanol by means of cloud-point and microcalorimetric measurements. They observed that the solution behavior was quite distinct from that of poly(vinyl methyl ether) (PVME) in the same binary solvent. Other co-nonsolvents, including dioxane and tetrahydrofuran, yielded phase diagrams similar to those of PNIPAAm in water–methanol mixtures. The authors used Flory–Huggins ternary solution theory for explaining the observed results only in terms of a change in the solvent–solvent interaction parameter (χ_{12}) [45] for PNIPAAm solutions. Pagonis and Bokias [115] showed the swelling behavior of hydrogels containing NIPAAm and *N,N*-dimethylacrylamide (DMAAm) in water–dioxane mixtures as a function of dioxane content and temperature. The composition of hydrogels in terms of DMAAm varied from 0 to 100%. The hydrogel containing only PNIPAAm significantly deswelled in the water-rich region, while the hydrogel containing only PDMAAm presented pronounced deswelling in the dioxane-rich region. The deswelling became less significant with decreasing the PDMAAm content in poly(NIPAM-co-DMAAm). Such a behavior was attributed to a combination of the LCST-type co-nonsolvency behavior of PNIPAAm and the upper critical solution temperature (UCST)-type co-nonsolvency behavior of PDMAAm in water–dioxane mixtures. Sakurai and Sato [139] investigated the swelling behavior of poly(4-vinyl phenol) (P4VPh) hydrogel in various kinds of mixed solvents (methanol, ethanol, 2-propyl alcohol, t-butyl alcohol, acetone, acetonitrile, dioxane, THF, and DMSO). The hydrogels firstly presented significant deswelling as the organic solvent content increased and then reswelled via lowest swelling points (10–60 vol% depending on the solvent), and finally showed a second deswelling in the highest content region (typically $\geq$ 80 vol%). According to those authors [139], this swelling-deswelling sequence was not reproduced when the organic solvent content was decreased. This remarkable hysteresis as well as the reentrant swelling with the extra second deswelling was justified in terms of the intermolecular interactions, i.e., H bonds between the phenol –OH groups and the Π-Π stacking of the phenol rings [139].

9.2.5 *Other Responsive Hydrogels*

Beyond temperature, pH, ionic strength, and solvent, there are several other ways of external stimuli [49]—magnetic field, light (UV, visible), electric voltage, mechanical stress, sound waves, and others—for inducing changes in hydrogels. As already mentioned, some systems have been developed to combine two or more stimuli-responsive mechanisms into a given polymer system. As this field is explosively growing, there is a trend in developing new smart hydrogels to be responsive not only to a conventional stimuli, but as actuator and biosensor having multi-response features. For instance, temperature-sensitive polymers may also respond to pH changes or solvent mixtures [19]. For deeper details the reader is addressed to reviews on this subject [49, 63, 132, 187].

9.3 Characteristic of Stimuli-Responsive Polysaccharide-Based Hydrogels

The development of stimuli-responsive polysaccharides-based hydrogels has been encouraged due to the several advantages of these materials over the synthetic ones. Biocompatibility, biodegradability, low or nontoxicity, hydrophilicity, similarity to biological environments, biological functionality at molecular level, cell signaling scheme, and immune recognition are good examples of the desirable properties attributed to the stimuli-responsive hydrogels based on polysaccharides [80, 118, 164]. Furthermore, the use of polysaccharides is also encouraged in respect of environmental concerns. Polysaccharides are the cheapest and most abundant, available, and renewable organic materials on earth. Their use is a good alternative to replace the petroleum-based polymers. Cellulose [93, 164], starch [4, 176], alginate [114], chitosan [177], and glycosaminoglycans [32, 38, 126] have been frequently used to prepared hydrogels matrices that provides some response to internal/external stimuli. Polysaccharides have been used in their natural or chemically modified forms to provide a specific characteristic/property. The great number of functional groups (e.g., –OH, –COOH, $-NH_2$, $-OSO_3H$) presented by most part of polysaccharides structure offers suitable chemical sites for modification leading to linear polymer and/or to polymeric network [30].

The chemical modification of functional groups allows combining polysaccharides with synthetic polymers to form stimuli-responsive hydrogels [119, 133]. In the last decades different types of crosslinking procedures have been developed. Generally, the reported reactions related to chemically modified polysaccharides are held in two main groups: graft copolymerization of suitable synthetic monomer(s) on polysaccharide backbone in the presence of a crosslinker, and direct crosslinking of polysaccharide [30]. Cellulose, for instance, could be grafted with some synthetic monomers, such as NIPAAm, forming a thermoresponsive hydrogel [119]. Furthermore, the functional groups present on the polysaccharide structure enable the formation of polyelectrolyte complexes, which are responsible to external stimuli. Polyelectrolyte complexes are formed by macromolecules that have charged functional groups or that acquire charge according to environmental changes. Chitosan is a good example of polysaccharide that shows chargeable functional groups allowing its combination with other opposite charged polysaccharides/polymer/ions [40]. The interactions between polysaccharides with oppositely charged moieties occur in mild conditions, resulting in materials with hydrophilic character and responsive to some properties (e.g., pH, ionic strength) [40]. Polysaccharides with polyelectrolyte features, such as chitosan and alginate, can be induced to undergo reversible sol–gel transition forming hydrogels or films by changing chemical or pH gradients [154]. These materials offer a versatile platform and customizable functionality to interface and interact with labile biologics (biomolecules, proteins, enzymes, cells) in aqueous environment. They are widely used in composite film coating, tissue engineering and drug delivery for creating functional materials with desired physical, chemical properties and even biological functionalities. Fajardo et al. [41] developed a hydrogel based on chitosan/chondroitin sulfate formed by polyelectrolyte complexation able to self-reorganizing its structure when swollen in basic conditions. This property allows tailoring the hydrogel structure to show a specific feature/property (e.g., crystallinity, higher thermal stability, porosity, and water uptake capacity) [41].

9.4 Strategies Used for Synthesis of Stimuli-Responsive Polysaccharide-Based Hydrogels

The most available biomolecules on earth are the carbohydrates, which present the empirical formula $(CH_2O)_n$ (although other atoms such as phosphorous, sulfur and nitrogen could also be found in their structures) [108]. Carbohydrates are classified as monosaccharides (single sugar unit, 3 to 9 carbons), oligosaccharides (small chains, most common are those with 2 monosaccharide units) and polysaccharides (as much as thousands of units of monosaccharides) [97]. Polysaccharides could be a homopolymer or heteropolymer according to the presence of single or different kinds of sugar residues, respectively. Besides, the several functional groups present in polysaccharide structure behave as active sites for preparing hydrogels. As earlier mentioned, polysaccharide-based hydrogels have been prepared by either chemical (covalent bonds link the chains) or physical (electrostatic interactions, H bond, etc.) crosslinking approach. Apart from crosslinking, chemical modification on side groups of carbohydrate polymers has also been employed to provide required functionalities and responsiveness necessary for specific application. Therefore, the rational design of hydrogels, the choice of the polysaccharide, the modification on side groups, the advantages and drawbacks of each crosslinking method are of paramount importance to the final properties and destination of smart hydrogels. Review papers focusing on different aspects of polysaccharide-based hydrogels are available elsewhere [11, 28, 48].

9.4.1 *Polysaccharide-Based Chemical Hydrogels*

Chemically crosslinked polysaccharide-based hydrogels have been prepared by either graft copolymerization reaction with suitable monomers or a proper crosslinker or by direct crosslinking through their functional groups using the well-known polysaccharide chemistry. The introduction of vinyl groups (carbon–carbon double bonds) in the polysaccharide backbone enabling the radical crosslinking and/or graft reaction of acrylate monomers is a common procedure. For instance, Zhang et al. [183] prepared dextran-based hydrogels responsive to pH and temperature using such pathway. The hydroxyl groups on dextran were reacted with maleic anhydride to add double bonds and carboxylic acid functionalities to dextran. Then, the modified dextran was photopolymerized/crosslinked in presence of NIPAAm generating temperature sensitive hybrid hydrogel. It was demonstrated that the magnitude of response to environmental changes is dependent on ratio of modified dextran/NIPAAm as well as the temperature in which the hydrophilic/hydrophobic phase transition occurs. Leach et al. [83] added photopolymerizable methacrylate groups to hyaluronic acid (HA) by reacting HA with glycidyl methacrylate (GMA) in aqueous medium. Sets of glycidyl methacrylate-HA (GMHA) conjugates were prepared and photopolymerized to form hydrogels targeting applications in tissue engineering as wound healing systems. It was showed that increasing the number of methacrylate groups, and therefore the crosslinking density, decreased the degradation rates. However, no important effects on the cytocompatibility and proliferation of human aortic endothelial cell associated to this change were observed. Besides, subcutaneous implants of GMHA hydrogels in rat models presented good biocompatibility and little inflammatory response.

The use of "click" chemistry has also been reported in the preparation of polysaccharide hydrogels. Crescenzi et al. [29] described the hydrogel preparation by reacting azide-functionalized hyaluronic with alkyne-functionalized hyaluronan, by "click" chemistry approach, based on Cu(I)-catalyzed Huisgen 1,3-dipolar cycloaddition reaction at room temperature and aqueous medium. The HA-based hydrogels presented great potential to be applied as carrier for drug delivery and scaffolds for tissue engineering. Mergy et al. [99] added double bonds to dextran through esterification of hydroxyl groups using pentenoic anhydride, followed by crosslinking using a difunctional mercaptan molecule, poly(ethylene glycol)-dithiol, through radical thiol-ene reaction. It was showed that the thiol-ene reaction is nontoxic and can be used to prepare cell-responsive polysaccharide hydrogels.

Many researchers took advantage of the existing functional groups in the polysaccharides backbone to prepare hydrogels, without the need of previous modification, by just choosing appropriated crosslinker agents such as sodium trimetaphosphate (STMP) [82, 173], glutaraldehyde [68, 127, 172], genipin [109, 149], diisocyanate [92, 168], divinylsulfone [65, 94], epichlorohydrin [20, 21], citric acid [16, 23], ionic species like metal ions [79, 98, 107], and borax [27], among others.

9.4.2 *Polysaccharide-Based Physical Hydrogels (Polyelectrolyte Complexes)*

Polyelectrolyte complexes (PECs), or physical hydrogels, present many advantages over chemical hydrogels since extra chemicals like initiator, crosslinker, or residual monomers as well as by-products are not present [180]. The pK_a of the ionizable groups on the polyelectrolytes determines whether such groups are dissociated or not in a specific pH condition. Therefore, the water uptake of polyelectrolyte complexes is dependent on pH and ionic strength (as pointed out in sections 9.2.2 and 9.2.3) since such parameters influence the interaction among chains (electrostatic repulsion or attraction) [10]. Such characteristics allow physical hydrogels to be applied in different fields. For instance, polyelectrolyte complexes based on chitosan and sodium alginate were prepared for vaginal delivery of antifungal and antibacterial chemicals such as chlorhexidine digluconate [1]. The adherence on the vaginal mucosa, the hydration and the release profile were dependent on complex formulation. It was showed that the ratio 1/9 (chitosan/alginate) was the most effective against the pathogens *Candida albicans* and *Escherichia coli*. Polyelectrolyte complexation has been also employed to enhance the mechanical properties of electrospun nanofibers [175]. The model system used was chitosan/gelatin. It was proved that the mechanical properties improved 10^3-fold in the polyelectrolyte system as the ratio between glucosamine units and carboxylic acid groups was 1/1 in comparison to pure chitosan or gelatin nanofibers. Besides, the annealing process could further improve the tensile properties. Polysaccharide-based polyelectrolyte complexes hydrogel have also been used in the separation process.

PECs based on chitosan (cationic polymer) and sodium alginate (anionic polymer) were prepared for separation of methyl ter-butyl ether/methanol mixtures through pervaporation technique [75]. The complex showed high selectivity and permeability due to the polarity and high free volume. Besides, the increase in temperature provided an increase in separation and permeation rate due to changes in the effective free volume in the membranes.

9.4.3 *Characterization of Stimuli-Responsive Polysaccharide-Based Hydrogels*

In order to characterize the morphological, physical and chemical aspects of polysaccharide-based hydrogels several techniques have been used. It is important to have in mind that the main properties of hydrogels are (i) swelling behavior, (ii) molecular structure, (iii) mechanical features, and (iv) morphology. Details of properties characterization required for specific application are not presented in this chapter but some discussion in this direction is given as follow. The water uptake capacity or degree of swelling (S) at equilibrium condition of hydrogels is usually expressed by the following equation:

$$S = \frac{W_s - W_d}{W_d} \tag{9.2}$$

where, W_s is the weight of swollen hydrogel and W_d is the weight of dry hydrogel. The swelling capacity in terms of percentage (%) is obtained by multiplying the S value for 100. The mechanism in which the hydrogel swells can be either diffusional- or chain-relaxation-controlled. Different mathematical models have been developed to predict the swelling and solute release behavior from hydrogels matrix. The semi-empirical equation described by Peppas et al. [77, 123, 124] has been largely employed; however it is only valid for the initial 60% of maximum absorbed fluid. Alternative model that can fit 100% of the released fraction has also been proposed [135]. The molecular structure (e.g., functional groups present) as well the molecular interaction among species (H bond, ionic interactions, etc.) of both physical and chemical hydrogel have been studied by FTIR [47], Raman [128] and NMR [146] spectroscopic techniques. The thermal behavior and crystallinity of hydrogels has been assessed mostly by DSC and TGA [9, 166] as well as XRD [41]. The mechanical features of polysaccharide-based hydrogels may be a key factor driven to specific application. Such properties can be evaluated by different means. For instance, the elastic modulus (E) can be obtained using data collected in a texturometer that correlates the necessary force (*stress*) to induce some deformation (*strain*) to a specimen through compression assays (e.g., compression stress–strain measurements) [42]. The morphology and surface topography of hydrogels have been evaluated by microscopy approaches such as SEM [76], TEM [144], and AFM [95].

9.5 Some Applications of Stimuli-Responsive Polysaccharide-Based Hydrogels

Polysaccharides-based stimuli-responsive hydrogels have been designed for several applications and the use as biomaterials should be highlighted. Polysaccharides, in general, show very interesting and desirables physicochemical and biological properties to form hydrogels for such purpose [186]. Hydrogels formed by polysaccharides are able to show high biocompatibility with the human biological system, which avoid rejection and maximize specific applications, such as drug release in specific body regions [113]. In addition, nowadays it is possible to form polysaccharides-based hydrogels with controlled biological activity and biodegradability to be applied in cell signaling [5] and growth schemes [155], immune recognition, wound healing [66], bone repair [154], etc. The promising polysaccharide-based hydrogels have become more attractive when exhibits sensitive features. Such a kind of material could be tailored to give an efficient response according to some external or internal stimulus as described in the previous sections.

Generally the choice of the polysaccharide to form a hydrogel for a specific application has a critical role. For example, a polysaccharide that easily degraded in stomach region (at acid conditions) is not a good candidate to form a colon-specific drug release system. Among the most common polysaccharides chosen to prepare biomaterial based on hydrogels includes the chitosan [184], alginates [42], pectin [104], starch [96], cellulose [22], dextran [89], and glycosaminoglycans (hyaluronic acid, chondroitin sulfate, dermantan sulfate, heparin, etc.) [62, 140, 153]. Take into account these polysaccharides, a great number of interesting studies about the formation of polysaccharide-based stimuli responsive hydrogels and their application has been published.

9.5.1 *Drug Delivery Carriers*

The general requirement about drug carriers is that they are substances or materials that show the ability to improve the site-specific delivery and the effectiveness of drugs [14]. It must be mentioned that the carriers are designed to prolong in vivo drug actions, to decrease the drug metabolism rate, to reduce some possible drug toxicity, and to target the drug at sites of pharmacological activity [14, 165]. One of the major benefits results from the decrease of daily doses administrated by the patient reducing the variation of drug concentration in the bloodstream and as a consequence the collateral effects.

In the last decades huge progresses have been made in the use of polysaccharides to form stable and versatile hydrogels to be applied as carriers for drug delivery. The success of such proposal allows encapsulating species more complex than drugs such living cell and biological agents, such as proteins, enzymes, growth factors, etc. [78, 106, 111]. Concerning the last topic, physically crosslinked hydrogels are of great interest, particularly because the gel formation can be often carried out under mild conditions and in the absence of organic solvents or cross-linkers [41]. In addition, most physically crosslinked hydrogels shows pH-responsive characteristic. One of the most common example is the hydrogels based on alginate physically crosslinked with divalent ions (e.g., Ca^{2+} ions). Lin et al. [91] prepared beads based on alginate blended with a water-soluble chitosan and crosslinked in a Ca^{2+}-aqueous solution to be tested as a pH-sensitive system for delivery of a model protein drug (bovine serum albumin, BSA). Release assays showed that the BSA releasing is clearly pH-dependent and the better results were observed for the assays carried out in neutral conditions (pH 7.4). The main advantage of this type of physically crosslinked hydrogel is that in general all formation procedures are performed in aqueous medium at neutral environment, which may preserve the bioactivity of proteins, enzymes, and drugs. Also, it is possible to obtain a dual-stimuli-responsive (temperature and pH) hydrogel from the association of alginate and some thermo-responsive polymer/copolymer. Shi et al. [147] mixed poly(NIPAAm-co-AAm) with alginate-Ca^{2+} to prepare inorganic-organic hybrid biomineralized beads. Release assays performed with *Indomethacin*, a model drug, showed very promising results. The potential carrier described in that work could prevent the permeability of the encapsulated drug and reduced the drug release rate effectively.

The formation of physically or chemically crosslinked hydrogels based on polysaccharides is a well-known fashion to obtain efficient carriers for drug delivery systems. The huge number of studies published regarding to this subject confirms such statement. On the other hand, the claims for new and enhanced materials induced the association of traditional polysaccharides-based networks and micro- and nanoparticles, making this procedure important methodology for this promising and still open research area. One example is the incorporation of iron/magnetite nanoparticles in the polysaccharide-based hydrogels networks [88, 156, 184]. Liang et al. [88] prepared a biocompatible polysaccharide-based hydrogel of maleilated-carboxymethyl chitosan grafted with PNIPAAm filled with magnetic iron oxide nanoparticles. By investigating the partition coefficients of BSA, as a model protein, this magnetic hydrogel was found to hold a potential application in magnetically assisted bioseparation. Similarly, Paulino et al. [121] related that external magnetic fields were responsible for increase the release rate of Curcumin from different polysaccharide-based hydrogel containing magnetite (i.e., $FeO{:}Fe_2O_3$).

9.5.2 *Scaffolds for Cell Growth*

Polysaccharide-based stimuli-responsive hydrogels as scaffolds for cell culturing is a promising field. Scaffolds can mimic the natural structure and are capable of supporting three-dimensional cell growth and tissue formation. Therefore, cells can be seeded in such artificial structures to grow. In general lines the scaffolds allow cell attachment, proliferation, and migration; enable the diffusion of vital cell nutrients; deliver and retain cells and biochemical factors; and exert certain mechanical and biological influences to modify the behavior of the cell phase [125, 150]. Hydrogels present a very suitable 3D network able to support cells and bioactive molecules, and their swollen state allows an efficient mass transfer [150]. Recently, polysaccharides-based stimuli-responsive hydrogels have been applied as efficient scaffolds because they can provide tailored biofunctions at molecular level, adjustable mechanical properties and morphologies according to external/internal stimuli as well as biocompatibility and biodegradability. The use of polysaccharides found in the extracellular matrix (i.e., the glycosaminoglycans) in the hydrogel formation allows obtaining a promising microenvironment for cell growth and tissue formation.

Hyaluronic acid is a key component of the extracellular matrix and is one of nature's most versatile and fascinating macromolecule. Due to its properties, hyaluronic acid has been considered an attractive starting material for the construction of hydrogels with desired morphology, stiffness and bioactivity to be applied as scaffolds. Several studies have attested that hyaluronic based hydrogels are promising materials for tissue repair and regeneration [17, 167].

Polysaccharide-based nanofibers hydrogels are an innovative method to obtain biomimetic materials that show the microstructure and porosity of extracellular matrices. For this finality, electrospinning has showed great applicability because of its efficiency and simplicity in fabricating of nanofibrous structure [24]. Polysaccharides such as gelatin, chitosan, and hyaluronic acid have been largely employed for electrospinning process.

9.5.3 *Other Technological Applications*

The high hydrophilic characteristic of polysaccharides has been exploited to form novel superabsorbent materials to be applied as efficient soil conditioner, water remediation, fertilizer release, heavy metals removal, etc. [55, 56, 90, 187]. In addition, several hydrogels composites in which the networks are formed by grafting vinyl and acrylic monomers onto polysaccharides backbone and then filled with inorganic or organic micro- and nanoparticles have been synthesized. Some studies demonstrate that the incorporation of such sized-particles (i.e., clays, iron, and silver nanoparticles, ashes, cellulose and chitin whiskers, etc.) is an advantageous method to form superabsorbent hydrogels with unique environmental and commercial advantages (e.g., biodegradability, low cost, and bioavailability) [53, 136, 152]. Furthermore, in some cases the incorporation of these particles enhances some response associated to a specific property or even creates a new one. The crescent environmental concerns on the formation of nature-friendly materials and the necessity to find substitutes for the oil-based polymers demonstrate that the use of polysaccharides could be such a promising solution to solve current environmental problems.

9.6 Perspectives of Stimuli-Responsive Polysaccharide-Based Hydrogels

This chapter demonstrates that polysaccharide-based hydrogels are versatile materials because their properties, beyond the biocompatibility and low cytotoxicity, are not found in other class of materials, allowing polysaccharides to be widely applied in several fields. It is also important to point out that the availability, renewable sources, broad chemical structures, hydrophilic properties, and the possibility of chemical modification make them a very important starting material for preparing hydrogels. The possibility in inserting groups in the polysaccharide main chain providing specific properties such external stimuli (e.g., temperature, pH, and magnetic) give rise to a plenty of possibilities to produce polysaccharide-based stimuli-responsive hydrogels with a great variety of properties. Therefore, there are still a lot of subjects to be investigated and improved in this type of materials. More comprehensive studies will broaden the understanding of structure–property relationship. Some future trends in this field can be highlighted, among others, as

(i) chemical modification of polysaccharides, in a specific point of polymer chain, for further click chemistry reaction targeting hydrogels preparation [81];

(ii) synthesis of block copolymers based on polysaccharides to obtain block with hydrophilic and block with hydrophobic characters targeting the self-assembling in certain conditions) allowing the drug encapsulating/release;
(iii) synthesis of polysaccharides in controlled conditions allowing control of chemical structure and molecular mass. For instance the obtaining bacterial cellulose scaffolds and cellulose nanowhiskers for tissue engineering [34];
(iv) nanohydrogels is a hot topic in science and technology of soft materials. In spite of lot of different nanogels have been prepared, they are considered still in naissance and very import for the future in hydrogel field, for instance in medicine applications, mainly those materials constituted by polysaccharides [131, 165].

9.7 Conclusions

Polysaccharide-based hydrogels are very important due to the set of properties not found in other class of materials. The number of publications announcing novel polysaccharide-based hydrogel with new structures and/or properties has increased year by year. The stimuli-responsive hydrogels, as smart material, occupy important place due to the versatile and wide applications. As pointed out in this chapter, however, there are challenges to be surpassed. More comprehensive studies will broaden the understanding of structure–property relationship in this very important class of materials allowing new materials and new applications. Thus, the future of polysaccharide-based stimuli-responsive hydrogels should remain in continuous growing, in a fast fashion, for the next decades.

Acknowledgments

The authors are grateful for the financial support from Fundação para a Ciência e Tecnologia (Portugal) and Coordenação para Aperfeiçoamento de Pessoal de Nível Superior (Brazil), or

FCT/CAPES application (Proc. No. 329/13), allowing scientific cooperation between Brazilian and Portuguese researches groups.

References

1. Abruzzo, A., F. Bigucci, T. Cerchiara, B. Saladini, M. C. Gallucci, F. Cruciani, B. Vitali, and B. Luppi (2013). Chitosan/ alginate complexes for vaginal delivery of chlorhexidine digluconate. *Carbohydrate Polymers* (**91**), pp. 651–658.
2. Aizawa, Y., S. C. Owen, and M. S. Shoichet (2012). Polymers used to influence cell fate in 3D geometry: new trends. *Progress in Polymer Science* (**37**), pp. 645–658.
3. Amiya, T., Y. Hirokawa, Y. Hirose, Y. Li, and T. Tanaka (1987). Reentrant phase-transition of *N*-isopropylacrylamide gels in mixed-solvents. *Journal of Chemical Physics* (**86**), pp. 2375–2379.
4. Athawale, V. D., and V. Lele (2001). Recent trends in hydrogels based on starch-graft-acrylic acid: a review. *Starch-Starke* (**53**), pp. 7–13.
5. Ayame, H., N. Morimoto, and K. Akiyoshi (2008). Self-assembled cationic nanogels for intracellular protein delivery. *Bioconjugate Chemistry* (**19**), pp. 882–890.
6. Bajpai, A. K., and R. Saini (2006). Preparation and characterization of novel biocompatible cryogels of poly(vinyl alcohol) and egg-albumin and their water sorption study. *Journal of Materials Science: Materials in Medicine* (**17**), pp. 49–61.
7. Bearat, H. H., B. H. Lee, and B. L. Vernon (2012). Comparison of properties between NIPAAm-based simultaneously physically and chemically gelling polymer systems for use in vivo. *Acta Biomaterialia* (**8**), pp. 3629–3642.
8. Becker, A. L., A. P. R. Johnston, and F. Caruso (2010). Layer-by-layer-assembled capsules and films for therapeutic delivery. *Small* (**6**), pp. 1836–1852.
9. Bellich, B., M. Borgogna, M. Cok, and A. Cesaro (2011). Water evaporation from gel beads: a calorimetric approach to hydrogel matrix release properties. *Journal of Thermal Analysis and Calorimetry* (**103**), pp. 81–88.
10. Berger, J., M. Reist, J. M. Mayer, O. Felt, and R. Gurny (2004). Structure and interactions in chitosan hydrogels formed by complexation or aggregation for biomedical applications. *European Journal of Pharmaceutics and Biopharmaceutics* (**57**), pp. 35–52.

11. Bhattarai, N., J. Gunn, and M. Zhang (2010). Chitosan-based hydrogels for controlled, localized drug delivery. *Advanced Drug Delivery Reviews* (**62**), pp. 83–99.
12. Billmeyer, F. W. (1984). *Textbook of Polymer Science*. New York: John Wiley & Sons, Inc.
13. Birdi, G., R. H. Bridson, A. M. Smith, S. P. M. Bohari, and L. M. Grover (2012). Modification of alginate degradation properties using orthosilicic acid. *Journal of the Mechanical Behavior of Biomedical Materials* (**6**), pp. 181–187.
14. Bourzac, K. (2012). Nanotechnology carrying drugs. *Nature* (**491**), pp. S58–S60.
15. Buenger, D., F. Topuz, and J. Groll (2012). Hydrogels in sensing applications. *Progress in Polymer Science* (**37**), pp. 1678–1719.
16. Bueno, V. B., R. Bentini, L. H. Catalani, and D. F. S. Petri (2013). Synthesis and swelling behavior of xanthan-based hydrogels. *Carbohydrate Polymers* (**92**), pp. 1091–1099.
17. Burdick, J. A., and G. Vunjak-Novakovic (2009). Engineered microenvironments for controlled stem cell differentiation. *Tissue Engineering Part A* (**15**), pp. 205–219.
18. Candido, J. D., R. C. F. Leitao, N. Ricardo, J. P. A. Feitosa, E. C. Muniz, and F. H. A. Rodrigues (2012). Hydrogels composite of poly(acrylamide-co-acrylate) and rice husk ash. I. Synthesis and characterization. *Journal of Applied Polymer Science* (**123**), pp. 879–887.
19. Casolaro, M., and I. Casolaro (2012). Multiple stimuli-responsive hydrogels for metal-based drug therapy. *Polymers* (**4**), pp. 964–985.
20. Chang, C., B. Duan, J. Cai, and L. Zhang (2010). Superabsorbent hydrogels based on cellulose for smart swelling and controllable delivery. *European Polymer Journal* (**46**), pp. 92–100.
21. Chang, C., B. Duan, and L. Zhang (2009). Fabrication and characterization of novel macroporous cellulose-alginate hydrogels. *Polymer* (**50**), pp. 5467–5473.
22. Chang, C., K. Han, and L. Zhang (2011). Structure and properties of cellulose/poly(*N*-isopropylacrylamide) hydrogels prepared by IPN strategy. *Polymers for Advanced Technologies* (**22**), pp. 1329–1334.
23. Chang, C., and L. Zhang (2011). Cellulose-based hydrogels: present status and application prospects. *Carbohydrate Polymers* (**84**), pp. 40–53.

24. Chen, J., X. Chen, X. Yang, C. Han, C. Gao, and Z. Gou (2013). Bioactive glasses-incorporated, core-shell-structured polypeptide/polysaccharide nanofibrous hydrogels. *Carbohydrate Polymers* (**92**), pp. 612–620.

25. Chetty, A., J. Kovacs, Z. Sulyok, A. Meszaros, J. Fekete, A. Domjan, et al. (2013). A versatile characterization of poly(*N*-isopropylacrylamideco-*N*,*N*′-methylene-bis-acrylamide) hydrogels for composition, mechanical strength, and rheology. *Express Polymer Letters* (**7**), pp. 95–105.

26. Costa, D., A. J. M. Valente, M. G. Miguel, and J. Queiroz (2011). Gel network photodisruption: a new strategy for the codelivery of plasmid DNA and drugs. *Langmuir* (**27**), pp. 13780–13789.

27. Coviello, T., P. Matricardi, A. Balena, B. Chiapperino, and F. Alhaique (2010). Hydrogels from scleroglucan and ionic crosslinkers: characterization and drug delivery. *Journal of Applied Polymer Science* (**115**), pp. 3610–3622.

28. Coviello, T., P. Matricardi, C. Marianecci, and F. Alhaique (2007). Polysaccharide hydrogels for modified release formulations. *Journal of Controlled Release* (**119**), pp. 5–24.

29. Crescenzi, V., L. Cornelio, C. Di Meo, S. Nardecchia, and R. Lamanna (2007). Novel hydrogels via click chemistry: synthesis and potential biomedical applications. *Biomacromolecules* (**8**), pp. 1844–1850.

30. d'Ayala, G. G., M. Malinconico, and P. Laurienzo (2008). Marine derived polysaccharides for biomedical applications: chemical modification approaches. *Molecules* (**13**), pp. 2069–2106.

31. de Moura, M. R., F. A. Aouada, M. R. Guilherme, E. Radovanovic, A. F. Rubira, and E. C. Muniz (2006). Thermo-sensitive IPN hydrogels composed of PNIPAAm gels supported on alginate-Ca2+ with LCST tailored close to human body temperature. *Polymer Testing* (**25**), pp. 961–969.

32. Ding, J., R. He, G. Zhou, C. Tang, and C. Yin (2012). Multilayered mucoadhesive hydrogel films based on thiolated hyaluronic acid and polyvinylalcohol for insulin delivery. *Acta Biomaterialia* (**8**), pp. 3643–3651.

33. Drury, J. L., and D. J. Mooney (2003). Hydrogels for tissue engineering: scaffold design variables and applications. *Biomaterials* (**24**), pp. 4337–4351.

34. Dugan, J. M., J. E. Gough, and S. J. Eichhorn (2013). Bacterial cellulose scaffolds and cellulose nanowhiskers for tissue engineering. *Nanomedicine* (**8**), pp. 287–298.

35. Durst, C. A., M. P. Cuchiara, E. G. Mansfield, J. L. West, and K. J. Grande-Allen (2011). Flexural characterization of cell encapsulated PEGDA hydrogels with applications for tissue engineered heart valves. *Acta Biomaterialia* (**7**), pp. 2467–2476.

36. Ekenseair, A. K., K. W. M. Boere, S. N. Tzouanas, T. N. Vo, F. K. Kasper, and A. G. Mikos (2012). Synthesis and characterization of thermally and chemically gelling injectable hydrogels for tissue engineering. *Biomacromolecules* (**13**), pp. 1908–1915.

37. Elias, H.-G. (1977). *Macromolecules* (1). Weinheim: Wiley-VCH Verlag GmbH.

38. Fajardo, A. R., L. C. Lopes, A. O. Caleare, E. A. Britta, C. V. Nakamura, A. F. Rubira, et al. (2013). Silver sulfadiazine loaded chitosan/chondroitin sulfate films for a potential wound dressing application. *Materials Science and Engineering: C* (**33**), pp. 588–595.

39. Fajardo, A. R., L. C. Lopes, A. G. B. Pereira, A. F. Rubira, and E. C. Muniz (2012). Polyelectrolyte complexes based on pectin-NH2 and chondroitin sulfate. *Carbohydrate Polymers* (**87**), pp. 1950–1955.

40. Fajardo, A. R., L. C. Lopes, A. J. M. Valente, A. F. Rubira, and E. C. Muniz (2011). Effect of stoichiometry and pH on the structure and properties of Chitosan/Chondroitin sulfate complexes. *Colloid and Polymer Science* (**289**), pp. 1739–1748.

41. Fajardo, A. R., J. F. Piai, A. F. Rubira, and E. C. Muniz (2010). Time- and pH-dependent self-rearrangement of a swollen polymer network based on polyelectrolytes complexes of chitosan/chondroitin sulfate. *Carbohydrate Polymers* (**80**), pp. 934–943.

42. Fajardo, A. R., M. B. Silva, L. C. Lopes, J. F. Piai, A. F. Rubira, and E. C. Muniz (2012). Hydrogel based on an alginate-Ca2+/chondroitin sulfate matrix as a potential colon-specific drug delivery system. *Rsc Advances* (**2**), pp. 11095–11103.

43. Feil, H., Y. H. Bae, J. Feijen, and S. W. Kim (1991). Molecular separation by thermosensitive hydrogel membranes. *Journal of Membrane Science* (**64**), pp. 283–294.

44. Fisher, O. Z., A. Khademhosseini, R. Langer, and N. A. Peppas (2010). Bioinspired materials for controlling stem cell fate. *Accounts of Chemical Research* (**43**), pp. 419–428.

45. Flory, P. J. (1953). *Principles of Polymer Chemistry*. New York: Cornell University.

46. Ganji, F., S. Vasheghani-Farahani, and E. Vasheghani-Farahani (2010). Theoretical description of hydrogel swelling: a review. *Iranian Polymer Journal* (**19**), pp. 375–398.
47. Garcia, H., A. S. Barros, C. Goncalves, F. M. Gama, and A. M. Gil (2008). Characterization of dextrin hydrogels by FTIR spectroscopy and solid state NMR spectroscopy. *European Polymer Journal* (**44**), pp. 2318–2329.
48. George, M., and T. E. Abraham (2006). Polyionic hydrocolloids for the intestinal delivery of protein drugs: alginate and chitosan–a review. *Journal of Controlled Release* (**114**), pp. 1–14.
49. Gil, E. S., and S. M. Hudson (2004). Stimuli-reponsive polymers and their bioconjugates. *Progress in Polymer Science* (**29**), pp. 1173–1222.
50. Guilherme, M. R., G. M. Campese, E. Radovanovic, A. F. Rubira, E. B. Tambourgi, and E. C. Muniz (2006). Thermo-responsive sandwiched-like membranes of IPN-PNIPAAm/PAAm hydrogels. *Journal of Membrane Science* (**275**), pp. 187–194.
51. Guilherme, M. R., R. da Silva, A. F. Rubira, G. Geuskens, and E. C. Muniz (2004). Thermo-sensitive hydrogels membranes from PAAm networks and entangled PNIPAAm: effect of temperature, cross-linking and PNIPAAm contents on the water uptake and permeability. *Reactive & Functional Polymers* (**61**), pp. 233–243.
52. Guilherme, M. R., M. R. de Moura, E. Radovanovic, G. Geuskens, A. F. Rubira, and E. C. Muniz (2005). Novel thermo-responsive membranes composed of interpenetrated polymer networks of alginate-Ca2+ and poly(*N*-isopropylacrylamide). *Polymer* (**46**), pp. 2668–2674.
53. Guilherme, M. R., A. R. Fajardo, T. A. Moia, M. H. Kunita, M. d. C. Goncalves, A. F. Rubira, et al. (2010). Porous nanocomposite hydrogel of vinyled montmorillonite-crosslinked maltodextrin-co-dimethylacrylamide as a highly stable polymer carrier for controlled release systems. *European Polymer Journal* (**46**), pp. 1465–1474.
54. Guilherme, M. R., T. A. Moia, A. V. Reis, A. T. Paulino, A. F. Rubira, L. H. C. Mattoso, et al. (2009). Synthesis and water absorption transport mechanism of a ph-sensitive polymer network structured on vinyl-functionalized pectin. *Biomacromolecules* (**10**), pp. 190–196.
55. Guilherme, M. R., A. V. Reis, A. T. Paulino, A. R. Fajardo, E. C. Muniz, and E. B. Tambourgi (2007). Superabsorbent hydrogrel based on modified polysaccharide for removal of Pb^{2+} and Cu^{2+} from water with excellent performance. *Journal of Applied Polymer Science* (**105**), pp. 2903–2909.

56. Guilherme, M. R., A. V. Reis, A. T. Paulino, T. A. Moia, L. H. C. Mattoso, and E. B. Tambourgi (2010). Pectin-based polymer hydrogel as a carrier for release of agricultural nutrients and removal of heavy metals from wastewater. *Journal of Applied Polymer Science* (**117**), pp. 3146–3154.

57. Guilherme, M. R., A. V. Reis, S. H. Takahashi, A. F. Rubira, J. P. A. Feitosa, and E. C. Muniz (2005). Synthesis of a novel superabsorbent hydrogel by copolymerization of acrylamide and cashew gum modified with glycidyl methacrylate. *Carbohydrate Polymers* (**61**), pp. 464–471.

58. Guilherme, M. R., R. Silva, E. M. Girotto, A. F. Rubira, and E. C. Muniz (2003). Hydrogels based on PAAm network with PNIPAAm included: hydrophilic-hydrophobic transition measured by the partition of orange II and methylene blue in water. *Polymer* (**44**), pp. 4213–4219.

59. Hennink, W. E., and C. F. van Nostrum (2002). Novel crosslinking methods to design hydrogels. *Advanced Drug Delivery Reviews* (**54**), pp. 13–36.

60. Hirokawa, Y., and T. Tanaka (1984). Volume phase-transition in a nonionic gel. *Journal of Chemical Physics* (**81**), pp. 6379–6380.

61. Hirotsu, S., Y. Hirokawa, and T. Tanaka (1987). Volume phase-transition of ionized *N*-isopropylacrylamide gels. *Journal of Chemical Physics* (**87**), pp. 1392–1395.

62. Hu, C.-H., X.-Z. Zhang, L. Zhang, X.-D. Xu, and R.-X. Zhuo (2009). Temperature- and pH-sensitive hydrogels to immobilize heparin-modified PEI/DNA complexes for sustained gene delivery. *Journal of Materials Chemistry* (**19**), pp. 8982–8989.

63. Hu, J., H. Meng, G. Li, and S. I. Ibekwe (2012). A review of stimuli-responsive polymers for smart textile applications. *Smart Materials and Structures* (**21**), 053001.

64. Huttermann, A., M. Zommorodi, and K. Reise (1999). Addition of hydrogels to soil for prolonging the survival of pinus halepensis seedlings subjected to drought. *Soil & Tillage Research* (**50**), pp. 295–304.

65. Ichibouji, T., T. Miyazaki, E. Ishida, A. Sugino, and C. Ohtsuki (2009). Apatite mineralization abilities and mechanical properties of covalently cross-linked pectin hydrogels. *Materials Science & Engineering C-Materials for Biological Applications* (**29**), pp. 1765–1769.

66. Ishihara, M., K. Nakanishi, K. Ono, M. Sato, M. Kikuchi, Y. Saito, et al. (2002). Photocrosslinkable chitosan as a dressing for wound occlusion and accelerator in healing process. *Biomaterials* (**23**), pp. 833–840.

67. Jamshidi, A., F. A. K. Beigi, K. Kabiri, and M. J. Zohuriaan-Mehr (2005). Optimized HPLC determination of residual monomer in hygienic SAP hydrogels. *Polymer Testing* (**24**), pp. 825–828.
68. Ji, C., A. Khademhosseini, and F. Dehghani (2011). Enhancing cell penetration and proliferation in chitosan hydrogels for tissue engineering applications. *Biomaterials* (**32**), pp. 9719–9729.
69. Jones, C. D., and L. A. Lyon (2000). Synthesis and characterization of multiresponsive core-shell microgels. *Macromolecules* (**33**), pp. 8301–8306.
70. Kaith, B. S., Saruchi, R. Jindal, and M. S. Bhatti (2012). Screening and RSM optimization for synthesis of a gum tragacanth-acrylic acid based device for in situ controlled cetirizine dihydrochloride release. *Soft Matter* (**8**), pp. 2286–2293.
71. Kang, S. I., K. Na, and Y. H. Bae (2001). Sulfonamide-containing polymers: a new class of pH-sensitive polymers and gels. *Macromolecular Symposia* (**172**), pp. 149–156.
72. Kang, S. I., K. Na, and Y. H. Bae (2003). Physicochemical characteristics and doxorubicin-release behaviors of pH/temperature-sensitive polymeric nanoparticles. *Colloids and Surfaces a-Physicochemical and Engineering Aspects* (**231**), pp. 103–112.
73. Karadag, E., O. B. Uzum, and D. Saraydin (2002). Swelling equilibria and dye adsorption studies of chemically crosslinked superabsorbent acrylamide/maleic acid hydrogels. *European Polymer Journal* (**38**), pp. 2133–2141.
74. Katayama, S., Y. Hirokawa, and T. Tanaka (1984). Reentrant phase-transition in acrylamide-derivative copolymer gels. *Macromolecules* (**17**), pp. 2641–2643.
75. Kim, S. G., G. T. Lim, J. Jegal, and K. H. Lee (2000). Pervaporation separation of MTBE (methyl tert-butyl ether) and methanol mixtures through polyion complex composite membranes consisting of sodium alginate/chitosan. *Journal of Membrane Science* (**174**), pp. 1–15.
76. Kim, S. H., and C. C. Chu (1999). Synthesis and characterization of dextran-methacrylate hydrogels and structural study by SEM. *Journal of Biomedical Materials Research* (**49**), pp. 517–527.
77. Korsmeyer, R. W., R. Gurny, E. Doelker, P. Buri, and N. A. Peppas (1983). Mechanisms of solute release from porous hydrophilic polymers. *International Journal of Pharmaceutics* (**15**), pp. 25–35.
78. Koutsopoulos, S., and S. Zhang (2012). Two-layered injectable self-assembling peptide scaffold hydrogels for long-term sustained release

of human antibodies. *Journal of Controlled Release* (**160**), pp. 451–458.

79. Kramer, J., R. K. Prudhomme, P. Wiltzius, P. Mirau, and S. Knoll (1988). Comparison of galactomannan crosslinking with organotitanates and borates. *Colloid and Polymer Science* (**266**), pp. 145–155.
80. Kuang, J., K. Y. Yuk, and K. M. Huh (2011). Polysaccharide-based superporous hydrogels with fast swelling and superabsorbent properties. *Carbohydrate Polymers* (**83**), pp. 284–290.
81. Kushwaha, D., P. Dwivedi, S. K. Kuanar, and V. K. Tiwari (2013). Click reaction in carbohydrate chemistry: recent developments and future perspective. *Current Organic Synthesis* (**10**), pp. 90–135.
82. Lavergne, M., M. Derkaoui, C. Delmau, D. Letourneur, G. Uzan, and C. Le Visage (2012). Porous polysaccharide-based scaffolds for human endothelial progenitor cells. *Macromolecular Bioscience* (**12**), pp. 901–910.
83. Leach, J. B., K. A. Bivens, C. W. Patrick, and C. E. Schmidt (2003). Photocrosslinked hyaluronic acid hydrogels: natural, biodegradable tissue engineering scaffolds. *Biotechnology and Bioengineering* (**82**), pp. 578–589.
84. Lessard, D. G., Y. F. Li, and X. X. Zhu (2012). Thermosensitive superabsorbents based on poly(*N*,*N*-diethylacrylamide-co-sodium acrylate). *Chinese Journal of Polymer Science* (**30**), pp. 873–878.
85. Lessard, D. G., M. Ousalem, and X. X. Zhu (2001). Effect of the molecular weight on the lower critical solution temperature of poly(*N*,*N*-diethylacrylamide) in aqueous solutions. *Canadian Journal of Chemistry-Revue Canadienne De Chimie* (**79**), pp. 1870–1874.
86. Lessard, D. G., M. Ousalem, X. X. Zhu, A. Eisenberg, and P. J. Carreau (2003). Study of the phase transition of poly(*N*,*N*-diethylacrylamide) in water by rheology and dynamic light scattering. *Journal of Polymer Science Part B-Polymer Physics* (**41**), pp. 1627–1637.
87. Li, H., and F. K. Lai (2010). Multiphysics modeling of responsive characteristics of ionic-strength-sensitive hydrogel. *Biomedical Microdevices* (**12**), pp. 419–434.
88. Liang, Y.-Y., L.-M. Zhang, W. Jiang, and W. Li (2007). Embedding magnetic nanoparticles into polysaccharide-based hydrogels for magnetically assisted bioseparation. *Chemphyschem* (**8**), pp. 2367–2372.
89. Lima, A. C., W. Song, B. Blanco-Fernandez, C. Alvarez-Lorenzo, and J. F. Mano (2011). Synthesis of temperature-responsive dextran-

MA/PNIPAAm particles for controlled drug delivery using superhydrophobic surfaces. *Pharmaceutical Research* (**28**), pp. 1294–1305.

90. Lin, L., and Q. Wu (2012). Improved conductivity of polysaccharide-co-polyacrylate/polyaniline conducting hydrogels. *Polymers & Polymer Composites* (**20**), pp. 377–386.
91. Lin, Y. H., H. F. Liang, C. K. Chung, M. C. Chen, and H. W. Sung (2005). Physically crosslinked alginate/*N*,O-carboxymethyl chitosan hydrogels with calcium for oral delivery of protein drugs. *Biomaterials* (**26**), pp. 2105–2113.
92. Lin-Gibson, S., H. J. Walls, S. B. Kennedy, and E. R. Welsh (2003). Reaction kinetics and gel properties of blocked diisocyinate crosslinked chitosan hydrogels. *Carbohydrate Polymers* (**54**), pp. 193–199.
93. Lindman, B., G. Karlstrom, and L. Stigsson (2010). On the mechanism of dissolution of cellulose. *Journal of Molecular Liquids* (**156**), pp. 76–81.
94. Lionetto, F., A. Sannino, G. Mensitieri, and A. Maffezzoli (2003). Evaluation of the degree of cross-linking of cellulose-based superabsorbent hydrogels: a comparison between different techniques. *Macromolecular Symposia* (**200**), pp. 199–207.
95. Luo, X. L., J. J. Xu, Q. Zhang, G. J. Yang, and H. Y. Chen (2005). Electrochemically deposited chitosan hydrogel for horseradish peroxidase immobilization through gold nanoparticles self-assembly. *Biosensors & Bioelectronics* (**21**), pp. 190–196.
96. Mahkam, M. (2010). Starch-based polymeric carriers for oral-insulin delivery. *Journal of Biomedical Materials Research Part A* (**92A**), pp. 1392–1397.
97. Mayers, R. A. (2001). *Encyclopedia of Physical Science and Technology: Biochemistry*. Amsterdam, Elsevier Science Ltd.
98. McConaughy, S. D., S. E. Kirkland, N. J. Treat, P. A. Stroud, and C. L. McCormick (2008). Tailoring the network properties of Ca2+ crosslinked aloe vera polysaccharide hydrogels for in situ release of therapeutic agents. *Biomacromolecules* (**9**), pp. 3277–3287.
99. Mergy, J., A. Fournier, E. Hachet, and R. Auzely-Velty (2012). Modification of polysaccharides via thiol-ene chemistry: a versatile route to functional biomaterials. *Journal of Polymer Science Part a-Polymer Chemistry* (**50**), pp. 4019–4028.
100. Molina, M. J., M. R. Gomez-Anton, and I. F. Pierola (2007). Determination of the parameters controlling swelling of chemically cross-linked pH-Sensitive Poly(*N*-vinylimidazole) hydrogels. *Journal of Physical Chemistry B* (**111**), pp. 12066–12074.

101. Morimoto, N., T. Endo, M. Ohtomi, Y. Iwasaki, and K. Akiyoshi (2005). Hybrid nanogels with physical and chemical cross-linking structures as nanocarriers. *Macromolecular Bioscience* (**5**), pp. 710–716.

102. Muniz, E. C., and G. Geuskens (2000). Influence of temperature on the permeability of polyacrylamide hydrogels and semi-IPNs with poly(*N*-isopropylacrylamide). *Journal of Membrane Science* (**172**), pp. 287–293.

103. Muniz, E. C., and G. Geuskens (2001). Compressive elastic modulus of polyacrylamide hydrogels and semi-IPNs with poly(*N*-isopropylacrylamide). *Macromolecules* (**34**), pp. 4480–4484.

104. Munjeri, O., P. Hodza, E. E. Osim, and C. T. Musabayane (1998). An investigation into the suitability of amidated pectin hydrogel beads as a delivery matrix for chloroquine. *Journal of Pharmaceutical Sciences* (**87**), pp. 905–908.

105. Na, K., E. S. Lee, and Y. H. Bae (2007). Self-organized nanogels responding to tumor extracellular pH: pH-dependent drug release and in vitro cytotoxicity against MCF-7 cells. *Bioconjugate Chemistry* (**18**), pp. 1568–1574.

106. Nakai, T., T. Hirakura, Y. Sakurai, T. Shimoboji, M. Ishigai, and K. Akiyoshi (2012). Injectable hydrogel for sustained protein release by salt-induced association of hyaluronic acid nanogel. *Macromolecular Bioscience* (**12**), pp. 475–483.

107. Narayanan, R. P., G. Melman, N. J. Letourneau, N. L. Mendelson, and A. Melman (2012). Photodegradable iron(III) cross-linked alginate gels. *Biomacromolecules* (**13**), pp. 2465–2471.

108. Nelson, D. L., and M. M. C. Lehninger (2004). *Principles of Biochemistry*. New York: W. H. Freeman and Company.

109. Nickerson, M. T., R. Farnworth, E. Wagar, S. M. Hodge, D. Rousseau, and A. T. Paulson (2006). Some physical and microstructural properties of genipin-crosslinked gelatin-maltodextrin hydrogels. *International Journal of Biological Macromolecules* (**38**), pp. 40–44.

110. Nicolson, P. C., and J. Vogt (2001). Soft contact lens polymers: an evolution. *Biomaterials* (**22**), pp. 3273–3283.

111. Nie, T., A. Baldwin, N. Yamaguchi, and K. L. Kiick (2007). Production of heparin-functionalized hydrogels for the development of responsive and controlled growth factor delivery systems. *Journal of Controlled Release* (**122**), pp. 287–296.

112. Noble, L., A. I. Gray, L. Sadiq, and I. F. Uchegbu (1999). A non-covalently cross-linked chitosan based hydrogel. *International Journal of Pharmaceutics* (**192**), pp. 173–182.

113. Oh, J. K., D. I. Lee, and J. M. Park (2009). Biopolymer-based microgels/nanogels for drug delivery applications. *Progress in Polymer Science* (**34**), pp. 1261–1282.

114. Omidian, H., J. G. Rocca, and K. Park (2006). Elastic, superporous hydrogel hybrids of polyacrylamide and sodium alginate. *Macromolecular Bioscience* (**6**), pp. 703–710.

115. Pagonis, K., and G. Bokias (2006). Simultaneous lower and upper critical solution temperature-type co-nonsolvency behaviour exhibited in water-dioxane mixtures by linear copolymers and hydrogels containing *N*-isopropylacrylamide and *N,N*-dimethylacrylamide. *Polymer International* (**55**), pp. 1254–1258.

116. Papancea, A., A. J. M. Valente, S. Patachia, M. G. Miguel, and B. Lindman (2008). PVA-DNA cryogel membranes: characterization, swelling, and transport studies. *Langmuir* (**24**), pp. 273–279.

117. Pasparakis, G., and M. Vamvakaki (2011). Multiresponsive polymers: nano-sized assemblies, stimuli-sensitive gels and smart surfaces. *Polymer Chemistry* (**2**), pp. 1234–1248.

118. Pasqui, D., M. De Cagna, and R. Barbucci (2012). Polysaccharide-based hydrogels: the key role of water in affecting mechanical properties. *Polymers* (**4**), pp. 1517–1534.

119. Paulino, A. T., G. M. Campese, S. L. Favaro, M. R. Guilherme, E. B. Tambourgi, and E. C. Muniz (2007). Water absorption profile of PAAm-co-PNIPAAm/chitosan hydrogel with sandwich-like morphology. *E-Polymers* (**122**).

120. Paulino, A. T., A. R. Fajardo, A. P. Junior, E. C. Muniz, and E. B. Tambourgi (2011). Two-step synthesis and properties of amagnetic-field-sensitive modified maltodextrin-based hydrogel. *Polymer International* (**60**), pp. 1324–1333.

121. Paulino, A. T., A. G. B. Pereira, A. R. Fajardo, K. Erickson, M. J. Kipper, E. C. Muniz, et al. (2012). Natural polymer-based magnetic hydrogels: potential vectors for remote-controlled drug release. *Carbohydrate Polymers* (**90**), pp. 1216–1225.

122. Pawar, S. N., and K. J. Edgar (2012). Alginate derivatization: a review of chemistry, properties and applications. *Biomaterials* (**33**), pp. 3279–3305.

123. Peppas, N. A., and P. Colombo (1997). Analysis of drug release behavior from swellable polymer carriers using the dimensionality index. *Journal of Controlled Release* (**45**), pp. 35–40.

124. Peppas, N. A., B. V. Slaughter, and M. A. Kanzelberger (2012). 9.20 - Hydrogels, *in* M. Krzysztof and M. Martin, eds., *Polymer Science: A Comprehensive Reference*. Amsterdam: Elsevier, pp. 385–395.

125. Peran, M., M. A. Garcia, E. Lopez-Ruiz, M. Bustamante, G. Jimenez, R. Madeddu, et al. (2012). Functionalized nanostructures with application in regenerative medicine. *International Journal of Molecular Sciences* (**13**), pp. 3847–3886.

126. Piai, J. F., A. F. Rubira, and E. C. Muniz (2009). Self-assembly of a swollen chitosan/chondroitin sulfate hydrogel by outward diffusion of the chondroitin sulfate chains. *Acta Biomaterialia* (**5**), pp. 2601–2609.

127. Pichayakorn, W., and P. Boonme (2013). Evaluation of cross-linked chitosan microparticles containing metronidazole for periodontitis treatment. *Materials Science and Engineering: C* (**33**), pp. 1197–1202.

128. Pielesz, A., and M. K. K. Bak (2008). Raman spectroscopy and WAXS method as a tool for analysing ion-exchange properties of alginate hydrogels. *International Journal of Biological Macromolecules* (**43**), pp. 438–443.

129. Pissis, P., and A. Kyritsis (2013). Hydration studies in polymer hydrogels. *Journal of Polymer Science Part B-Polymer Physics* (**51**), pp. 159–175.

130. Pourjavadi, A., M. Sadeghi, and H. Hosseinzadeh (2004). Modified carrageenan. 5. Preparation, swelling behavior, salt- and pH-sensitivity of partially hydrolyzed crosslinked carrageenan-graft-polymethacrylamide superabsorbent hydrogel. *Polymers for Advanced Technologies* (**15**), pp. 645–653.

131. Qian, Z. Y., S. Z. Fu, and S. S. Feng (2013). Nanohydrogels as a prospective member of the nanomedicine family. *Nanomedicine* (**8**), pp. 161–164.

132. Qiu, Y., and K. Park (2001). Environment-sensitive hydrogels for drug delivery. *Advanced Drug Delivery Reviews* (**53**), pp. 321–339.

133. Qiu, Y., and K. Park (2012). Environment-sensitive hydrogels for drug delivery. *Advanced Drug Delivery Reviews* (**64**), pp. 49–60.

134. Reis, A. V., M. R. Guilherme, O. A. Cavalcanti, A. F. Rubira, and E. C. Muniz (2006). Synthesis and characterization of pH-responsive hydrogels based on chemically modified Arabic gum polysaccharide. *Polymer* (**47**), pp. 2023–2029.

135. Reis, A. V., M. R. Guilherme, A. F. Rubira, and E. C. Muniz (2007). Mathematical model for the prediction of the overall protile of in vitro solute release from polymer networks. *Journal of Colloid and Interface Science* (**310**), pp. 128–135.

136. Rodrigues, F. H. A., A. R. Fajardo, A. G. B. Pereira, N. M. P. S. Ricardo, J. P. A. Feitosa, and E. C. Muniz (2012). Chitosan-graft-poly(acrylic acid)/rice husk ash based superabsorbent hydrogel composite: preparation and characterization. *Journal of Polymer Research* (**19**).

137. Rowley, J. A., G. Madlambayan, and D. J. Mooney (1999). Alginate hydrogels as synthetic extracellular matrix materials. *Biomaterials* (**20**), pp. 45–53.

138. Sadeghi, M., F. Shafeie, and E. Mohammadinasab (2013). Partially hydrolyzed biocopolymer based on carboxymethylcellulose/alginate-graft-AAm and effect of pH, salinity onto swelling capacity. *Asian Journal of Chemistry* (**25**), pp. 1621–1624.

139. Sakurai, M., and M. Satoh (2012). Reentrant swelling behavior of poly(4-vinyl phenol) gel in aqueous organic solvent mixtures. *E-Polymers* (**12**).

140. Santos, J. R., N. M. Alves, and J. F. Mano (2010). New thermo-responsive hydrogels based on poly(*N*-isopropylacrylamide)/hyaluronic acid semi-interpenetrated polymer networks: swelling properties and drug release studies. *Journal of Bioactive and Compatible Polymers* (**25**), pp. 169–184.

141. Schild, H. G. (1992). Poly (*N*-isopropylacrylamide) - experiment, theory and application. *Progress in Polymer Science* (**17**), pp. 163–249.

142. Schild, H. G., M. Muthukumar, and D. A. Tirrell (1991). Cononsolvency in mixed aqueous-solutions of poly(*N*-isopropylacrylamide). *Macromolecules* (**24**), pp. 948–952.

143. Seo, S., C. S. Lee, Y. S. Jung, and K. Na (2012). Thermo-sensitivity and triggered drug release of polysaccharide nanogels derived from pullulan-g-poly(*L*-lactide) copolymers. *Carbohydrate Polymers* (**87**), pp. 1105–1111.

144. Serp, D., M. Mueller, U. von Stockar, and I. W. Marison (2002). Low-temperature electron microscopy for the study of polysaccharide ultrastructures in hydrogels. I. Theoretical and technical considerations. *Biotechnology and Bioengineering* (**79**), pp. 243–252.

145. Sershen, S. R., S. L. Westcott, N. J. Halas, and J. L. West (2000). Temperature-sensitive polymer-nanoshell composites for photother-

mally modulated drug delivery. *Journal of Biomedical Materials Research* (**51**), pp. 293–298.

146. Shapiro, Y. E. (2011). Structure and dynamics of hydrogels and organogels: an NMR spectroscopy approach. *Progress in Polymer Science* (**36**), pp. 1184–1253.
147. Shi, J., L. H. Liu, X. P. Liu, X. M. Sun, and S. K. Cao (2008). Inorganic-organic hybrid alginate beads with LCST near human body temperature for sustained dual-sensitive drug delivery. *Polymers for Advanced Technologies* (**19**), pp. 1467–1473.
148. Shu, X. Z., K. J. Zhu, and W. H. Song (2001). Novel pH-sensitive citrate cross-linked chitosan film for drug controlled release. *International Journal of Pharmaceutics* (**212**), pp. 19–28.
149. Silva, S. S., A. Motta, M. T. Rodrigues, A. F. M. Pinheiro, M. E. Gomes, J. F. Mano, et al. (2008). Novel genipin-cross-linked chitosan/silk fibroin sponges for cartilage engineering strategies. *Biomacromolecules* (**9**), pp. 2764–2774.
150. Slaughter, B. V., S. S. Khurshid, O. Z. Fisher, A. Khademhosseini, and N. A. Peppas (2009). Hydrogels in regenerative medicine. *Advanced Materials* (**21**), pp. 3307–3329.
151. Spagnol, C., F. H. A. Rodrigues, A. G. V. C. Neto, A. G. B. Pereira, A. R. Fajardo, E. Radovanovic, et al. (2012). Nanocomposites based on poly(acrylamide-co-acrylate) and cellulose nanowhiskers. E*uropean Polymer Journal* (**48**), pp. 454–463.
152. Spagnol, C., F. H. A. Rodrigues, A. G. B. Pereira, A. R. Fajardo, A. F. Rubira, and E. C. Muniz (2012). Superabsorbent hydrogel composite made of cellulose nanofibrils and chitosan-graft-poly(acrylic acid). *Carbohydrate Polymers* (**87**), pp. 2038–2045.
153. Strehin, I., Z. Nahas, K. Arora, T. Nguyen, and J. Elisseeff (2010). A versatile pH sensitive chondroitin sulfate-PEG tissue adhesive and hydrogel. *Biomaterials* (**31**), pp. 2788–2797.
154. Suh, J. K. F., and H. W. T. Matthew (2000). Application of chitosan-based polysaccharide biomaterials in cartilage tissue engineering: a review. *Biomaterials* (**21**), pp. 2589–2598.
155. Sun, G., Y.-I. Shen, S. Kusuma, K. Fox-Talbot, C. J. Steenbergen, and S. Gerecht (2011). Functional neovascularization of biodegradable dextran hydrogels with multiple angiogenic growth factors. *Biomaterials* (**32**), pp. 95–106.
156. Takahashi, H., S.-I. Sawada, and K. Akiyoshi (2011). Amphiphilic polysaccharide nanoballs: a new building block for nanogel biomedical

engineering and artificial chaperones. *ACS Nano* (**5**), pp. 337–345.

157. Takigawa, T., T. Yamawaki, K. Takahashi, and T. Masuda (1997). Change in Young's modulus of poly(*N*-isopropylacrylamide) gels by volume phase transition. *Polymer Gels and Networks* (**5**), pp. 585–589.
158. Taylor, L. D., and L. D. Cerankowski (1975). Preparation of films exhibiting a balanced temperature-dependence to permeation by aqueous-solutions - study of lower consolute behavior. *Journal of Polymer Science Part a-Polymer Chemistry* (**13**), pp. 2551–2570.
159. Uliniuc, A., M. Popa, T. Hamaide, and M. Dobromir (2012). New approaches in hydrogel synthesis – Click chemistry: a review. *Cellulose Chemistry and Technology* (**46**), pp. 1–11.
160. Valente, A. J. M., S. M. A. Cruz, M. C. Moran, D. B. Murtinho, E. C. Muniz, and M. G. Miguel (2010). Release of DNA from cryogel PVA-DNA membranes. *Express Polymer Letters* (**4**), pp. 480–487.
161. Valente, A. J. M., S. M. A. Cruz, D. M. B. Murtinho, M. G. Miguel, and E. C. Muniz (2013). DNA-poly(vinyl alcohol) gel matrices: release properties are strongly dependent on electrolytes and cationic surfactants. *Colloids and Surfaces B-Biointerfaces* (**101**), pp. 111–117.
162. Van den Bulcke, A. I., B. Bogdanov, N. De Rooze, E. H. Schacht, M. Cornelissen, and H. Berghmans (2000). Structural and rheological properties of methacrylamide modified gelatin hydrogels. *Biomacromolecules* (**1**), pp. 31–38.
163. Van Vlierberghe, S., P. Dubruel, and E. Schacht (2011). Biopolymer-based hydrogels as scaffolds for tissue engineering applications: a review. *Biomacromolecules* (**12**), pp. 1387–1408.
164. Vieira, E. F. S., A. R. Cestari, C. Airoldi, and W. Loh (2008). Polysaccharide-based hydrogels: preparation, characterization, and drug interaction behaviour. *Biomacromolecules* (**9**), pp. 1195–1199.
165. Wang, M., and M. Thanou (2010). Targeting nanoparticles to cancer. *Pharmacological Research* (**62**), pp. 90–99.
166. Wang, T., and S. Gunasekaran (2006). State of water in chitosan-PVA hydrogel. *Journal of Applied Polymer Science* (**101**), pp. 3227–3232.
167. Wang, X., J. He, Y. Wang, and F.-Z. Cui (2012). Hyaluronic acid-based scaffold for central neural tissue engineering. *Interface Focus* (**2**), pp. 278–291.
168. Welsh, E. R., and R. R. Price (2003). Chitosan cross-linking with a water-soluble, blocked diisocyanate. 2. Solvates and hydrogels. *Biomacromolecules* (**4**), pp. 1357–1361.

169. Wenceslau, A. C., F. G. dos Santos, E. R. F. Ramos, C. V. Nakamura, A. F. Rubira, and E. C. Muniz (2012). Thermo- and pH-sensitive IPN hydrogels based on PNIPAAm and PVA-Ma networks with LCST tailored close to human body temperature. *Materials Science & Engineering C-Materials for Biological Applications* (**32**), pp. 1259–1265.

170. Wenceslau, A. C., A. F. Rubira, and E. C. Muniz (2012). IPN hydrogels based on PNIPAAm and PVA-Ma networks: characterization through measure of LCST, swelling ratio and mechanical properties. *Acta Scientiarum-Technology* (**34**), pp. 233–241.

171. Wichterle, O., and D. Lim (1960). Hydrophilic gels for biological use. *Nature* (**185**), pp. 117–118.

172. Wilson, L. D., D. Y. Pratt, and J. A. Kozinski (2013). Preparation and sorption studies of beta-cyclodextrin-chitosan-glutaraldehyde terpolymers. *Journal of Colloid and Interface Science* (**393**), pp. 271–277.

173. Woo, K., and P. A. Seib (1997). Cross-linking of wheat starch and hydroxypropylated wheat starch in alkaline slurry with sodium trimetaphosphate. *Carbohydrate Polymers* (**33**), pp. 263–271.

174. Xiao, C. M., H. Q. Li, and Y. K. Gao (2009). Preparation of fast pH-responsive ferric carboxymethylcellulose/poly(vinyl alcohol) double-network microparticles. *Polymer International* (**58**), pp. 112–115.

175. Xu, J., N. Cai, W. Xu, Y. Xue, Z. Wang, Q. Dai, and F. Yu (2013). Mechanical enhancement of nanofibrous scaffolds through polyelectrolyte complexation. *Nanotechnology* (24), 025701.

176. Yan, Q. Z., W. F. Zhang, G. D. Lu, X. T. Su, and C. C. Ge (2005). Frontal copolymerization synthesis and property characterization of starch-graft-poly(acrylic acid) hydrogels. *Chemistry-a European Journal* (**11**), pp. 6609–6615.

177. Yin, L., X. Zhao, L. Cui, J. Ding, M. He, C. Tang, and C. Yin (2009). Cytotoxicity and genotoxicity of superporous hydrogel containing interpenetrating polymer networks. *Food and Chemical Toxicology* (**47**), pp. 1139–1145.

178. Yu, J., Y. P. Pan, Q. F. Lu, W. Yang, J. Z. Gao, and Y. Li (2013). Synthesis and swelling behaviors of P(AMPS-co-AAc) superabsorbent hydrogel produced by glow-discharge electrolysis plasma. *Plasma Chemistry and Plasma Processing* (**33**), pp. 219–235.

179. Zaroslov, Y. D., O. E. Philippova, and A. R. Khokhlov (1999). Change of elastic modulus of strongly charged hydrogels at the collapse transition. *Macromolecules* (**32**), pp. 1508–1513.

180. Zhang, H., F. Zhang, and J. Wu (2013). Physically crosslinked hydrogels from polysaccharides prepared by freeze–thaw technique. *Reactive and Functional Polymers* (**73**), pp. 923–928.

181. Zhang, L., E. J. Choi, M. H. Kim, G. M. Lee, H. Suh, and I. Kim (2012). pH-reversible supramolecular hydrogels based on aminoalkyl phosphoamide compounds. *Supramolecular Chemistry* (**24**), pp. 189–196.

182. Zhang, T., G. Y. Li, L. Guo, and H. Chen (2012). Synthesis of thermo-sensitive CS-g-PNIPAM/CMC complex nanoparticles for controlled release of 5-FU. *International Journal of Biological Macromolecules* (**51**), pp. 1109–1115.

183. Zhang, X. Z., D. Q. Wu, and C. C. Chu (2004). Synthesis and characterization of partially biodegradable, temperature and pH sensitive Dex-MA/PNIPAAm hydrogels. *Biomaterials* (**25**), pp. 4719–4730.

184. Zhang, Y., Y. Thomas, E. Kim, and G. F. Payne (2012). pH- and voltage-responsive chitosan hydrogel through covalent cross-linking with catechol. *Journal of Physical Chemistry B* (**116**), pp. 1579–1585.

185. Zhu, C. H., Y. Lu, J. Peng, J. F. Chen, and S. H. Yu (2012). Photothermally sensitive poly(*N*-isopropylacrylamide)/graphene oxide nanocomposite hydrogels as remote light-controlled liquid microvalves. *Advanced Functional Materials* (**22**), pp. 4017–4022.

186. Zhu, J. M., and R. E. Marchant (2011). Design properties of hydrogel tissue-engineering scaffolds. *Expert Review of Medical Devices* (**8**), pp. 607–626.

187. Zohuriaan-Mehr, M. J., and K. Kabiri (2008). Superabsorbent polymer materials: a review. *Iranian Polymer Journal* (**17**), pp. 451–477.

188. Zohuriaan-Mehr, M. J., H. Omidian, S. Doroudiani, and K. Kabiri (2010). Advances in non-hygienic applications of superabsorbent hydrogel materials. *Journal of Materials Science* (**45**), pp. 5711–5735.

Chapter 10

Properties and Biomedical Applications of Gellan Gum Hydrogels

Cencetti Claudia,[a,b] Pietro Matricardi,[a] Franco Alhaique,[a] and Davide Bellini[b]

[a] *Department of Drug Chemistry and Technologies, University of Rome "La Sapienza", Piazzale Aldo Moro 5, 00185 Rome, Italy*
[b] *Novagenit s.r.l., Viale Trento, 115/117, 38017 Mezzolombardo (TN), Italy*
claudia.cencetti@gmail.com

Gellan gum is a bacterial exopolysaccharides (EPS) of the "sphingans" family, a group of structurally related EPS secreted by members of the genus Sphingomonas.

Gellan is a multifunctional gelling agent based on a tetrasaccharide repeating unit composed of two molecules of D-glucose, one of L-rhamnose and one of D-glucuronic acid; in its native form, gellan is partially esterified with acyl substituents. Due to its unique characteristics, gellan gum is used for a great variety of applications, particularly in the fields of food, pharmaceutics, and biomedicine.

Polysaccharide Hydrogels: Characterization and Biomedical Applications
Edited by Pietro Matricardi, Franco Alhaique, and Tommasina Coviello

ISBN 978-981-4613-61-3 (Hardcover), 978-981-4613-62-0 (eBook)
www.panstanford.com

Table 10.1 Sphingomonas bacteria and related EPS

Sphingomonas genus	Sphingan Former name	Sphingan Current name
Sphingomonas elodea ATCC 31461	S-60	Gellan
Sphingomonas sp. ATCC 31555	S-130	Welan
Sphingomonas sp. ATCC 31961	S-194	Rhamsan
Sphingomonas sp. ATCC 53159	S-657	Diutan
Sphingomonas sp. ATCC 31554	S-88	—
Sphingomonas sp. ATCC 31853	S-198	—
Sphingomonas sp. ATCC 21423	S-7	—

10.1 Origin and Biosynthesis

10.1.1 *Sphingans*

The "sphingans" are a family of structurally related exopolysaccharides (EPS) secreted by members of the genus Sphingomonas [54]. Unlike other gram-negative bacteria, they have glycosphingolipids in their outer membranes instead of lipopolysaccharides.

Due to their biodegradative and biosynthetic capabilities, several Sphingomonas have been studied for the production of this new class of EPS: a few of these bacteria, and the related EPS, are listed in Table 10.1.

Sphingans are characterized by the same linear tetrasaccharide back bone structure (–X–glucose–glucuronic acid–glucose–X, where X is either L-rhamnose or L-mannose) to which distinct side groups are attached [54].

10.1.2 *Gellan Biosynthesis*

Gellan gum is a high molecular weight EPS aerobically produced by *Sphingomonas elodea* ATCC 31461: this nonpathogenic strain was originally classified as Pseudomonas elodea, and only later defined as Sphingomonas by Pollock [54].

S. elodea is aerobic, rod-shaped with a single flagellum, and forms round mucoid and yellow pigmented colonies, and it was isolated from the surface of a plant from Elodea genus as a result of an extensive screening effort by the Kelco Company, a division of Merck [27], and thereafter improved for food applications [12].

The gellan biosynthetic pathway, extensively described by Fialho et al. [18], is a multi-sequential-step process:

(i) intracellular synthesis of sugar-activated precursors
(ii) assembly of the tetrasaccharide repeating units linked to the inner membrane
(iii) translocation of the repeating units to the periplasmic space
(iv) polymerization of the repeating units
(v) translocation of the polysaccharidic chain through the outer membrane.

10.2 Structure and Physical Properties

Gellan gum has a backbone molecular structure based on a tetrasaccharide repeating unit composed of (1-3)-β-D-glucose, (1-4)-β-D-glucuronic acid, (1-4)-β-D-glucose, and (1-4)-α-L-rhamnose; the native polysaccharide is partially esterified with acyl substituents of L-glycerate (approximately 1 per repeating unit) and acetate (approximately 0.5 per repeating unit) at the C-2 and C-6 positions of the (1-3)-linked D-glucose (Fig. 10.1a).

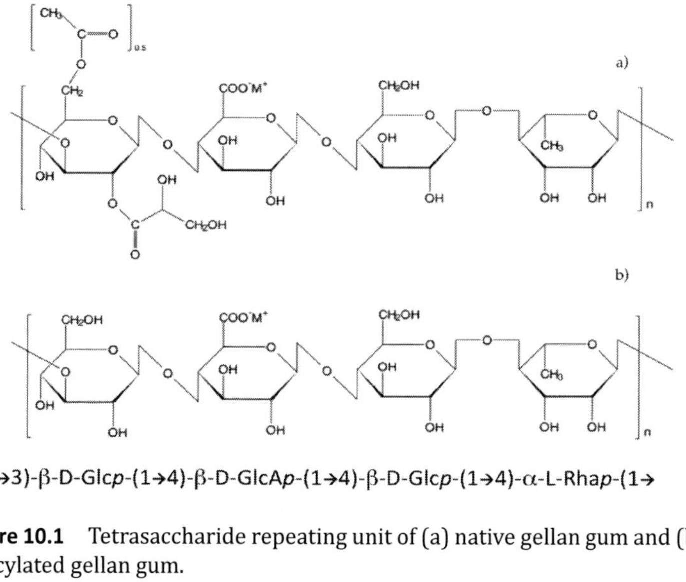

→3)-β-D-Glc*p*-(1→4)-β-D-GlcA*p*-(1→4)-β-D-Glc*p*-(1→4)-α-L-Rha*p*-(1→

Figure 10.1 Tetrasaccharide repeating unit of (a) native gellan gum and (b) deacylated gellan gum.

The acetyl groups in native gellan gum can be removed by alkaline treatment to produce deacetylated gellan gum (Fig. 10.1b). The acyl substituents affect the rheological behavior of the polysaccharide, and deacetylation of native gellan leads to a significant change: from soft, elastic, thermoreversible gels to harder, more brittle gels with higher thermal stability. The steps involved in deacetylation of native gellan gum include hot treatment of the fermentation broth, followed by alkaline treatment up to pH $= 10$ with 1 M NaOH (10 min at 80°C). The broth is then neutralized and centrifuged, and the supernatant is finally added to ice-cold ethanol to precipitate the deacetylated gellan [26].

10.2.1 *Structure of Gellan in Solid State*

A first attempt to understand the gellan molecular structure was performed by Upstill et al. [67] by means of X-ray diffraction on fibers of gellan (Li^+ salt form), but the proposed model was unconvincing; later studies by Chandrasekaran et al. [8] (Fig. 10.2) proved that the molecular structure of gellan was a double helix, formed by two intertwined left-handed, threefold helical chains with a pitch of 5.64 nm): the two chains run parallel to one another and are exactly half-staggered, so that the repeat distance of the double helix is one half of that of the individual strands (2.82 nm).

This helical geometry is promoted by the (1→3) linkage in the gellan repeating unit. Three of the four glycosidic linkages in the tetrasaccharide repeating sequence of gellan (Fig. 10.1) involve equatorial bonds at C(1) and C(4) of the participating residues: this type of linkage promotes the formation of flat, ribbon-like structures in the solid state; on the other side, the (1→3) linkage introduces a systematic "twist" in the direction of the chain and promotes the helix formation.

10.2.2 *Conformational Transition of Gellan in Solution*

In order to understand the behavior of gellan in solution, avoiding complications due to gel formation, the adopted procedure in numerous investigations is the conversion of the native polysaccharide

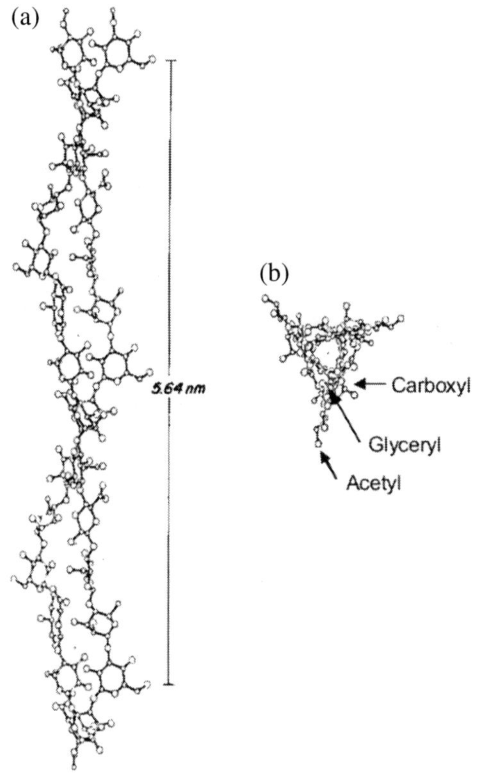

Figure 10.2 Double helix structure of deacylated gellan viewed (a) perpendicular to the helix axis and (b) along the helix axis. The sites of attachment of glyceryl and acetyl substituents in high acyl gellan, relative to the position of the carboxyl group on the neighboring glucuronate residue, are indicated in (b). Reprinted with permission from Ref. 8. Copyright © 1988, Elsevier.

in tetramethylammonium salt form (Me_4N^+), since in such form gellan gelation occurs only at very high salt concentrations.

Crescenzi et al. [15] monitored the conformational transition of gellan by means of optical rotation, as a function of temperature, for dilute solutions of Me_4N^+ gellan (1.2 mM $\approx$ 0.086 wt%) in water and in the presence of Me_4NCl at different concentrations. As reported in Fig. 10.3, the plot of the specific rotation for the solution in water is featureless, showing only a slight, approximately linear, increase in absolute value of optical rotation with decreasing temperature;

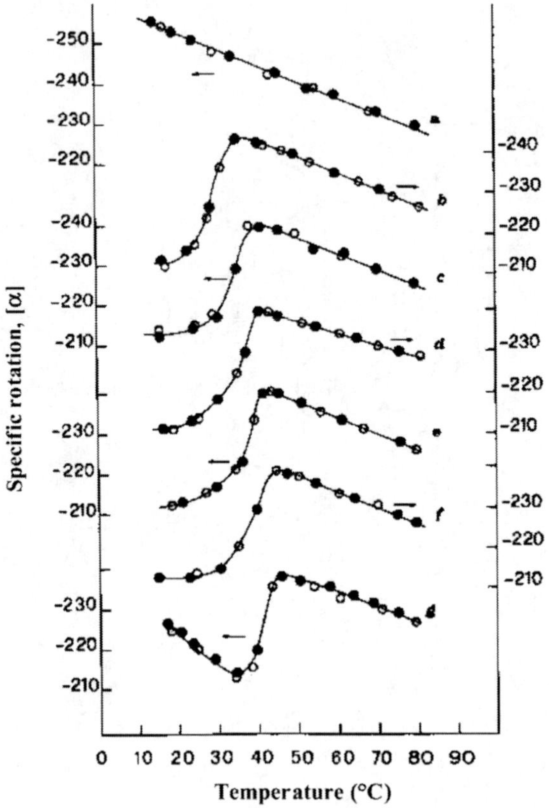

Figure 10.3 Temperature dependence of optical rotation (302 nm) for 1.2 mM (0.086 wt%) Me4N$^+$ gellan on heating (○) and cooling (•) in the presence of Me4NCl at concentrations (mM) of (a) 0, (b) 30, (c) 75, (d) 92, (e) 120, (f) 150, and (g) 250. Reprinted with permission from Ref. 15. Copyright © 1987, Elsevier.

while a well-defined sigmoidal trend occurs for the solutions with added Me_4NCl.

The detected shifts show no thermal hysteresis between cooling and heating and they occur at progressively higher temperatures as salt concentration increases. These sigmoidal trends of optical rotation can be considered as the evidence of a conformational transition between an ordered structure at low temperatures and a disordered coil state at high temperatures.

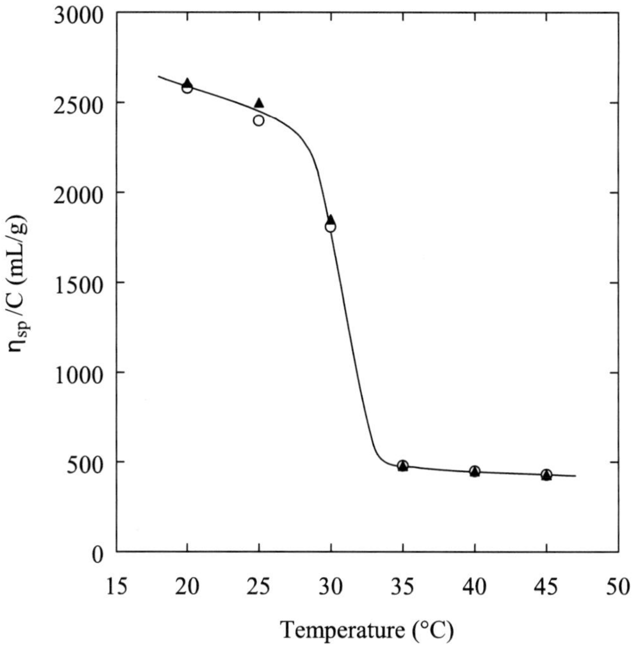

Figure 10.4 Temperature dependence of reduced specific viscosity on heating (σ) and cooling ($\circ$) for 0.05 wt% Me_4N^+ gellan in 30 mM Me_4NCl. Reprinted with permission from Ref. 15. Copyright © 1987, Elsevier.

Crescenzi et al. [15] showed that this conformational transition is accompanied by an important increase in reduced specific viscosity (η_{sp}/c, where c is polymer concentration) (Fig. 10.4); since at the low experimental gellan concentration ($c = 0.05$ wt%), the reduced specific viscosity closely approximates the intrinsic viscosity $[\eta]$, which is directly related to hydrodynamic volume in solution: the significantly higher values obtained at low temperatures demonstrate that the ordered structure of Me_4N^+ gellan is stiffer, and has a much bigger hydrodynamic volume, than that of the disordered state.

10.2.3 *Gelation of Gellan*

Gellan gum can form gels according to the following mechanism: at high temperatures, gellan gum is in its coil form; as temperature decreases, a thermally reversible coil to double-helix transition occurs, which is a prerequisite for gel formation. However, conformational ordering does not lead, per se, to a cohesive network: formation of stable hydrogels requires the arrangement (sol–gel transition) of an ordered structure composed of antiparallel double helixes self assembled to form oriented bundles, called junction zones. Untwined regions of polysaccharide chains, in the form of extended helical chains, link the junction zones, leading to the formation of a three dimensional network, that creates the gel (Fig. 10.5).

The sol–gel transition is inhibited by electrostatic repulsion among the helixes. Such repulsion can be suppressed, thus allowing gel formation, by lowering the pH value which reduces the charge on the helixes by conversion of glucuronate carboxyl groups from the negatively charged COO— form to the uncharged COOH form.

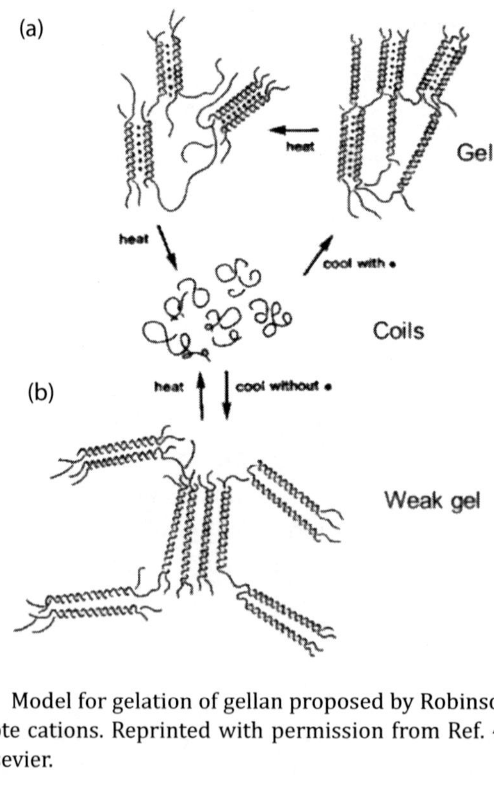

Figure 10.5 Model for gelation of gellan proposed by Robinson et al. Filled circles denote cations. Reprinted with permission from Ref. 46. Copyright © 2012, Elsevier.

Aggregation can also be promoted by salts: anions and cations from dissolved salts are capable to screen the electrostatic repulsion among the gellan helixes; moreover, cations may cause further reduction of the repulsion by clustering around the helixes and thus lowering their effective negative charge.

10.2.3.1 Monovalent cation-induced gelation of gellan

In the presence of cations, gellan gelation occurs even at low polymer concentrations, because they shield the electrostatic repulsion of helixes, promoting a more stable aggregation.

Site binding is triggered initially by the electrostatic attraction of cations to the carboxylate groups of the polymer, but is increased and stabilized by the formation of a coordination complex with appropriately spaced oxygen atoms from both strands of the double helix.

Coil–helix and sol–gel transition of gellan gum were widely studied by means of differential scanning calorimetry (DSC) [37, 41, 56].

Robinson et al. [56] performed DSC on Na^+ gellan gum in the presence of NaCl at different concentrations ($0 \div 130$ mM) (Fig. 10.6).

The DSC cooling scans showed only a single exotherm which moved to progressively higher temperature as the concentration of Na^+ was raised. Heating of Na^+ gellan gum in water (Fig. 10.6A) gave a single endotherm, essentially equal and opposite to the exotherms observed for the same samples upon cooling.

When salt was added (50 mM NaCl, Fig. 10.6B) two different thermal processes became evident in the DSC heating plots, with the first one at approximately the same position as the exotherm in the corresponding cooling curves.

By further increase in salt concentration (100 mM NaCl, Fig. 10.6C), the first endotherm in the heating curves became progressively smaller, but remained at approximately the same position as the corresponding exotherm. On the other side, the second endotherm moved to much higher temperatures and increased in size, until at the highest NaCl concentration (130 mM NaCl, Fig. 10.6D) it became the only detectable process. The second

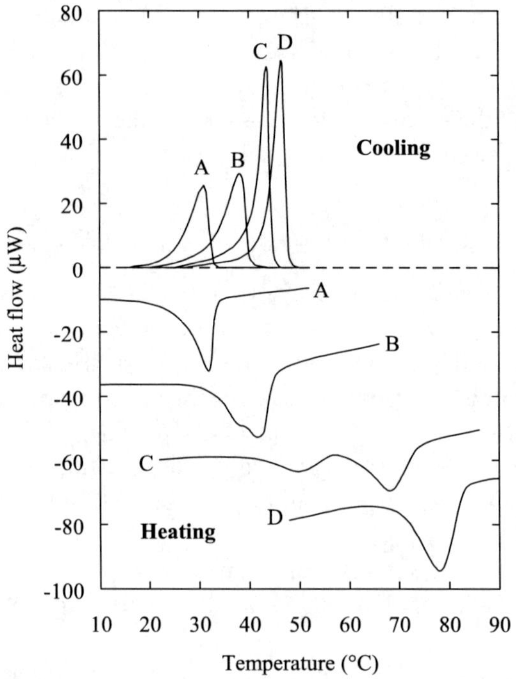

Figure 10.6 DSC cooling and heating plots for 1 wt% (15 mN) Na^+ gellan at total concentrations (mM) of Na^+ (counterions to the polymer plus added NaCl) of (A) 15, (B) 65, (C) 115, and (D) 145. Baselines were subtracted from the cooling curves; the heating curves are displaced vertically by arbitrary units to avoid overlapping [35].

thermal process, observed upon heating, is caused by the melting of the junction zones (gel–sol transition), while the first endotherm can be ascribed to the helix–coil transition of the residual unaggregated helixes, which melt at the same temperature at which they were formed during cooling [56].

The increase in the size of the second endotherm is correlated to the increase in the extent of aggregations, while, as salt concentration increases, the progressive increase in thermal hysteresis between the formation of individual helixes on cooling and melting of helix–helix aggregates on heating indicates that true gels are formed by association of gellan double helixes into stable aggregates. These conclusions are also confirmed by the rheological

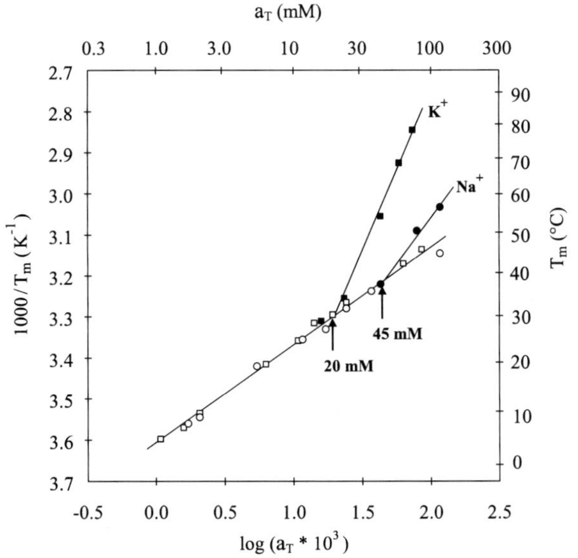

Figure 10.7 Dependence of transition temperature (T_m) from optical rotation on cooling (open symbols) and heating (filled symbols) with activity of Na^+ (circles) and K^+ (squares) in solutions of gellan. Reprinted with permission from Ref. 38. Copyright © 1996, Elsevier.

measurements performed by Morris et al. [45] and by optical rotation measurements carried out by Milas and Rinaudo [38], as described in Fig. 10.7.

Ogawa et al. [49] studied, by means of circular dichroism (CD) and viscosity measurements, the influence of different types of monovalent cations on the conformational transition of gellan gum; the order of effectiveness of the cationic species in promoting the partial aggregation was $K^+ >> Li^+ > Na^+$: the author's interpretation of this behavior was that cations such as Na^+ and Li^+ are the structure-ordering ions for water, while cations such as K^+ belong to the structure-disordering ions. For this reason, cations as K^+ can shield the electrostatic repulsion of carboxyl groups in the gellan molecules more directly than Na^+ and Li^+. These results are confirmed by means of the rheological and DSC studies by Miyoshi et al. [42] and by the optical rotation measurements performed by Milas and Rinaudo [38]; furthermore, Grasdalen et al. [22], by means of light scattering, NMR and gel strength measurements, demonstrated that the order of monovalent cation effectiveness was

$$Cs^+ > K^+ > Na^+ > Li^+ > TMA^+.$$

10.2.3.2 Divalent cation-induced gelation of gellan

Gelation with divalent cations is qualitatively different: these ions promote aggregation by site binding between pairs of carboxylate groups on neighboring helixes, rather than by suppressing electrostatic repulsion by binding to individual helixes.

Morris et al. [46] showed that the thermal hysteresis associated with stable aggregation starts immediately (Fig. 10.8) by progressive incorporation of $CaCl_2$ in solutions of Me_4N^+ gellan, and is

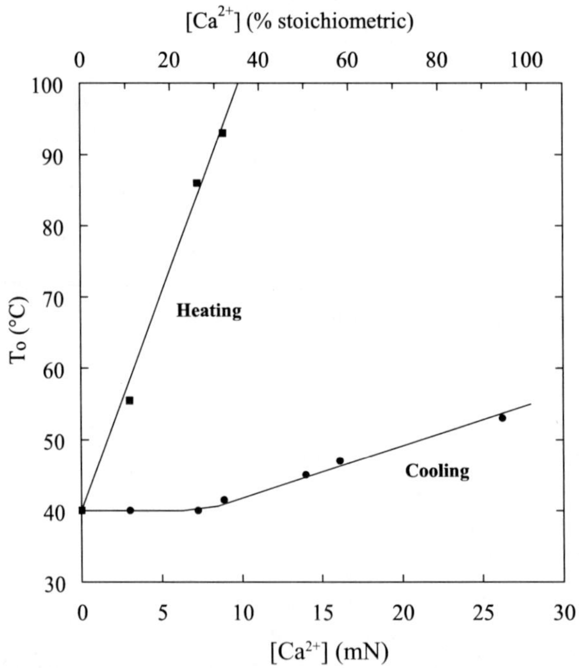

Figure 10.8 Temperature (T_0) at the onset of gelation on cooling (•) and completion of gel melting on heating (■) for 2.0 wt% (27.8 mM) $Me4N^+$ gellan by progressive addition of $CaCl_2$. Reprinted with permission from Ref. 46. Copyright © 2012, Elsevier.

accompanied by immediate massive increase in storage and loss moduli (G' and G'').

Kapasis et al. [29] studied, by means of rheological temperaure sweep, the mechanical properties of gellan solutions in NaCl or $CaCl_2$ at different concentrations: as reported in Fig. 10.9, qualitatively, addition of sodium or calcium chloride to the gellan system leads

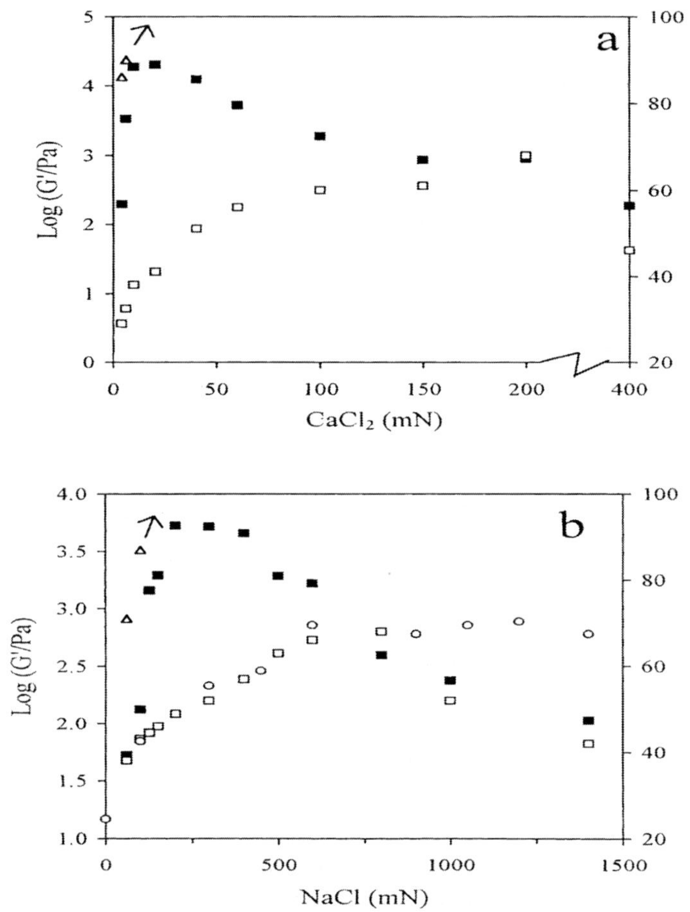

Figure 10.9 Storage modulus (G') at 20°C (■), rheological onset of network formation (□) and melting temperatures (△), and onset of conformational transitions using polarimetry (○) for 0.5% deacylated gellan with increasing levels of (a) calcium and (b) sodium ions. G' values are on the left ordinate. Reprinted with permission from Ref. 29. Copyright © 1999, Elsevier.

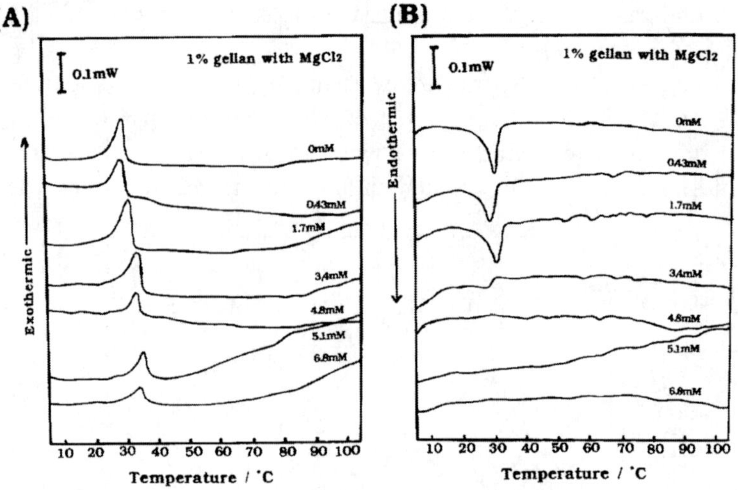

Figure 10.10 Cooling and heating DSC curves for 1% gellan gum solutions containing $MgCl_2$ at various concentrations. Reprinted with permission from Ref. 41. Copyright © 1995, Elsevier.

to a similar profile, but the maximum values in the structural properties and network formation profile are observed at much lower concentrations of divalent salt. The authors explained that this behavior was related to the strong specific interaction between calcium and carboxyl groups, whereas sodium ions were seen only as a surrounding "cloud" capable to neutralize the gellan chains.

Miyoshi et al. [41, 42] performed DSC measurements on gellan gum (1% w/v) in solutions containing $MgCl_2$ and $CaCl_2$ at various concentrations. As shown in Fig. 10.10, up to a salt concentration of 1.7 mM, melting temperature shifts to higher temperatures with the cation concentration, with increasing enthalpies increasing; this occurs both for endothermic and esothermic peaks. With higher salt concentrations (i.e., above 1.7 mM) the endothermic curves are characterized by a number of small peaks, too broad to be resolved. These results suggested that a sufficient amount of divalent cations was capable to induce unhomogenization in the gellan gum solution, leading to a strongly thermally stable network, formed by junction zones of different thermal stability.

As already demonstrated for monovalent cations, Grasdalen et al. [22] showed, by means of light scattering, NMR and gel strenght measurements, that also for divalent cations an order of effectiveness can be proposed: $Pb^{++} > Cu^{++} > Zn^{+} > Mg^{++}$, Ca^{++}, Sr^{++}, Ba^{++}. Actually, differently from other uronic acid containing polysaccharides, such as alginate and pectin, gellan exhibits lack of specificity among the alkaline earth cations and selectivity among the transition elements. This order seems to be closely related to the coordination number of the cation–water interaction: cations, when in aqueous solution, become hydrated and, therefore, form hydration shells, according to their hydration number, thus significantly reducing their shielding ability.

10.2.3.3 Gelation of gellan without added salts

In the absence of added salts, gellan gelation occurs only at high polymer concentrations: at low gellan gum concentrations the coil–helix transition actually occurs, but the low number of helical aggregates is not capable to give a continuous network in the whole bulk of the system.

Miyoshi et al. [41] studied the gelation of gellan in the absence of added salts by means of rheological measurements of thermal sweeps: they found that at low polymer concentrations ($c_P < 2\%$ w/v) the viscoelastic behavior of gellan gum are characteristic of dilute or concentrated solution, with a single step-like shift of G' and G'' corresponding to the coil–helix transition ($T_{c\text{-}h}$). By increasing the polymer concentration, the system viscoelastic behavior becomes that of a weak gel, with a crossover temperature (corresponding to the sol–gel transition $T_{s\text{-}g}$) and two step-like shifts of G' and G'' (corresponding to the two transitions $T_{c\text{-}h}$ and $T_{s\text{-}g}$).

The same behavior was observed by Miyoshi et al. [42] and by Miyoshi and Nishinari [43] by means of DSC measurements on Na^{+} gellan in water at different polymer concentrations (Fig. 10.11). The DSC cooling scans showed only a single exotherm, which moved to progressively higher temperatures as Na^{+} gellan concentration increased (corresponding to the coil–helix transition). The DSC heating scans gave single endotherms for samples at gellan concentrations lower than 3% w/v, essentially equal and

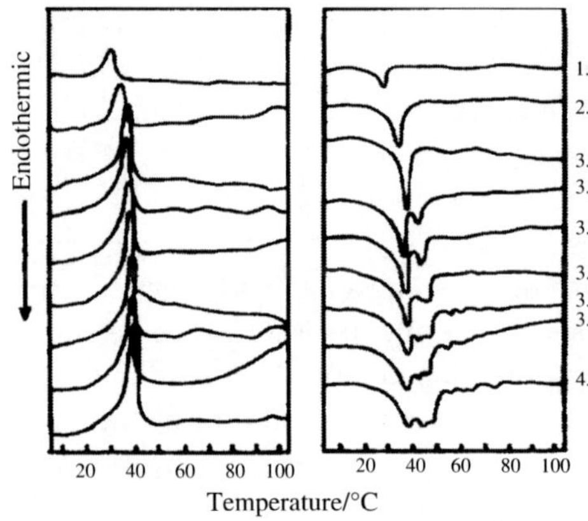

Figure 10.11 DSC plots recorded on cooling (left) and heating (right) at 0.5°C/min for Na^+ gellan in water at the concentrations shown to the right of the heating curves (1.0–4.2 wt%). Reprinted with permission from Ref. 42. Copyright © 1996, Elsevier.

opposite to the exotherms observed for the same samples on cooling. When the polymer concentration increases, a split in the heating curve becomes detectable, but this phenomenon must be attributed more to the counterions concentration, rathen than to the polymer concentration; actually, a gellan concentration of 3.5% w/v is equivalent to a Na^+ counterion concentration of 52.5 mM of, close to the value, observed by Robinson et al. ([56], Fig. 10.6), at which a second endotherm occurs.

10.2.3.4 Effect of acyl substituents

In water, native (i.e., acylated) gellan gum forms more elastic gels than low acyl or deacetylated gellan gum: in the absence of added salts, electrostatic repulsion between the helixes occurs, and the acyl groups decrease the charge density within the molecular backbone, thus promoting the stabilization of the double helix (particularly via the glycerate group) and leading to increased elasticity of the gelled system. On the other side, in the presence of salt, hydrogels of native gellan gum are less elastic than those obtained with the deacetylated polysaccharide: in this case acyl groups destibilize the formation of junction zones via steric hindrance.

Rheological characteristics of native gellan ($C_p = 10$ g/l) in presence of different salt contents[a]

C_s (mol/l)	G' at 20°C (N/m^2)	G'' at 20°C (N/m^2)	G'/G'' at 20°C	T_m (°C)
0.01	28	18	1.55	60
0.05	82	28	2.92	63
0.1	140	42	3.3	75

Rheological characteristics of deacylated gellan ($C_p = 10$ g/l) in the presence of different salt contents[a]

C_s (mol/l)	G' at 20°C (N/m^2)	G'' at 20°C (N/m^2)	G'/G'' at 20°C	T_m (°C)
0.01	10	12.5	0.8	
0.05	200	35	3.71	40
0.1	8200	1050	7.8	80

[a] T_m corresponds to the cross-over G'–G'' vs T. Strain 10%; $\omega = 1$ Hz.

Figure 10.12 Rheological characteristics of native and deacylated gellan gum in presence of different NaCl contents. Reprinted with permission from Ref. 37. Copyright © 1999, Elsevier.

Kapasis et al. [29] studied, by means of rheological temperature sweep, the mechanical properties of native gellan solutions (c_P = 0.5% w/v) in water, and in NaCl or $CaCl_2$ at different concentrations: network formation was possible in the absence of added counterions. In this case (i.e., high acyl), no thermal hysteresis was observed, suggesting the absence of aggregation, and the presence of counterions does not influence the onset of network can be observed.

Mazen et al. [37] compared the mechanical properties, obtained by means of rheological characterization, of native and deacylated gellan gum in the presence of NaCl at various concentrations: the native gellan formed weak gels at all NaCl concentrations in the studied range (0.01 ÷ 0.1 M), and the gelation temperature (T_m) was almost independent of salt concentration, while NaCl heavily affected both gel strength and T_m of the networks from deacylated gellan. Moreover, by means of DSC, Mazen et al. confirmed the

Table 10.2 Various gellan gum products available in market

Gellan Gum Product	Company	Acyl Content	Application
Kelcogel	CP Kelco	Low Acyl	Food
Kelcogel F	CP Kelco	Low Acyl	Food
Kelcogel LT100	CP Kelco	High Acyl	Food
Kelcogel AFT	CP Kelco	Low Acyl	Industrial
Kelcogel CG-LA	CP Kelco	Low Acyl	Pharmaceutical, oral care, and personal care
Kelcogel CG-HA	CP Kelco	High Acyl	Pharmaceutical and personal care
Gelzan CM (Gelrite)	CP Kelco	No Acyl	Microbiological media and plant tissue culture
Phytagel	Sigma-Aldrich	No Acyl	Microbiological media and plant tissue culture
Gel-Gro	ICN Biochemical	No Acyl	Microbiological media and plant tissue culture

results obtained by Kapasis et al. [29]: while the deacylated gellan gum systems are characterized by a large thermal hysteresis, which improves with salt concentrations, the native gellan gum shows no thermal hysteresis, both in water and in saline solutions.

10.3 Applications

Gellan gum was approved in the United States and the European Union for food and medical applications as a gelling, stabilizing, and suspending agent [20]. Gellan is characterized by an excellent thermal and acid stability, adjustable mechanical properties, high transparency, and good flavor release. Gellan is commercially available in three forms: high, low, and no acyl content. In Table 10.2, trademarks and applications of different gellan gum products are reported.

Deacetylated gellan gum is mostly used as a substitute of agar for the culture of thermophilic bacterial species and in plant tissue culture media. Moslemy et al. described potential environmental applications for gellan gum, such as in the gasoline biodegradation process [48] and in gel encapsulation of bacteria for the bioaugmentation of contaminated aquifers [47].

Food-grade partially acetylated gellan gum (both low and high acyl form, and identified as E418), is widely used as gelling, stabilizer and suspending agent in foods (dessert gels, puddings, sauces, ice-creams, and microwavable foods) and personal care products (lotions, creams and toothpastes).

In the last 15 years the capability in obtaining a highly purified gellan gum, with low endotoxin levels, increased significantly, consequently it has been possible to utilize this polysaccharide also in pharmaceutics [52]. Gellan applications in the field of pharmaceutical formulations include its use as a pharmaceutical excipient for nasal, ocular, gastric, and colonic drug delivery [2, 25, 65].

In 2000, Monsanto filed a patent for the purification of different polysaccharides, including gellan gum, finalized to the parenteral use of these macromolecules [6]; finally, in the last 10 years, gellan gum was tested in tissue engineering and surgical applications [7, 52], being proved to be a suitable material for the construction of 3D scaffolds [5, 11, 64].

10.3.1 *Gellan for Drug Delivery System*

Due to the peculiar property of cation-induced gelation, gellan gum has been widely used in the formulation of in situ gelling ophthalmic preparations [4, 53, 57, 59] as well as in situ gelling oral controlled release formulations [39, 40, 55].

According to Paulsson et al. [53], instillation of low-viscosity solutions of gellan gum into the conjunctival sac of the eye, leads to a conformational change in the polymer, producing a gel because of the saline composition of the tear fluid, thus increasing the residence time of a drug solution. Cao et al. [4] exploited the saline composition of nasal mucosa to prepare an in situ gelling gellan gum formulation for nasal administration of mometasone furoate.

Miyazaki et al. [39] produced in situ gelling gellan formulations as vehicles for oral drug delivery: the proposed formulation was a gellan solution containing calcium chloride and sodium citrate, which complexes the free Ca^{++} ions and releases them only in the highly acidic environment of the stomach. According to this approach, the formulation remains in liquid form until it reaches the stomach, where gelation takes place almost instantaneously.

Following a similar method, Rajinikanth et al. [55] developed a floating in situ gelling system of amoxicillin for eradication of *Helicobacter pylori*, with solutions of gellan gum and amoxicillin containing sodium citrate and calcium carbonate, as gas-forming agents.

The cation-induced ionotropic gelation was also exploited for the preparation of hydrogels, beads, microparticles, and tablets based on gellan gum, alone or in association with other polymers.

For the preparation of sustained release dosage forms, Santucci et al. [58] prepared gellan gum beads by dropping a polymer solution containing clay, oil, and a model molecule into a calcium chloride solution. Kedzierewicz [31], following a similar dropping technique, prepared gellan gum beads for the release of propranolol; Agnihotri et al. [2] produced beads of gellan gum, by dropping gellan gum solutions containing cephalexin into a counterion solution containing calcium chloride and zinc sulfate.

The opposite approach, i.e., dropping a viscous calcium chloride solution containing the active substance into a gellan solution, was proposed by Alhaique et al. [3] for the preparation of sustained release gel capsules.

Mahajan and Gattani [34] prepared, via spray drying, gellan gum microparticles containing metoclopromide hydrochloride for intranasal delivery, providing a longer contact time with the mucosa and consequently an improvement of drug transport across the nasal membrane, because of the significantly slowed clearance by the mucocilliary mechanism.

Chemically crosslinked gellan hydrogels were also studied as drug delivery systems, in order to tune the release profile for different drugs.

Coviello et al. [14] prepared a co-crosslinked matrix, chemically crosslinking a physical hydrogel of gellan and scleroglucan, studying the effect of calcium content of the release profile of a model drug.

Matricardi et al [36] obtained chemical gellan gum hydrogels with different content of L-lysine ethil ester, characterized by much slower release profiles, compared to the corresponding physical hydrogels.

Chemical modification, i.e., addition of functional group, was performed onto the gellan gum chains in order to modify the physicochemical behavior of the polymer.

Mahajan et al. [34] prepared an in situ gelling system for nasal administration of dimenhydrinate, enhancing the mucoadhesiveness and the permeation profile of gellan gum by addition of thiol functional groups, with no modifications of viscosity and in vitro drug diffusion with respect to gellan gum.

D'Arrigo et al. [16] functionalized the gellan gum with prednisolone, introducing a hydrophobic moiety in the polysaccharidic chain, thus allowing a self-assembly of the gellan gum in nanohydrogels, that could penetrate within the cells, releasing the prednisolone by slow degradation. In this case, prednisolone simultaneously acts as a pharmacological agent and a promoter of the nanohydrogel formation.

10.3.2 *Gellan in Tissue Engineering*

Gellan gum was shown to be very versatile in terms of processing, which can be controlled by both temperature and pH, forming structures with different shapes using simple technologies. As a demonstration of the wide range of possible applications for this biomaterial, discs, membranes, fibers, particles, and scaffolds have been produced [50].

Recent investigations by Smith et al. [64] showed that various cell types immobilized into or onto gellan gel-based matrices could maintain high viability and appropriate functionalities, implying the potential use of gellan gum for tissue engineering purposes. Oliveira et al. [51] demonstrated that gellan gum solutions containing autologous cells could be injected in rabbit knee defects, efficiently regenerating hyaline-like cartilage tissue.

Nevertheless, in physiological conditions (i.e., salts concentration and body temperature) the commercially available gellan gum gels in seconds, leaving no window of time for manipulations.

In order to optimize the gelation temperature for injection, Gong et al. [21] degraded the high molecular weight gellan gum (via oxidative cleavage), and demonstrated the potential use of this material for therapeutic chondrocyte delivery in the long-term cartilage regeneration.

With the same purpose, Du et al. [17] prepared an injectable, in situ crosslinkable gellan-based hydrogel, via chemical functionalization of the gellan backbone with L-cysteine: thiolated gellan have a decreased gelation temperature, while keeping its physical gelation capacity. A chemical crosslinking occurred while the thiol groups underwent air oxidation in physiological conditions, leading to a stable chemical hydrogel.

It should be pointed out that physically crosslinked gellan hydrogels tend to loose their stability in vivo after implantation due to the exchange of divalent cations with monovalent ones that are present in higher concentrations in the physiological environment. In order to overcome this drawback, several chemical modifications to the gellan gum backbone have been proposed.

Shin et al. [61] prepared, by a two-step photocrosslinking, a double network (DN) hydrogel based on gellan gum methacrylate, for a rigid and brittle first network, and gelatine methacrylamide, for a soft and ductile second network. These DN showed the ability to encapsulate cells, and were characterized by high mechanical strength, thus being potentially useful for the regeneration of load-bearing tissues.

Silva et al. [60] demostrated that gellan gum, chemically modified with a peptide in order to enhance cell adhesion, could be efficiently associated with olfactory ensheathing glia (OEG) for the proliferation of neural stem/progenitor cell (NSPC), thus obtaining a potential application in the regeneration of injured adult central nervous system (CNS) tissues.

Coutinho et al. [13] studied the effect of different degrees of methacrylation on the physical and chemical gelation properties of gellan gum, in order to obtain hydrogels with tunable mechanical properties, applicable for a wide range of tissue engineering approaches.

In the tissue engineering applications, hydrogels of gellan gum or of modified gellan gum has been reported for their capability of act as a physical barrier against vascular or fibroblast invasion, due to their negative charge.

Silva-Correia et al. [62, 63] studied the biocompatibility and the angiogenic potential of methacrylated gellan gum hydrogels for the application in nucleus pulposus regeneration, demonstrating that they act as a physical barrier for vascular invasion in the treatment of degenerative intervertebral disc.

Chang et al. [10] prepared sponges of chemically crosslinked gellan gum as dental filling to maintain alveolar space, indicating that these materials could effectively act as a barrier, inhibiting the growth of fibroblasts at wound sites.

Cencetti et al. [7] prepared an hydrogel of gellan gum and sulphated hyaluronic acid for potential epidural scar prevention, where gellan gum can act as both "structural" and "barrier" component, and, at the same time, sulphated hyaluronic acid enhances the electrostatic repulsion in the presence of fibroblasts, thus inhibiting cell invasion of the damaged spinal tissue.

References

1. Agnihotri, S. A., and Aminabhavi, T. M. (2005). Development of novel interpenetrating network gellan gum-poly(vinyl alcohol) hydrogel microspheres for the controlled release of carvediloly, *Drug Dev. Ind. Pharm.*, **31**, pp. 491–503.
2. Agnihotri, S. A., Jawalkar, S. S., and Aminabhavi, T. M. (2006). Controlled release of cephalexin through gellan gum beads: effect of formulation parameters on entrapment efficiency, size, and drug release, *Eur. J. Pharm. Biopharm.*, **63**, pp. 249–261.
3. Alhaique, F., Santucci, E., Carafa, M., Coviello, T., Murtas, E., Riccieri, F. M. (1996). Gellan in sustained release formulations: preparation and release studies, *Biomaterials*, **17**, pp. 1981–1986.
4. Cao, S., Ren, X., Zhang, Q., Chen, E., Xu, F., Chen, J., Liu, L., and Jiang, X. (2009). In situ gel based on gellan gum as new carrier for nasal administration of mometasone furoate, *Int. J. Pharm.*, **365**, pp. 109–115.
5. Bertoni, F., Barbani, N., Giusti, P., and Ciardelli, G. (2006). Transglutaminase reactivity with gelatine: perspective applications in tissue engineering, *Biotechnol. Lett.*, **28**, pp. 697–702.
6. Bousman, W. S., Colegrove, G. T., Montana, M. A., Raczkowski, R., and Holahan, J. (2000). Biopolymers salts with low endotoxin levels,

biopolymer compositions thereof and methods of making the same, *WO0029449.*

7. Cencetti, C., Bellini, D., Longinotti, C., Martinelli, A., and Matricardi, P. (2011). Preparation and characterization of a new gellan gum and sulphated hyaluronic acid hydrogel designed for epidural scar prevention, *Mater. Sci. Mater. Med.*, **22**, pp. 263–271.
8. Chandrasekaran, R., Millane, R. P., Arnott, S., and Atkins, E. D. T. (1988). The crystal structure of gellan, *Carbohydr. Res.*, **175**, pp. 1–15.
9. Chandrasekaran, R., Puigjaner, L. C., Joyce, K. L., and Arnott, S. (1988). Cation interactions in gellan: an X-ray study of the potassium salt, *Carbohydr. Res.*, **181**, pp. 23–40.
10. Chang, S. J., Huang, Y., Yang, S., Kuo, S., and Lee, M. (2012). In vitro properties of gellan gum sponge as the dental filling to maintain alveolar space, *Carbohydr. Pol.*, **88**, pp. 684–689.
11. Ciardelli, G., Chiono, V., Vozzi, G., Pracella, M., Ahluwalia, A., Barbani, N., Cristallini, C., and Giusti, P. (2005). Blends of poly-(epsilon-caprolactone) and polysaccharides in tissue engineering applications, *Biomacromolecules*, **6**, pp. 1961–1976.
12. Cleary, J. M., Coleman, R. J., Harding, N. E., and Patel, Y. N. (2006). Genetically purified gellan gum, *WO2006009938.*
13. Coutinho, D. F., Sant, S. V., Shin, H, Oliveira, J. T., Gomes, M. E., Neves, N. M., Khademhosseini, A., and Reis, R. L. (2010). Modified gellan gum hydrogels with tunable physical and mechanical properties, *Biomaterials*, **31**, pp. 7494–7502.
14. Coviello, T., Dentini, M., Rambone, G., Desideri, P., Carafa, M., Murtas, E., Riccieri, F. M., and Alhaique, F. (1998). A novel co-crosslinked polysaccharide: studies for a controlled delivery matrix, *J. Control. Release*, **55**, pp. 57–66.
15. Crescenzi, V., Dentini, M., and Dea, I. C. M. (1987). The infuence of side-chains on the dilute-solution properties of three structurally related bacterial anionic polysaccharides, *Carbohyd. Res.*, **160**, pp. 283–302.
16. D'Arrigo, G., Di Meo, C., Gaucci, E., Chichiarelli, S., Coviello, T., Capitani, D., Alhaique, F., and Matricardi, P. (2012). Self-assembled gellan-based nanohydrogels as a tool for prednisolone delivery, *Soft Matter*, **8**, pp. 11557–11564.
17. Du, H., Hamilton, P., Reilly, M., and Ravi, N. (2012). Injectable in situ physically and chemically crosslinkable gellan hydrogel, *Macromol. Biosci.*, **12**, pp. 952–961.

18. Fialho, A. M., Moreira, L. M., Granja, A. T., Popescu, A. O., Hoffmann, K., and Sá-Correia, I. (2008). Occurrence, production, and applications of gellan: current state and perspectives, *Appl. Microbiol. Biotechnol.*, **79**, pp. 889–900.
19. Gal, A., and Nussinovitch, A. (2007). Hydrocolloid carriers with filler inclusion for diltiazem hydrochloride release. *J. Pharm. Sci.*, **96**, pp. 168–178.
20. Giavasis, I., Harvey, L. M., and McNeil, B. (2000). Gellan gum. *Crit. Rev. Biotechnol.*, **20**, pp. 177–211.
21. Gong, Y., Wang, C., Lai, R. C., Su, K., Zhang, F., and Wang, D. (2009). An improved injectable polysaccharide hydrogel: modified gellan gum for long-term cartilage regeneration in vitro, *J. Mater. Chem.*, **19**, pp. 1968–1977.
22. Grasdalen, H., and Smidsrød, O. (1987). Gelation og gellan gum. *Carbohyd. Pol.*, **7**, pp. 371–393.
23. Huang, Y., Singh, P. P., Tang, J., and Swanson, B. J. (2004). Gelling temperatures of high acyl gellan as affected by monovalent and divalent cations with dynamic rheological analysis, *Carbohyd. Pol.*, **56**, pp. 27–33.
24. Jampen, S., Britt, I. J., Tung, M. A. (2000). Gellan polymer solution properties: dilute and concentrated regimes, *Food Res. Int.*, **33**, pp. 579–586.
25. Jansson, B., Hägerström, H., Fransén, N., Edsman, K., and Björk, E. (2005). The influence of gellan gum on the transfer of fluorescein dextran across rat nasal epithelium in vivo, *Eur. J. Pharm. Biopharm.*, **59**, pp. 557–564.
26. Kang, K. S., Veeder, G. T., and Kaneko, T. (1982). Agar like polysaccharide produced by pseudomona species: Production and basic properties, *Appl. Environ. Microbiol.*, **43**, 1086–1091.
27. Kang, K.S., and Veeder, G.T. (1983). Fermentation process for preparation of polysaccharide S-60, *US4377636*
28. Kani, K., Horinaka, J., Maeda, S. (2005). Effects of monovalent cation and anion species on the conformation of gellan chains in aqueous systems, *Carbohydr. Pol.*, **61**, pp. 168–173.
29. Kapasis, S., Giannouli, P., Hember, M. W. N., Evageliou, V., Poulard, C., Tort-Bourgeois, B., and Sworn G. (1999). Structural aspects and phase behaviour in deacylated and high acyl gellan systems, *Carbohyd. Pol.*, **38**, pp. 145–154.
30. Kawasaki, S.R., Moriguchi, R., Sekiya, K., Nakai, T., Ono, E., Kume, K., and Kawahara, K. (1994). The cell envelope structure of the

lipopolysaccharide-lacking gram-negative bacterium *Sphingomonas paucimobilis*, *J. Bacteriol.*, **176**, pp. 284–290.

31. Kedzierewicz, F., Lombry, C., Rios, R., Hoffman, M., and Maincent, P. (1999). Effect of the formulation on the in-vitro release of propranolol from gellan beads, *Int. J. Pharm.*, **178**, pp. 129–136.
32. Li, J., Kamath, K., and Dwivedi, C. (2001). Gellan film as an implant for insulin delivery, *J. Biomater. Appl.*, **15**, pp. 321–343.
33. Mahajan, H. S., Shaikh, H., Gattani, S., and Nerkar, P. (2009). In-situ gelling system based on thiolated gellan gum as new carrier for nasal administration of dimenhydrinate, *Int. J. Pharm. Sci. Nanotech*, **2**, pp. 544-550.
34. Mahajan, H. S., and Gattani, S. G. (2009). Gellan gum based microparticles of metoclopromide hydrochloride for intranasal delivery: development and evaluation, *Chem. Pharm. Bull.*, **4**, pp. 388–392.
35. Manning, C.E. (1992). Formation and melting of gellan polysaccharide gels. PhD thesis, Cranfield Institute of Technology, Silsoe College, Silsoe, Bedford, UK.
36. Matricardi, P., Cencetti, C., Ria, R., Alhaique, F., and Coviello, T. (2009). Preparation and characterization of novel gellan gum hydrogels suitable for modified drug release, *Molecules*, **14**, pp. 3376–3391.
37. Mazen, F., Milas, M., and Rinaudo, M. (1999). Conformational transition of native and modified gellan, *Int. J. Biol. Macromol.*, **26**, pp. 109–118.
38. Milas, M., and Rinaudo, M. (1996). The gellan sol-gel transition, *Carbohydr. Pol.*, **30**, pp. 177–184.
39. Miyazaki, S., Aoyama, H., Kawasaki, N., Kubo, W., and Attwood, D. (1999). In situ-gelling gellan formulations as vehicles for oral drug delivery, *J. Control. Release*, **60**, pp. 287–295.
40. Miyazaki, S., Kawasaki, N., Kubo, W., Endo, K., and Attwood, D. (2001). Comparison of in situ gelling formulations for the oral delivery of cemetidine, *Int. J. Pharm.*, **220**, pp. 161–168.
41. Miyoshi, E., Takaia, T., and Nishinari, K. (1995). Effect of salts on the sol-gel transition of gellan gum by differential scanning calorimetry and thermal scanning rheology, *Thermochim. Acta*, **267**, pp. 269–287.
42. Miyoshi, E., Takaia, T., and Nishinari, K. (1996). Rheological and thermal studies of gel-sol transition in gellan gum aqueous solution, *Carbohydr. Pol.*, **30**, pp. 109–119.
43. Miyoshi, E., and Nishinari, K. (1999a). Rheological and thermal properties near the sol-gel transition of gellan gum aqueous solutions, *Prog. Coll. Pol. Sci.*, **114**, pp. 68–82.

44. Miyoshi, E., and Nishinari, K. (1999b). Non-Newtonian flow behaviour of gellan gum aqueous solutions, *Coll. Pol. Sci.*, **277**, pp. 727–734.

45. Morris, E. R., Richardson, R. K., and Whittaker, L. E. (1999). Rheology and gelation of deacylated gellan polysaccharide with Na+ as the sole counterion, *Prog. Coll. Pol. Sci.*, **114**, pp. 109–115.

46. Morris, E. R., Nishinari, K., and Rinaudo, M. (2012). Gelation of gellan - A review, *Food Hydrocoll.*, **28**, pp. 373–411.

47. Moslemy, P., Neufeld, R. J., Millette, D., and Guiot S. R. (2003). Transport of gellan gum microbeads through sand: an experimental evaluation for encapsulated cell bioaugmentation, *J. Environ. Manage.*, **69**, pp. 249–259.

48. Moslemy, P., Guiot, S. R., and Neufeld, R. J. (2004). Activated sludge encapsulation in gellan gum microbeads for gasoline biodegradation, *Bioprocess. Biosyst. Eng.*, **26**, pp. 197–204.

49. Ogawa, E., Matsuzawa, H., and Iwahashi M. (2002). Conformational transition of gellan gum of sodium, lithium, and potassium types in aqueous solutions, *Food Hydrocoll.*, **16**, pp. 1–9.

50. Oliveira, J. T., Martins, L., Picciochi, R., Malafaya, P. B., Sousa, R. A., Neves, N. M., Mano, J. F., and Reis, R. L. (2010). Gellan gum: a new biomaterial for cartilage tissue engineering applications, *J. Biomed. Mater. Res. A*, **93**, pp. 852–863.

51. Oliveira, J. T., Gardel, L. S., Rada, T., Martins, L., Gomes, M. E., and Reis, R. L. (2010). Injectable gellan gum hydrogels with autologous cells for the treatment of rabbit articular cartilage defects, *J. Orthop. Res.*, **28**, pp. 1193–1199.

52. Osmałek, T., Froelich, A., Tasarek, S. (2014). Application of gellan gum in pharmacy and medicine, *Int. J. Pharm.*, **466**, pp. 328–340.

53. Paulsson, M., Hägerström, H., and Edsman, K. (1999). Rheological studies of the gelation of deacetylated gellan gum (Gelrite®) in physiological conditions, *Eur. J. Pharm. Sci.*, **9**, pp. 99–105.

54. Pollock, T. J. (1993). Gellan-related polysaccharides and the genus Sphingomonas, *J. Gen. Microbiol.*, **139**, pp. 1939–1945.

55. Rajinikanth, P. S., Balasubramaniam, J., and Mishra, B. (2007). Development and evaluation of a novel floating in situ gelling system of amoxicillin for eradication of *Helicobacter pylori*, *Int. J. Pharm.*, **335**, pp. 114–122.

56. Robinson, G., Manning, C. E., and Morris, E. R. (1991). Conformation and physical properties of the bacterial polysaccharides gellan, welan and rhamsan. In E. Dickinson (ed.), *Food Polymers, Gels and Colloids* (pp. 22e33). Cambridge, UK: Royal Society of Chemistry.

57. Rozier, A., Mazuel, C., Grove, J., and Plazonnet, B. (1997). Functionality testing of gellan gum, a polymeric excipient material for ophthalmic dosage forms, *Int. J. Pharm.*, **153**, pp. 191–198.
58. Santucci E. Alhaique F. Carafa M. Coviello T., Murtas E. Riccieri F.M. (1996). Gellan beads for sustained delivery, *J. Control. Release*, **42**, pp. 157–164.
59. Sanzgiri, Y. D., Maschi, S., Crescenzi, V., Callegaro, L., Topp, E. M., and Stella, V. J. (1993). Gellan-based systems for ophthalmic sustained delivery of methylprednisolone, *J. Control. Release*, **26**, pp. 195–201.
60. Silva, N. A., Cooke, M. J., Tam, R. Y., Sousa, N., Salgado, A. J., Reis, R. L., and. Shoichet M. S. (2012). The effects of peptide modified gellan gum and olfactory ensheathing glia cells on neural stem/progenitor cell fate, *Biomaterials*, **33**, pp. 6345–6354.
61. Shin, H., Olsenc, B. D., and Khademhosseini, A. (2012). The mechanical properties and cytotoxicity of cell-laden double-network hydrogels based on photocrosslinkable gelatin and gellan gum biomacromolecules, *Biomaterials*, **33**, pp. 3143–3152.
62. Silva-Correia, J., Zavan, B., Vindigni, V., Silva, T. H., Oliveira, J. M., Abatangelo, G., and Reis, R. L. (2012), Biocompatibility evaluation of ionic- and photo-crosslinked methacrylated gellan gum hydrogels: in vitro and in vivo study, *Adv. Healthcare Mater.*, doi: 10.1002/adhm.201200256
63. Silva-Correia, J., Miranda-Gonçalves, V., Salgado, A. J., Sousa, N., Oliveira, J. M., Reis, R. M., and Reis, R. L. (2012). Angiogenic potential of gellan-gum-based hydrogels for application in nucleus pulposus regeneration: in vivo study, *Tissue Eng. A*, **11–12**, pp. 1203–1212.
64. Smith, A. M., Shelton, R. M., Perrie, Y., and Harris, J. J. (2007). An initial evaluation of gellan gum as a material for tissue engineering applications, *J. Biomater. Appl.*, **22**, pp. 241–254.
65. Sultana, Y., Aqil, M., and Ali, A. (2006). Ion-activated, Gelrite-based in situ ophthalmic gels of pefloxacin mesylate: comparison with conventional eye drops, *Drug. Deliv.*, **13**, pp. 215–219.
66. Sutherland, I. W. (2001). Microbial polysaccharides from Gram-negative bacteria, *Int. Dairy J.*, **11**, pp. 663–674.
67. Upstill, C., Atkins, E. D. T., and Atwool, P. T. (1986). Helical conformations of gellan gum, *Int. J. Biol. Macromol.*, **8**, pp. 275–288.

Chapter 11

Polysaccharide Hydrogels with Magnetic Nanoparticles

Serena Fedi[a] and Rolando Barbucci[b]

[a]*University of Siena, Department of Biotechnology, Chemistry and Pharmacy, Via Aldo Moro 2, 53100 Siena, Italy*
[b]*CRISMA Advanced Medical Systems Interuniversity Research Centre, Viale Giacomo Matteotti 15/16, 53034 Colle Val d'Elsa, Siena, Italy*
barbucci@unisi.it

The concept of using external magnetic fields to achieve a pulsatile release from polymer composites was first pursued by Langer and coworkers in 1981 [1]. They incorporated magnetic steel beads in ethylene vinyl acetate polymers and showed that the application of low-frequency oscillating magnetic fields can be used to control release rates of different compounds like bovine serum albumin and insulin. In recent years, stimuli-responsive polymers, which can respond to external stimuli, such as pH, temperature, light, ultrasound, and the presence of magnetic or electric fields, have attracted a great deal of interest [2–4].

This chapter presents an overview of polysaccharide magnetic hydrogels with a particular interest in their reactions to external magnetic fields, which lead to their potential biomedical applications in externally actuated intelligent drug delivery systems.

Polysaccharide Hydrogels: Characterization and Biomedical Applications
Edited by Pietro Matricardi, Franco Alhaique, and Tommasina Coviello

ISBN 978-981-4613-61-3 (Hardcover), 978-981-4613-62-0 (eBook)
www.panstanford.com

11.1 Introduction

Polysaccharides are widely used in pharmaceutical formulations thanks to their biocompatibility, biodegradability, and versatile chemistry. The polysaccharides usually employed for the preparation of magnetic hydrogels are hyaluronic acid, cellulose, guar gum, chitosan, gelatin, collagen, and alginate. Combining the properties of polysaccharide hydrogels with those of magnetic nanoparticles (NPs), magnetic hybrid hydrogels are obtained. These new composites are of special interest because they maintain the peculiarities of both components, but what are the properties of magnetic NPs? First, they have controllable sizes ranging from a few nanometers up to hundreds of nanometers, which places them at dimensions that are smaller than or comparable to those of a cell (10–100 μm), a virus (20–450 nm), a protein (5–50 nm), or a gene (2 nm wide and 10–100 nm long). This means that they can "get close" to a biological entity of interest. Second, the NPs are magnetic, which means that they can be manipulated by an external magnetic field gradient. This "action at a distance," combined with the intrinsic penetrability of magnetic fields into human tissue, opens up many applications involving the transport and/or immobilization of magnetic NPs, or of magnetically tagged biological entities in the site of interest. In this way they can be prepared to deliver a package, such as an anticancer drug to a targeted region of the body. Third, the magnetic NPs can be prepared to respond to a time-varying magnetic field, with advantageous results related to the transfer of energy from the field to the NPs [5].

The NPs usually employed in the preparation of magnetic hydrogels are composed of iron oxides, and this is the reason that they are also called ferrogels. The most used magnetic NP is magnetite, Fe_3O_4, composed of iron II–iron III oxides, followed by cobalt ferrite, $CoFe_2O_4$, for its enhanced magnetic properties and maghemite, γ-Fe_2O_3.

Magnetic materials show different magnetic properties that depend on chemical components, size, and shape. Magnetic NPs with a typical size equal to or less than that of the monodomain exhibit superparamagnetic behavior. An estimate of monodomain diameters for cobalt or iron NPs, spherical in shape, is 70 and 14 nm,

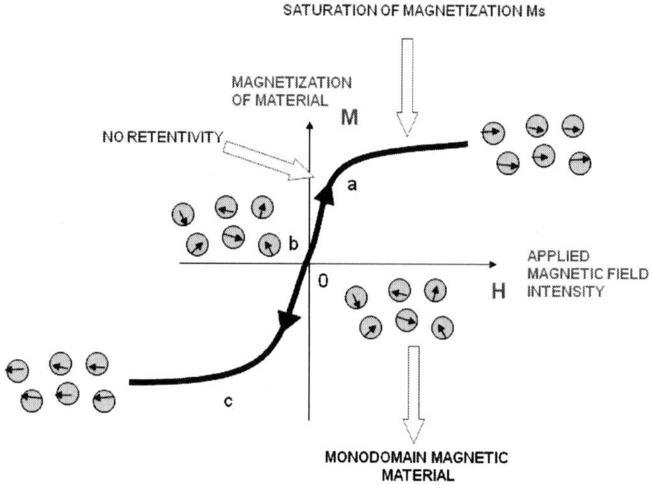

Figure 11.1 Magnetization curve of monodomain material. (a) The material is magnetized starting from a zero field value; (b) the material loses its magnetization as the applied field comes back to zero and (c) toward saturation in the opposite direction.

or less, respectively. Nevertheless, the NP shape is important; in fact, the monodomain dimensions of anisotropic NPs are larger than the spherical ones. In Fig. 11.1 the arrows represent the magnetic moment of a monodomain particle. In the absence of a static external magnetic field they are randomly oriented due to thermal agitation, and thus there isn't a net bulk magnetization.

As soon as an external field is applied, the magnetic moments tend to align with the field and produce a bulk magnetic moment. As the strength of the field increases, more particles align their moments along the direction of the field, until it reaches a saturation of magnetization to a value $\mathbf{M_s}$. If the applied field is turned off, the particles randomize through two relaxation mechanisms, having different characteristic times, which strongly depend on particle core size, particle anisotropy, and hydrodynamic volume, and as a result $\mathbf{M_s}$ again reduces to zero. This means that the magnetization curve shows no hysteresis at all.

NPs with a typical size larger than that of the monodomain are multidomain and exhibit a hysteresis loop on the magnetization

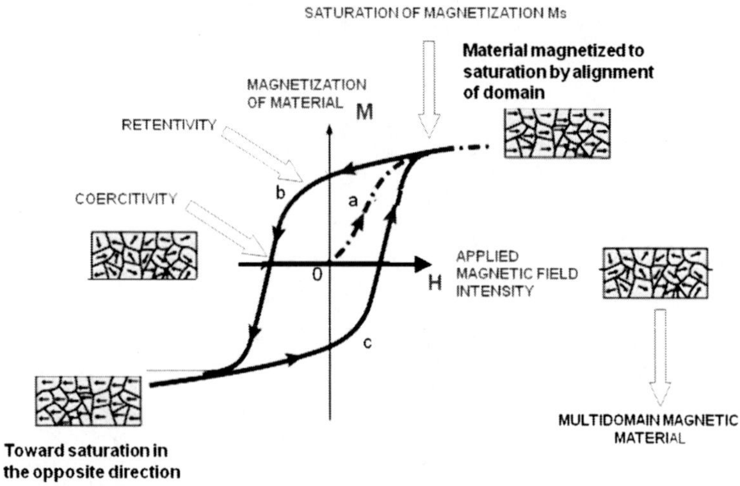

Figure 11.2 Hysteresis loop of a multidomain material. (a) The material follows a nonlinear magnetization curve when magnetized starting from a zero field value. (b) The material retains its magnetization even if the applied field is in the opposite direction. (c) The material, after saturation in the opposite direction, retains again its magnetization.

curve. As illustrated in Fig. 11.2, the hysteresis loop shows the "history-dependent" nature of magnetization of a ferromagnetic material. Once the material has been driven to saturation, the external magnetizing field can then be dropped to zero and the material will retain most of its magnetization (it remembers its history). Hysteresis makes permanent magnetism possible.

Because of the significant difference between monodomain and multidomain magnetic NPs incorporated into a gel matrix, the magnetic properties of the gel will be also different [6].

One way to classify the hydrogel nanocomposites is based on their structure: bulk nanocomposites, particle nanocomposites, and core–shell nanocomposites (Fig. 11.3). The bulk hydrogel nanocomposites are predominantly synthesized in the form of mass hybrid hydrogels or thin films. The nanocomposites in particle form can range from tens of nanometers to micrometers in diameter and the magnetic NPs can be dipped into the gel network or cover the hydrogel beads (colloidosome). The core–shell nanocomposites, on

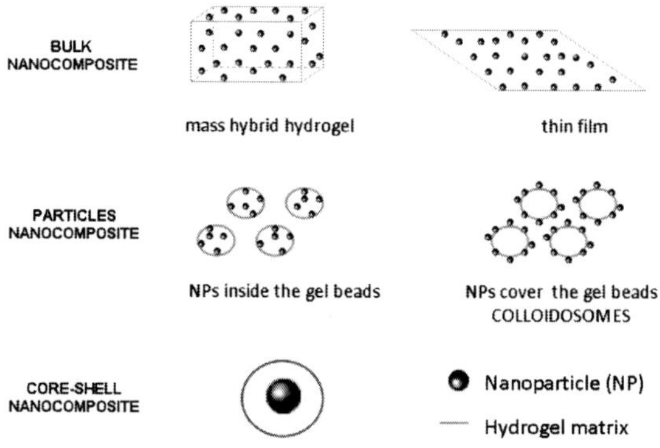

Figure 11.3 Classification of magnetic hydrogels.

the other hand, consist of a single NP or small agglomerate of NPs surrounded by a hydrogel shell.

11.2 Methods of Preparation of Magnetic Hydrogels

To date, a number of approaches for the preparation of hybrid hydrogels have been suggested and developed. They can be divided into the five methods basically illustrated in Figs. 11.4 and 11.5.

The first one is the incorporation of preformed magnetic NPs into the polysaccharide solution by mixing and subsequent in situ gelation (Fig. 11.4a). This was the method of synthesizing copolymers of chitosan, maltodexrin, and gum arabic [7], and hydrogels based on gelatin [4] and collagen [8] containing preformed Fe_3O_4 NPs.

The second method is the simultaneous formation of both hydrogel matrix and inorganic NPs in the reaction system (Fig. 11.4b). As an example, Hu in 2007 prepared magnetic ferroscaffolds by an in situ co-precipitation process adding the appropriate amount of $FeCl_2$ and $FeCl_3$ with a molar ratio of 2:1 to a gelatin solution to form hybrid sols which were rapidly cooled to 4°C to gel the gelatin. Subsequently they immersed the gels in a water solution of NH_4OH to start the formation of iron oxide [9].

The third one is the formation of inorganic NPs in the preformed hydrogel matrix. In this way the polymeric gel acts as a spatial framework to control the size and distribution of the iron oxide NPs

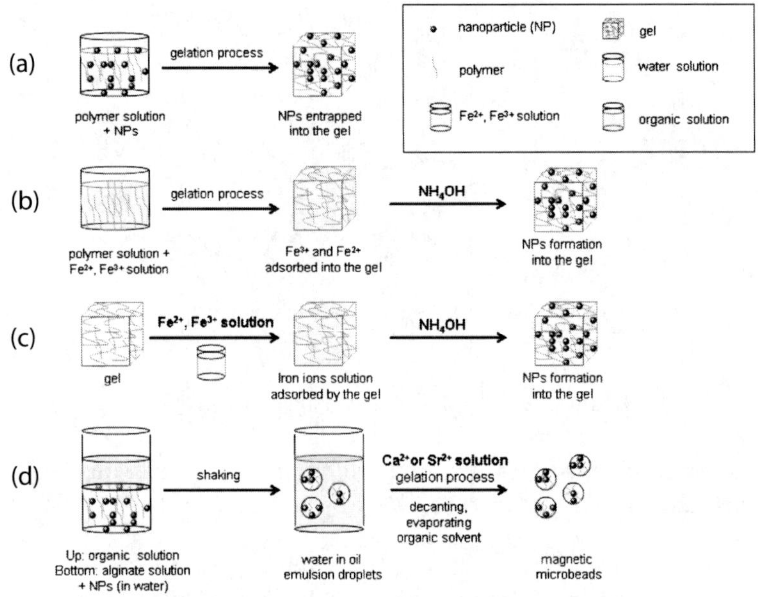

Figure 11.4 Methods of preparation of magnetic hydrogels.

(Fig. 11.4c). Reddy prepared polyacrylamide/gelatin copolymers by free radical polymerization and immersed these hydrogels in a solution containing iron(II) chloride tetrahydrate and iron(III) chloride hexahydrate. Then they placed the gels in an ammonium hydroxide solution to obtain brown- or black-colored hydrogel magnetic nanocomposites [10, 11]. In a similar manner Hernandez and coworkers synthesized semi-interpenetrating polymer networks constituted of alginate and poly(*N*-isopropylacrylamide). These gels were immersed in an acid solution containing a mixture of Fe^{3+}/Fe^{2+} (2:1 molar ratio), after which the authors immersed the loaded gels in an alkaline solution to transform the iron cations to iron oxide NPs [12, 13].

The fourth method is the fabrication of well-defined hybrid hydrogel beads (Fig. 11.4d). This method is based on the formation of a microemulsion of a water solution containing the polymer (alginate) and the NPs in an organic solvent. After achieving stable water-in-oil emulsion droplets a crosslinking agent is added to perform the gelation process.

Saslawski reported the preparation of alginate–strontium ferrite microspheres [14] and Liu H. prepared colloidosome hydrogel beads with alginate gel cores and shells of maghemite nanoparticles [15]. Liu J. synthesized magnetic γ-Fe_2O_3 alginate microspheres [16] and Brulé prepared magnetic microbeads encapsulating a concentrated magnetic fluid composed of iron oxide NPs (diameter 8 nm) with a maghemite core into alginate [17].

All these methods share the lack of a specific interaction between the NPs and the polymer matrix allowing the movement of the NPs inside the hydrogels. The NPs can attract each other with the consequent formation of big aggregates, which inhibit the magnetic properties of the gel. In some cases, depending on the size of the NPs and the gel mesh size, magnetic NPs can lift out the hydrogel when exposed to an external magnetic field [18]. In order to overcome these problems, polysaccharides such as carboxymethylcellulose and hyaluronic acid were crosslinked with magnetic NPs. The NPs are crosslinker agents and therefore bind to the polysaccharide by covalent bonds [19–21].

To obtain these crosslinked hydrogels a previous functionalization of the NPs is necessary. The chemistry of 3-aminopropyltrimethoxysilane is used to insert on the NP surface a $-NH_2$ group that in a subsequent reaction is bonded to the carboxylic group of carboxymethylcellulose or hyaluronic acid (Fig. 11.5). Three-dimensional structures are then obtained which overcome the problem of undesired NPs movement in the gel or dispersion outside of the gel [19–21].

11.3 Methods of Characterization

Magnetic hydrogels can be characterized by their physicochemical, morphological, and rheological properties.

11.3.1 *Physicochemical Studies*

The physicochemical studies usually carried out for hybrid hydrogel characterization are infrared spectroscopy, Raman spectroscopy, potentiometry, thermogravimetric analysis (TGA), and water uptake.

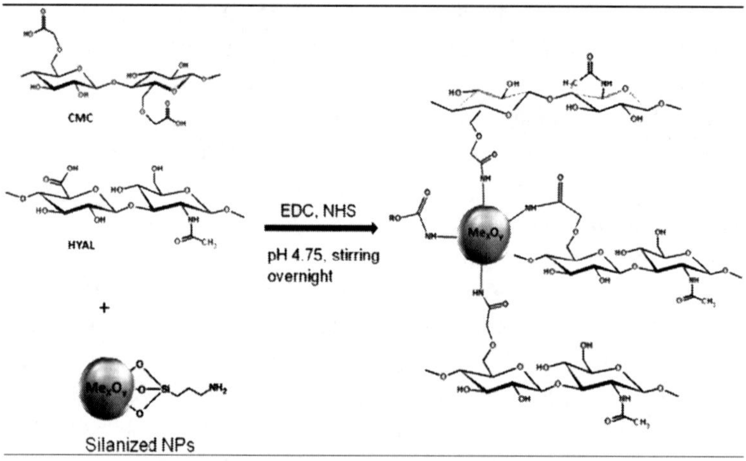

Figure 11.5 Reaction scheme of the formation of the HYAL-NP hybrid hydrogel. The reaction involves the formation of an amide bond between the carboxylic groups of the polymer and the amine groups of the NP-NH_2 (the crosslinker agent) in the presence of *N*-(3-dimethylaminopriopyl)-*N*-ethylcarbodiimide hydrochloride (EDC) and *N*-hydroxysuccinimide (NHS). Reprinted from Ref. 20, © 1996–2015 MDPI AG (Basel, Switzerland).

FTIR spectroscopy is a useful tool to determine the hydrogel characteristics (because polysaccharide rings present common adsorption peaks in the 900–1200 cm^{-1} range) as well as the chemical route pursued to synthesize the hydrogel. As an example in the case of functionalized NPs with 3-aminopropyl trimethoxysilane, bound to some polysaccharides via EDC and NHS, by FTIR it was possible to ascertain that some carboxylic groups of carboxymethylcellulose, or hyaluronic acid, were effectively bonded to NH_2 groups to form amide groups showing a characteristic peak at ~1640 cm^{-1} relative to amidic C=O stretching [19, 21]. The presence of amino groups on the surface of the NPs was previously determined by XPS analysis and cyclic voltammetry [19, 21].

Using infrared spectroscopy, it is also possible to verify the presence of iron oxide NPs and the silane coating because of the occurrence of peaks at near 600 cm^{-1} due to the stretching vibrations of the Fe–O bond and at ~1100 and ~1040 cm^{-1} due to the Si–O and Si–O–Si bonds [10, 11, 16, 21].

Confocal Raman spectroscopy is a way to determine the iron oxide composition in the preformed hydrogel matrix because the low-frequency region (300–800 cm^{-1}) contains the most characteristic Raman bands of iron oxides. Magnetite and maghemite, in fact, give rise to the occurrence of two peaks at 720 and 670 cm^{-1}, respectively [12, 13].

To solve the same problem X-ray diffraction is sometimes utilized [11]. Nevertheless, the wide-angle X-ray diffractograms corresponding to magnetite and maghemite are almost identical, and therefore this technique does not allow for a detailed characterization of the molecular structure of iron oxide NPs synthesized inside a preformed hydrogel.

Occasionally, potentiometric titration becomes a useful tool to determine the crosslinking degree of a given hydrogel that contains amine or carboxylic groups (more generally acid or base groups) which are involved in the crosslinking reaction. As an example, the crosslinking degree in the carboxymethylcellulose hydrogel reticulated with NH_2-crosslinker agents is calculated by potentiometric titration of the remaining free carboxylic groups and then by relating this value to the total carboxylic groups present in the polymeric chain [22].

The knowledge of the quantity of NPs present in the hydrogel is mandatory to interpret the physicochemical hydrogel properties. A method to determine the iron oxide content in the hydrogel is by thermogravimetric analysis. A typical thermogram of a lyophilized hybrid hydrogel shows two or more weight loss values in the range of 90–600°C attributed to water loss and polymer chain degradation. The remained residue trespassing 600°C is the iron oxide content [10, 11, 16].

The water uptake or degree of swelling of the hydrogel depends on pH, ionic strength, and temperature and is usually set starting from the dried hydrogel of known mass and placing it in contact with a specified volume of water for various contact times. For each contact time, the mass of the swollen hydrogel is measured to calculate the water uptake according to Eq. 11.1:

$$\text{Water uptake} = \frac{W_\text{s} - W_\text{d}}{W_\text{d}} \times 100 \qquad (11.1)$$

where W_s is the weight of the swollen hydrogel at equilibrium and W_d is the initial weight of the dried hydrogel.

The water uptake depends on the presence of hydrophilic groups on the polymer chain, on the crosslinking degree, and on the NPs content and distribution, which in turn affect the extent of the magnetic hydrogel response upon a magnetic stimulus. Generally, more pronounced magnetic responses are observed for less rigid matrixes [2, 4, 23, 24].

11.3.2 *Rheological Properties*

Stress vs. strain is another parameter employed for hydrogel characterization. In fact, the deformation of swollen hydrogels resulting from suitable/applied forces is a parameter that depends on hydrogel softness or roughness and can be correlated to the water uptake, the crosslinking degree and the NPs' content [25, 26].

The rheological measurements give the values of the storage modulus (G') and the loss modulus (G''). The storage modulus (or elasticity modulus) G' is a measure of the gel's elastic behavior associated with energy storage, and the loss modulus (or viscosity modulus), G'' is a measure of the gel's viscous behavior associated with energy dissipation. $G' < G''$ indicates liquid-like behavior, and $G' > G''$ indicates more of a solid-like behavior [21].

Table 11.1 shows the G' and G'' values of different carboxymethyl-cellulose (CMC) hydrogels prepared as in Fig. 11.5. CMC-NPs hydrogels show values of the storage modulus G' higher than the

Table 11.1 G' and G'' values for CMC-DAP and CMC- NPs hydrogels (50 and 70 $CoFe_2O_4$%). The G' and G'' for the CMC polymer solution at a concentration of 1.5% w/v are 15±2 Pa and 25±1 Pa, respectively. Data are reported as a mean of the three measurements ± standard error

Sample	G' (Pa)	G'' (Pa)
CMC-DAP hydrogel	500 ± 50	60 ± 30
CMC-NPs-50 hydrogel	3300 ± 300	90 ± 20
CMC-NPs-70 hydrogel	2000 ± 200	110 ± 20

Source: Modified from [20]

loss modulus G'', demonstrating the gel-like nature of the hydrogels formed by using magnetic NPs as crosslinking agents. Moreover, the values of G' of the hydrogels with NPs (CMC-NPs-50 and CMC-NPs-70 NPs=$CoFe_2O_4$) are greater than that of CMC crosslinked with diaminopropane (CMC-DAP), demonstrating that the presence of NPs in the gel matrix causes the stiffening of the hydrogel. Another obvious consequence of the insertion of NPs in the gel is the decrease in the water uptake [20, 21]. The comparison between G' values of CMC-NPs-50 and CMC-NPs-70 reveals that the NP concentration and distribution is another important parameter to be taken into account: if the NPs are too crowded, as in the case of CMC-NPs-70, the formation of NP aggregates produces big pores in the hydrogel network and the hydrogel becomes softer; consequently the G' value decreases but the magnetic properties also decrease [24].

Additional hydrogel parameters that are investigated through the rheological analysis are the thixotropic properties. Thixotropy is a characteristic of some pseudoplastic materials that leads to a progressive decrease in their viscosity when these materials are subjected to a specific mechanical stress, followed by the recovery of its rheological properties after a suitable period of rest. This peculiar property was found for some polysaccharide-based hydrogels [2, 8, 19]. The hydrogel injectability through a syringe is proof of its thixotropic nature and mandatory in the biomedical field, in order to facilitate the introduction of such materials at the site of application. For this reason the thixotropic nature of the CMC-NPs-50 hydrogel was investigated by steady state flow tests and stress sweep tests before and after the passage through a syringe. The rheological measurements have demonstrated the thixotropic nature of CMC-NPs-50 hydrogel: the material has characteristics of a liquid after a period of increasing applied stress ($G'' > G'$) and recovers the initial characteristic after a suitable period of time. Moreover, the infrared spectroscopy measurements showed that the chemical structure of the magnetic hydrogel was not changed after the passage through a syringe [19]. These results demonstrate that CMC-NPs-50 can be used for biomedical applications.

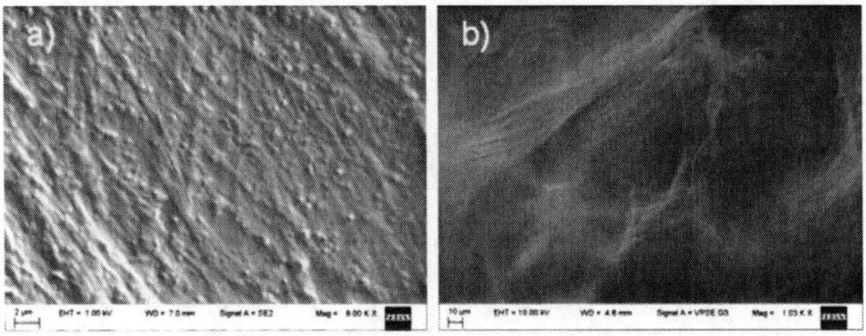

Figure 11.6 FESEM images of (a) lyophilized and (b) wet CMC-$CoFe_2O_4$ hydrogels [38].

11.3.3 *Morphological Studies*

The morphological studies of magnetic hydrogels are of crucial interest for the visualization of NPs' dimension and dispersibility, which in turn are strictly connected to the magnetic properties of the gel. Giani et al. demonstrated that the sonication of an NP solution before the gelation process is essential to have good dispersibility of the magnetic NPs inside the polysaccharide gel [20]. The size of the magnetic NPs and their dispersibility in the gel network can be determined by transmission electron microscopy (TEM).

The surface morphology of the magnetic hydrogel can be visualized by a scanning electron microscope SEM or field emission scanning electron microscope (FESEM). Figure 11.6 illustrates the FESEM morphological structure of dried (a) and hydrated (b) CMC-$CoFe_2O_4$ hydrogels. It is possible to observe NPs of about 100 nm at a higher magnification for the dried hydrogel, while the hydrated hydrogel shows a regular surface. FESEM-EDX studies on the hydrated gel have indicated that the $CoFe_2O_4$ NPs are inside the hydrogel and not on the surface.

SEM analysis can give information about modifications of the hydrogel surface upon introduction of magnetic NPs and, most importantly, after magnetic field exposition. As an example, De Paoli demonstrated through SEM analysis a reversible physical deformation of the collagen hydrogel matrix following the application of an oscillating magnetic field for 12 hours [8].

Figure 11.7 (a) Image showing the attraction of the CMC–NP hydrogel by a magnet. (b, c) Images showing the attraction of CMC–NP hydrogels loaded with toluidine blue after being squeezed through a syringe. The magnet is capable of attracting the material from the center to the wall of the beaker. Reproduced from Ref. 19 with permission from The Royal Society of Chemistry.

11.4 Magnetic Hydrogel Responses to Magnetic Stimuli

The incorporation of magnetic NPs inside the gel matrix provides enhanced properties: the resulting hybrid hydrogel reacts to external magnetic stimuli (Fig. 11.7).

To test the magnetic properties of hybrid hydrogels a stationary magnetic field (SMF) or an alternating magnetic field (AMF) is usually applied. Normally, an increase in the amplitude of the applied SMF conducts improved responses of a sample containing magnetic NPs (until it reaches a saturation of the magnetization) [2].

A magnetic hydrogel can be exposed to two different SMFs: uniform and non-uniform. If the field is non-uniform, the particles (monodomain or multidomain) inside the gel are subjected to field-particle interactions. If monodomain NPs are present within the hydrogel matrix, their magnetic moments, which without the field are randomly oriented, tend to align with the field. If the gel contains

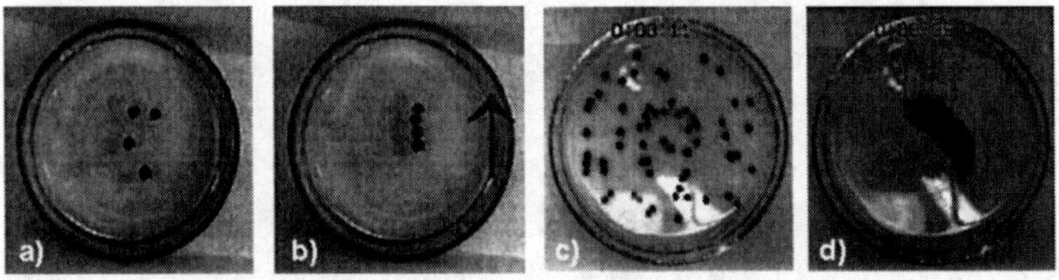

Figure 11.8 Magnetophoresis of the poly(*N*-isopropylacrilamide) gel beads: (a) and (c) no magnetic field is applied; (b) needle-like aggregation of gel beads in a uniform magnetic field. The arrow indicates the direction of the field, which has an intensity of 300 mT. (d) Aggregation of the gel beads in a non-uniform magnetic field. Reprinted with permission from Ref. 6. Copyright © 2000, American Chemical Society.

multidomain particles they are magnetized by the external field and again tend to orientate with its direction. In both cases, the presence of magnetic field gradients makes the displacement of the particles possible. In the uniform magnetic fields no particle displacement can be observed. In the event that the particles approach a close proximity to one another, they can interact. This interaction can be very strong, leading to a significant change in the structure of the particle ensembles. The particles attract each other and a pearl-chain structure develops.

The exposition of a hydrogel containing NPs to these different types of SMF induces macroscopic effects: Zrìni and coworkers prepared magnetic poly(*N*-isopropylacrilamide) or poly(vinylalcohol) gel beads. As illustrated in Fig. 11.8, the gel beads experience different magnetophoretic forces as a consequence of uniform (Fig. 11.8b) or non-uniform (Fig. 11.8d) SMF application [6, 27].

A uniform SMF of 400 Oe was used by Chen and coworkers to induce magnetic NPs aggregation in gelatin-based hydrogels [4, 9]. The NPs' aggregation reduces the pore size and increases the tortuosity of the pore channels across the hydrogel. When the field is turned off, the pores reopen [28].

The macroscopic effect due to nonuniform SMF application on the CMC-$CoFe_2O_4$ hydrogel exposed to two permanent magnets of 0.48 T is illustrated in Fig. 11.9. When the magnetic hydrogel is subjected to non-uniform SMF, the gel stretches. Even in this case, the result is probably a reduction of the pore dimensions.

Figure 11.9 Carboxymethylcellulose-$CoFe_2O_4$ hydrogel macroscopic modification upon 12 h non-uniform SMF (0.48 T) [38].

The AMFs that can be applied can be divided in two types: low-frequency magnetic fields (0.3–10 Hz) and high-frequency magnetic fields (up to 300 kHz). The frequencies of generally utilized oscillating magnetic fields pass harmlessly through the body. Nevertheless, there is a limit on the strength of an applied field, which is about 8–16 kAm^{-1} (10–20 mT) [29].

For the alternating magnetic fields manipulation it is advantageous to use NPs that show a very low coercivity (superparamagnetic properties) in order to maximize the energy that the external AMF can transfer to the magnetic particles.

The choice of the frequency at which the external magnetic field causes an effect on the magnetic NPs depends on the core size, anisotropy, and hydrodynamic volume of the NPs inside the gel. If the NPs have superparamagnetic cores, are spherical, and have a small hydrodynamic volume, the magnetic hydrogel, in general, will respond at high-frequency magnetic fields. If the NPs have core dimensions higher than the superparamagnetic one, or are highly anisotropic and have a large hydrodynamic volume, the hydrogel will respond at rather low-frequency magnetic fields.

One other important parameter is the distribution of the magnetic core and the hydrodynamic volume even if it is not an easy task to obtain a high degree of monodispersibility. If the hydrogel contains NPs with various core-size, shape, and hydrodynamic volume, it will respond to a wide band of frequency of the applied magnetic field and thus when a monochromatic AMF is applied only a small amount of the magnetic NPs will effectively interact with the magnetic field. Ultimately, the system responds weakly and is not used for biomedical application.

Low-frequency AMFs were used to enhance the release rate of drugs from polymer gels [8, 30].

High-frequency AMFs were usually utilized in the remote-controlled heating of the magnetic hydrogel and consequently led to an increase in the release of a loaded drug. Magnetic NPs, in fact, have the property to generate heat if subjected to a high-frequency AMF due to Neel and Brownian relaxations [31]. The magnetic NPs inside the gel matrix give rise to a temperature increase of several degrees (up to 40°C) [32]. This approach offers the possibility of multiple modality treatment because it couples the controlled drug release with hyperthermia therapy: the heating of cancer tissue to between 41°C and 45°C [3, 12, 13].

11.5 Drug Release Applications of Magnetic Hydrogels

In the previous section we have shown the ability of magnetic hydrogels to respond to external magnetic fields. This ability has attracted a great deal of interest due to their potential applications as externally controlled drug delivery systems.

Before showing several examples of magnetic polysaccharide drug delivery systems it is opportune to consider their biocompatibility. Parameters that determine the biocompatibility of bionanocomposites are the nature of the magnetically responsive component and the nature of the polymer. There is much research regarding citotoxicity of bare or coated magnetic nanoparticles [33–35], but there is a lack of reasearch on the biocompatibility of polysaccharide hybrid hydrogels even if the polysaccharide presence is a guarantee of biocompatibility. Reddy studied the blood compatibility of gelatin-based magnetic hydrogel nanocomposites by using thrombus formation and hemolysis tests. It was evident that the nanocomposites with a greater magnetite percentage and a greater gelatin amount exhibit greater blood compatibility. The blood compatible nature of the magnetic hydrogel was principally due to the iron oxide compatibility [11].

Polysaccharide-based magnetic hydrogels have been proposed by Huang and Yang for the fabrication of a sandwich-like structure of magnetic gelatin hydrogel coated stainless steel for drug eluting

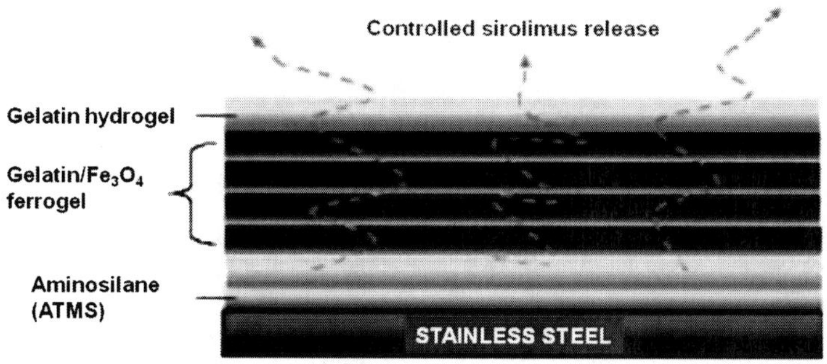

Figure 11.10 Illustration of the sandwich-like structure of stainless-steel-gelatin-ferrogel stent and sirolimus delivery. Reprinted with permission from Ref. 36. Copyright © 2007, Elsevier.

stents (Fig. 11.10). Using a SMF of 400 Oe they decreased the sirolimus (an immunosuppressant drug used to prevent rejection in organ transplantation) release of about 2/3 compared to the normal diffusive release. The gelatin-based magnetic hydrogel is the key part of the proposed device because it triggers the sirolimus delivery by switching the magnetic field "on" or "off" [36].

Chen and coworkers synthesized several types of gelatin hydrogels that contained magnetite and/or maghemite nanoparticles for the controlled release of vitamin B_{12} (taken as a model drug). One type was exposed to a STM of 400 Oe, which stops the drug release [4], and the other type was exposed to pulsed AMF (50–100 kHz), which enhances the drug release [37].

De Paoli and coworkers found differences in the dextrane-rhodamine release properties of two kinds of collagen gels (filled with nanoparticles or microparticles) upon application of an oscillating magnetic field (0.14 T; 0.3 Hz). The oscillating magnetic field affects mainly the magnitude of the initial release burst in the gel-containing nanoparticles and increases regularly and constantly the amount of released dextrane-rhodamine from the gel-containing magnetic microparticles [8].

Recently, Brulé demonstrated the magnetic properties of alginate microbeads loaded with NPs and doxorubicine. The combination of the magnetic control of heating and drug delivery triggered a substantial reduction of live cancer cells [17].

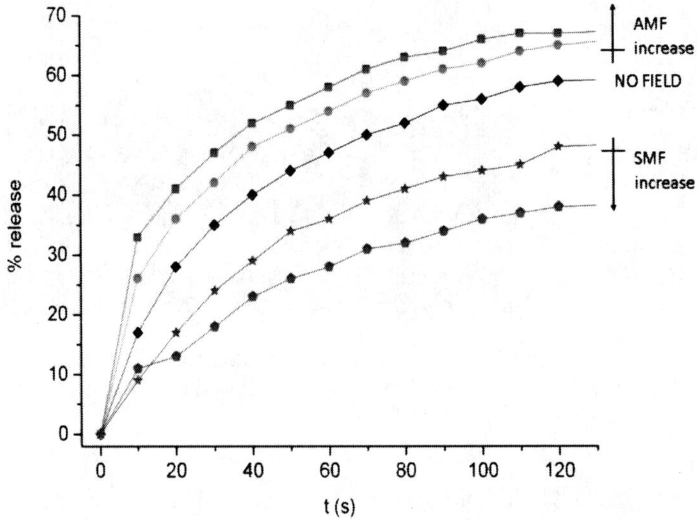

Figure 11.11 Release of methylene blue from carboxymethylcellulose-$CoFe_2O_4$ with the application of static or alternating magnetic field [38].

Finally CMC-based magnetic hydrogels containing $CoFe_2O_4$ NPs were subjected to an AMF and a STM to investigate the possibility of tuning the release of a model molecule, methylene blue (MB), by changing the condition of the applied magnetic field.

As shown in Fig. 11.11, when AMF (1 mT or 2 mT at 40 kHz) is applied on the system the amount of MB released from the CMC-NP is greater than that released by the same CMC-NP not exposed to magnetic stimuli. By applying a SMF (0.23 T or 0.48 T), the release of the MB is slowed down compared to that obtained without the application of any magnetic field.

The novelty of this system is the ability to control an increase or a decrease in the release of MB dye using the same hydrogel.

Acknowledgments

The authors thank the MIUR (Ministero Istruzione, Università, Ricerca) for financial support for the Project "Strategies for engineering ferromagnetic nanoparticles as crosslinkers of polymer chains: application as a targeted drug delivery system in primary and secondary bone tumors," FIRB Project RBAP11ZJFA, 2010.

The authors also thank Prof. Valerio Biancalana and Dr. Yordanka Dancheva for their good advice regarding physical topics.

References

1. Hsieh, D. S. T., Langer, R., and Folkman, J. (1981). Magnetic modulation of release of macromolecules from polymers. *Proc. Natl. Acad. Sci. USA*, **78**, pp. 1863–1867.
2. Kost, J., and Langer, R. (2012). Responsive polymeric delivery systems. *Adv. Drug Delivery Rev.*, **64**, pp. 327–341.
3. Satarkar, N. S. and Hilt, J. Z. (2008). Hydrogel nanocomposites as remote-controlled biomaterials. *Acta Biomater.*, **4**, pp. 11–16.
4. Liu, T. Y., Hu, S. H., Liu K. H., Liu, D. M., and Chen, S. Y. (2006). Preparation and characterization of smart magnetic hydrogels and its use for drug release. *J. Magn. Magn. Mater.*, **304**, pp. e397–e399.
5. Pankhurst, Q. A., Connolly, J., Jones, S. K., and Dobson, J. (2003). Application of magnetic nanoparticles in biomedicine. *J. Phys. D Appl. Phys.*, **36**, pp. R167–R181.
6. Xulu, P. M., Filipcsei, G., and Zrínyi, M. (2000). Preparation and responsive properties of magnetically soft poly(*N*-isopropylacrylamide) gels. *Macromolecules*, **33**, pp. 1716–1719.
7. Paulino, A. T., Pereira, A. G. B., Fajardo, A. R., Erickson, K., Kipper, M. J., Muniz, E. C., Belfiore, L. A., and Tambourgi, E. B. (2012). Natural polymer-based magnetic hydrogels: potential vectors for remote-controlled drug release. *Carbohydr. Polym.*, **90**, pp. 1216–1225.
8. De Paoli, V. M., De Paoli Lacerda, S. h., Spinu, L., Ingber, B., Rosenzweig, Z., and Rosenzweig, N. (2006). Effect of an oscillating magnetic field on the release properties of magnetic collagen gel. *Langmuir*, **22**, pp. 5894–5899.
9. Hu, S. H., Liu, T. Y., Tsai, C. H., and Chen, S. Y. (2007). Preparation and characterization of magnetic ferroscaffolds for tissue engineering. *J. Magn. Magn. Mater.*, **310**, pp. 2871–2873.
10. Reddy, N. N., Mohan, Y. M., Varaprasad, K., Ravindra, S., Joy, P. A., and Raju, K. M. (2011). Magnetic and electric responsive hydrogel-magnetic nanocomposites for drug-delivery application. *J. Appl. Polymer Sci.*, **122**, pp. 1364–1375.

11. Reddy, N. N., Varaprasad, K., Ravindra S., Reddy, G. V. S., Reddy, K. M. S., Reddy, K. M. M., and Raju, K. M. (2011). Evaluation of blood compatibility and drug release studies of gelatine based magnetic hydrogel nanocomposites. *Colloids Surf. A*, **385**, pp. 20–27.

12. Hernández, R., Sacristán, J., Nogales, A., Ezquerra, T. A., and Mijangos, C. (2009). Structural organization of iron oxide nanoparticles synthesizedinside hybrid polymer gels derived from alginate studied with small-angle X-ray scattering. *Langmuir*, **25**, pp. 13212–13218.

13. Hernández, R., Sacristán, J., Asín, L., Torres, T. E., Ibarra, M. R., Goya, G. F., and Mijangos, C. (2010). Magnetic hydrogels derived from polysaccharides with improved specific power absorption: potential devices for remotely triggered drug delivery. *J. Phys. Chem. B*, **114**, pp. 12002–12007.

14. Saslawski, O., Weigarten, C., Benoit, J. P., and Couvreur, P. (1988). Magnetically responsive microspheres for pulsed delivery of insulin. *Life Sci.*, **42**, pp. 1521–1528.

15. Liu, H., Wang, C., Gao, Q., Chen, J., ren, B., Liu, X., and Tong, Z. (2009). Facile fabrication of well-defined hydrogel beads with magnetic nanocomposite shells. *Int. J. Pharm.*, **376**, pp. 92–98.

16. Liu, J., Zhang, Y., Wang, C., Xu, R., Chen, Z., and Gu, N., (2010). Magnetically sensitive alginate-templated polyelectrolyte multilayer microcapsules for controlled release of doxorubicin. *J. Phys. Chem. C*, **114**, pp. 7673–7679.

17. Brulé, S., Levy, M., Wilhelm, C., Letourneur, D., Gazeau, F., Ménager, C., and Le Visage, C. (2011). Doxorubicin release triggered by alginate embedded magnetic nanoheaters: a combined therapy. *Adv. Mater.*, **23**, pp. 787–790.

18. Galicia, J. A., Cousin, F., Dubois, E., Sandre, O., Cabuil, V., and Perzynski, R. (2009). Static and dynamic structural probing of swollen polyacrylamide ferrogels. *Soft. Matter.*, **5**, pp. 2614–2624.

19. Barbucci, R., Pasqui, D., Giani, G., De Cagna, M., Fini, M., Giardino, R., and Atrei, A. (2011). A novel strategy for engeering hydrogels with ferromagnetic nanoparticles as crosslinkers of the polymer chains. Potential application as a targeted drug delivery system. *Soft Matter*, **7**, pp. 5558–5565.

20. Giani, G., Fedi, S., and Barbucci, R., (2012). Hybrid magnetic hydrogel: a potential system for controlled drug delivery by means of alternating magnetic fields. *Polymers,* **4**, pp. 1157–1169.

21. Barbucci , R., Giani, G., Fedi, S., Bottari, S., and Casolaro, M. (2012). Biohydrogels with magnetic nanoparticles as crosslinker: characteristics and

potential use for controlled antitumor drug-delivery. *Acta Biomater.*, **8**, pp. 4244–4252.

22. Barbucci, R., Magnani, A., and Consumi, M. (2000). Swelling behaviour of carboxymethylcellulose hydrogels in relation to cross-linking, pH, and charge density. *Macromolecules*, **33**, pp. 7475–7480.
23. Saslawski, O., Couvreur, P., and Peppas, N. (1988). Alginate magnetic release systems: crosslinked structure, swelling and release studies. In J. Heller, F. Harris, D. Lohmann, H. Merkle, and I. Robinson (eds.), Proceedings of the International Symposium on Controlled Release of Bioactive Materials, Basel, p. 26.
24. Liu, T. Y., Hu, S. H., Liu, K. H., Liu D. M., and Chen, S. Y. (2008). Study on controlled drug permeation of magnetic-sensitive ferrogel: effect of Fe_3O_4 and PVA. *J. Controlled Release*, **126**, pp. 228–236.
25. Tjong, S. C., (2006). Structural and mechanical properties of polymer nanocomposites. *Mater. Sci. Eng. Rep.*, **53**, pp. 73–197.
26. Markides, H., Rotherham, M., and El Haj, J. (2012). Biocompatibility and toxicity of magnetic nanoparticles in regenerative medicine. *J. Nanomater.*, **2012**, pp. 1–11.
27. Zrínyi, M. (2000). Intelligent polymer gels controlled by magnetic fields. *Colloid Polym. Sci.*, **278**, pp. 98–103.
28. Liu, T. Y., Hu, S. H., Liu, T. Y., Liu, D. M., and Chen, S. Y. (2006). Magnetic-sensitive behaviour of intelligent ferrogel for controlled release of drug. *Langmuir*, **22**, pp. 5974–5987.
29. Andra, W. (1998). In Andra, W., and Novak, H. (eds.), *Magnetism in Medicine*, Wiley-VCH, Berlin, pp. 455–470.
30. Edelman, E., Kost, J., Bobeck, H., and Langer, R. (1985). Regulation of drug release from polymer matrices by oscillating magnetic fields. *J Biomed. Mater. Res.* **19**, pp. 67–83.
31. Erné, B. H., Butter, K., Kuipers, B. W. M., and Vroege, G. J. (2003). Rotational diffusion in iron ferrofluids. *Langmuir*, **19**, pp. 8218–8225.
32. Satarkar, N.S., and Hilt, J. Z. (2008). Magnetic hydrogel nanocomposites for remote controlled pulsatile drug release. *J. Controlled Release*, **130**, pp. 246–251.
33. Schlachter, E. V., Widmer, H. R., Bregy, A., Lönnfors-Weitzel, T., Vajtai, I., Corazza, N., Bernau, V. J. P., Weitzel, T., Mordasini, P., Slotboom, J., Herrmann, G., Bogni, S., Hofmann, H., Frenz, M., and Reinert, M. (2011). *Int. J. Nanomed.*, **6**, pp. 1793–1800.
34. Singh, R. K., Kim, T. H., Patel, K. D., Knowles, J. C., and Kim, H. W. (2012). Biocompatible magnetite nanoparticles with varying silica-coating layer

for use in biomedicine: physicochemical and magnetic properties, and cellular compatibility. *J. Biomed. Mater. Res. Part A*, **100A**, pp. 1734–1742.

35. Kim, D. H., Kim, K. N., Kim, K. M., Shim, I. B., and Lee, Y. K. (2007). In vitro and in vivo toxicity of $CoFe_2O_4$ for application to magnetic hyperthermia. *NSTI-Nanotech 2007*, **2**, pp. 748–751.
36. Huang, L. Y., and Yang, M. C. (2007). Behaviours of controlled drug release of magnetic- gelatine hydrogel coated stainless steel for drug-eluting-stents application. *J. Magn. Magn. Mater.*, **310**, pp. 2874–2876
37. Hu, S. H., Liu, T. Y., Liu, D. M., and Chen, S. Y. (2007). Controlled pulsatile drug release from a ferrogel by a high frequency magnetic field. *Macromolecules*, **40**, pp. 6786–6788.
38. Uva, M., Pasqui, D., Mencuccini, L., Fedi, S., and Barbucci, R. (2014). Influence of alternating and static magnetic fields on drug release from hybrid hydrogels containing magnetic nanoparticles. *J. Biomater. Nanobiotechnol.*, **5**, pp. 116–127.

Chapter 12

Physical and Chemical Hyaluronic Acid Hydrogels and Their Biomedical Applications

Assunta Borzacchiello,[a] Luisa Russo,[a] Sabrina Zaccaria,[a] and Luigi Ambrosio[a,b]

[a] *Institute for Polymers, Composites and BioMaterials IMCB-CNR, Italy*
[b] *Department of Chemical Science and Materials Technology DCSMT- CNR, Italy*
luigi.ambrosio@cnr.it

Hyaluronic acid (HA), or hyaluronan, is a naturally occurring polysaccharide, discovered in the bovine vitreous humor by Meyer and Palmer in 1934 [1]. It is ubiquitous in the human body and its molecular structure is highly conserved among the mammalian species. Thanks to its versatile properties, such as biocompatibility, non-immunogenicity, biodegradability, and viscoelasticity, it has been used for medical, pharmacological, and cosmetic applications. More recently, with the development of different strategies for the modification of its structure, HA has become an important building block for the production of new biomaterials to be used in regenerative medicine and drug delivery.

To prolong the in vivo residence time and improve the mechanical properties of HA, a common method is to form hydrogels by

Polysaccharide Hydrogels: Characterization and Biomedical Applications
Edited by Pietro Matricardi, Franco Alhaique, and Tommasina Coviello

ISBN 978-981-4613-61-3 (Hardcover), 978-981-4613-62-0 (eBook)
www.panstanford.com

crosslinking the HA polysaccharide chains into a three-dimensional network. HA hydrogels, as well as the HA, have become important for many biomedical applications.

The present chapter discusses the main features of hyaluronic acid and focuses on the formation of HA hydrogels through physical or chemical interactions and their applications in the biomedical field.

12.1 Hyaluronic Acid

12.1.1 *General Features and Biodistribution*

Hyaluronic acid is a linear polysaccharide composed of repeating disaccharide unit of D-glucuronic acid (GlcUA) and *N*-acetyl-D-glucosamine (GlcNAc), which are linked together through alternating β-1,4 and β-1,3 glycosidic bonds [1]. The carboxyl groups of HA are fully ionized at physiological pH. Thus the molecule is negatively charged and it is present as sodium salt (also known as sodium hyaluronate).

Its osmotic activity is non-ideal and disproportionately high in relation to its molecular weight. For this and other reasons it is capable of deep effects on the distribution and movements of water and plays a major part in water homeostasis [2].

HA belongs to an important group of biomacromolecules known as glycosaminoglycans (GAGs), being the most simple among them. Indeed it is the only GAG not covalently associated with a core protein, non-sulfated, and unlike all the other GAGs. HA is synthesized in the plasma membrane rather than the endoplasmic reticulum or in the Golgi apparatus [3]. In addition, although it consists of a single polysaccharide chain like other glycosaminoglycans, its molecular weight (MW) usually reaches millions. HA is also one of the most important GAGs in all vertebrates, mainly present in the extracellular matrix (ECM) and in connective tissues, and it can be found also in the capsules of some bacteria (e.g., strains of Streptococci) [4]. In the human body, HA is found in high concentrations; it is known that in a 70 kg individual there is around 15 g of hyaluronan, 5 g of which is turned over (degraded and

Table 12.1 HA occurrence in different animal tissues and fluids

Tissue or fluid	Concentration (μg/ml)
Rooster comb	7500
Human umbilical cord	4100
Human synovial fluid	1400–3600
Bovine nasal cartilage	1200
Human vitreous body	140–340
Human dermis	200–500
Human epidermis	100
Rabbit brain	65
Rabbit heart	27
Human urine	0.1–0.3
Human plasma	0.01–0.1

synthesized) every day. The highest concentrations of HA in the human body are found in connective tissues such as the umbilical cord, synovial fluid, skin, and the vitreous body (Table 12.1). Considerable amounts are also present in the lungs, kidney, brain, and muscle, but very little in the liver. The lowest concentration is found in blood serum, while the human umbilical cord is so rich in HA that it was the main source of this polysaccharide for the early studies [5]. Currently, HA is commercially produced from microbial fermentations [6] or from animal tissues such as cock's comb [7].

The biological function of HA has been widely investigated, at the beginning, outside the field of matrix biology. Most workers have regarded the polymer as a space filler without interest. Then attention toward HA progressively increased when it was shown that cells have specific receptors (cluster determinant 44 [CD44] [8], receptor for hyaluronate-mediated motility [RHAMM] [9], HA receptor for endocytosis [HARE] [10, 11], and lymphatic vessel endothelial hyaluronan receptor-1 [LYVE-1] [12]) that can recognize and bind HA selectively. Now it is known that HA plays a pivotal role in many biological functions.

Concerning the skin, the largest organ of the human body, HA plays a series of important functions: it can immobilize water in tissue and thereby change dermal volume and compressibility, act as a scavenger of free radicals generated by the ultraviolet rays from sunlight, and influence cell proliferation, differentiation, and tissue repair. Changes in HA observed with aging, wound healing, and degenerative diseases further highlight its importance in the skin [13].

In cartilage, despite its relatively low content, HA functions as an important structural constituent of the matrix, forming an aggregation center for Aggrecan, a large chondroitin sulfate proteoglycan that retains its macromolecular assembly in the matrix due to specific HA–protein interactions [14].

In the synovial fluid, there is an elevated concentration of high molecular weight HA which serves as shock absorber and provides necessary lubrication for the joints and reduces friction between the moving bones, thus diminishing wear. Under inflammatory conditions, such as osteoarthritis or rheumatoid arthritis, high MW HA is degraded by reactive oxygen species (ROS), which reduces its viscosity and its lubricant and shock-absorbing properties, leading to deteriorated joint movement and pain [15].

HA is also involved in embryogenesis, signal transduction, and cell motility, and is associated with cancer invasiveness and metastasis [16]. Moreover, HA has a wide molecular weight, ranging from 1000 to 10,000,000 Da, and it plays different roles in the body depending on its MW. For example, high MW HA, is space-filling, anti-angiogenic, and immunosuppressive, whereas the intermediate-sized polymers comprising 25–50 disaccharides are inflammatory, immunostimulatory, and highly angiogenic. Smaller HA oligosaccharides seem to function as endogenous danger signals. Some of the variably sized fragments trigger different signal transduction pathways. A recent review summarizes the vast range of size-specific activities of HA polymers [17].

12.1.2 *Physical and Chemical Properties of HA*

The exact chemical structure of HA (see Fig. 12.1) was first determined by Meyer and coworkers in 1954 [18, 19]. They found that the uronic acid and amino sugar are linked by β-1,4 glycosidic bond while β 1-3 bonds link the disaccharides to form the HA chain [20]. The number of repeating disaccharide units in a HA molecules

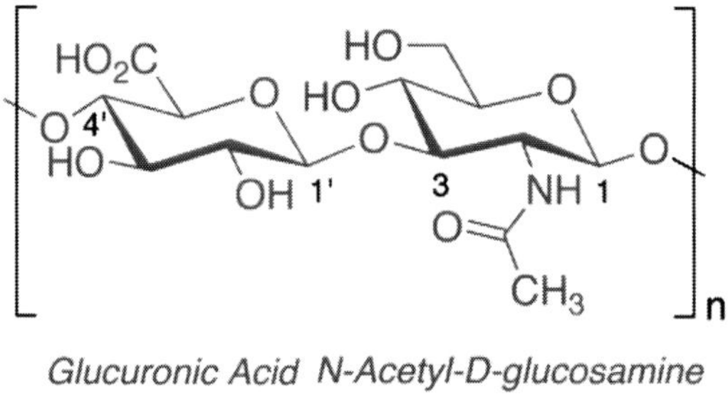

Figure 12.1 Chemical structure of the repeating disaccharide units in HA.

can reach 10,000 or more, which generate large molecules of up to several million daltons in weight and up to 10 μm in length.

In contrast with other biopolymers, such as nucleic acids and proteins, the structural chemistry of hyaluronan is much less developed. In the solid state, the HA conformation is helical and depends on pH, counterion type, temperature, and degree of hydratation [21]. Experiments of intrinsic viscosity [22], light scattering [23], and ultracentrifugation [20] suggested that HA chains in solution adopt an expanded random coil structure. On the other hand, it should be taken into account that these techniques provide only a measure of the bulk properties and thus are indicative of the mean contribution of a large number of molecules within a polydispersive distribution but do not necessarily provide molecular information. Other studies based on nuclear magnetic resonance spectroscopy (NMR), X-ray diffraction, and molecular modeling provide evidence of the existence of ordered secondary structures in HA solutions due to the formation of hydrogen bonds between two neighboring disaccharide units [20, 24–26]. These impart some rigidity and create a twist in the chain that adopts a tape-like two-fold helical structure [27]. In addition, the axial hydrogen atoms of about 8 CH groups on the alternating sides of the secondary structure create secondary hydrophobic faces that energetically favor the formation a meshwork-like β-sheet tertiary structure as a result of molecular aggregation. The hydrophobic and hydrogen bonding interactions, countering electrostatic repulsion, allow HA molecules to aggregate, leading to the formation of molecular networks (matrices) of HA. This structure is rigid and very stable, even if it has some degree of flexibility and a large number of conformations are possible. It is also to be noted that the tertiary structure of hyaluronan is sensitively dependent on the environment: it varies with pH and ionic strength, a behavior typical of a flexible polyelectrolytes. In particular, it was demonstrated that HA chains contract with increasing ionic strength and decreasing pH [20].

Significant information about the conformation of HA molecules in water can be obtained by dynamical mechanical analysis. This allows to obtain the viscoelastic properties of the materials by measuring the response of a sample when it is deformed under small periodic oscillation in shear [28].

Moreover, the knowledge of HA solution's viscoelastic properties allows to get information on the solution structure, thus giving the possibility to engineer the solution performance for a specific biomedical application.

From this analysis it is possible to obtain the dependence of the elastic modulus $G'(\omega)$ and the viscous modulus $G''(\omega)$ upon frequency, the so-called mechanical spectra. In particular G' gives information about the elasticity or the energy stored in the material during deformation, whereas G'' describes the viscous character or the energy dissipated as heat.

The ratio between the viscous modulus and the elastic modulus is expressed by the loss tangent:

$\tan\delta = \frac{G''}{G'}$ where δ is the phase angle. The loss tangent is a measure of the ratio of energy lost to energy stored in the cyclic deformation. The phase angle, δ, is equal to 90° for a purely viscous material, 0° for a pure elastic material, and $0^\circ < \delta < 90^\circ$ for viscoelastic materials [29]. The information on the HA structural properties can be obtained by analyzing the mechanical spectra of HA solution [30].

HA-based materials exhibit different viscoelastic properties. They can indeed behave as dilute solutions (Fig. 12.2) or entangled solutions (Fig. 12.3) depending on the concentration and the molecular weight of HA.

Low molecular weight HA solutions (150 KDa) behave as dilute solutions, as reported in Fig. 12.2. The mechanical spectra are characterized by a loss modulus greater than elastic modulus in all frequency range; these results suggest that no topological

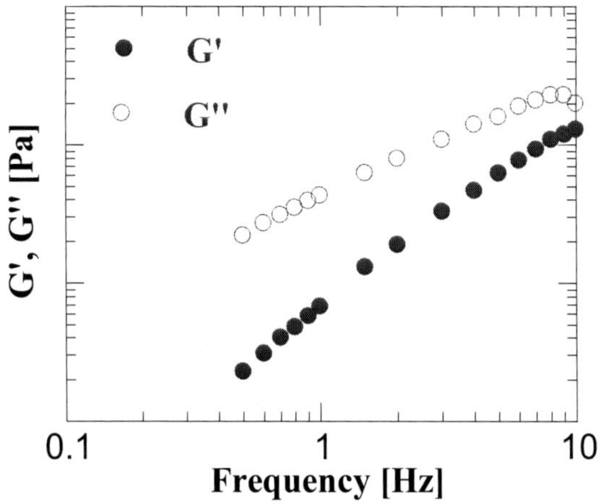

Figure 12.2 Mechanical spectra of HA solutions at low molecular weight.

interaction, such as entanglement, occurs among HA chains, so HA molecules can flow as individual flow unit.

Due to the expansion of HA random coil in water, the molecular weight increases also at low concentration, leading to a different rheological behavior. Indeed, even at low-concentration, high-molecular weight solutions exhibit entangled network.

The solution shows a viscous behaviour at low frequency ($G'' > G'$) and a prevalently elastic one at high frequencies ($G' > G''$), as reported in Fig. 12.3. The limit between the two regions is represented by the crossover frequency, usually expressed as ω_c.

This behavior can be explained by considering that at low frequency, entanglements among the chains occur and the molecular chains can release stress by disentanglement and molecular re-arrangement during the period of oscillation, and hence the solution shows viscous behavior ($G'' > G'$); instead at high frequency, molecular chains cannot disentangle during this short period of oscillation, and therefore they behave as a temporary crosslinked network, and the elastic behavior ($G' > G''$) is prevalent [31–33].

By increasing the HA concentration, the viscoelasticity of the solutions increases [34–37]. The viscoelastic properties of HA in aqueous solution are also influenced by the presence of small molecular agents [38–40]. For example, the presence of phospholipids,

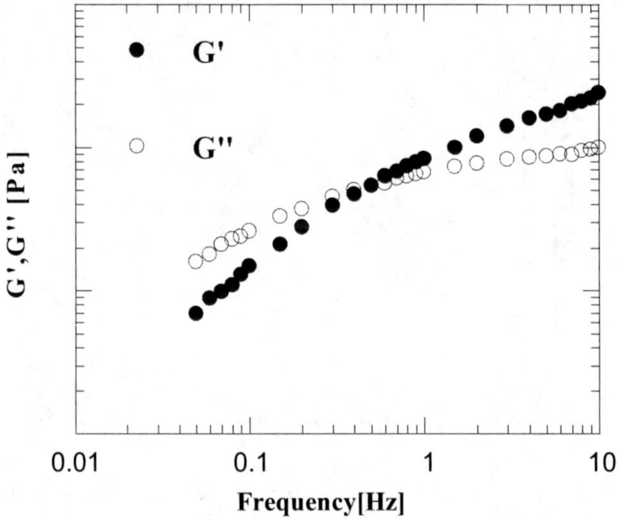

Figure 12.3 Mechanical spectra of HA solutions at high molecular weight.

guanidine, and sodium chloride leads to decreases in both elastic and viscous moduli, while the addition of sugars appears to promote structure. The effect of these compounds on the viscoelasticity is due to the disruption of intermolecular interactions and reduction in the entanglement between HA molecules [41].

12.2 Physical and Chemical HA-Hydrogels

12.2.1 *Chemical Modifications of HA: Synthesis of New HA Derivatives*

Thanks to its previously mentioned physical-chemical properties and to its large-scale production, HA has been used in numerous applications in the last decades. At the same time, however, the enlarged use of this biopolymer has led to consider some disadvantages such as its short in vivo residence time, due to the degradation catalyzed by hyaluronidase [42], and its poor mechanical properties. To overcome these drawbacks, various chemical modifications of the HA structure have been developed to synthesize new HA derivatives

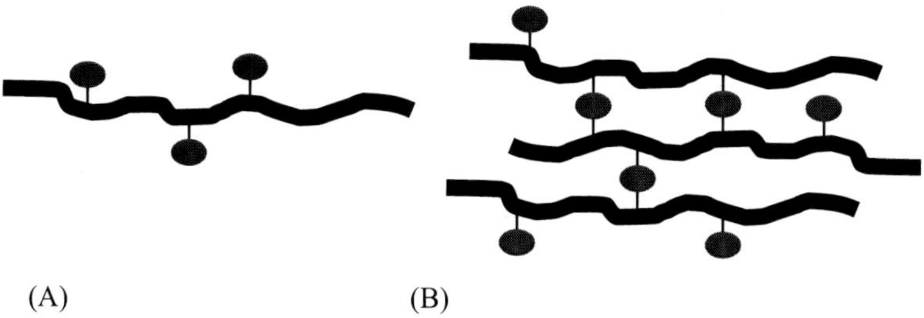

Figure 12.4 Chemical conjugation (A) and chemical crosslinking (B) of a polymer.

which are less susceptible to chemical and enzymatic hydrolysis and show enhanced mechanical properties [43]. Chemical modifications of HA target three different functional groups: the primary and secondary hydroxyl groups, the glucuronic acid carboxylic group, and the *N*-acetyl group (after deacetylation). Usually hydroxyls are modified through etherification, divynilsulfone crosslinking, esterification, and bis-epoxide crosslinking, whereas carboxylates are modified by carbodiimide-mediated reactions, esterification and amidation. When modifying HA, it has to be taken into account that modifications carried on the —COOH can alter the HA biological behavior in the body because the carboxylic groups are the recognition sites for hyaluronidase and HA receptors [44]. For this reason, in many cases low degree of modification are preferred to preserve the HA biological characteristic.

HA can be modified through two main ways, conjugation or crosslinking; in the first case, a compound is grafted onto HA chain, whereas in the second case different HA chains are linked together (Fig. 12.4). Crosslinked HA are also commonly referred to as hylans and can reach very high molecular weight (up to 23×10^6 Da) and prolonged in vivo residence time (1.5–9 days) [20, 45].

To crosslink or conjugate HA polymers, numerous methods have been developed, both in water or in organic solvents, such as dimethylformamide (DMF) or dimethylsulfoxide (DMSO). In this case, the native HA, which is in the form of a sodium salt, needs to be in its acidic form or a tetra-butylammonium salt (HA-TBA) for solubilization in organic solvents. However, this requires additional steps of physical and chemical treatments that can induce HA chain

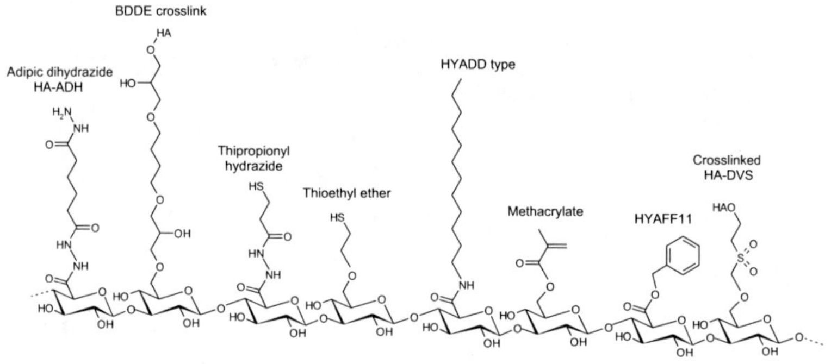

Figure 12.5 An hypothetical HA octasaccharide showing some chemical modifications.

hydrolysis [46, 47]. On the other hand, even reaction performed in water can led to HA chain fragmentation if acidic or alkaline conditions are needed [48].

Recently, a new appealing classification for HA derivatives has been proposed. It considers two different categories, "monolithic" and "living" HA derivatives [49, 50]. The first arises when the chemical modification gives a final structure that cannot form new covalent bonds in the presence of cells or tissues and can only be prefabricated in different forms. On the contrary, living HA derivatives can be further modified or crosslinked in the presence of cells or tissues enabling a change in their physical form during in vivo or in vitro use.

Figure 12.5 shows an hypothetical structure of an HA octasaccharide, containing some chemical modification targeting the carboxylic acid or the primary hydroxyl group of the *N*-acetylglucosamine. Some of these modifications are monolithic, like the benzyl ester, the HYADD-type amide, the DVS, and the butane-1,4-diol diglycidyl ether (BDDE), while the others are examples of living HA derivatives.

Concerning the modification of the carboxylic group of the glucuronic acid, one of the most widely used reaction is amidation [51, 52], which usually needs an activator agent. EDC (1-ethyl-3-[3-(dimethylamino)-propyl]-carbodiimide) is preferred among the carbodiimide-based activator agents because of its water solubility. The EDC reacts with the HA carboxylic acid forming an *O*-acyl isourea-HA intermediate, which in turn reacts with the chosen amine to form the amide bond. However the *O*-acyl isourea intermediate is very reactive and can hydrolyze and then rearrange into a stable *N*-acyl urea by-product. In this regard, Bullpitt reported the use of *N*-hydroxsuccinimide (NHS) and 1-hydroxybenzotriazole (HOBt) with EDC to form a less hydrolyzable intermediate thus preventing the formation of the *N*-acylurea by-product [52]. One of main advantages of the amidation using EDC is that it can be performed in water using the native HA. However, some authors have developed DMSO (dimethyl sulfoxide) methodologies on the acidic form of the HA to minimize the EDC hydrolysis, thus reaching substitution degrees of up to 60–80% [51, 53]. Kuo et al. demonstrated that carbodiimides can also be used as the reagent, not only as the activator agent, and using biscarbodiimides managed to crosslink HA and form stable bis(*N*-acylurea) crosslinked gels [54]. Nonetheless, using dihydrazides, like the adipic hydrazide (ADH), no crosslinking was observed, but only single functionalization (see Fig. 12.5 for HA-ADH) [55].

EDC-hydrazide chemistry has also been to introduce thiol groups on the HA chain (see Fig. 12.5). This result was accomplished by EDC-mediated reaction of HA with disulfide containing dihydrazide, followed by reduction with dithiothreitol (DTT) [56].

Amidation using CMPI (2-chloro-1-methylpyridinium iodide) as the activator agent is usually performed in dimethylformamide (DMF). Thus HA have to be converted into tetrabutylammonium salt (HA-TBA) through additional steps of preparation and purification [57]. With this method, high degree of substitution of up to 100% can be achieved. Therefore, it is highly effective. In addition, if no amine is added to the reaction medium, the CMPI-activated HA reacts with its own hydroxyl groups, forming an auto-crosslinked HA gel through ester crosslink between the HA chains [58]. These crosslinked HA products have been prepared by Fidia Advanced Biopolymers S.p.A. (FAB) and are referred to as ACP (auto-crosslinked polysaccharide) gels. In gels of this kind, no bridge molecules are present, and thus during their in vivo degradation, only natural components of HA are released [58].

The amidation of the –COOH of glucuronic acid with alkylamides having long carbon chain (see Fig. 12.5) has led also to a class of amphiphilic HA amide derivatives called HYADD and patented by Fidia [59].

A further chemical modification that involves carboxylic group is the esterification; this can be performed using epoxides, diazomethane, or alkylation using alkyl halides or tosylate activation. These reactions are usually performed in an organic solvent, such as DMF, on the TBA salt of HA. Among the ester derivatives of HA, HYAFF® is particularly interesting. These are a class of hyaluronan derivative polymer, patented by Fidia S.p.A., obtained by esterifying the free carboxyl group with different types of alcohol (aliphatic, aryliphatic, cycloaliphatic, and others). These derivatives are characterized by a reduced hydrophilic character that results in a decreased solubility in water and a slight increase in the structural rigidity of the macromolecule, caused by the interaction of hydrophobic group organized in hydrophobic patches. The benzyl ester of hyaluronan (see Fig. 12.5), called HYAFF11®, has been well characterized from both the chemical-physical and the biological point of view [60]. It has been showed that the numerous hydrophobic benzyl groups along the polymeric chain are probably organized in hydrophobic patches. This gives the molecule a more rigid structure and reduces its ability to interact freely with water molecules. Amphiphilic ester derivatives have also been synthesized in which a rather small proportion of the carboxylic groups (~10% mol) was esterified by long alkyl chains ($C_{12}H_{25}$ or $C_{18}H_{37}$) [47]. In this case an alkyl halide (dodecyl or octadecyl bromide) reacts with the HA carboxylic groups, preliminarily transformed into its tetrabutylammonium (TBA) salts to be solubilized in DMSO.

For what concerns chemical modifications involving the –OH groups, one of the most common is the formation of ether using epoxides, divinyl sulfone (DVS), or ethylene sulfide. Among the epoxides, the most used is the butanediol-diglycidyl ether (BDDE, see Fig. 12.5), which was patented for this reaction in 1986 by Malson and Lindqvist [61]. Nowadays, BDDE is widely used for crosslinked HA hydrogels; the reaction consists of the epoxide ring opening to form ether bonds with the HA hydroxyl groups. Crosslinked HA hydrogels are also obtained using DVS, as reported in Fig. 12.5. In this case, it has been demonstrated through histological assay that even if the starting material is highly toxic, HA-DVS hydrogels are biocompatible [62].

Thiol groups can be added to the HA chain not only by modifying the –COOH, as previously mentioned, but also through reaction at the hydroxyl group. Indeed, Serban, Yang, and Prestwich have used ethylene sulfide to synthesize 2-thioethyl ether HA derivatives (see Fig. 12.5) performing the reaction at pH 10 overnight and at pH 8.5 for 24 hours after addiction of DTT.

The –OHs can also be converted into esters by reaction with activated compounds. For example, it is possible to graft acyl-chloride activated carboxylate compound onto the hydroxyl groups of HA by chloroacylation with thionyl chloride. This reaction is performed at room temperature in an organic solvent, such as DMSO. Thus it is also possible to graft poly(lactic acid) (PLA) oligomers to the HA chain by first converting the HA to a more hydrophobic salt like the cetyltrimethyl-ammonium bromide (CTA) salt, which is easier to prepare than HA–TBA [63].

Methacrylated HA is another type of esterification of the –OH groups of HA (see Fig. 12.5). It is usually performed using methacrylic anhydride in ice cold water for 12 hours at pH 8–10. The obtained derivatives can be crosslinked by free-radical polymerization using UV light (365 nm) and a photo-initiator to obtain HA hydrogels.

HA can also be modified by oxidation with sodium periodate of the hydroxyl groups of the glucuronic acid moiety of HA to dialdehydes, thereby opening the sugar ring. This derivative has been employed to graft peptides to the HA by reaction with the aldehyde groups [64], or to crosslink with HA–hydrazide derivatives [65].

Eventually, even the *N*-acetyl group of the *N*-acetyl glucosamine can be a modification site after deacetylation [59]. In this way an amino group is recovered that can then react with an acid using the same amidation methods described before. However, deacetylation can induce severe chain fragmentation. Platt and Szoka mention the possibility of using enzymes for HA deacetylation, which was previously performed on the *N*-acetylglucosamine moiety of heparin and heparan sulfate [66].

12.2.2 *From HA Derivatives to HA Physical and Chemical Hydrogels*

HA derivatives are used to prepare hydrogels for several applications. Some of them, like amphipilic derivatives or thiol-modified HA, can spontaneously form gels, whereas others need to be crosslinked.

As well as for hydrogels in general, HA hydrogels can be broadly classified into two categories, chemical and physical, according to the kind of interaction between the HA chains. Chemical, or permanent, hydrogels are covalently crosslinked networks. They attain an equilibrium swelling state which depends on the HA–water interaction and on the crosslink density. Physical, or reversible, hydrogels are instead held together by molecular entanglements, and/or secondary forces including ionic, hydrogen bonding or hydrophobic interactions. All of these interactions are reversible and can be disrupted by changes in physical conditions or application of stress. That is why they are called reversible. In this kind of hydrogels, dissolution is prevented by the physical interactions between different HA chains.

In the last decades many different strategies have been developed for the production of chemical crosslinked HA hydrogels. The important ones are reported in Table 12.2, and some of them have been already mentioned in the previous section and shown in Fig. 12.5. It is also to be noticed that there are different crosslinking procedures: direct crosslinking of native HA, crosslinking of an HA derivative, and crosslinking of different HA derivatives.

As previously pointed out bis-carbodiimides can be used to chemically crosslink HA. Carbodiimides react with the HA carboxylic groups to give bis *N*-acyl-urea intermediate, which in turn can react with free HA carboxylates, leading to gelation [56]. The acyl-urea moiety can also react with polyvalent hydrazides, affording crosslinked HA gels with a range of mechanical properties that can be tuned simply by varying the molar ratios of the reagent [67]. Divynil sulfone crosslinking has been employed to prepare hydrogels like Hylan B, which is a highly swollen gel composed of a three-dimensional network of divinylsulfone crosslinks. It is worth noting that Shu, Liu, Luo, Roberts, and Prestwich [56] have obtained hydrogels from HA disulfide derivative, simply by air

Table 12.2 Some of the main strategies for HA crosslinking

HA group involved	Crosslinking agent	HA product
–COOH	bis-carbodiimide	crosslinked via *N*-acyl-urea or *O*-acyl-isourea groups
	polyvalent hydrazide coupled with carbodiimide and co-activator	crosslinked via hydrazide bonds
	carbodiimide and co-activators or CMPI	auto-crosslinked via ester bonds
	ditiobishydrazide coupled with carbodiimide	disulfide-crosslinked via air oxidation
	methacrylating agent coupled with carbodiimide	photocrosslinked after exposition to UV
–OH	bis-epoxide	crosslinked via ether bonds
	divinyl sulfone	crosslinked via ether bonds

oxidation. Indeed they need a reduction step with DTT to have the HA derivatives solution because they spontaneously crosslink. Therefore, this method is interesting since it does not rely on synthetic crosslinkers and allow in situ crosslinking of the gel. Auto-crosslinked HA (ACP, see section 2.1) can instead be prepared by activation of HA carboxylic groups with CMPI. This adduct can form internal ester bonds, leading to the formation of a gel network [58]. HA has also been crosslinked by diglycidyl ether under alkaline conditions to give a biocompatible hydrogel [68]. HA can also be crosslinked twice with the same bis-epoxide, first under basic conditions and then under acidic conditions. This double corsslinked HA is less susceptible to hyaluronidase degradation compared with single crosslinked intermediate [69]. Photocrosslinking, a technique that provides advantages such as increased spatial and temporal control over crosslinking and biocompatibility with in situ polymerization, has been used to polymerize acrylate-HA. Metacrylated, methacrylamide, and glycidylmethacrylated HA derivatives have been photocrosslinked into hydrogels with either UV (365 nm) or visible light and a photo-initiator such as 2-oxo-ketoglutaric acid or 4-(2-hydroxyethoxy)phenyl-(2-hydroxy-2-propyl)ketone (marketed as Irgacure 2959 by Ciba Specialty Chemicals, Switzerland) [70–72].

Although chemical modifications lead to an improvement of mechanical properties of HA and thus to the formation of more stable hydrogels, they may also induce severe reduction in the HA biocompatibility and can alter its biological activity. Several concerns have been expressed for the potential toxicity of the crosslinking agents and of the degradation products [73].

Thus, in order to enhance the mechanical performance of HA without compromising its biology activity, various strategies have been adopted to induce the formation of HA physical hydrogels, where they are not chemical crossliking.

For instance, HA ester solutions, such as HYAFF® (see the previous paragraph) solutions, can form a weak gel network without any further chemical process. In this case the gelification is due to the interactions arising among the aliphatic or aromatic alcohol linked to the HA backbone [74].

Evidence of the formation of a physically crosslinked hydrogel was provided also for some amphiphilic HA derivatives, prepared by esterification or amidation of the polysaccharide backbone with long alkyl chains, which exhibit the typical rheological properties of hydrophobically associating polymers [48, 75, 76]. Physical hydrogels have been prepared using HYADD4® (see also section 2.1), in turn synthesized by amidation of the —COOH of the glucuronic acid (2% degree of derivatization) of an HA of 500–730 kDa with an hexadecylamide group [77]. This amphiphilic derivative is able to associate into aqueous solution, and its rheological behavior is typical of a weak gel. Indeed, in this kind of hydrogels, a simple and short chemical modification allows the formation of hydrophobic interactions that can improve the rheological properties of the material [66].

Another kind of amphiphilic HA derivatives, able to hydrophobically associate in an aqueous medium giving a physical HA hydrogel, have been synthesized using HA as a hydrophilic backbone and polylactic acid (PLA) as an aliphatic polyester [78]. PLA is a biocompatible and biodegradable synthetic polymer, approved by the FDA (Food and Drug Administration) and used for various biomedical applications; this polymer can be a profitable pendant chain to increase the HA lipophilicity. HA-PLA graft copolymers show in aqueous medium a clear tendency for hydrophobic interaction dependent on the amount of PLA chains linked to the HA backbone.

Physical hydrogels can also be prepared by interpenetrating low molecular weight HA (LMW-HA) with a fibrillar collagen scaffold [79]. In this case the HA form a matrix that is intimately interdispersed in the fibrillar collagen, which, in turn, acts as a binding cross element and a mechanical support. As the collagen-HA structure strongly resembles the organization of the connective tissues, this kind of hydrogels have been used in soft tissue regeneration.

Eventually, physical hydrogels capable of gelifying at temperatures near the body temperature have also been prepared using poloxamers (or Pluronics), which are triblock copolymers of poly(ethylene oxide)–poly(propylene oxide)–poly(ethylene oxide) (PEO–PPO–PEO) [80]. By formulating properly poloxamers (F127/F68)/HA platforms at specific concentrations, it was possible to obtain a thermoreversible gel with a T_{gel} close to body temperature. In this case, HA presence led to a strong increase in the poloxamer rheological properties, thus indicating possible HA interactions with poloxamers micelles through secondary bonds, such as hydrogen ones, which reinforce the gel structure.

12.3 Biomedical Applications

HA-based hydrogels, thanks to their unique properties such as excellent biocompatibility, high water content, and capacity to degrade in safe product, have found a broad range of biomedical applications.

As mentioned in the previous paragraph, HA hydrogels can be divided into chemical and physical gels. The formation of the gel by both chemical and physical methods allows to obtain a suitable biomaterial, which, being less susceptible to chemical and enzymatic hydrolysis, shows a prolonged in vivo persistence [81].

12.3.1 *Biomedical Applications of Chemical HA Hydrogels*

Chemical HA-based gels find application especially in aesthetic medicine, in the treatment of osteoarthritis, in vocal fold augmentation, and in drug delivery.

In the last decade, HA-based dermal fillers (DFs) have become the most successful response to the current massive demand for nonsurgical soft tissue augmentation. HA is one of the main structural elements in the skin; its concentration in the dermis decreases with age, promoting the formation of wrinkles. By means of an injection of DFs it is possible to fill wrinkles and to augment the volume of soft tissues [81]. There are several kinds of HA-based DFs in commerce that differ in their physical and chemical properties such as HA concentration, the crosslinking agent, the crosslinking degree, the particle size, and the swelling capacity [82, 83].

These properties can significantly influence the final clinical performance and the desired properties of DFs for a particular application. For example, HA concentration is one of the factors that influences the persistence and stability of a filler, particle size affects the degree of correction and volume filling, and swelling capacity is related to the degree of inflammation and induration [84]. Moreover, it was demonstrated that the performance of a dermal filler in vivo seems to be directly correlated not only with the polymer concentration but also with the product between the concentration and the percent elasticity [85].

Other parameters that can strongly influence the performance of DFs are viscosity, extrusion force, and the stress applied. For example, DFs, during their use, are subjected to high and low stress. In fact, a high stress is applied when a DF is injected through the small bore needle, so in this case the DF must have a low viscosity; on the other hand, when the materials are at the site of application, there is a low stress environment, and the DF must be predominately elastic [86].

Crosslinking with both BDDE and divinyl sulfone is used to realize gel for DF application.

HA-based DF in which BDDE is used as crosslinker is Restylane, producted by Q-Med (Upsalla, Sweden) [83, 87, 88]; an example of HA hydrogel crosslinked with divinyl sulfone is Hylaform, producted by Genzyme Corp. (Boston, MA) [83, 87, 88].

Crosslinked HA-based hydrogels are also used as viscosupplementation products in the joints affected by arthritis [89]. A joint affected by osteoarthritis and rheumatoid arthritis is characterized by a synovial fluid in which both HA molecular weight [90, 91] and concentration [92] are reduced, and so a decline in its viscoelastic properties occurs [93]. The purpose of viscosupplementation by intra-articular injection of HA hydrogels is to restore the rheological properties (viscosity and elasticity) of synovial fluids [20, 76], by normalizing the concentration and molecular weight of HA [94]. Furthermore, viscosupplementation not only allows to restore SF mechanical properties, but it has been also been seen to allow an induction of the synthesis of high molecular weight (HMW) HA by synoviocytes [95].

An example of HA-based chemical hydrogel used as viscosupplementation product is Synvisc (Biomatrix), made of Hylan GF-20, an HA derivative crosslinked with DVS [96].

Chemical HA-based hydrogels have been also used in vocal fold (VF) augmentation. VF augmentation is used in the therapy of scarring and paralysis of the vocal folds. In healthy conditions, the vocal folds are brought together near the center of the larynx by the muscle attached to the vocal folds' base. They are able to vibrate and produce sound when the air is forced through these folds. When the vocal folds do not move well enough to be close each other, a gap exists between them and the air leaks out too quickly, thus causing the voice to sound breathy and weak.

In this case, to fill the gap, an augmentation substance is injected in the VF to achieve complete glottal closure.

For vocal fold regeneration, HA hydrogels at the microscale were prepared by crosslinking HA derivatives carrying hydrazide (HA-ADH) and aldehyde (HA-ALD) functionalities within the inverse emulsion droplets. Macroscopic hydrogels with tunable viscoelasticity can be readily obtained by further crosslinking with soluble macromolecules through the residual functional groups on microgel surfaces, leading to a doubly crosslinked network (DXN) with hierarchical organization [65].

Scaffolds for tissue engineering were also prepared by combining HA with other molecules. In particular, composite matrixes containing collagen and HA were realized by means of the freeze-drying technique and crosslinked with 1-ethyl-3-(3-dimethyl aminopropyl) carbodiimide (EDC) in order to improve its mechanical stability [97].

For the regeneration of skin and cartilage, gelatin-chondroitin 6 sulfate-hyaluronic acid scaffolds were realized. Chondroitin-6-sulfate and hyaluronic acid were incorporated within the gelatin membrane to mimic skin composition and create an appropriate microenvironment for cell proliferation, differentiation, and migration. For the skin regeneration, the lower layer of the membrane was seeded with dermal fibroblasts so that it acted as the feeder layer for keratinocyte inoculation. Meanwhile, the upper layer was seeded with keratinocytes for epidermalization [98].

Gelatin/hyaluronic acid/chondroitin-6-sulfate scaffolds were also prepared for cartilage tissue engineering. These composite systems showed very good results, but also problems of cell seeding and cell distribution. It was demonstrated that gelatin/hyaluronic acid/chondroitin-6-sulfate particles added into glue provide better mechanical strength, better cell distribution, and easier cell seeding, which would be expected to improve cartilage regeneration in vitro [99].

Furthermore, HA crosslinked hydrogels were used to incorporate drug molecules and to release them in a controlled kinetic. For example, Sei K. Hahn et al. synthesized two different types of HA hydrogels by crosslinking HA with divinyl sulfone (DVS) and poly(ethylene glycol)-divinyl sulfone (VS-PEG-VS) [100]. These HA hydrogels were loaded with an anti-inflammatory drug, vitamin E succinate (VES), and with a model of anti-inflammatory protein drugs, bovine serum albumin (BSA), by means of a swelling process; in particular dehydrated gels were left to swell in VES (0.4 mg/ml) or BSA (2 mg/ml) PBS solutions for a day. HA hydrogels, loaded with VES, release about 50–60% of the VES in the first few hours, while most of the remainder gradually released during the next several days, obtaining a plateau in the curve release after about four days. In the case of hydrogel loaded with BSA, the release kinetic was similar to that of VES, reaching a plateau after about two days.

12.3.2 *Biomedical Application of Physical HA Hydrogels*

Physical HA-based hydrogels are widely used in the field of tissue engineering and drug delivery. The benzyl ester of hyaluronan, HYAFF®, is the most utilized HA-based product for tissue engineering application. In particular, HYAFF11® was widely used for skin, cartilage and fibro-cartilaginous tissue engineering. In the case of skin tissue engineering, HYAFF11® was used to form scaffolds cultured in vitro with keratinocytes and fibroblasts [101, 102] in order to obtain epidermal and dermal-like tissues layers. HYAFF11® were also used as scaffold for the culture of chondrocytes in tissue-engineering procedures of cartilage reconstruction [103]. The cultured cells expressed cartilage-specific collagen type II, indicating that they were able to redifferentiate within the scaffold of HYAFF11® and were able to retain a chondrocyte phenotype even after a long period of in vitro conditions. Nevertheless, the expression of collagen type I, which was produced by dedifferentiated or incompletely redifferentiated chondrocytes, was noticeable. Additional data were obtained by subcutaneous implantation of samples seeded with human cells in the in vivo model of the athymic nude mouse. The results after one month revealed the development of tissue similar to hyaline cartilage [104].

Composite scaffolds based on HYAFF11® reinforced with polycaprolactone (PCL) were produced for engineering fibro-cartilaginous tissues such as the central part of the meniscus. The scaffolds were realized by phase separation and seeded with human chondrocytes [105].

HYADD3®, a dodecylamide of HA, and HYAFF120®, a crosslinked HA ester, have been also used as scaffolds for the regeneration of nucleus pulposus (NP) [106].

Physical HA-based hydrogels are also used as therapeutic drug delivery systems. In particular, injectable temperature-sensitive systems based on PNIPAAm (poly(*N*-isopropylacrylamide))-grafted chitosan and/or HA were used for the controlled release of bioactive molecules. These hydrogels were loaded with nalbuphine that is a narcotic analgesic used in the treatment of both acute and chronic pain. The presence of both synthetic and natural polymers in these systems allow to maintain the temperature sensitivity of PNIPAAM on the one hand, and to improve its biocompatibility and biomedical properties on the other [107]. When the hydrogels were injected in vivo, they formed a non-flowing gel at body temperature; upon incorporation of pharmaceutical agents, the hydrogel systems could act as sustained drug release depot in situ.

Furthermore, thermosensitive gels made of poloxamers (or Pluronics) and HA are also successful platforms for drug delivery [108]. Poloxamer thermoresponsive gels were widely used as drug delivery systems. However, they show inadequate mechanical properties, poor bioadhesiveness, and high permeability to water. It was demonstrated that HA hampers water–poloxamer interactions, thereby strongly influencing the physicochemical properties of poloxamer gels. Improving the mechanical properties of these gels allowed to obtain diffusion kinetics of macromolecular active molecules that are drastically slowed down [109]. The optimized systems were loaded with insulin and acyclovir and the platforms were able to sustain a prolonged and controlled release of the loaded drug [80].

References

1. Meyer, K., and Palmer, J. W. (1934). The polysaccharide of vitreous humor, *J. Biomol. Chem.* **107**, 629–634.
2. Laurent, T. C., and Fraser, J. R., (1992). Hyaluronan, *FASEB J.l*, **6**, (7), 2397–2404.
3. Kogan, G., Soltes, L., Stern, R., and Gemeiner, P. (2007). Hyaluronic acid: a natural biopolymer with a broad range of biomedical and industrial applications, *Biotechnol. lett.*, **29**, 17–25.
4. Nogovitsin, E. A., and Budkov, Y. A. (2012). Self consistent field theory investigation of the behavior of hyaluronic acid chains in aqueous salt solutions, *Physica A,* **391**, 2507–2517.
5. Kuo, J. W. (2006). *Practical Aspects of Hyaluronan-Based Medical Products,* (CRC Press, New York).
6. Kendall, F. E., Heidelberger, M., and Dawson, M. H. (1937). A serologically inactive polysaccharide elaborated by mucoid strains of group A hemolytic Streptococcus, *J. Biol. Chem.*, **118**, 61–69.
7. Boas, N. F. (1949). Isolation of hyaluronic acid from the cock's comb, *J. Biol. Chem.*, **181**, 573–575.
8. Aruffo, A., Stamenkovic, I., Melnick, M., Underhill, C. B., and Seed, B. (1990). CD44 is the principal cell surface receptor for hyaluronate, *Cell*, **61**, 1303–1313.

9. Entwistle, J., Hall, C. L., and Turley, E. A. (1996). HA receptors: regulators of signaling to the cytoskeleton, *J. Cell. Biochem.*, **61**, 569–577.
10. Asayama, S., Nogawa, M., Takei, Y., Akaike, T., and Maruyama, A. (1998). Synthesis of novel polyampholyte comb-type copolymers consisting of a poly(L-lysine) backbone and hyaluronic acid side chains for a DNA carrier, *Bioconj. Chem.*, **9**, 476–481.
11. Takei, Y., Maruyama, A., Ferdous, A., Nishimura, Y., Kawano, S., Ikejima, K., Okumura, S., Asayama, S., Nogawa, M., Hashimoto, M., Makino, Y., Kinoshita, M., Watanabe, S., Akaike, T., Lemasters, J. J., and Sato, N. (2004). Targeted gene delivery to sinusoidal endothelial cells: DNA nanoassociate bearing hyaluronan–glycocalyx, *FASEB J.*, **18**, 699–701.
12. Schledzewski, K., Falkowski, M., Moldenhauer, G., Metharom, P., Kzhyshkowska, J., Ganss, R., Demory, A., Falkowska-Hansen, B., Kurzen, H., Ugurel, S., Geginat, G., Arnold, B., and Goerdt, S. (2006). Lymphatic endothelium-specific hyaluronan receptor LYVE-1 is expressed by stabilin-1^+, F4/80^+, CD11b^+ macrophages in malign ant tumors and wound healing tissue in vivo and in bone marrow cultures in vitro: implications for the assessment of lymphangiogenesis, *J. Pathol.*, **209**, 67–77.
13. Juhlin, L. (1997). Hyaluronan in skin, *J. Intern. Med.*, **242**, 61–66.
14. Prehm, P. (2000). Hyaluronan. In: Vandamme, E. J., De Baets, S., Steinbüchel, A. (eds.), *Biopolymers: Biology, Chemistry, Biotechnology, Applications*, vol. 5, *Polysaccharides* I. *Polysaccharides from Prokaryotes*. Wiley-VCH, Weinheim, pp. 379–404.
15. Soltes, L., and Mendichi, R. (2003). Molecular characterization of two host–guest associating hyaluronan derivatives, *Biomed. Chromatogr.*, **17**, 376–384.
16. Kogan, G., Soltes, L., Stern, R., Schiller, J., and Mendichi, R. (2006). *Hyaluronic Acid: Its Function and Degradation in in vivo Systems.* In: Rahman, A. (ed.), *Studies in Natural Products Chemistry*, vol. 35, *Bioactive Natural Products*, Part D. Elsevier, Amsterdam.
17. Stern, R., Asari, A. A., and Sugahara, K. N. (2006). Hyaluronan fragments: an information-rich system, *Eur. J. Cell. Biol.*, **85**, 699–715.
18. Weissman, B., and Meyer, K. (1954). The structure of hyalobiuronic acid and of hyaluronic acid from umbilical cord, *J. Am. Chem. Soc.*, **76**, 1753–1757.
19. Linker, A., and Meyer, K. (1954). Production of unsaturated uronides by bacterial hyaluronidases, *Nature*, **174**, 1192–1194.

20. Lapcik, L., De Smedt, S., Demeester, J., and Chabrecek, P. (1998). Hyaluronan: preparation, structure, properties and applications, *Chem. Rev.*, **98**, 2663–2684.
21. Hargittai, I., and Hargittai, M. (2008). Molecular structure of hyaluronan: an introduction, *Struct. Chem.*, **19**, 697–717.
22. Laurent, T. C., Ryan, M., and Pietruszkiewicz, A. (1960). Fractionation of hyaluronic acid the polydispersity of hyaluronic acid from the bovine vitreous body, *Biochim. Biophys. Acta*, **42**, 476–485.
23. Laurent, T. C., and Gergely, J. (1955). Light-scattering studies in hyaluronic acid, *J. Biol. Chem.*, **212**, 325–333.
24. Scott, J. E. (1984). The periphery of the developing collagen fibril-quantitative relationships with dermatan sulfate and other surface-associated species. *Biochem. J.*, **218**, 229–233.
25. Scott, J. E., and Heatley, F. (1999). Hyaluronan forms specific stable tertiary structures in aqueous solution: a C-13 NMR study, *Proc. Natl. Acad. Sci. USA*, **96**, 4850–4855.
26. Day, A. J., and Sheehan, J. K. (2001). Hyaluronan: polysaccharide chaos to protein organization, *Curr. Opin. Struct. Biol.*, **11**, 617–622.
27. Scott, J. E., (1992) Supramolecular organization of extracellular matrix glycosaminoglycans, in vitro and in the tissues, *FASEB J.*, **6**, 2639–2645.
28. Ferry, J. D. (1970) *Viscoelastic Properties of Polymers* (Wiley, New York).
29. D'Errico, G., De Lellis, M., Mangiapia, G., Tedeschi, A., Ortona, O., Fusco, S., Borzacchiello, A., and Ambrosio, L. (2008). Structural and mechanical properties of UV-photo-cross-linked poly(*N*-vinyl-2-pyrrolidone) hydrogels, *Biomacromolecules*, **9**, 231–240.
30. Ambrosio, L., Borzacchiello, A., Netti, P. A., and Nicolais, L. (1999). Rheological study on hyaluronic acid and its derivative solutions, *J. Mater. Sci. Pure Appl. Chem.*, **A36**(7 and 8), 991–1000.
31. Borzacchiello, A., Netti, P. A., Ambrosio, L., and Nicolais, L. (2000). Hyaluronic acid derivatives mimic the rheological properties of vitreous body, *New Frontiers in Medical Sciences: Redefining Hyaluronan*, 195–202.
32. Borzacchiello, A., and Ambrosio, L. (2001). Network formation of low molecular weight hyaluronic acid derivatives, *J. Biomater. Sci. Polym. Edn.*, **12**(3), 307–316.
33. Carlforsa, J., Edsmana, K., Peterssonb, R., and Jörnvingb, K. (1998). Rheological evaluation of Gelrite® in situ gels for ophthalmic use, *Eur. J. Pharm. Sci.*, **6**, 113–119.

34. Gibbs, D. A., Merrill, E. W., Smith, K. A., and Balazs, E. A. (1968). Rheology of hyaluronic acid, *Biopolymers*, **6**, 777–791.

35. Kobayashi, Y., Okamoto, A., and Nishinari, K. (1994). Viscoelasticity of hyaluronic-acid with different molecular weights, *Biorheology*, **31**, 235–244.

36. Brown, M. B., Siligardi, G., Karim, S., Bennet, F. C., Martin, G. P., and Marriott, C. (1996). The determination of the viscosity-average molecular weight of hyaluronan by capillary viscosity, *Fourth International Workshop on Hyaluronan in Drug Delivery*, Willoughby, D., Ed.; RSM Press: London, pp. 79.

37. Fujii, K., Kawata, M., Kobayashi, Y., Okamoto, A., and Nishinari, K. (1996). Effects of the addition of hyaluronate segments with different chain lengths on the viscoelasticity of hyaluronic acid solutions, *Biopolymers*, **38**, 583–591.

38. Pasquali-Ronchetti, I., Quaglino, D., Mori, G., Bacchelli, B., and Ghosh, P. (1997). Hyaluronan-phospholipid interactions, *J. Struct. Biol.*, **120**, 1–10.

39. Mo, Y., Takaya, T., Nishinari, K., Kubota, K., and Okamoto, A. (1999). Effects of sodium chloride, guanidine hydrochloride, and sucrose on the viscoelastic properties of sodium hyaluronate solutions, *Biopolymers*, **50**, 23–34.

40. Ghosh, P., Hutadilok, N., Adam, N., and Lentini, A. (1994). Interactions of hyaluronan (hyaluronic-acid) with phospholipids as determined by gelpermeation chromatography, multi-angle laser-light-scattering photometry and h-1-nmr spectroscopy, *Int. J. Biol. Macromol.*, **16**, 237–244.

41. Liao, Y.-H. (2005). Hyaluronan: pharmaceutical characterization and drug delivery, *Drug Delivery*, **12**, 327–342,

42. Stern, R. (2004). Hyaluronan catabolism: a new metabolic pathway, *Eur. J. Cell Biol.*, **83**, 317–325.

43. Kuo, J. W., and Prestwich, G. D. (2010). *Materials of Biological Origin–Materials Analysis and Implant Uses, Comprehensive Biomaterials* (Ducheyne, P., Healy, K., Hutmacher, D., and Kirkpatrick, J., eds.). Elsevier.

44. Banerji, S., Wright, A. J., Noble, M., Mahoney, D. J., Campbell, I. D., Day, A. J., and Jackson, D. G. (2007). Structures of the Cd44-hyaluronan complex provide insight into a fundamental carbohydrate-protein interaction, *Nat. Struct. Mol. Biol.*, **14**(3), 234–239.

45. Reichenbach, S., Blank, S., Rutjes, A. W. S., Shang, A., King, E. A., Dieppe, P. A., Juni, P., and Trelle, S. (2007). Hylan versus hyaluronic acid for osteoarthritis of the knee: a systematic review and meta-analysis, *Arthritis Rheum.*, **57**(8), 1410–1418.
46. Bergman, K., Elvingson, C., Hilborn, J., Svensk, G., and Bowden, T. (2007). Hyaluronic acid derivatives prepared in aqueous media by triazine-activated amidation, *Biomacromolecules*, **8**(7), 2190–2195.
47. Pelletier, S., Hubert, P., Lapicque, F., Payan, E., and Dellacherie, E. (2000). Amphiphilic derivatives of sodium alginate and hyaluronate: Synthesis and physico-chemical properties of aqueous dilute solutions, *Carbohydr. Polym.*, **43**(4), 343–349.
48. Maleki, A., Kjøniksen, A., and Nyström, B. (2008). Effect of pH on the behavior of hyaluronic acid in dilute and semidilute aqueous solutions, *Macromol. Symp.*, **274** (1), 131–140.
49. Prestwich, G. D., and Kuo, J. W. (2008). Chemically-modified HA for therapy and regenerative medicine, *Curr. Pharm. Biotechnol.*, **9**, 242.
50. Burdick, J., and Prestwich, G. D. (2011). Hyaluronic acid hydrogels for biomedical applications, *Adv. Mater.*, **23**, H41–H56.
51. Oh, E., Park, K., Kim, K., Kim, J., Yang, J., Kong, J., et al. (2010). Target specific and long- acting delivery of protein, peptide, and nucleotide therapeutics using hyaluronic acid derivatives, *J. Controlled Release*, **141**(1), 2–12.
52. Bulpitt, P., and Aeschlimann, D. (1999). New strategy for chemical modification of hyaluronic acid: preparation of functionalized derivatives and their use in the formation of novel biocompatible hydrogels, *J. Biomed. Mater. Res.*, **47**(2), 152–169.
53. Schneider, A., Picart, C., Senger, B., Schaaf, P., Voegel, J., and Frisch, B. (2007). Layer-by-layer films from hyaluronan and amine-modified hyaluronan, *Langmuir*, **23**(5), 2655–2662.
54. Kuo, J., Swann, D., and Prestwich, G. (1991). Chemical modification of hyaluronic acid by carbodiimides, *Bioconjugate Chem.*, **2**(4), 232–241.
55. Pouyani, T., and Prestwich, G. (1994). Functionalized derivatives of hyaluronic acid oligosaccharides: drug carriers and novel biomaterials, *Bioconjugate Chem.*, **5**(4), 339–347.
56. Shu, X. Z., Liu, Y., Roberts, M. C., and Prestwich, G. D. (2002) *Biomacromolecules*, **3**, 1304–1311
57. Magnani, A., Rappuoli, R., Lamponi, S., and Barbucci, R. (2000). Novel polysaccharide hydrogels: Characterization and properties, *Polym. Adv. Technol.*, **11**(8–12), 488–495.

58. Della Valle, F. (1994). Crosslinked carboxy polysaccharides, EP341745.
59. Bellini, D., and Topai, A. (1999). Amides of hyaluronic acid and the derivatives thereof and a process for their preparation, WO2000001733 A1.
60. Iannace, S., Ambrosio, L., Nicolais, L., Rastrelli, A., and Pastorello, A. (1992). Thermomechanical properties of hyaluronic acid derived products, *J. Mater. Sci. Mater. Med.*, **3**(1), 59–64.
61. Malson, T., and Lindqvist, B. (1986). Gels of crosslinked hyaluronic acid for use as a vitreous humor substitute, WO1986000079.
62. Eun, J., Kang, S., Kim, B., Jiang, G., Il, H., and Sei, K. (2008). *J. Biomed. Mater. Res. Part A*, **86**, 685–693.
63. Pravata, L., Braud, C., Boustta, M., El Ghzaoui, A., Tømmeraas, K., Guillaumie, F., Schwach-Abdellaoui, K., and Vert, M. (2008). New amphiphilic lactic acid oligomer–hyaluronan conjugates: Synthesis and physicochemical characterization, *Biomacromolecules*, **9**(1), 340–348.
64. Glass, J., Dickerson, K., Stecker, K., and Polarek, J. (1996). Characterization of a hyaluronic acid-Arg-Gly-Asp peptide cell attachment matrix, *Biomaterials*, **17**(11), 1101–1108.
65. Jia, X., Colombo, G., Padera, R., Langer, R., and Kohane, D. (2004). Prolongation of sciatic nerve blockade by in situ cross-linked hyaluronic acid, *Biomaterials*, **25**(19), 4797–4804.
66. Platt, V., and Szoka, F., Jr. (2008). Anticancer therapeutics: targeting macromolecules and nanocarriers to hyaluronan or CD44, a hyaluronan receptor, *Mol. Pharm.*, **5**(4), 474–486.
67. Prestwich, G. D., Marecak, D. M., Marecek, J. F., Vercruysse, K. P., and Ziebell, M. R. (1998). Controlled chemical modification of hyaluronic acid: synthesis, applications, and biodegradation of hydrazide derivatives, *J. Control Rel.*, **53**(1–3), 93–103.
68. Lindqvist, U., Tolmachov, V., Kaircmo, K., Astrom, G., Jonsson, E., and Lundqvist, H. (2002). Elimination of stabilised hyaluronan from the knee joint in healthy men, *Clin. Pharmacokinet.*, **41**, 603–613.
69. Zhan, X. B., Fraser, J. E., Alexander, C., Lockett, C., and White, B. J. (2002). Synthesis and characterization of a novel double crosslinked hyaluronan hydrogel, *J. Mater. Sci. Mater. Med.*, **13**, 11–16.
70. Park, Y. D., Tirelli, N., and Hubbell, J. A. (2003). Photopolymerized hyaluronic acid-based hydrogels and interpenetrating networks, *Biomaterials*, **24**(6), 893–900.

71. Leach, J. B., Bivens, K. A., Patrick, C. W., Jr., and Schmidt, C. E. (2003). Photocrosslinked hyaluronic acid hydrogels: natural, biodegradable tissue engineering scaffolds, *Biotechnol. Bioeng.*, **82**(5), 578–589.
72. Smeds, K. A., Pfister-Serres, A., Hachell, D. L., and Grinastaff, M. W. (2001). Photocrosslinkable polysaccharides for in situ hydrogel formation, *J. Biomed. Mater. Res.*, **54**(1), 115–121.
73. Campoccia, D., Doherty, P., Radice, M., Brun, P., Abatangelo, G., and Williams, D. F. (1998). Semisynthetic resorbable materials from hyaluronan esterification, *Biomaterials*, **19**, 2101–2127.
74. Borzacchiello, A., and Ambrosio, L. (2001). Network formation of low molecular weight hyaluronic acid derivatives, *J. Biomater. Sci. Polym. Edn.*, **12**, 307–316.
75. Pelletier, S., Hubert, P., Payan, E., Marchal, P., Choplin, L., and Dellacherie, E. (2001). Amphiphilic derivatives of sodium alginate and hyaluronate for cartilage repair: rheological properties, *J. Biomed. Mater. Res.*, **54**, 102–108.
76. Borzacchiello, A., Mayol, L., Schiavinato, A., and Ambrosio, L. (2010). Effect of hyaluronic acid amide derivative on equine synovial fluid viscoelasticity, *J. Biomed. Mater. Res. Part A*, **92A**(3), 1162–1170.
77. Bellini, D., and Schiavinato, A. (2006). Amide derivatives of hyaluronic acid in osteoarthrosis, US7863256 B2.
78. Palumbo, F. S., Pitarresi, G., Mandracchia, D., Tripodo, G., and Gimmona, G. (2006). New graft copolymers of hyaluronic acid and polylactic acid: Synthesis and characterization, *Carbohydr. Polym.*, **66**, 379–385.
79. Xin, X., Borzacchiello, A., Netti, P. A., Ambrosio, L., and Nicolais, L. (2004). Hyaluronic-acid-based semi-interpenetrating materials, *J. Biomater. Sci. Plym. Edn.*, **15**, 1223–1236.
80. Mayol, L., Quaglia, F., Borzacchiello, A., Ambrosio, L., and La Rotonda, M. I. (2008). A novel poloxamers/hyaluronic acid in situ forming hydrogel for drug delivery: rheological, mucoadhesive and in vitro release properties, *Eur. J. Pharm. Biopharm.*, **70**, 199–206.
81. Brown, M. B., and Jones, S. A. (2005). Hyaluronic acid: a unique topical vehicle for the localized delivery of drugs to the skin, *J. Eur. Acad. Dermatol. Venereol.*, **19**(3), 308–318.
82. Allemann, I. B., and Baumann, L. (2008). Hyaluronic acid gel (Juvederm) preparations in the treatment of facial wrinkles and folds, *Clin. Interv. Aging*, **3**(4): 629–634.
83. Beasley, K. L., Weiss, M. A., and Weiss, M. D. (2009). Hyaluronic acid fillers: a comprehensive review. *Facial. Plast. Surg.*, **25**, 86–94.

84. Monheit, G. D., and Coleman, K. M. (2006). Hyaluronic acid fillers, *Dermatol. Ther.*, **19**, 141–150.

85. Santoro, S., Russo, L., Argenzio, V., and Borzacchiello, A. (2011). Rheological properties of cross-linked hyaluronic acid dermal fillers, *J. Appl. Biomater. Biomech.*, **9**(2), 127–136.

86. Falcone, S. J., and Berg, R. A. (2009). temporary polysaccharide dermal fillers: a model for persistence based on physical properties, *Dermatol. Surg.*, **35**, 1238–1243.

87. Matarasso, S. L. (2004). Understanding and using hyaluronan, *Aesthetic Surg.*, **J 24**, 361–364.

88. Manna, F., Dentini, M., De Pità, O., Mortilla, E., and Maras, B. (1999). Comparative chemical evaluation of two commercially available derivatives of hyaluronic acid (Hylaform® from rooster combs and Restylane® from streptococcus) used for soft tissue augmentation, *J. Eur. Acad. Dermatol. Venereol.*, **13**, 183–192.

89. Dahlqvist, A., Garskog, O., Laurent, C., Hertegard, S., Ambrosio, L., and Borzacchiello, A. (2004). Viscoelasticity of rabbit vocal folds after injection augmentation, *Laryngoscope*, **114**(1), 138–142.

90. Balazs, E. A., and Denlinger, J. L. (1993). Viscosupplementation: a new concept in the treatment of osteoarthritis, *J. Rheumatol.*, **20** (Supp 39), 3–9.

91. Leach, J. B., and Schmidt, C. E. (2004). *Hyaluronan. Encyclopedia of Biomaterials and Biomedical Engineering* (Marcel Dekker, New York). pp. 779–789.

92. Laurent, T. C., Fraser, J. R., Laurent, U. B., and Engstrom-Laurent, A. (1995). Hyaluronan in inflammatory joint disease, *Acta. Orthop. Scand.*, **66**(Supp 226), 116–120.

93. Kirwan, J. R., and Rankin, E. (1997). Intra-articular therapy in osteoarthritis, *Baillieres Clin. Rheumatol.*, **11**, 769–795.

94. Barbucci, R., Lamponi, S., Borzacchiello, A., Ambrosio, L., Fini, M., Torricelli, P., and Giardino, R. (2002). Hyaluronic acid hydrogel in the treatment of osteoarthritis, *Biomaterials*, **23**(23), 4503–4513.

95. Bellamy, N., Campbell, J., Robinson, V., Gee, T., Bourne, R., and Wells, G. (2006). Viscosupplementation for the treatment of osteoarthritis of the knee, *Cochrane Database Syst. Rev.*, 19:CD005321; Review.

96. Conrozier, T., and Chevalier, X. (2008). Long-term experience with hylan GF-20 in the treatment of knee osteoarthritis, *Expert Opin. Pharmacother.*, **9**(10), 1797–1804.

97. Park, S.-N., Park, J.-C., Kim, H. O., Song, M. J., and Suh, H. (2002). Characterization of porous collagen/hyaluronic acid scaffold modified by 1-ethyl-3-(3-dimethylaminopropyl)carbodiimide cross-linking, *Biomaterials*, **23**, 1205–1212.
98. Wang, T.-W., Wu, H.-C., Huang, Y.-C., Sun, J.-S., and Lin, F.-H. (2006). Biomimetic bilayered gelatin-chondroitin 6 sulfate-hyaluronic acid biopolymer as a scaffold for skin equivalent tissue engineering, *Artif. Organs*, **30**(3), 141–149.
99. Chou, C.-H., Cheng, W. T. K., Kuo, T.-F., Sun, J.-S., Lin F.-H., and Tsai, J.-C. (2007). Fibrin glue mixed with gelatin/hyaluronic acid/chondroitin-6-sulfate tri-copolymer for articular cartilage tissue engineering: The results of real-time polymerase chain reaction, *J. Biomed. Mater. Res. Part A*, **82A**(3), 757–767.
100. Hahn, S. K., Jelacic, S., Maier, R. V., Stayton, P. S., and Hoffman, A. S. (2004). Anti-inflammatory drug delivery from hyaluronic acid hydrogels, *J. Biomater. Sci. Polym. Edn.*, **15**(9), 1111–1119.
101. Necas, J., Bartosikova, L., Brauner, P., and Kolar, J. (2008). Hyaluronic acid (hyaluronan): a review Veterinarni Medicina, **53**(8), 397–411.
102. Weigel, P. H., Fuller, G. M., and LeBoeuf, R. D. (1986). A model for the role of hyaluronic acid and fibrin in the early events during the inflammatory response and wound healing, *J. Theor. Biol.*, **119**(2), 219–234.
103. Borzacchiello, A., Gloria, A., Mayol, L., Dickinson, S., Miot, S., Martin, I., and Ambrosio, L. (2011). Natural/synthetic porous scaffold designs and properties for fibro-cartilaginous tissue engineering, *J. Bioact. Compat. Poly.*, **26**(5), 437–451.
104. Aigner, J., Tegeler, J., Hutzler, P., Campoccia, D., Pavesio, A., Hammer, C., Kastenbauer, E., and Naumann, A. (1998). Cartilage tissue engineering with novel nonwoven structured biomaterial based on hyaluronic acid benzyl ester, *J. Biomed. Mater. Res.*, **42**(2), 172–181.
105. Hemmrich, K., von Heimburg, D., Rendchen, R., Di Bartolo, C., Milella, E., and Pallua, N. (2005). Implantation of preadipocyte-loaded hyaluronic acid-based scaffolds into nude mice to evaluate potential for soft tissue engineering, *Biomaterials*, **26**(34), 7025–7037.
106. Pereira, D. R., Silva-Correia, J., Oliveira, J. M., and Reis, R. L. (2013). Hydrogels in acellular and cellular strategies for intervertebral disc regeneration, *J. Tissue Eng. Regener. Med.*, **7**(2), 85–98.
107. Fang, J.-Y., Chen, J.-P., Leu, Y.-L., and Hu, J.-W. (2008). Temperature-sensitive hydrogels composed of chitosan and hyaluronic acid as

injectable carriers for drug delivery, *Eur. J. Pharm. Biopharm.*, 68, 626–636.

108. Xiong, X. Y., Tam, K. C., and Gan, L. H. (2006). Polymeric nanostructures for drug delivery applications based on pluronic copolymer systems, *J. Nanosci. Nanotechnol.*, **6**(9–10), 2638–50.

109. Mayol, L., Biondi, M., Quaglia, F., Fusco, S., Borzacchiello, A., Ambrosio, L., and La Rotonda, M. I. (2011). Injectable thermally responsive mucoadhesive gel for sustained protein delivery, *Biomacromolecules*, **12**, 28–33.

Chapter 13

Alginate Hydrogels: Properties and Applications

Gudmund Skjåk-Bræk,[a] Ivan Donati,[b] and Sergio Paoletti[b]

[a]*Institute of Biotechnology, Norwegian University of Science and Technology (NTNU), Sem Sælands v. 6/8, N-7491 Trondheim, Norway*
[b]*Department of Life Sciences, University of Trieste, Via Licio Giorgieri 5, I-34127 Trieste, Italy*
gudmund.skjak-brak@ntnu.no

13.1 Sources

Alginic acid was first discovered by Stanford and recognized as the structural component of marine brown algae (*Phaeophyceae*) [136] where it constitutes, as an insoluble mixture of calcium, magnesium, potassium and sodium salts [62], up to 40% of the dry matter. In the first half of the twentieth century, the repeating units constituting the polysaccharide backbone were identified with mannuronic and guluronic acids [7, 47, 120]. The content of the two different uronic groups in alginate samples was found to vary over a wide range depending on their source. Nowadays, to underline the chemical inhomogeneity of this polysaccharide, the term *alginate* is used to describe a family of polysaccharides

Polysaccharide Hydrogels: Characterization and Biomedical Applications
Edited by Pietro Matricardi, Franco Alhaique, and Tommasina Coviello

ISBN 978-981-4613-61-3 (Hardcover), 978-981-4613-62-0 (eBook)
www.panstanford.com

[113] produced by brown algae and bacteria. In seaweeds, alginates contribute the strength and flexibility of the plant. Indeed, alginates have the same morphophysiological properties in brown algae as those of cellulose and pectins in terrestrial plants [38]. The large-scale production of alginates is exclusively based on the harvesting of brown seaweeds belonging to the family of *Laminaria* (of the species *hyperborea, digitata, japonica*), to *Macrocystis pyrifera*, and to *Durvillea antarctica*. Alginate composition depends on the seasonal and growth conditions as well as on the parts of the plant that the alginate is isolated from [3, 73].

Alginates are also produced by several bacteria belonging to the genera *Pseudomonas* and *Azotobacter* [19, 51, 52]. In bacteria, alginate can be regarded as a virulence factor mediating the biofilm growth which causes colonization of the lung. Indeed, alginate overproduction is the major pathological feature of the opportunistic human pathogen *Pseudomonas aeruginosa*, which infects the lungs of patients suffering from cystic fibrosis [20, 53, 97]. Although the precise role of alginate in biofilm formation is not yet completely clear [161], it is believed that by forming a slimy capsule around the pathogen, it suppresses phagocytosis [121] and impedes host immune clearance [81]. The main peculiar feature of bacterial alginates is the presence of O-acetyl groups which are completely absent in algal ones. In alginates from *A. vinelandii* and *P. aeruginosa*, the acetyl groups distribution among C-2 and C-3 in the mannuronic acid residues is approximately 3:2 and even some di-O-acetyl modification on uronic moieties are encountered [127]. Although the role of the acetyl groups in bacterial alginate in *P. aeruginosa* has not been elucidated, they seem to mediate cell aggregation into microcolonies in the early stage of biofilm development [152]. In *Azotobacter*, where alginate forms the outer protective walls in the metabolic dormant cysts, the acetyl substitution controls C-5 epimerization and indirectly facilitates the reswelling of the cysts [129].

Considering that applications of alginates from seaweed origin by far overcome those of material from bacterial origin, we will focus on the former group in this chapter.

13.2 Chemical Structure

Alginates have a heterogeneous structure with no regular repeating unit. From the chemical standpoint, alginate is a linear copolymer composed of 1→4 linked β-D-mannuronic acid (M) and its C-5 epimer, α-L-guluronic acid (G). Its block structure was recognized after acid hydrolysis and fractionation of the whole polysaccharide [59]. Along alginate chain, the G and M residues are arranged in pattern showing homopolymeric regions of G residues (G-blocks) and homopolymeric regions of M residues (M-blocks) interspersed by regions in which the two groups coexist in a strictly alternating sequence (MG-blocks) [60, 61] (Fig. 13.1).

The wide variability of the composition and sequential structure of alginate is a key functional attribute of the polysaccharide (Table 13.1) [29, 38, 73, 124, 132]. The epimerization of the C-5 residue from M to G imparts different macromolecular characteristics to the uronic blocks. While the di-equatorial glycosidic bonds lead to a ribbon-like structure for M blocks, G blocks are buckled and shorter than the previous one due to the diaxial linkages. This different spatial arrangements is reflected on the physical properties of the blocks: the linear distance between fixed charges projected along the polyelectrolyte axis equals 4.35 Å and 5.17 Å for G and M residues, respectively [5, 6], and the associated linear charge density of the homopolymeric blocks is 1.64 and 1.38, respectively [90–92]. Hence,

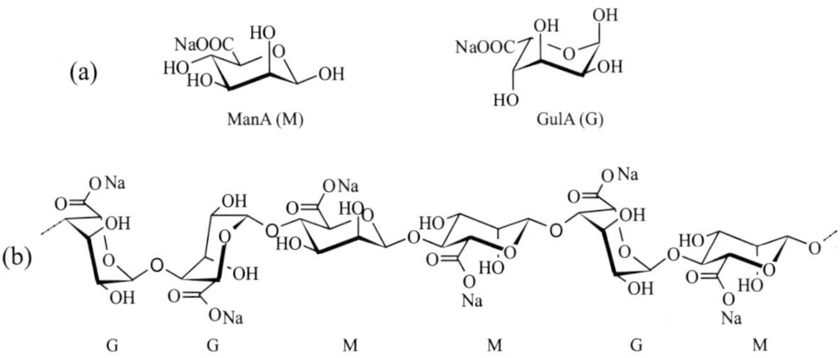

Figure 13.1 (a) The $^4C_1$ conformation of β-D-mannuronic acid (M) sodium salt and the $^1C_4$ conformation of α-L-guluronic acid (G) sodium salt. (b) Block composition of alginate.

Table 13.1 Chemical composition of some algal alginates determined by high-resolution ^{1}H-NMR

Source	F_G	F_M	F_{GG}	F_{MM}	$F_{GM,MG}$	F_{GGG}	F_{MGM}	$F_{GGM,MGG}$	$N_{G>1}$
Laminaria hyperborea (stipe)	0.63	0.37	0.52	0.26	0.11	0.48	0.07	0.05	15
Laminaria hyperborea (leaf)	0.49	0.51	0.31	0.32	0.19	0.25	0.13	0.05	8
Macrocystis pyrifera	0.42	0.58	0.20	0.37	0.21	0.16	0.17	0.04	6
Laminaria digitata	0.41	0.59	0.25	0.43	0.16	0.20	0.11	0.05	6
Lessonia nigrescens	0.41	0.59	0.22	0.40	0.19	0.17	0.14	0.05	6
Ascophyllum nodosum	0.41	0.59	0.22	0.38	0.21	0.13	0.14	0.07	5
Laminaria japonica	0.35	0.65	0.18	0.48	0.17				
Durvillaea antarctica	0.32	0.68	0.16	0.51	0.17	0.11	0.12	0.05	4

F_G denotes the fraction of alginate consisting of guluronic acid; F_{GG} and F_{GGG} indicate the fraction of alginate consisting of guluronic acid in blocks of dimers and trimers, respectively, whereas F_M and F_{MM} indicate the fraction of alginate consisting of mannuronic acid and its dimers; $F_{GGM,MGG}$ indicates the fraction of alginate which starts or ends with a block of guluronic acid; $F_{GM,MG}$ indicates the fraction of alginate consisting of mixed sequences of guluronic and mannuronic acid, with F_{MGM} denoting the fraction of alginate consisting of two mannuronic acids interspaced with guluronic acid; $N_{G>1}$ denotes the average length of guluronic acid blocks.

composition has consequences on alginate as a polyelectrolyte as to counterion condensation and, in turn, on its physical-chemical properties [28, 31, 55, 90, 91, 92, 114]. As MG blocks alternate axial and equatorial bonds, it has intermediate properties with respect to the homopolymeric blocks and it is expected to be more flexible. Indeed, the stiffness of the chain blocks has been found to increase in the order MG < MM < GG [25, 133, 142].

The molar fractions of guluronic and mannuronic acid in an alginate sample are indicated as F_G and F_M, respectively. Typically, algal alginates display a fraction of guluronic acid (F_G) ranging from 0.2 to 0.85. In general terms, *L. hyperborea* plants exhibit a higher content of guluronic acid than the other alginates from seaweed such as those isolated from *M. pyrifera*, *A. nodosum*, and *L. digitata*.

Considering the block-like structure of alginate, the monomeric fraction is not sufficient to give an overall picture of the physical-chemical characteristics of the polysaccharide. To further frame the compositional characteristics of alginate samples and correlate those with its functional properties, the dyadic composition is usually reported, expressed as fractions F_{GG}, F_{GM}, F_{MG}, and F_{MM}. The first two stand for the fraction of a G residue neighboring another G residue or an M residue, respectively. At variance, F_{MG} and F_{MM}

indicate the fractions of M residues neighboring a G residue or another M residue, respectively. For sufficiently long polymer chains in which the end-effects can be neglected, the following relations hold (Eq. 13.1):

$$\begin{aligned} &F_G + F_M = 1 \\ &F_G = F_{GG} + F_{GM} = F_{GG} + F_{MG} \\ &F_M = F_{MM} + F_{GM} = F_{MM} + F_{MG} \\ &F_{GM} = F_{MG} \\ &F_{GG} + F_{GM} + F_{MG} + F_{MM} = 1 \end{aligned} \tag{13.1}$$

A deeper characterization, although more demanding, of the block structure of alginate can be accomplished by indicating the residues that precede and follow a specific monomer. This structural characterization is based on the determination of the G-centered (F_{GGG}, F_{GGM}, F_{MGG}, F_{MGM}) or M-centered (F_{MMM}, F_{MMG}, F_{GMM}, F_{GMG}) triads. In this case, it holds (Eq. 13.2):

$$\begin{aligned} &F_G = F_{GGG} + F_{GGM} + F_{MGG} + F_{MGM} \\ &F_M = F_{MMM} + F_{MMG} + F_{GMM} + F_{GMG} \\ &F_{GGM} = F_{MGG} \\ &F_{MMG} = F_{GMM} \\ &F_{MG} = F_{GM} = F_{GGM} + F_{MGM} = F_{MMG} + F_{GMG} \end{aligned} \tag{13.2}$$

All alginate samples are polydisperse as to their molecular weight. This stems from two main reasons: (i) the production of the polysaccharide is enzymatically controlled and not gene-encoded; (ii) extraction process leads to partial depolymerization of the chain. The number-average molecular weight ($\overline{M}_N$) and the weight-average molecular weight ($\overline{M}_W$) represent the most common averages used to characterize the polysaccharide and their ratio indicates the so-called polydispersity index, PI $= \overline{M}_W/\overline{M}_N$ [38]. The polydispersity of the alginate sample, much like its number- and weight-average molecular weight, has a strong impact on the functional properties of the alginate based hydrogels and materials, especially when biomedical and pharmaceutical applications are considered. The selection of an alginate sample with proper average molecular weight and polydispersity index is then considered of great importance. In general terms, commercial alginates have a

$\overline{M}_w$ of approximately 200,000, but alginates with values as high as 400,000–500,000 are available. Typically, alginate samples have a PI ranging from 1.5 to 3, although values as high as 6 have been reported [38].

13.3 Biosynthesis and Sequence Engineering

Although initially studied on the marine brown algae *Fucus gardnerii* [84], most of the existing knowledge on alginate biosynthetic pathway derives from the work on bacteria such as *P. aeruginosa* [86, 162] and *A. vinelandii* [18, 115]. Despite the conclusions drawn from the early studies [83], it has been established that the biosynthesis of alginate proceeds through the formation of a homopolymer of mannuronic acid (mannuronan) [16]. The biosynthesis of mannuronan starts from fructuose-6-phosfate which is converted into GDP-mannose prior to its oxidation to GDP-mannuronic acid and eventual polymerization. The guluronic moieties are then introduced by mannuronan C-5 epimerases which catalyze, through a NAD+/NADH independent mechanism, the conversion of mannuronic residues into their C-5 epimers, i.e., guluronic residues. The post-polymerization epimerization of mannuronan mirrors the one found for heparin in mammalian cells where a glucuronan C-5-epimerase converts some of the D-glucuronic acid residues into L-iduronic acid ones [85].

Epimerase activity was first spotted by Haug and Larsen from an *A. vinelandii* culture [64]. The complex dependence of the enzyme activity and sequence selectivity from the concentration Ca^{2+} led Ofstad and Larsen to propose the existence of different mannuronan C-5 epimerases with different calcium optima; their relative amounts vary among batch cultivations of the bacterium [43]. The screening of a genomic bacteriophage λ library of *A. vinelandii* allowed identifying a family of eight homologous mannuronan C-5-epimerase genes, indicated as algE genes; seven of these, once expressed in *Escherichia coli*, led to proteins with mannuronan C-5 epimerase activity and they were indicated as AlgE epimerases [42, 44, 147].

The seven C-5 epimerases (AlgE1–AlgE7) are modular enzymes containing repeats of two types of structural units, indicated as A and R (Table 13.2). Within each module type, the homology of epimerases ranges from 50% to 100%. Although the A module contains the catalytic site, the presence of the R module regulates the activity of the enzyme by a tenfold increase [130]. Both A and R modules contain Ca^{2+}-binding motifs.

The seven epimerases reported in Table 13.2 produce alginates with different epimerization patterns. While AlgE2 and AlgE5 catalyze the production of relatively short G blocks, AlgE1 and AlgE6 introduce long G blocks in the polysaccharide chain. In addition to G blocks, AlgE1 and AlgE3 introduce also alternating blocks in the chain (for an extensive description see Ref. 43). The catalytic properties of AlgE4 are of special interest as it introduces strictly MG blocks along the chain [69]. All these epimerases differ further as to the substrate that is accepted by the enzyme for the epimerization and the amount of calcium required for epimerization [14, 70].

As to the mode of action of epimerases, different techniques pointed out that AlgE4 acts in a processive manner epimerizing, on the average, ten M residues before the enzyme-substrate complex dissociates. In the case of AlgE4, the minimum substrate to support epimerase activity was estimated to be an hexameric mannuronan oligomer [14, 58]. It has been further demonstrated that also AlgE1 and AlgE6 introduce G-blocks in a processive manner when acting on polyMG as a substrate, with a seven to eight and eight to ten residues, respectively, as the minimum oligosaccharide sequence required to support epimerase activity [70].

Mannuronan C-5 epimerase activity has been described in brown algae although the properties of algal epimerases, such as specificity and mode of action, are still elusive. Literature reports indicate that *L. digitata* encodes a family of at least 21 mannuronan C-5 epimerases, which likely enable the seaweed with a high tailoring potential for the alginate molecule [107].

The intracellular overexpression of mannuronan C-5 epimerases from *A. vinelandii* in different *E. coli* has allowed producing large amounts of these enzymes that have been used to *in vitro* modify natural alginate and build-up novel polymer sequences. The latter

Table 13.2 Modular structure, molecular mass and enzyme activities of the extracellular mannuronan C-5 epimerase AlgE1-AlgE7 from *A. vinelandii*. Each mannuronan C-5 epimerase is composed of different numbers of A modules (385 amino acids) and R modules (153 amino acids), designed A_1-A_2 and R_1-R_7. Closely related A modules are indicated with identical shading. Closely related R modules are indicated with identical Greek letter ($\alpha, \beta, \gamma, \delta, \varepsilon, \zeta, \eta$)

Epimerase	Module structure	[kDa]	Products
AlgE1	α β γ ε A_1 R_1 R_2 R_3 A_3 R_4	*147.2*	*G and MG blocks*
AlgE2	α β γ δ A_1 R_1 R_2 R_3 R_4	*103.1*	*G blocks (short)*
AlgE3	α β γ α β γ δ A_1 R_1 R_2 R_3 A_3 R_4 R_5 R_6 R_7	*191.1*	*G and MG blocks*
AlgE4	ε A_1 R_1	*57.5*	*MG blocks*
AlgE5	α β γ δ A_1 R_1 R_2 R_3 R_4	*103.7*	*G blocks (medium)*
AlgE6	ε ε ε A_1 R_1 R_2 R_3	*90.2*	*G blocks (long)*
AlgE7	ζ η δ A_1 R_1 R_2 R_3	*90.4*	*G blocks, MG blocks and lyase activity*

was also based on the availability, in kilogram scale, of mannuronan ($F_M = 1$) from epimerase negative ($AlgG^-$) mutants of *P. aeruginosa* [45] and *P. fluorescens* [50]. The combined biotechnological approach for the production of mannuronan and for in vitro epimerization using recombinant C-5 epimerases has allowed to obtain, for the first time, the production of engineered alginates with one largely prevailing block sequence such as polyMG with $F_G = 0.47$ completely lacking of G blocks and polyG with $F_G = 0.97$ (13.2) [70].

Indeed, the production of a polyG is very challenging as the calcium ions, necessary for the epimerase activity to take place, are gradually depleted by the binding to the G blocks. In addition, since three to four residues at the reducing and nonreducing ends are not epimerized, an alginate with $F_G = 0.97\%$ is the highest achievable by epimerization reactions. The use of mutants allows for a potential industrial production of mannuronan opening up to the production of tailor-made engineered alginates. Indeed, although efforts have been successfully carried out to reach an automated solid-phase synthesis of mannuronan oligosaccharides [26, 158], an efficient and cost-effective complete synthesis of a high-molecular weight polysaccharide is still far to come.

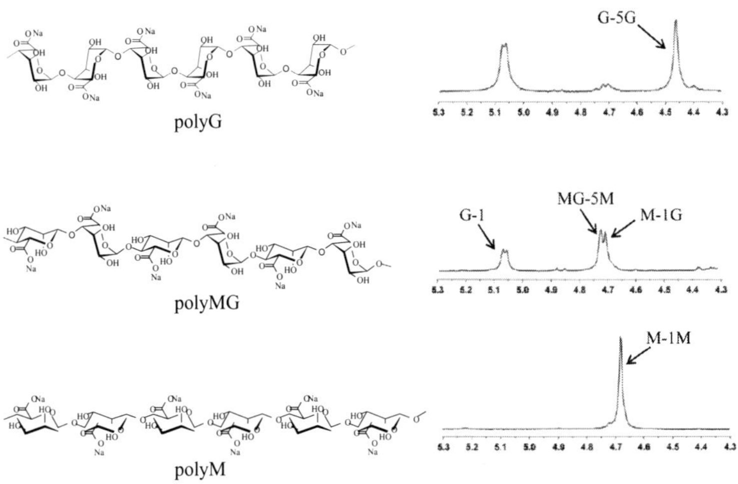

Figure 13.2 Chemical structure and ^{1}H-NMR (anomeric region) of polyG, polyMG and polyM (mannuronan).

13.4 Sequence Determination

Given the details on alginate biosynthetic pathway reported in the previous paragraph, it can be clearly understood how a purely Bernoullian statistics cannot be used to predict the monomer distribution along the alginate chain [112, 133]. Similarly, the use of a more complex second-order Markov model provides only an approximate (and unsatisfactory) description of the block composition in natural alginates [80]. Because of the functional implications of the block structure in alginates, the development of suitable analytical techniques for its determination are of fundamental importance.

A basic, nondestructive and very accessible characterization of alginate samples, in terms of determination of monomer and diad molar fraction, is based on the use of circular dichroism (CD). This relies on the availability of the molar ellipticity in the UV wavelength range, $[\theta](\lambda)$, for each of the different sequential structures that compose alginate, i.e., $[\theta]_{GG}(\lambda)$ for the G blocks, $[\theta]_{MM}(\lambda)$ for the M blocks, and $[\theta]_{MG,GM}(\lambda)$ for the MG blocks. The dyadic composition of an alginate sample (F_{GG}, F_{MM}, $F_{MG,GM}$) can then be readily determined from its CD spectrum ($[\theta]_{Alg}(\lambda)$) (Eq. 13.3)

$$[\theta]_{A\lg}(\lambda) = F_{GG}[\theta]_{GG}(\lambda) + F_{MM}[\theta]_{MM}(\lambda) + (1 - F_{GG} - F_{MM})[\theta]_{MG,GM}(\lambda) \quad (13.3)$$

with

$$F_{MG,GM} = 1 - F_{GG} - F_{MM}$$

The method has been originally proposed in 1980 by Morris and coworkers [103]. However, the availability of alginate composed of an (almost) pure block structure (polyM, polyG, and polyMG) allowed to increase the accuracy in the determination [32].

A complete characterization of the composition, sequential structure and (average) block length is accomplished by means of high resolution ^{1}H- and ^{13}C-NMR. Fig. 13.3 reports the anomeric region of a ^{1}H-NMR spectrum for an alginate sample with the assignations of the various peaks.

The analysis of the anomeric region allows to calculate, by means of the relations reported in equations 1, the composition in terms

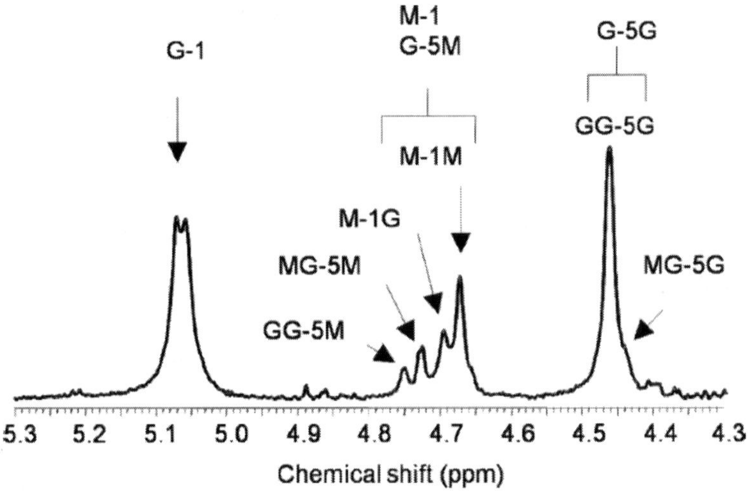

Figure 13.3 ^{1}H-NMR spectrum (anomeric region) of an alginate isolated from *Laminaria hyperborea*. M-1M and M-1G represent the anomeric proton of an M residue neighboring another M residue or a G residue, respectively. MG-5M, GG-5M, and MG-5G refer to the H-5 proton of the central G residue in an MGM, GGM, or MGG triad, respectively. G-1 refers to the anomeric proton of G residues and GG-5G refers to the anomeric proton of G residues in G blocks.

of monomers, diads and G-centered triads and, from the latter, the average length of the G blocks, $N_{G>1}$, as (Eq. 13.4):

$$N_{G>1} = \frac{F_G - F_{MGM}}{F_{GGM}} \tag{13.4}$$

The determination of the M-centered triads requires ^{13}C-NMR data [56].

The determination of the composition by ^{1}H-NMR analysis allows also to get a relative estimation of the length of the alternating blocks, N_{GM}, by means of Eq. 13.5 [56], although for a quantitative calculation the information from the M centered triads are needed.

$$N_{MG} \propto \frac{F_{MGM}}{F_{GGM}} \tag{13.5}$$

$N_{G>1}$ is very often used to interpret the structure-functional properties relationship in alginates. It should be underlined, however, that the ^{1}H-NMR analysis provides an evaluation of the block length averaged over the whole G-block population. As the average molecular weight of a polymer provides only a partial information on its potential performances which strongly depend on the polydispersity, the information provided by $N_{G>1}$ is only partial as the real length of the various blocks and its distribution might play a fundamental role.

An important improvement along this line has been recently accomplished and it is based on alginate sequencing using a combination of different analytical techniques such as ^{1}H-NMR, high-performance anion exchange chromatography–pulsed amperometric detection (HPAED-PAD) and size exclusion chromatography–multi-angle laser light scattering (SEC-MALLS), and alginate lyases (Fig. 13.4) [2]. These enzymes are endolyases that have been isolated from many sources [160], including marine algae [88], marine mollusks [12, 67, 105], and various microorganisms like *Pseudomonas aeruginosa* [40] and *Azotobacter vinelandii* [74].

Alginate lyases target the glycosidic bond on the polysaccharide and catalyze the cleavage of the chain through a β-elimination that leaves an unchanged saturated uronate on the reducing end and a 4-deoxy-L-erythro-hex-4-enepyranosyluronate at the non-reducing end [60]. The reaction mechanism of the alginate lyases is analogous

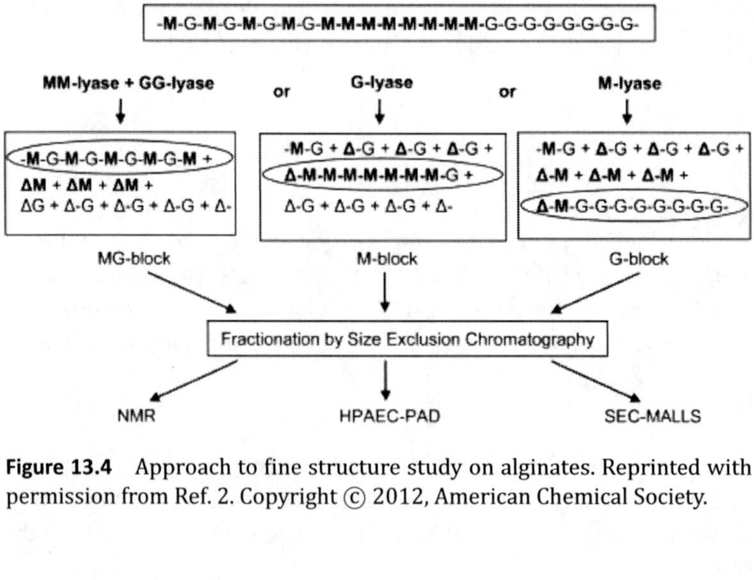

Figure 13.4 Approach to fine structure study on alginates. Reprinted with permission from Ref. 2. Copyright © 2012, American Chemical Society.

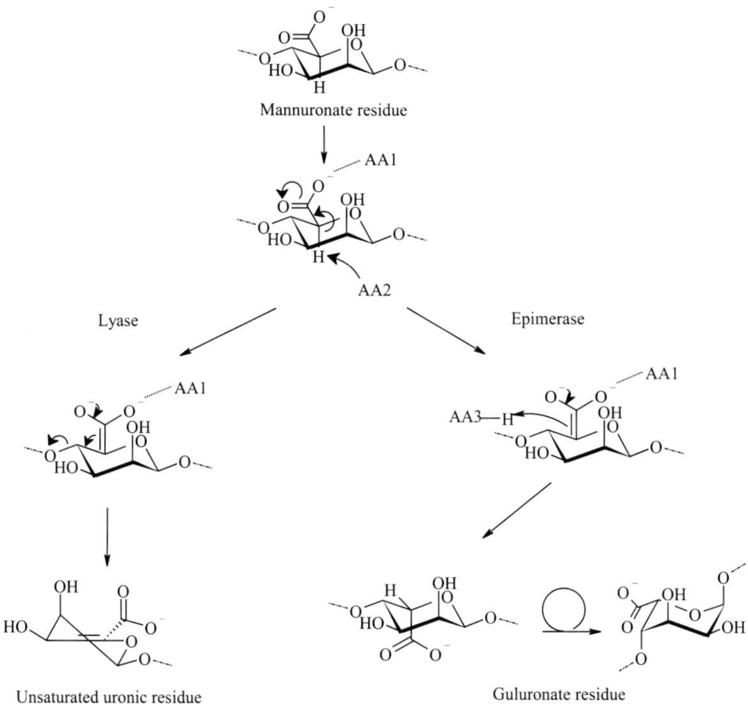

Figure 13.5 Unified reaction mechanism for mannuronan C-5 epimerases and lyases.

to the one proposed for other polysaccharide lyases [116, 117] and it is closely related to the epimerization mechanism for mannuronan C-5 epimerases (Fig. 13.5) [48]. The specificity of alginate lyases is expressed toward one of the residues in the glycosidic linkage. For example, the lyase from *A. vinelandii* is specific for M residues in MM or MG diads. Alginate lyase from *Klebsiella aerogenes* [11, 65] is specific for G residues followed by either another G or an M [11, 65]. At variance, lyase from *P. alginovora* cleaves only MM diads [15].

Aarstad et al. used four different lyases with appropriate specificity, to obtain small alginate fragments which were then characterized and the results interpreted in statistical terms [2]. This novel approach showed already its enormous potential allow-

Table 13.3 Block length in alginates has determined by a combined enzymatic and analytical approach

Alginate	Lyase	Block	DP_n (SEC-MALLS)	DP_n (^{1}H-NMR)
L. hyp.	M-	G	120	117
M. pyr.	M-	G	132	105
M. pyr.	MM- + GG-	MG		39

ing to unveil the processive character of the epimerization by C-5 epimerases [14, 70]. As to the alginate structure, the combination of enzymatic treatment and analytical analyses revealed the presence, in algal alginates isolated from *L. hyperborea* and *M. pyrifera*, of extremely long G blocks composed of more than 100 G residues (Table 13.3). Although these data are not directly comparable with the $N_{G>1}$ values obtained from the NMR analysis, which requires a prior hydrolysis to the analysis, still these recent findings seem to indicate that real block lengths in natural algal alginates are much longer than expected. Although the functional role of such long G blocks in natural alginates remains an open question, it can be expected that the recent data will have an impact on further refinements of the model for alginate gel formation.

Although the present alginate lyases available do not allow to determine accurately the length of the alternating sequences in different alginates, the data so far collected seem to indicate that alginate from *M. pyrifera* is characterized by the longest MG blocks, in line with what previously proposed [33]. The determination of the real length of MG blocks is likely to contribute in clarifying the differences in behavior and in the mechanical properties of natural alginate from different sources.

13.5 Gel Formation and Ion Binding Properties

13.5.1 *Ionic Gels*

Present and future applications of alginate, in particular for the food industry and in the biotechnological and biomedical field, are linked to its ability to efficiently bind divalent cations, such as Ca^{2+}, Sr^{2+}, and Ba^{2+} to name a few, leading eventually to hydrogel formation. Upon treating an alginate solution with a polymer concentration around 10–20 g/L with a proper divalent cation, a rapid gel formation occurs. The gelation mechanism is based on the formation of a physical network where ions hold together polysaccharide chains forming a continuous swollen network. The driving force for the hydrogel formation is the markedly higher affinity of alginate for some divalent ions, like Ca^{2+}, than for monovalent ions (typically Na^{+}). In addition, the affinity for a specific ion increases with its concentration in a near-neighbor autocooperative process [131].

Alginate ionic hydrogels can be prepared through controlled setting or diffusion setting. In the former case, an inactivated form of the cross-linking ion is mixed with the alginate solution. The divalent ion is then mobilized by lowering the pH with inorganic acids or slowly hydrolyzing lactones. In a typical experiment of calcium-alginate gel formation, $CaCO_3$ is used in combination with δ-gluconolactone (GDL) in a molar ratio 1:2, which guarantees only a slightly acidic pH in the final hydrogel [37]. The slow release of calcium from the inactivated form accounts for the formation of a homogeneous network where gel formation kinetics can be efficiently controlled by alginate and calcium carbonate concentration as well as by the particle size of the latter salt. In addition, the composition of the alginate sample used influences the gel kinetics. Indeed, the presence of a higher amount of MG blocks leads to a faster gel setting [35, 36].

The preparation of inhomogeneous alginate gels is, at variance, accomplished through a diffusion setting approach. This is based on the diffusion of the cross-linking ion (e.g., Ca^{2+}) from a large reservoir into the alginate solution. The chelation of the divalent ion by the alginate chains causes an immediate gelation at the polysaccharide-external solution boundary. Combined with the rapid diffusion towards the boundary of the alginate, driven by the Na^{+} counterion, this leads to a highly inhomogeneous hydrogel. This has been verified by means of several methods, including synchrotron radiation induced X-ray emission, magnetic resonance imaging, confocal microscopy, and Raman spectroscopy [66, 144, 150].

Since the distribution in mass of the polysaccharide within the hydrogel is determined by the coupled diffusion of Na^+ and alginate towards the gelling zone and of the divalent ion towards the inner part of the alginate solution, the suppression of such driving force acts reducing the inhomogeneity of the system. This can be accomplished by adding non gelling Na^+ ions in both the calcium reservoir and alginate solution. In such a way, the diffusion of alginate is slowed and the cross-linking ion can diffuse throughout the alginate solution [126, 144].

13.5.2 *Ion Binding Properties and Sequence Relationship*

13.5.2.1 Sequence specificity: concepts

The highly selective binding ability of alginate towards different alkaline-earth (and transition metal) divalent ions is known since 1967 and it was found to increase in the order $Mg^{2+} \ll Mn^{2+} < Ca^{2+} < Sr^{2+} < Ba^{2+} < Cu^{2+} < Pb^{2+}$ [62, 63]. The selectivity in ion binding by alginate was found to depend on the composition of the polysaccharide, in particular on the presence of G residues. In other terms, the selectivity of binding was allocated to the specific feature of divalent ions to induce a strong interchain interaction between stretches of G blocks. This idea led to the so-called "egg-box" model to describe the interchain associations in alginates which underlines the formation of junction zones holding the polysaccharide network together in a swollen network [54, 104, 131].

According to this model, two facing ordered stretches of G sequences bind the divalent ion in a chelate type of binding and the geometrical characteristics of the cavity formed by the diaxially linked G residues determines the selectivity in the cation binding (Fig. 13.6). The formation of the egg-box junctions in alginate is a cooperative process that takes place according to the "zipper" mechanism: the binding of the first cross-linking ion is thermodynamically less favorable than the binding of the following ones. This process requires a minimum length of the G stretch which depends upon the affinity for the divalent ion and that, in the case of Ca^{2+}, has been reported to range from 8 to 20 adjacent G residues [78, 141, 140].

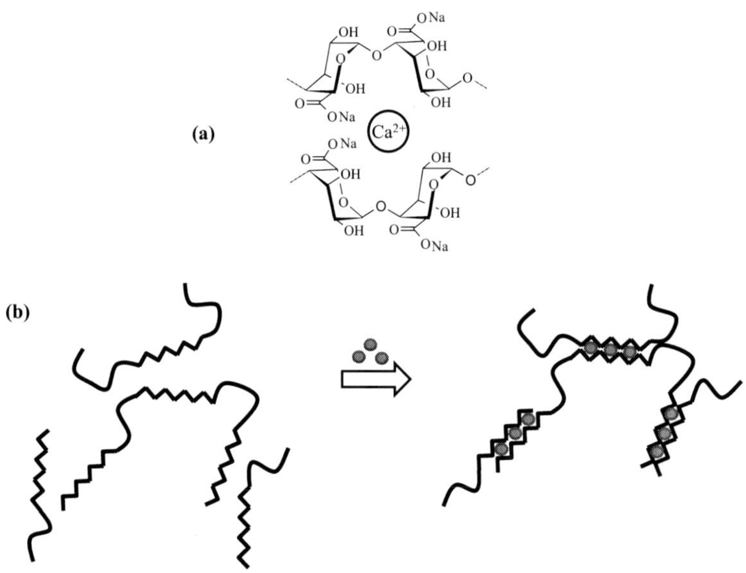

Figure 13.6 (a) Chelation of divalent cations by the GG sequences according to the egg-box model. (b) Interchain junction formation.

Over the recent years, several investigations have attempted to refine the description of the egg-box model by means of X-ray diffraction, NMR spectroscopy and molecular modeling [4, 125, 137]. They all showed that the egg-box model for alginate junctions as originally proposed should be regarded as basically correct. Some of the features of the junctions within the hydrogel, however, have been revised. In particular, a large-scale lateral association of junctions, induced by the presence of an excess of calcium ions, has been proposed by Stokke *et al.* on the basis of SAXS measurements [139]. However, recent X-ray analyses have indicated that this lateral association occurs through unspecific electrostatic interactions and have confirmed that the junctions within the gel closely resemble the egg-box model [125]. Recently, viscometry, light scattering, circular dichroism, and fluorescence quenching studies were performed on both Ca^{2+} and Mg^{2+} alginate systems to focus on the initial binding of calcium ions by alginate chains in dilute solution. The first mode of bonding was proposed to be formed by four G residues from facing chains in a conformation which, albeit ordered, is different from the "egg-box" one. This was indicated as a locally tilted conformation (tilted egg-box). The addition of further bonding ions caused the well-known cooperative egg-box conformation with notable variation in the physical-chemical properties of the dilute solution [10].

As anticipated, the selective binding of alginate towards divalent cations is strongly dependent on its sequence structure. Experiments involving equilibrium dialysis showed that the binding strength of divalent ions to the three different components of alginate follows the order:

G blocks: $Ba^{2+} > Sr^{2+} > Ca^{2+} \gg Mg^{2+}$
M blocks: $Ba^{2+} > Sr^{2+} \approx Ca^{2+} \approx Mg^{2+}$
MG blocks: $Ba^{2+} \approx Sr^{2+} \approx Ca^{2+} \approx Mg^{2+}$

This overall picture has been, to some extent, revised. Indeed, the large scale availability of mannuronan and of the whole set of AlgE epimerases makes it possible to engineer alginates with designed sequential structures. The latter has allowed a better understanding of the sequence–functional properties relationships and revisiting some of the concepts of ion binding and gel formation. In particular, the availability of an alginate composed exclusively of MG blocks has shown that the latter sequences are able to bind calcium ions. Specifically, treatment of a polyMG alginate in calcium limiting conditions ends up in gel formation (Fig. 13.7).

This finding led to propose the existence of gel junctions involving MG blocks, namely MG/MG junctions and MG/GG junctions (Fig. 13.8). In particular the latter can be encountered in those situation in which alginate presents quite long G blocks and long MG blocks. Since the length of the former blocks in alginate can be only estimated as an average value from ^{1}H-NMR and the one of the MG blocks is not easily accessible, novel analytical methods, which are now being reported in the literature and have been summarized in the previous paragraph, are expected to bring about a notable development in the understanding of the molecular aspects of junction formation in alginate.

Along the same line, the ion selectivity of the different block sequences present in alginate have been revised using alginates produced through the use of specific epimerases. The analysis

Figure 13.7 Hydrogel formed by polyMG upon treatment with calcium ions. Reprinted with permission from Ref. 33. Copyright © 2005, American Chemical Society.

revealed that Ca^{2+} binds to both G and MG blocks, Ba^{2+} to G and M blocks, and Sr^{2+} to G blocks solely [100]. These considerations led to re-interpret some results on natural alginates from *L. hyperborea* and *M. pyrifera* in view of the possible presence, in particular for the latter sample, of junctions which are not composed exclusively by the G-blocks.

13.5.2.2 Sequence specificity: macroscopic consequences

The elastic modulus (or Young's modulus, E) is often used as a measure of gel strength for alginate gels. This is normally measured in the small deformation regime where a linear relationship holds between the strain ($\varepsilon = \Delta l/l$) on the network and the stress exerted ($\sigma = F/A$) (Hooke's law) (Eq. 13.6).

$$\sigma = E\varepsilon \tag{13.6}$$

In general, the Young's modulus recorded for alginate hydrogels depends on the deformation rate. The elastic modulus is also dependent on the amount of cross-linking ion and on the concentration of alginate. As to the former effect, the higher is the cross-linking ion concentration the higher is the number of junctions formed within the hydrogel. It follows that the calcium saturated alginates gels are

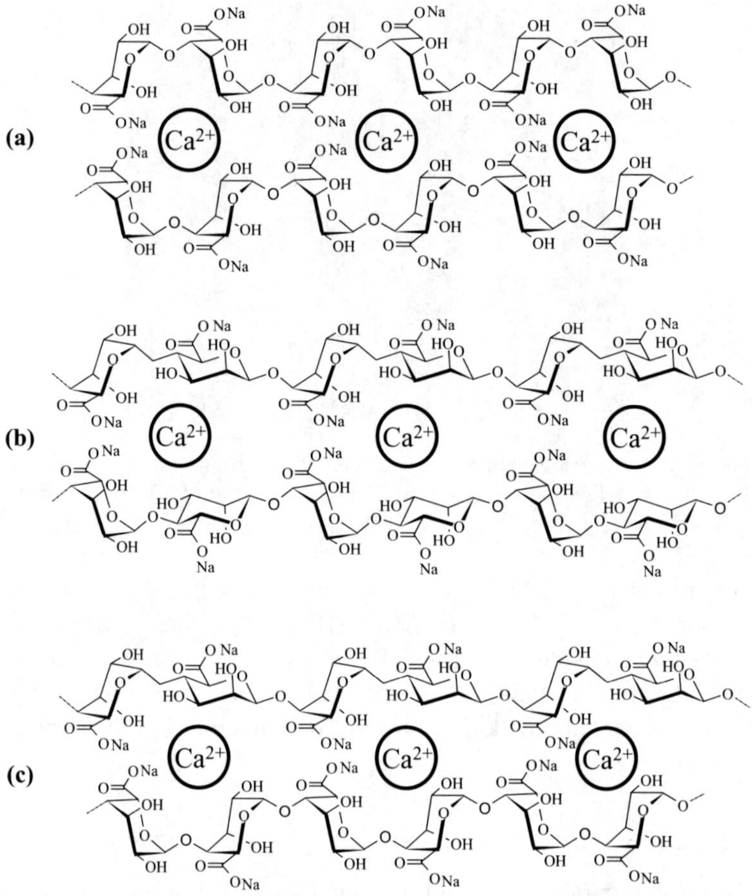

Figure 13.8 Three possible junctions in alginate hydrogels. (a) GG/GG junction. (b) MG/MG junction. (c) MG/GG junction.

better performing than the calcium-limited ones. Apart from this effect, the Young's modulus of alginate hydrogels has been reported to depend on the affinity between the polymer and the cross-linking ion [131]. Along this line, it was recently found that the addition of small amounts of barium are beneficial to increase the gel strength both for high-G and high-M alginate [100].

Rheological experiments are also very often used for the determination of the mechanical performance of alginate hydrogels.

In this case, the viscoelastic nature of the systems is explored through the determination of the elastic (G') and loss (G'') moduli. The data from mechanical spectroscopy of calcium gels can be interpreted in terms of the generalized Maxwell model composed of a sequence of elements in parallel (spring and dashpot) to which an additional spring has been added (Eq. 13.7).

$$G' = G_e + \sum_{i=1}^{n} G_i \frac{(\lambda_i 2\pi \nu)^2}{1 + (\lambda_i 2\pi \nu)^2}; \quad G_i = \frac{\eta_i}{\lambda_i}$$

$$G'' = \sum_{i=1}^{n} G_i \frac{(\lambda_i 2\pi \nu)^2}{1 + (\lambda_i 2\pi \nu)^2} \tag{13.7}$$

where n is the number of Maxwell elements considered, G_i, η_I, and λ_I represent the spring constant, the dashpot viscosity and the relaxation time of the ith Maxwell element, respectively. G_e is the spring constant of the (n+ 1)th Maxwell element which is supposed to be purely elastic [154]. The use of these constitutive equations allow to determine the shear modulus, G, as (Eq. 13.8)

$$G = G_e + \sum_{i=1}^{n} G_i \tag{13.8}$$

which is correlated to the Young's modulus through the following equation (Eq. 13.9):

$$G = \frac{E}{2\,(1 + \theta)} \tag{13.9}$$

with θ the Poisson ratio of the hydrogel which, in the case of alginate, has been evaluated to be 0.5 [102].

From the theoretical point of view, the response of an alginate network to the application of a stress is commonly described in terms of the rubber elasticity theory. However, the specific features of alginate hydrogels fail to stringently meet the requirements of the theory developed by Flory. The presence of long restricted junctions rather than point-like crosslinks and the enthalpic contributions [154]—arising from junction deformation/un-zipping—suffice to consider the application of the rubber elasticity theory in the case of alginate hydrogels with some caution.

Nevertheless, since enthalpic contributions are negligible at small deformations where the entropic contribution is expected to prevail [27, 68], the concept of a direct correlation between the elastic modulus and the number (and stiffness) of the chains between crosslinks (elastically active chains) has been largely exploited [34, 154].

Focusing on the characteristics of the polysaccharide, the composition of alginate samples brings about a substantial effect on hydrogel stiffness. In general, G-rich samples form stronger gels than M-rich alginates. In particular, it was found that the mechanical strength of saturated alginate gels increases with the increase of the average length of G-blocks, $\overline{N}_{G>1}$ [128].

Besides the undoubtedly major role of G blocks in determining the gel strength, the important role of the alternating sequences has also been recently pointed out. In particular, the enzyme-based elongation of the MG blocks in natural alginate samples led to a notable increase in the Young's modulus, especially for the high-M alginates [33, 35]. In particular, in the case of *A. nodosum* the AlgE4 treatment leads to an approximately threefold increase in the gel strength. This was correlated with the possible existence of mixed junctions composed of GG and MG blocks. In alginate samples prepared by combination of AlgE4 (MG-block forming) and AlgE1 (MG and GG block forming) there is an approximately linear relationship between the amount of overall G residues and the Young's modulus in the range $F_G = 0.48 - 0.77$ (Fig. 13.9) [101].

Very recently, Aarstad has shown that the relation between $N_{G>1}$ and the Young's modulus does not hold in the case of AlgE6 epimerized samples [1]. Indeed, the presence of very long G blocks in epimerized alginates is detrimental as to the mechanical properties of the hydrogel when compared to those of a sample from a natural alginate with comparable F_G. These results, besides stressing on the importance of the block composition on the mechanical properties of alginate hydrogels underline once more the role of the real (not averaged) block length and the need of analytical methods to determine it.

Rupture strength is also quite often used to frame the mechanical properties of alginate gels. Typically, rupture strength is related to the energy (enthalpy) required to fracture the junctions within the network [8, 9]. In general, hydrogels from high-M alginates with low elastic modulus show a high rupture strength while gels from

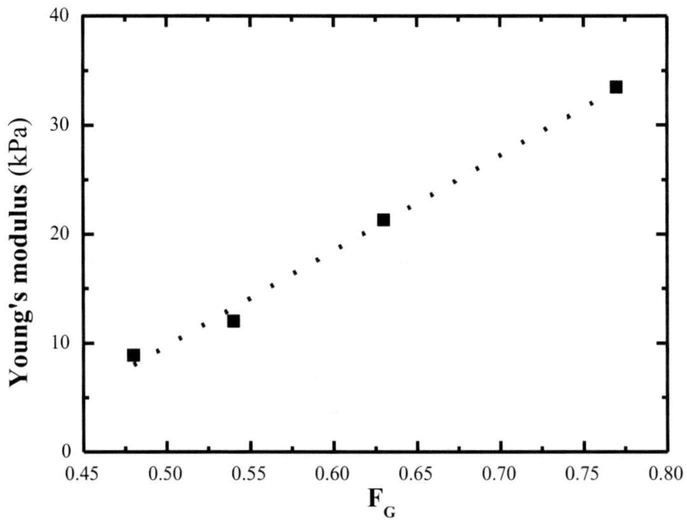

Figure 13.9 Dependence of the Young's modulus from F_G for alginate samples prepared by epimerizing mannuronan with AlgE4 and AlgE1. Reprinted with permission from Ref. 99. Copyright © 2007, American Chemical Society.

G-rich alginates are brittle [95]. Very recently, it has been pointed out that long MG blocks strongly affect the rupture strength. Indeed, a hydrogel based on an alginate sample composed exclusively of MG sequences shows an extraordinary high value of the rupture strength [101]. Along the same line, the elongation of MG sequences in natural alginates led to an increase of the energy required for gel rupture. Both cases have been traced back to the fact that, upon compression, Ca^{2+} bond MG blocks "slide" along each other, unzipping the MG junctions and releasing part of the energy applied (Fig. 13.10) [34].

This seems supported by the comparison between a natural alginate and an epimerized one, both at $F_G = 0.66$, where the presence of long MG-blocks in the former accounts for a twofold increase in rupture strength [101]. Indeed, in the engineered alginate samples, produces by means of two successive epimerizations with AlgE4 and AlgE1, the elasticity of the hydrogel decreases almost linearly with the increase of F_G (Fig. 13.11).

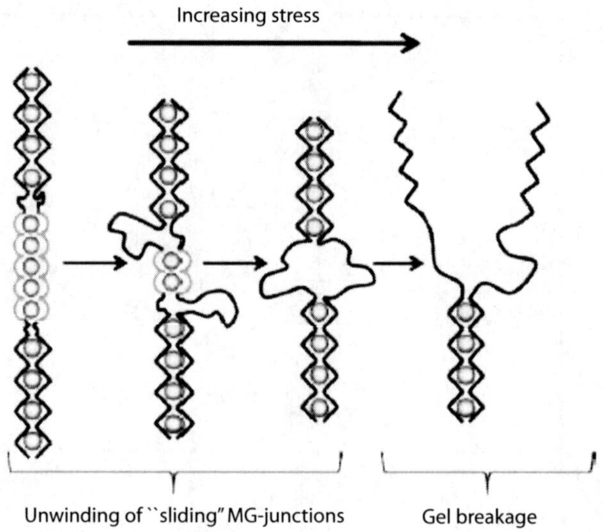

Figure 13.10 Graphical representation of the sliding of MG/MG junctions upon application of a stress.

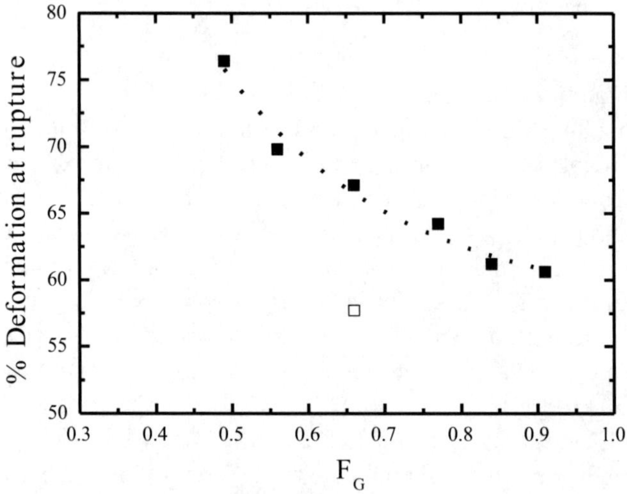

Figure 13.11 (■) Dependance of the elasticity of gel cylinders on F_G for alginates prepared from mannuronan epimerized with AlgE4 and AlgE1. (□) Gel cylinders from a natural alginate isolated from *L. hyperborea* (F_G = 0.66). Reprinted with permission from Ref. 99. Copyright © 2007, American Chemical Society.

Although based on different physical concepts, stability, syneresis and permeability of alginate hydrogels are highly correlated and strongly dependent on alginate block structure. The term *stability* refers to the resistance of the gel to swelling and dissolution upon treatment with ions which compete for the uronate residues and eventually displace Ca^{2+} ions from the network junctions. Alginate gels can be seen as osmotic swelling systems where an osmotic pressure, given by the mobile counterions of the alginate polyanion chain, is exerted against an elastic retraction force given by the network junctions which opposes to the swelling. In terms of chemical potential, the osmotic contribution to swelling of an ionic network is composed by the polymer-solvent mixing term ($\Delta\mu_{\text{mix}}$) and an ionic term ($\Delta\mu_{\text{ion}}$) arising from the uneven distribution of ions between the inside and the outside of the gel. At equilibrium, these two terms are counterbalanced by the elastic resistance of the network, $\Delta\mu_{\text{elastic}}$ (Eq. 13.10).

$$\Delta\mu_{\text{mix}} + \Delta\mu_{\text{ion}} + \Delta\mu_{\text{elastic}} = 0 \qquad (13.10)$$

For these reasons, the stability of the alginate hydrogels depends on the number and strength of the junctions. Consequently, the presence of long G blocks [95, 149] and/or the use of stronger binding ions such as Ba^{2+} and Sr^{2+} increases the stability of alginate network [21, 95]. However, it has been shown that the effect of the cross-linking ion strongly depends on alginate composition, with the same amount of Ba^{2+} playing a notable stabilization role for alginate from *L. hyperborea* while having negligible effect on *M. pyrifera* (Fig. 13.12) [100].

The use of Ca^{2+} or Ba^{2+} ions leads also to marked differences as to the inhomogeneity of the hydrogel (Fig. 13.13), although basically no effect was noticed on permeability of the gels. Alternating sequences play also a role in the case of alginate gel stabilization. Indeed, the elongation of the MG blocks by means of AlgE4 epimerization led to a general increase in the stability of the hydrogels from natural alginates [34, 146]. To an extreme point, an alginate prepared by an epimerization approach and characterized by a $F_G = 0.64$, distributed in very long MG blocks and a few G stretches, showed an extreme stability towards swelling with a dimensional variation of less than 10% after 7 shifts of saline

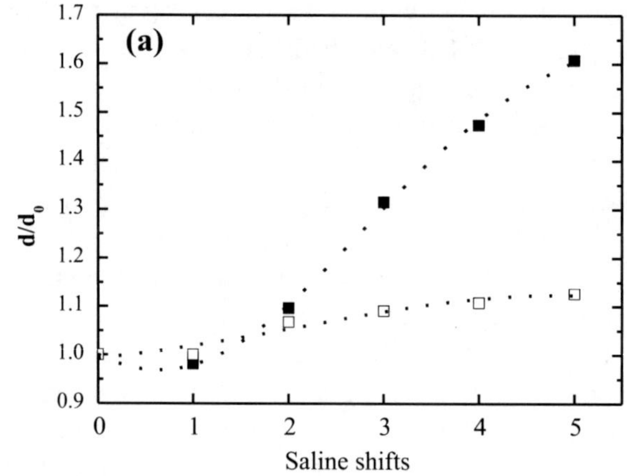

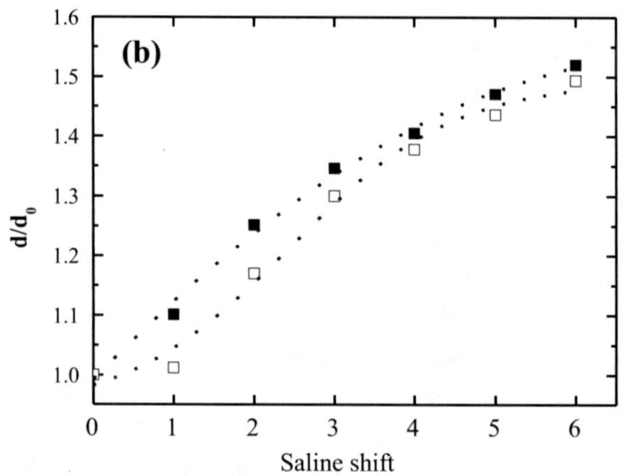

Figure 13.12 Swelling of alginate beads in saline solution (NaCl 0.9%) prepared using in the gelling bath $CaCl_2$ 50 mM (■) or 10 mM $BaCl_2$ (□) for high-G alginate (**a**, $F_G = 0.66$) and high-M alginate (**b**, $F_G = 0.40$). Lines are drawn to guide the eye. Reprinted with permission from Ref. 100. Copyright © 2006, American Chemical Society.

solution [101]. For comparison, a dimensional increase of more than 60% was found for a natural alginate with $F_G = 0.66$ (Fig. 13.14).

Syneresis is defined as the shrinkage of hydrogels with release of water which leads to a smaller gel with a higher polysaccharide concentration. In general, alginate gel syneresis increases with increasing cross-linking ion (Ca^{2+}) concentration, gelling time

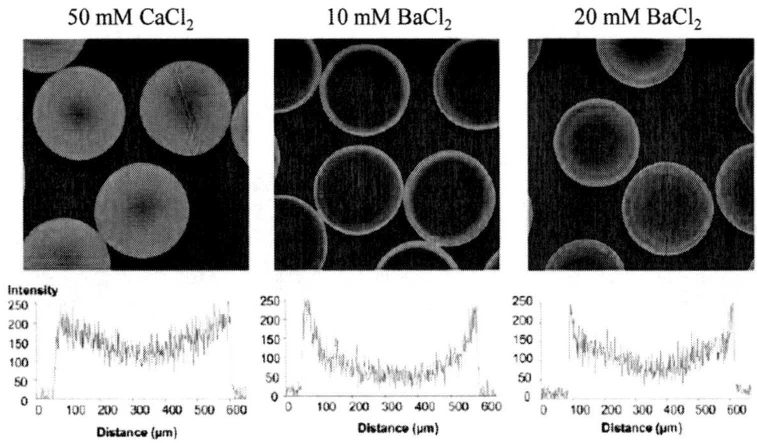

Figure 13.13 Effect of the cross-linking ion on the inhomogeneity of alginate hydrogels as detected by means of confocal laser scanning microscopy. Reprinted with permission from Ref. 100. Copyright © 2006, American Chemical Society.

and polysaccharide molecular weight [36, 95]. However, the fine composition of the polysaccharide brings about a very strong effect on this characteristic of alginate gels. In general, long G blocks will shrink less than samples with short G blocks due probably to the lack of the reorganization of the network structure [95]. However, a fundamental role in syneresis is also played by the alternating sequences. In particular, the elongation of the alternating blocks in natural alginates increases the syneresis of the gels [33, 34, 101]. In particular, for alginate samples enzymatically synthesized containing long MG blocks the syneresis of the samples was found to decrease in a linear fashion from the overall G content, F_G (Fig. 13.15). In addition, the epimerized sample showed a higher syneresis than the natural ones with the same F_G [99].

A property strictly connected with syneresis is the permeability of the hydrogels. The (average) pores size of alginate hydrogels have been evaluated to be between 5 and 150 nm from microscopic analyses [3]. Recent relaxometric determinations have indeed shown an average mesh ratio of around 12 nm for calcium saturated alginate gels at 1% w/V [154]. Permeability of alginate hydrogels depends on the composition of the polysaccharide; the higher the

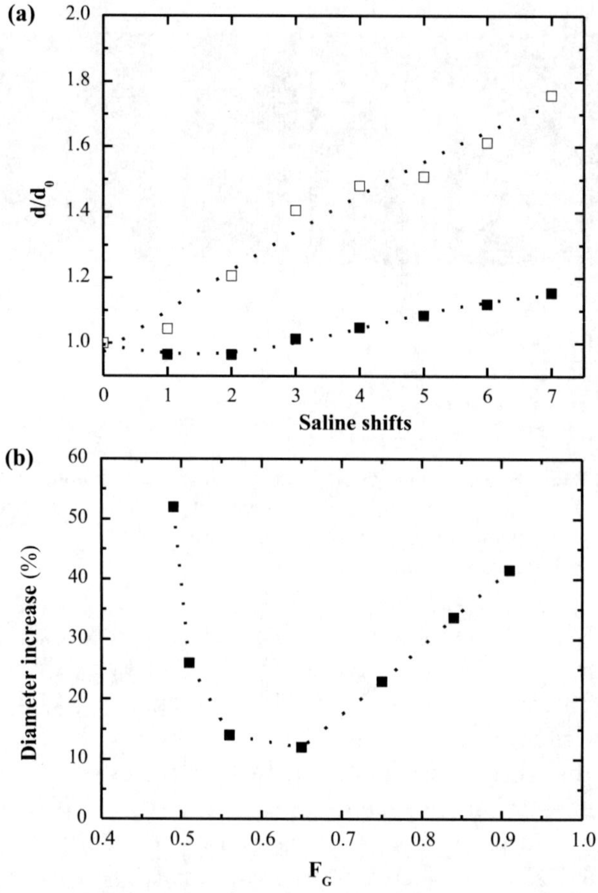

Figure 13.14 (a) Relative diameter increase (d/d_0) for alginate beads treated with saline solution for a natural high-G alginate (□, $F_G = 0.66$) and for an epimerized alginate (■, $F_G = 0.64$). Lines are drawn to guide the eye. (b) Increase in diameter (%) with respect to the initial value after 7 saline shifts for beads obtained from epimerized alginates. Line is drawn to guide the eye. Reprinted with permission from Ref. 99. Copyright © 2007, American Chemical Society.

G content the larger are the pores within the gels. This has been explained considering the increase in stiffness of the network in the presence of long G blocks compared to the elastic structures obtained in the presence of short G block. In addition, the increase of

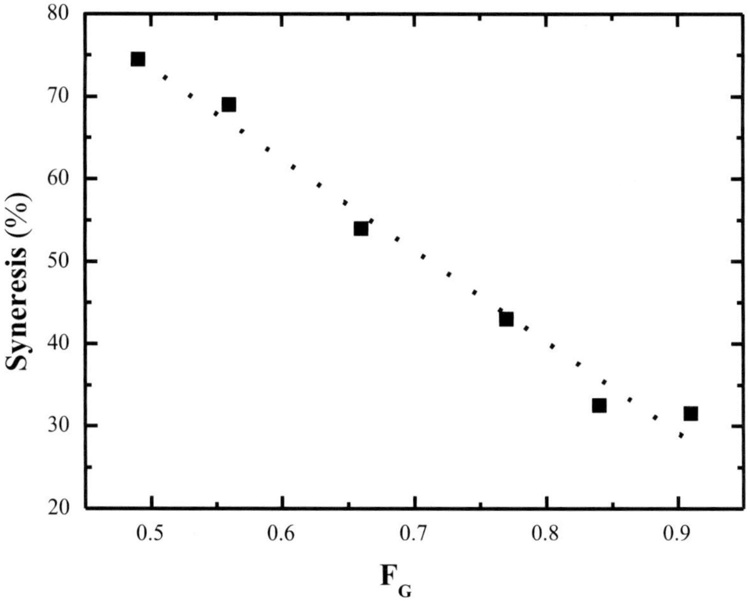

Figure 13.15 Dependence of the syneresis from F_G for alginate samples prepared by epimerizing mannuronan with AlgE4 and AlgE1. Reprinted with permission from Ref. 99. Copyright © 2007, American Chemical Society.

Ca^{2+} concentration will result in lateral association, which is favored by the length of the G blocks, leaving more space in between the junction zones [49, 95]. However, permeability of alginate hydrogels is influenced by the composition of alginate to a higher level. In particular, it has been shown that the presence of long MG blocks reduces the permeability of the hydrogel, in line with the results of syneresis. In particular, epimerized samples with high content of MG blocks displayed a lower permeability than natural alginate samples with similar F_G content [99].

The permeability of an alginate gel capsule can be, to some extent, controlled by alginate concentration and by the use of different divalent ions [111, 139], although in this latter case the macroscopic effect depends of the composition of the polysaccharide and of the specific ion–polysaccharide interactions. Indeed, it has been found that the presence of Ba^{2+} reduces notably the permeability of high-

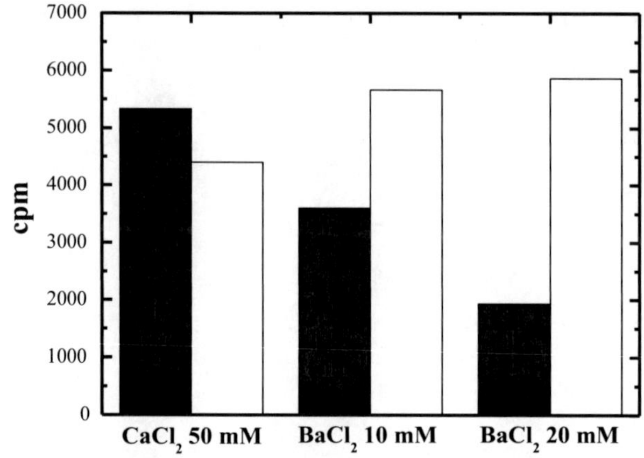

Figure 13.16 Permeability of [^{125}I]-labeled IgG into beads of high G (F_G = 0.66, filled columns) and low G (F_G = 0.40, open columns). Reprinted with permission from Ref. 100. Copyright © 2006, American Chemical Society.

G alginate but brings about the opposite effect on high-M alginates (Fig. 13.16) [100].

13.6 Applications

Alginates extracted from brown seaweeds are used for a variety of industrial processes including textile printing, paper and water treatments, production of food for humans and pets [110]. However, the availability on the market of biocompatible ultra-pure alginate samples with a very low content of pyrogens allows their use in the pharmaceutical and biotechnological industry, e.g., for wound healing, treatment of oesophageal reflux, as dental impression materials and in the immobilization of cells.

One of the most interesting and challenging uses of alginate is in the field of bioencapsulation. Indeed, alginate is by far the most studied and well characterized material for cell encapsulation technology [72, 106]. Immobilization of living cells encompasses applications in food industry, medicine, and agriculture. On the industrial scale, cell immobilization offers the advantage of an easier handling with respect to the free cells. Examples of industrial processes based on immobilized cells are the production of ethanol and beer from yeast cells and the synthesis of steroids and antibiotics [57, 157].

Bioencapsulation in alginate hydrogels has been proposed in the literature for regenerative medicine purposes. Alginate beads and capsules have been used for the regeneration of several different tissues. As an example, rat bone marrow cells and osteoblasts have been encapsulated in alginate for the regeneration of bone tissue [79, 134]. Alginate beads have also been used to encapsulate chondrocytes for cartilage regeneration [75]. In this case, the encapsulated primary culture maintained its phenotype without de-differentiation, although no significant cell proliferation was detected [30]. At variance, when another bioactive agent was included within the alginate capsule, specifically a bioactive chitosan derivative bearing galactose moieties flanking the polysaccharide backbone (chitlac), a notable improvement in the cell proliferation has been achieved [94]. This can be traced back to the presence of a galectin in the cartilage tissue which bridges the bioactive chitosan with one integrin of the cell surface [93]. In the field of cartilage regeneration, alginate is one of the components of Cartipatch, a three-dimensional scaffold that is used for treatment of cartilage defects [138]. It requires arthroscopic cartilage harvest, laboratory isolation and culture of the cells in the scaffold, and follow-up mini-arthrotomy. Alginate beads and capsules have been used for cell encapsulation of non-autologous cells for the treatment of different diseases [72]. For example, encapsulation of kidney cells has been exploited to produce endostatin for the treatment of brain tumor [118]. Similarly, myoblasts and fibroblasts have been encapsulated to produce human factor IX for the treatment of hemophilia [71, 87].

In the field of cell encapsulation, major interest has been devoted to the immobilization of insulin producing cells for the treatment of type 1 diabetes [22, 24, 135, 159]. Complete overviews on the use alginate for immunoprotection of Langerhans islets have been recently published [22, 24]. Insulin is a hormone controlling normal glucose homeostasis that is secreted by the β cells within the Langerhans islets in the pancreas. Diabetes is characterized by hyperglycemia resulting from either low levels of insulin or from abnormal resistance to its effects. Type 1 (insulin-dependent) diabetes results from an autoimmune destruction of β cells causing an insulin deficiency and the need of endogenous hormone [96]. The current administration of exogenous insulin do not prevent glycemic fluctuations and the complications associated with this condition (hypoglycemia, ketoacidosis, cardiovascular diseases, stroke, kidney failure, blindness) represent a real risk for the patients [148, 153]. A normoglycemic insulin-independent state can be accomplished by whole pancreas transplantation. However, combination of a major surgery procedure (with a simultaneous kidney transplantation) and the constant immunosuppression that is required, makes pancreas transplantation not sustainable for the vast majority of diabetic patients. Transplantation of Langerhans islets through the portal vein represents a much less invasive procedure than pancreas transplantation. In this field, the application of the so-called Edmonton protocol allowed achieving insulin independence with a steroid-free immunosuppressive therapy in diabetic patients [122]. However, the need of a quite high mass of freshly prepared Langerhans islets isolated from two or more pancreas donors and the lack of long term insulin independence in most of the patients 5 years after transplantations are the main limitations of such protocol [123].

Recent progresses in the development of pancreatic β cells from human stem cells address the possibility of performing transplantation without immunosuppression [89]. However, as diabetes is an autoimmune disease, there is the need of immunoprotection of the insulin-producing β cells from the host immune system.

It is in this framework that the use of alginate gels as an immune-protective system has gained the interest of the scientific community [22]. In fact, the availability of pure and very well characterized alginate samples make this material an ideal choice for microencapsulation of Langerhans islets. Along this line, Lin and Sun have used alginate based capsules for immunoprotection of Langerhans islets and they have shown to restore normoglycemia in diabetic dogs [82]. Similarly, allotransplantation in humans of pancreatic islets has been successfully performed [13].

The preparation of alginate beads can be simply accomplished by dripping a sodium alginate solution (1–4% w/V) containing the biological material into a solution of gelling ions, Ca^{2+} and/or Ba^{2+} as an example. (The risks related to the toxicity of Ba^{2+} are highly reduced by the extremely high affinity of the ion for alginate, so as to prevent barium displacement and release in physiological conditions).The immediate gelation of the alginate solution will cause an immobilization of the cells within the 3D network. The initial size of the bead is determined basically by the size of the droplet and different encapsulation techniques can be exploited to control bead mean diameter and size distribution [39]. One of most known and diffused techniques for cell encapsulation makes use of an electrostatic generator which creates an electrostatic potential (1–20 kV) between the gelling bath and the needle feeding the alginate solution. Needle diameter size and electrostatic potential, which pulls the droplet from the needle to the gelling bath, both control the initial size of the bead which is also influenced by the viscosity of alginate and by the distance between the needle and the gelling bath [143]. This approach allows obtaining alginate beads with an initial size ranging from 1 mm to 200 μm.

Concepts like biocompatibility, stability and permeability of the microcapsules are fundamental aspects in the development of a stable system for immunoprotection of cells. All the general considerations that have been discussed in the previous sections concerning alginate gels apply in the context of the microbead. In fact, syneresis, stability, mechanical resistance, which strongly depend on alginate macromolecular characteristics (molecular weight and distribution) and composition and on the divalent cation used, determine the success or failure of cell encapsulation [95].

Permeability is of fundamental importance when alginate gels are used for immunoprotection purposes. Given the limited effect of the cross-linking ion, an increase in stability and a decrease in permeability of alginate beads can be achieved by converting them into capsules by means of an adsorption, on the bead surface, of a polycation such as poly-lysine [77, 82, 149, 151, 156] or polyornithine [13].

The negative charges on alginate and the positive ones on the polycation lead to the formation of strong complexes which do not dissolve in the presence of non-gelling ions or calcium chelators and are very stable in physiological conditions. The polycation-polyanion complex on the surface of the alginate hydrogel increases the stability of the capsule by both increasing the elastic retraction force and discharging the polymer network [77]. The composition of alginate plays a role also on the binding of the polycation, with poly-L-lysine binding to a higher extent to high-M than to high-G alginate.

The use of poly-L-lysine (PLL) offers numerous advantages as to stability and permeability of alginate capsules. However, PLL has been reported to be toxic towards many cell lines [17]. In addition, coating the alginate beads with PLL has been reported to lead to an overgrowth of fibrous capsule [23, 76, 145, 155] which impairs the oxygen and nutrient supply to encapsulated cells [98]. These adverse effects are to some extent mitigated when an additional coating of alginate is performed on the capsule to mask the positive charges of the PLL. However, also in these cases, capsular overgrowth has been observed [145]. In addition, PLL adsorbed on alginate capsules was shown to induce the inflammatory cytokines (TNF, Il-1β, Il-6), the anti-inflammatory mediators (IL-1RA, IL-10), the chemokines (IL-8, MIP-1α, MCP-1) and the growth factor VEGF in a complement dependent manner [119]. Overall, concern on the use of a polycation for the immunoprotection of Langerhans islets has arisen and the actual trend considers that the presence of PLL brings about more adverse effects than benefits and that capsules based on alginate-PLL-alginate systems have low possibility of success [46]. At variance, other reports showed in vivo promising results when PLL is replaced with poly(L-ornithine) [13, 46] or poly(methylene-co-guanidine) [159].

These considerations gain new interest in view of the recent results that reported a normalization of blood glucose for 1 year in NOD mouse using allogenic islets embedded in barium beads [41, 108]. In such conditions and based on the molecular dimension, the beads are permeable to IgG and to the cytokines IL-1β and TNF-α. Although the same results have not been reached in the case of xenografts, [109] still the question on the requirement as to the permeability of the beads to grant sufficient immunoprotection to Langerhans islets remains open.

13.7 Conclusions and Future Perspectives

Twenty years ago it was generally believed that there was not much more to discover about alginates, neither on molecular structure nor on their structure function correlation. From a chemical point of view, the alginate molecule looked very simple containing only the two monomer units M and G linked by the same 1→4 linkages. This simplification of its chemical structure has also lead commercial users of alginate to treat alginate as a commodity like, e.g., many of the cellulose derivatives. Moreover alginate ionic gel formation was correlated to the content of G-blocks only, while MM and MG blocks were supposed to have a minor impact on the gelling properties due to their lack of cooperative ion binding. The breakthrough in alginate genetics and biosynthesis in bacteria which enabled manipulation of the composition and sequential structure of the polymer, have given a wide range of new functional polymers and, more important, added more knowledge to the link between polymer nanostructure and macroscopic properties. By having compositionally homogeneous alginates, namely poly M, poly G, and polyMG, it has been possible to sort out the contribution for the various block structure to the gel properties. The role of polyMG both in gel formation and syneresis is a good example. Although the molecular (primary) structure and average monomer distribution of alginates is well described, there is a need for further understanding, particularly of the absolute length and distribution of the various blocks. Along this line, it has been recently discovered that algal alginates, irrespectively on their monomer composition, contain a fraction of extremely long G blocks. These long G blocks apparently act as enforcement bars of the network contributing significantly the mechanical strength and volume stability. This finding suggests that in the future one should look at alginate gels as a nanocomposite material where compensation between length and overall amount of G residues takes place. It is also becoming clear that all the sequences and their length in algal alginate play an important role for the mechanical properties to be reached and thus they should all be properly considered and their effect accounted for. Nowadays, framing the properties of alginate hydrogels based on the alginate monomer composition appears to be an oversimplification.

When microbial alginate and epimerase-modified alginate will have entered into the marked in the near future, the possibility of tailor-making alginate to diverse applications will be increased even further. A future trend of growing applications (which actually has already started) can be easily predicted: the exploitation of alginate is gradually shifting from low-tech applications, facing the increasing competition from cheap alternatives, to more advanced knowledge-based applications in the food, pharmaceutical, and biomedical areas.

References

1. Aarstad, O., Strand, B. L., Klepp-Andersen, L. M., and Skjåk-Bræk, G. (2013). Analysis of G-blocks distributions and their impact on gel properties of in vitro epimerized mannuronan, *Biomacromolecules*, **14**, pp. 3409-3416.
2. Aarstad, O. A., Tøndervik, A., Sletta, H., and Skjåk-Bræk, G. (2011). Alginate sequencing: an analysis of block distribution in alginates using specific alginate degrading enzymes, *Biomacromolecules*, **13**, pp. 106–116.
3. Andersen, I.-L., Skipnes, O., Smidsrød, O., Ostgaard, K., and Hemmer, P. C. (1977). Some biological functions of matrix components in benthic algae in relation to their chemistry and the composition of seawater, in *Cellulose chemistry and technology* American Chemical Society, 361–381.
4. Arnott, S., Bian, W., Chandrasekaran, R., and Manis, B. R. (2000). Lessons for today and tomorrow from yesterday - the structure of alginaic acid, *Fibre Diffr. Rev.*, **9**, pp. 44–51.
5. Atkins, E. D. T., Nieduszynski, I. A., Mackie, W., Parker, K. D., and Smolko, E. E. (1973). Structural components of alginic acid. I. The crystalline structure of poly-α-D-mannuronic acid. Results of X-ray diffraction and polarized infrared studies, *Biopolymers*, **12**, pp. 1865–1878.
6. Atkins, E. D. T., Nieduszynski, I. A., Mackie, W., Parker, K. D., and Smolko, E. E. (1973). Structural components of alginic acid. II. The crystalline structure of poly-α-L-guluronic acid. Results of X-ray diffraction and polarized infrared studies, *Biopolymers*, **12**, pp. 1879–1887.

7. Atsuki, K., and Tomoda, Y. (1926). Studies on seaweeds of Japan I. The chemical constituents of *Laminaria*, *J. Soc. Chem. Ind. Japan*, **29**, pp. 509–517.
8. Baumberger, T., Caroli, C., and Martina, D. (2006). Solvent control of crack dynamics in a reversible hydrogel, *Nat. Mater.*, **5**, pp. 552–555.
9. Baumberger, T., and Ronsin, O. (2010). Cooperative effect of stress and ion displacement on the dynamics of cross-link unzipping and rupture of alginate gels, *Biomacromolecules*, **11**, pp. 1571–1578.
10. Borgogna, M., Sjkåk-Bræk, G., Paoletti, S., and Donati, I. (2013). On the initial binding of alginate by calcium ions. The tilted egg-box hypothesis, *J. Phys. Chem.*, **117**, pp. 7277–7282.
11. Boyd, J., and Turvey, J. R. (1978). Structural studies of alginic acid, using a bacterial poly-α-L-guluronate lyase, *Carbohyd. Res.*, **66**, pp. 187–194.
12. Boyen, C., Kloareg, B., Polne-Fuller, M., and Gibor, A. (1990). Preparation of alginases from marine molluscs for protoplast isolation in brown algae, *Phycologia*, **29**, pp. 173–181.
13. Calafiore, R., Basta, G., Luca, G., Lemmi, A., Montanucci, M. P., Calabrese, G., Racanicchi, L., Mancuso, F., and Brunetti, P. (2006). Microencapsulated pancreatic islet allografts into nonimmunosuppressed patients with type 1 diabetes: first two cases, *Diabetes Care*, **29**, pp. 137–138.
14. Campa, C., Holtan, S., Nilsen, N., Bjerkan, T. M., Stokke, B. T., and Skjåk-Bræk, G. (2004). Biochemical analysis of the processive mechanism for epimerization of alginate by mannuronan C-5 epimerase AlgE4, *Biochem. J.*, **381**, pp. 155–164.
15. Chavagnat, F., Duez, C., Guinand, M., Potin, P., Barbeyron, T., Henrissat, B., Wallach, J., and Ghuysen, J. M. (1996). Cloning, sequencing and overexpression in *Escherichia coli* of the alginate-lyase-encoding *aly* gene of *Pseudomonas alginovora*: identification of three classes of alginate lyases, *Biochem. J.*, **319**, pp. 575–583.
16. Chitnis, C. E., and Ohman, D. E. (1990). Cloning of *Pseudomonas aeruginosa* algG, which controls alginate structure, *J. Bacteriol.*, **172**, pp. 2894–2900.
17. Choksakulnimitr, S., Masuda, S., Tokuda, H., Takakura, Y., and Hashida, M. (1995). In vitro cytotoxicity of macromolecules in different cell culture systems, *J. Control. Release*, **34**, pp. 233–241.
18. Clementi, F. (1997). Alginate production by *Azotobacter Vinelandii*, *Crit. Rev. Biotechnol.*, **17**, pp. 327–361.

19. Cote, G. L., and Krull, L. H. (1988). Characterization of the exocellular polysaccharides from *Azotobacter chroococcum, Carbohyd. Res.*, **181**, pp. 143–152.

20. Damron, F. H., and Goldberg, J. B. (2012). Proteolytic regulation of alginate overproduction in *Pseudomonas aeruginosa, Mol. Microbiol.*, **84**, pp. 595–607.

21. Darrabie, M. D., Kendall, W. F., and Opara, E. C. (2006). Effect of alginate composition and gelling cation on micro-bead swelling, *J. Microencapsul.*, **23**, pp. 29–37.

22. de Vos, P., Bucko, M., Gemeiner, P., Navrátil, M., Švitel, J., Faas, M., Strand, B. L., Skjåk-Bræk, G., Mørch, Y. A., Vikartovská, A., Lacík, I., Kolláriková, G., Orive, G., Poncelet, D., Pedraz, J. L., and Ansorge-Schumacher, M. B. (2009). Multiscale requirements for bioencapsulation in medicine and biotechnology, *Biomaterials*, **30**, pp. 2559–2570.

23. de Vos, P., De Haan, B., and Van Schilfgaarde, R. (1997). Effect of the alginate composition on the biocompatibility of alginate-polylysine microcapsules, *Biomaterials*, **18**, pp. 273–278.

24. de Vos, P., Faas, M. M., Strand, B., and Calafiore, R. (2006). Alginate-based microcapsules for immunoisolation of pancreatic islets, *Biomaterials*, **27**, pp. 5603–5617.

25. Dentini, M., Rinaldi, G., Risica, D., Barbetta, A., and Skjåk-Bræk, G. (2005). Comparative studies on solution characteristics of mannuronan epimerized by C-5 epimerases, *Carbohyd. Polym.*, **59**, pp. 489–499.

26. Dinkelaar, J., van den Bos, L. J., Hogendorf, W. F. J., Lodder, G., Overkleeft, H. S., Codée, J. D. C., and van der Marel, G. A. (2008). Stereoselective synthesis of L-guluronic acid alginates, *Chem. Eur. J.*, **14**, pp. 9400–9411.

27. Doi, M., and Kuzuu, N. Y. (1980). Nonlinear elasticity of rodlike macromolecules in condensed state, *J. Polym. Sci. Polym. Phys. Ed.*, **18**, pp. 409–419.

28. Donati, I., Cesaro, A., and Paoletti, S. (2006). Specific interactions versus counterion condensation. 1. Nongelling ions/polyuronate systems, *Biomacromolecules*, **7**, pp. 281–287.

29. Donati, I., and Paoletti, S. (2009). Material properties of alginates, in *Alginates: Biology and Applications* Rhem, B. H. A., Springer-Verlag, Berlin Heidelberg, pp. 1–53.

30. Donati, I., Stredanska, S., Silvestrini, G., Vetere, A., Marcon, P., Marsich, E., Mozetic, P., Gamini, A., Paoletti, S., and Vittur, F. (2005). The

aggregation of pig articular chndrocyte and synthesis of extracellular matrix by lactose-modified chitosan, *Biomaterials*, **26**, pp. 987–998.

31. Donati, I., Asaro, F., and Paoletti, S. (2009). Experimental evidence of counterion affinity in alginates: the case of nongelling ion Mg^{2+}, *J. Phys. Chem. B*, **113**, pp. 12877–12886.
32. Donati, I., Gamini, A., Skjåk-Bræk, G., Vetere, A., Campa, C., Coslovi, A., and Paoletti, S. (2003). Determination of the diadic composition of alginate by means of circular dichroism: a fast and accurate improved method, *Carbohyd. Res.*, **338**, pp. 1139–1142.
33. Donati, I., Holtan, S., Mørch, Y. A., Borgogna, M., Dentini, M., and Skjåk-Bræk, G. (2005). New hypothesis on the role of alternating sequences in calcium–alginate gels, *Biomacromolecules*, **6**, pp. 1031–1040.
34. Donati, I., Mørch, Y. A., Strand, B. L., Skjåk-Bræk, G., and Paoletti, S. (2009). Effect of elongation of alternating sequences on swelling behavior and large deformation properties of natural alginate gels, *J. Phys. Chem. B*, **113**, pp. 12916–12922.
35. Draget, H. I., Strand, B. L., Hartmann, M., Valla, S., Smidsrød, O., and Skjåk-Bræk, G. (2000). Ionic and acid gel formation of epimerised alginates; the effect of AlgE4, *Int. J. Biol. Macromol.*, **27**, pp. 117–122.
36. Draget, K. I., Gåserød, O., Aune, I., Andersen, P. O., Storbakken, B., Stokke, B. T., and Smidsrød, O. (2001). Effects of molecular weight and elastic segment flexibility on syneresis in Ca-alginate gels, *Food Hydrocolloid.*, **15**, pp. 485–490.
37. Draget, K. I., Østagaard, K., and Smidsrød, O. (1991). Homogeneous alginate gels: a technical approach, *Carbohyd. Polym.*, **14**, pp. 159–178.
38. Draget, K. I., Smidsrød, O., and Skjåk-Bræk, G. (2005). Alginates from algae, in *Polysaccharides and Polyamides in the Food Industry. Properties, Production, and Patents.* Steinbüchel, A., and Rhee, S. K. eds, Wiley-VCH Verlag GmbH & Co., Weinheim, pp. 1–30.
39. Dulieu, C., Poncelet, D., and Neufeld, R. J. (1999). Encapsulation and immobilization techniques, in *Cell Encapsulation Technology and Therapeutics* Kuhtreiber, W. M., Lanza, R. P., and Chick, W. eds, Birkhauser Boston, New York, pp. 3–17.
40. Dunne, W. M., and Buckmire, F. L. (1985). Partial purification and characterization of a polymannuronic acid depolymerase produced by a mucoid strain of *Pseudomonas aeruginosa* isolated from a patient with cystic fibrosis, *Appl. Environ. Microbiol*, **50**, pp. 562–567.

41. Duvivier-Kali, V. F., Omer, A., Parent, R. J., O'Neil, J. J., and Weir, G. C. (2001). Complete protection of islets against allorejection and autoimmunity by a simple barium-alginate membrane, *Diabetes*, **50**, pp. 1698–1705.
42. Ertesvåg, H., Doseth, B., Larsen, B., Skjåk-Bræk, G., and Valla, S. (1994). Cloning and expression of an *Azotobacter vinelandii* mannuronan C-5-epimerase gene, *J. Bacteriol.*, **176**, pp. 2846–2853.
43. Ertesvåg, H., Valla, S., and Skjåk-Bræk, G. (2009). Enzymatic alginate modification, in *Alginates: Biology and Applications* Rehm, B. H. A. ed., Springer, Verlag Berlin Heidelberg, pp. 95–115.
44. Ertesvåg, H., Høidal, H. K., Hals, I. K., Rian, A., Doseth, B., and Valla, S. (1995). A family of modular type mannuronan C-5-epimerase genes controls alginate structure in *Azotobacter vinelandii*, *Mol. Microbiol.*, **16**, pp. 719–731.
45. Ertesvåg, H., and Valla, S. (1998). Biosynthesis and applications of alginates, *Polym. Degrad. Stabil.*, **59**, pp. 85–91.
46. Espevik, T., Rokstad, A. M., Kulseng, B., Strand, B. L., and Skjåk-Bræk, G. (2009). Mechanisms of the host immune response to alginate microcapsules, in *The Bioartificial Pancreas and Other Biohybrid Therapies* Hallé, J.-P., de Vos, P., and Rosenberg, L. eds, Research Science Post, 279–290.
47. Fischer, F. G., and Dörfel, H. (1955). Die Polyuronsäuren der Braunalgen (Kohlenhydrate der Algen I), *H. -S. Z. Physiol. Chem.*, **302**, pp. 186–203.
48. Gacesa, P. (1987). Alginate-modifying enzymes: A proposed unified mechanism of action for the lyases and epimerases, *FEBS Lett.*, **212**, pp. 199–202.
49. Gåserød, O., Sannes, A., and Skjåk-Bræk, G. (1999). Microcapsules of alginate–chitosan. II. a study of capsule stability and permeability, *Biomaterials*, **20**, pp. 773–783.
50. Gimmestad, M., Sletta, H., Ertesvåg, H., Bakkevig, K., Jain, S., Suh, S. J., Skjåk-Bræk, G., Ellingsen, T. E., Ohman, D. E., and Valla, S. (2003). The *Pseudomonas fluorescens* AlgG protein, but not its mannuronan C-5-epimerase activity, is needed for alginate polymer formation, *J. Bacteriol.*, **185**, pp. 3515–3523.
51. Gorin, P. A., and Spencer, J. F. T. (1966). Exocellular alginic acid from *Azotobacter vinelandii*, *Can. J. Chem.*, **44**, pp. 993–998.
52. Govan, J. R. W., and Fyfe, J. F. M. (1981). Isolation of alginate-producing mutants of *Pseudomonas fluorescens*, *Pseudomonas putida*, and *Pseudomonas mendocina*, *J. Gen. Microbiol.*, **125**, pp. 217–220.

53. Govan, J. R. W., and Harris, G. S. (1986). *Pseudomonas aeruginosa* and cystic fibrosis: unusual bacterial adaptation and pathogenesis, *Microbiol. Sci.*, **3**, pp. 302–308.

54. Grant, G. T., Morris, E. R., Rees, D. A., Smith, P. J. C., and Thom, D. (1973). Biological interactions between polysaccharides and divalent cations: the egg-box model, *FEBS Lett.*, **32**, pp. 195–198.

55. Grasdalen, H., and Kvam, B. J. (1986). Sodium-23 NMR in aqueous solutions of sodium polyuronates. Counterion binding and conformational conditions, *Macromolecules*, **19**, pp. 1913–1920.

56. Grasdalen, H., Larsen, B., and Smisrød, O. (1981). ^{13}C-n.m.r. studies of monomeric composition and sequence in alginate, *Carbohyd. Res.*, **89**, pp. 179–191.

57. Groboillot, A., Boadi, D. K., Poncelet, D., and Neufeld, R. J. (1994). Immobilization of cells for application in the food industry, *Crit. Rev. Biotechnol.*, **14**, pp. 75–107.

58. Hartmann, M., Holm, O. B., Johansen, G. A. B., Skjåk-Bræk, G., and Stokke, B. T. (2002). Mode of action of recombinant *Azotobacter vinelandii* mannuronan C-5 epimerases AlgE2 and AlgE4, *Biopolymers*, **63**, pp. 77–88.

59. Haug, A., Larsen, B., and Smidsrød, O. (1966). A study of the constitution of alginic acid by partial acid hydrolysis, *Acta Chem. Scand.*, **20**, pp. 183–190.

60. Haug, A., Larsen, B., and Smidsrød, O. (1967). Studies on the sequence of uronic acid residues in alginic acid, *Acta Chem. Scand.*, **21**, pp. 691–704.

61. Haug, A., and Smidsrød, O. (1965). Fractionation of alginate by precipitation with calcium and magnesium ions, *Acta Chem. Scand.*, **19**, pp. 1221–1226.

62. Haug, A., and Smidsrød, O. (1967). Strontium-calcium selectivity of alginates, *Nature*, **215**, pp. 757–757.

63. Haug, A., and Smidsrød, O. (1970). Selectivity of some anionic polymers for divalent metal ions, *Acta Chem. Scand.*, **24**, pp. 843–854.

64. Haug, A., and Larsen, B. (1969). Biosynthesis of alginate. Epimerisation of D-mannuronic to L-guluronic acid residues in the polymer chain, *BBA-Gen. Subjects*, **192**, pp. 557–559.

65. Haugen, F., Kortner, F., and Larsen, B. (1990). Kinetics and specificity of alginate lyases: Part I, a case study, *Carbohyd. Res.*, **198**, pp. 101–109.

66. Heinemann, M., Meinberg, H., Büchs, J., Koß, H. J., and Ansorge-Schumacher, M. B. (2005). Method for quantitative determination of spatial polymer distribution in alginate beads using Raman spectroscopy, *Appl. Spectrosc.*, **59**, pp. 280–285.

67. Heyraud, A., Colin-Morel, P., Girond, S., Richard, C., and Kloareg, B. (1996). HPLC analysis of saturated or unsaturated oligoguluronates and oligomannuronates. Application to the determination of the action pattern of *Haliotis tuberculata* alginate lyase, *Carbohyd. Res.*, **291**, pp. 115–126.

68. Higgs, P. G., and Ball, R. C. (1989). Some ideas concerning the elasticity of biopolymer networks, *Macromolecules*, **22**, pp. 2432–2437.

69. Høidal, H. K., Ertesvåg, H., Skjåk-Bræk, G., Stokke, B. T., and Valla, S. (1999). The recombinant *Azotobacter vinelandii* mannuronan C-5-epimerase AlgE4 epimerizes alginate by a nonrandom attack mechanism, *J. Biol. Chem.*, **274**, pp. 12316–12322.

70. Holtan, S., Bruheim, P., and Skjåk-Bræk, G. (2006). Mode of action and subsite studies of the guluronan block-forming mannuronan C-5 epimerases AlgE1 and AlgE6, *Biochem. J.*, **395**, pp. 319–329.

71. Hortelano, G., Al-Hendy, A., Ofosu, F. A., and Chang, P. L. (1996). Delivery of human factor IX in mice by encapsulated recombinant myoblasts: a novel approach towards allogeneic gene therapy of hemophilia B, *Blood*, **87**, pp. 5095–5103.

72. Hunt, N., and Grover, L. (2010). Cell encapsulation using biopolymer gels for regenerative medicine, *Biotechnol. Lett.*, **32**, pp. 733–742.

73. Indergaard, M., and Skjåk-Bræk, G. (1987). Characteristics of alginate from *Laminaria digitata* cultivated in a high-phosphate environment, *Hydrobiologia*, **151–152**, pp. 541–549.

74. Kennedy, L., Mcdowell, K., and Sutherland, I. W. (1992). Alginases from *Azotobacter* species, *J. Gen. Microbiol.*, **138**, pp. 2465–2471.

75. Khattak, S. F., Chin, K., Bhatia, S. R., and Roberts, S. C. (2007). Enhancing oxygen tension and cellular function in alginate cell encapsulation devices through the use of perfluorocarbons, *Biotechnol. Bioeng.*, **96**, pp. 156–166.

76. King, A., Sandler, S., and Andersson, A. (2001). The effect of host factors and capsule composition on the cellular overgrowth on implanted alginate capsules, *J. Biomed. Mater. Res.*, **57**, pp. 374–383.

77. King, A., Strand, B., Rokstad, A. M., Kulseng, B., Andersson, A., Skjåk-Bræk, G., and Sandler, S. (2003). Improvement of the biocompatibility of alginate/poly-L-lysine/alginate microcapsules by the use

of epimerized alginate as a coating, *J. Biomed. Mater. Res. A*, **64A**, pp. 533–539.

78. Kohn, R., and Larsen, B. (1972). Preparation of water-soluble polyuronic acids and their calcium salts, and the determination of calcium ion activity in relation to the degree of polymerization, *Acta Chem. Scand.*, **26**, pp. 2455–2468.
79. Kong, H. J., Smith, M. K., and Mooney, D. J. (2003). Designing alginate hydrogels to maintain viability of immobilized cells, *Biomaterials*, **24**, pp. 4023–4029.
80. Larsen, B., Smidsrød, O., Painter, T., and Haug, A. (1970). Calculation of the nearest-neighbour frequencies in fragments of alginate from the yields of free monomer after partial hydrolysis, *Acta Chem. Scand.*, **24**, pp. 726–728.
81. Leid, J. G., Willson, C. J., Shirtliff, M. E., Hassett, D. J., Parsek, M. R., and Jeffers, A. K. (2005). The exopolysaccharide alginate protects *Pseudomonas aeruginosa* biofilm bacteria from IFN-γ-mediated macrophage killing, *J. Immunol.*, **175**, pp. 7512–7518.
82. Lim, F., and Sun, A. M. (1980). Microencapsulated islets as bioartificial endocrine pancreas, *Science*, **210**, pp. 908–910.
83. Lin, T. Y., and Hassid, W. Z. (1966). Isolation of guanosine diphosphate uronic acids from a marine brown alga, *Fucus gardneri Silva*, *J. Biol. Chem.*, **241**, pp. 3283–3293.
84. Lin, T. Y., and Hassid, W. Z. (1966). Pathway of alginic acid synthesis in the marine brown alga, *Fucus gardneri Silva*, *J. Biol. Chem.*, **241**, pp. 5284–5297.
85. Lindahl, U., Bäckström, G., Malmström, A., and Fransson, L.-Å. (1972). Biosynthesis of L-iduronic acid in heparin: epimerization of D-glucuronic acid on the polymer level, *Biochem. Bioph. Res. Co.*, **46**, pp. 985–991.
86. Linker, A., and Jones, R. S. (1964). A polysaccharide resembling alginic acid from a *Pseudomonas* micro-organism, *Nature*, **204**, pp. 187–188.
87. Liu, H.-W., Ofosu, F. A., and Chang, P. L. (1993). Expression of human factor IX by microencapsulated recombinant fibroblasts, *Hum. Gene Ther.*, **4**, pp. 291–301.
88. Madgwick, J., Haug, A., and Larsen, B. (1973). Alginate lyase in the brown alga *Laminaria digitata* (Huds.) Lamour, *Acta Chem. Scand.*, **27**, pp. 711–712.
89. Madsen, O. D., and Serup, P. (2006). Towards cell therapy for diabetes, *Nat. Biotechnol.*, **24**, pp. 1481–1483.

90. Manning, G. S. (1969). Limiting laws and counterion condensation in polyelectrolyte solutions I. Colligative properties, *J. Chem. Phys.*, **51**, pp. 924–933.

91. Manning, G. S. (1978). Limiting laws and counterion condensation in polyelectrolyte solutions: V. Further development of the chemical model, *Biophys. Chem.*, **9**, pp. 65–70.

92. Manning, G. S., and Paoletti, S. (1987). Theoretical approach to ionic polysaccharide solutions., in *Industrial polysaccharides: the impact of biotechnology and advanced methodologies* Stivala, S. S., Crescenzi, V., and Dea, I. C. M. eds., Gordon and Breach Science, New York, pp. 305–324.

93. Marcon, P., Marsich, E., Vetere, A., Mozetic, P., Campa, C., Donati, I., Vittur, F., Gamini, A., and Paoletti, S. (2005). The role of galectin-1 in the interaction between chondrocytes and a lactose-modified chitosan, *Biomaterials*, **26**, pp. 4975–4984.

94. Marsich, E., Borgogna, M., Donati, I., Mozetic, P., Strand, B. L., Gomez Salvador, S., Vittur, F., and Paoletti, S. (2008). Alginate/lactose-modified chitosan hydrogels: a bioactive biomaterial for chondrocyte encapsulation, *J. Biomed. Mater. Res.*, **84A**, pp. 364–376.

95. Martinsen, A., Skjåk-Bræk, G., and Smidsrød, O. (1989). Alginate as immobilization material: I. Correlation between chemical and physical properties of alginate gel beads, *Biotechnol. Bioeng.*, **33**, pp. 79–89.

96. Mathis, D., Vence, L., and Benoist, C. (2001). β-Cell death during progression to diabetes, *Nature*, **414**, pp. 792–798.

97. May, T. B., and Chakrabarty, A. M. (1994). *Pseudomonas aeruginosa*: genes and enzymes of alginate synthesis, *Trends Microbiol.*, **2**, pp. 151–157.

98. Mazaheri, R., Atkinson, P., Stiller, C., Dupré, J., Vose, J., and O'Shea, G. (1991). Transplantation of encapsulated allogeneic islets into diabetic Bb/W rats: effects of immunosuppression, *Transplantation*, **51**, pp. 750–753.

99. Mørch, Y. A., Donati, I., Strand, B. L., and Skjåk-Bræk, G. (2007). Molecular engineering as an approach to design new functional properties of alginate, *Biomacromolecules*, **8**, pp. 2809–2814.

100. Mørch, Y. A., Donati, I., Strand, B. L., and Skjåk-Bræk, G. (2006). Effect of Ca^{2+}, Ba^{2+}, and Sr^{2+} on alginate microbeads, *Biomacromolecules*, **7**, pp. 1471–1480.

101. Mørch, Y. A., Holtan, S., Donati, I., Strand, B. L., and Skjåk-Bræk, G. (2008). Mechanical properties of C-5 epimerized alginates, *Biomacromolecules*, **9**, pp. 2360–2368.

102. Moresi, M., and Bruno, M. (2007). Characterisation of alginate gels using quasi-static and dynamic methods, *J. Food Eng.*, **82**, pp. 298–309.

103. Morris, E. R., Rees, D. A., and Thom, D. (1980). Characterisation of alginate composition and block-structure by circular dichroism, *Carbohyd. Res.*, **81**, pp. 305–314.

104. Morris, E. R., Rees, D. A., Thom, D., and Boyd, J. (1978). Chiroptical and stoichiometric evidence of a specific, primary dimerisation process in alginate gelation, *Carbohyd. Res.*, **66**, pp. 145–154.

105. Muramatsu, T., Hirose, S., and Katayose, M. (1977). Isolation and properties of alginate lyase from the mid-gut gland of wreath shell *Turbo cornutus*, *Agr. Biol. Chem. Tokyo*, **41**, pp. 1939–1946.

106. Murua, A., Portero, A., Orive, G., Hernáíndez, R. M., de Castro, M., and Pedraz, J. L. (2008). Cell microencapsulation technology: Towards clinical application, *J. Control. Release*, **132**, pp. 76–83.

107. Nyvall, P., Corre, E., Boisset, C., Barbeyron, T., Rousvoal, S., Scornet, D., Kloareg, B., and Boyen, C. (2003). Characterization of mannuronan C-5-epimerase genes from the brown alga *Laminaria digitata*, *Plant Physiol.*, **133**, pp. 726–735.

108. Omer, A., Duvivier-Kali, V., Fernandes, J., Tchipashvili, V., Colton, C. K., and Weir, G. C. (2005). Long-term normoglycemia in rats receiving transplants with encapsulated islets, *Transplantation*, **79**, pp. 52–58.

109. Omer, A., Keegan, M., Czismadia, E., de Vos, P., Van Rooijen, N., Bonner-Weir, S., and Weir, G. C. (2003). Macrophage depletion improves survival of porcine neonatal pancreatic cell clusters contained in alginate macrocapsules transplanted into rats, *Xenotransplantation*, **10**, pp. 240–251.

110. Onsøyen, E. (1996). Commercial applications of alginates, *Carbohyd. Eur.*, **14**, pp. 26–31.

111. Øyaas, J., Storrø, I., Svendsen, H., and Levine, D. W. (1995). The effective diffusion coefficient and the distribution constant for small molecules in calcium-alginate gel beads, *Biotechnol. Bioeng.*, **47**, pp. 492–500.

112. Painter, T., Smidsrød, O., Larsen, B., and Haug, A. (1968). A computer study of the changes in composition-distribution occurring during random depolymerization of a binary linear heteropolysaccharide, *Acta Chem. Scand.*, **22**, pp. 1637–1648.

113. Painter, T. J. (1983) Algal polysaccharides, in *The polysaccharides* Aspinall, G. O., Academic Press, New York, pp. 195–285.

114. Paoletti, S., Cesàro, A., Delben, F., Crescenzi, V., and Rizzo, R. (1985). Polyelectrolyte aspects of conformational transitions and interchain interactions in ionic polysaccharide solutions: comparison of theory and microcalorimetric data, in *Microdomains in polymer colutions* Dubin, P. L. ed, Plenum Press, New York, pp. 159–189.

115. Pindar, D. F., and Bucke, C. (1975). The biosynthesis of alginic acid by *Azotobacter vinelandii, Biochem. J.*, **152**, pp. 617–622.

116. Ponnuraj, K., and Jedrzejas, M. J. (2000). Mechanism of hyaluronan binding and degradation: structure of *Streptococcus pneumoniae* hyaluronate lyase in complex with hyaluronic acid disaccharide at 1.7 Å resolution, *J. Mol. Biol.*, **299**, pp. 885–895.

117. Rao, M. N., Kembhavi, A. A., and Pant, A. (1996). Role of lysine, tryptophan and calcium in the β-elimination activity of a low-molecular-mass pectate lyase from *Fusarium moniliformae, Biochem. J.*, **319**, pp. 159–164.

118. Read, T. A., Farhadi, M., Bjerkvig, R., Olsen, B. R., Rokstad, A. M., Huszthy, P. C., and Vajkoczy, P. (2001). Intravital microscopy reveals novel antivascular and antitumor effects of endostatin delivered locally by alginate-encapsulated cells, *Cancer Res.*, **61**, pp. 6830–6837.

119. Rokstad, A. M., Brekke, O. L., Steinkjer, B., Ryan, L., Kolláriková, G., Strand, B. L., Skjåk-Bræk, G., Lambris, J. D., Lacík, I., Mollnes, T. E., and Espevik, T. (2013). The induction of cytokines by polycation containing microspheres by a complement dependent mechanism, *Biomaterials*, **34**, pp. 621–630.

120. Schmidt, E., and Vocke, F. (1926). Zur kenntnis der polyglykuronsäuren (I.), *Ber. Dtsch. Chem. Ges.*, **59**, pp. 1585–1588.

121. Schwarzmann, S., and Boring, J. R. (1971). Antiphagocytic effect of slime from a mucoid Strain of *Pseudomonas aeruginosa, Infect. Immun.*, **3**, pp. 762–767.

122. Shapiro, A. M. J., Lakey, J. R. T., Ryan, E. A., Korbutt, G. S., Toth, E., Warnock, G. L., Kneteman, N. M., and Rajotte, R. V. (2000). Islet transplantation in seven patients with type 1 diabetes mellitus using a glucocorticoid-free immunosuppressive regimen, *New Engl. J. Med.*, **343**, pp. 230–238.

123. Shapiro, A. M. J., Ricordi, C., Hering, B. J., Auchincloss, H., Lindblad, R., Robertson, R. P., Secchi, A., Brendel, M. D., Berney, T., Brennan,

D. C., Cagliero, E., Alejandro, R., Ryan, E. A., DiMercurio, B., Morel, P., Polonsky, K. S., Reems, J. A., Bretzel, R. G., Bertuzzi, F., Froud, T., Kandaswamy, R., Sutherland, D. E. R., Eisenbarth, G., Segal, M., Preiksaitis, J., Korbutt, G. S., Barton, F. B., Viviano, L., Seyfert-Margolis, V., Bluestone, J., and Lakey, J. R. T. (2006). International trial of the Edmonton protocol for islet transplantation, *New Engl. J. Med.*, **355**, pp. 1318–1330.

124. Sherbrock-Cox, V., Russell, N. J., and Gacesa, P. (1984). The purification and chemical characterisation of the alginate present in extracellular material produced by mucoid strains of *Pseudomonas aeruginosa*, *Carbohyd. Res.*, **135**, pp. 147–154.

125. Sikorski, P., Mo, F., Skjåk-Bræk, G., and Stokke, B. T. (2007). Evidence for egg-box-compatible interactions in calcium–alginate gels from fiber X-ray diffraction, *Biomacromolecules*, **8**, pp. 2098–2103.

126. Skjåk-Bræk, G., Grasdalen, H., and Smidsrød, O. (1989). Inhomogeneous polysaccharide ionic gels, *Carbohyd. Polym.*, **10**, pp. 31–54.

127. Skjåk-Bræk, G., Grasdalen, H., and Larsen, B. (1986). Monomer sequence and acetylation pattern in some bacterial alginates, *Carbohyd. Res.*, **154**, pp. 239–250.

128. Skjåk-Bræk, G., Smidsrød, O., and Larsen, B. (1986). Tailoring of alginates by enzymatic modification in vitro, *Int. J. Biol. Macromol.*, **8**, pp. 330–336.

129. Skjåk-Bræk, G., Zanetti, F., and Paoletti, S. (1989). Effect of acetylation on some solution and gelling properties of alginates, *Carbohyd. Res.*, **185**, pp. 131–138.

130. Sletmoen, M., Skjåk-Bræk, G., and Stokke, B. T. (2005). Mapping enzymatic functionalities of mannuronan C-5 epimerases and their modular units by dynamic force spectroscopy, *Carbohyd. Res.*, **340**, pp. 2782–2795.

131. Smidsrød, O. (1974). Molecular basis of some physical properties of alginates in the gel state, *Faraday Discuss. Chem. Soc.*, **57**, pp. 263–274.

132. Smidsrød, O., and Skjåk-Bræk, G. (1990). Alginate as immobilization matrix for cells, *Trends Biotechnol.*, **8**, pp. 71–78.

133. Smidsrød, O., and Whittington, S. G. (1969). Monte Carlo investigation of chemical inhomogeneity in polymers, *Macromolecules*, **2**, pp. 42–44.

134. Smith, A. M., Harris, J. J., Shelton, R. M., and Perrie, Y. (2007). 3D culture of bone-derived cells immobilised in alginate following light-triggered gelation, *J. Control. Release*, **119**, pp. 94–101.

135. Soon-Shiong, P., Heintz, R. E., Merideth, N., Yao, Q. X., Yao, Z., Zheng, T., Murphy, M., Moloney, M. K., Schmehl, M., Harris, M., Mendez, R., Mendez, R., and Sandford, P. A. (1994). Insulin independence in a type 1 diabetic patient after encapsulated islet transplantation, *The Lancet*, **343**, pp. 950–951.

136. Stanford, E. C. C. (1883). On algin: a new substance obtained from some of the commoner species of marine algae, *Chem. News*, **47**, pp. 254–257.

137. Steginsky, C. A., Beale, J. M., Floss, H. G., and Mayer, R. M. (1992). Structural determination of alginic acid and the effects of calcium binding as determined by high-field n.m.r, *Carbohyd. Res.*, **225**, pp. 11–26.

138. Stein, S., Strauss, E., and Bosco, J. (2013). Advances in the surgical management of articular cartilage defects: autologous chondrocyte implantation techniques in the pipeline, *Cartilage*, **4**, pp. 12–19.

139. Stokke, B. T., Draget, K. I., Smidsrød, O., Yuguchi, Y., Urakawa, H., and Kajiwara, K. (2000). Small-Angle X-ray Scattering and rheological characterization of alginate gels. 1. Ca—Alginate gels, *Macromolecules*, **33**, pp. 1853–1863.

140. Stokke, B. T., Smidsrød, O., Bruheim, P., and Skjåk-Bræk, G. (1991). Distribution of uronate residues in alginate chians in relation to alginate gelling properties, *Macromolecules*, **24**, pp. 4637–4645.

141. Stokke, B. T., Smidsrød, O., Zanetti, F., Strand, W., and Skjåk-Bræk, G. (1993). Distribution of uronate residues in alginate chains in relation to alginate gelling properties. 2. Enrichment of β-D-mannuronic acid and depletion of α-L-guluronic acid in sol fraction, *Carbohyd. Polym.*, **21**, pp. 39–46.

142. Stokke, B. T., Smidsrød, O., and Brant, D. A. (1993). Predicted influence of monomer sequence distribution and acetylation on the extension of naturally occurring alginates, *Carbohyd. Polym.*, **22**, pp. 57–66.

143. Strand, B. L., Gåserød, O., Kulseng, B., Espevik, T., and Skjåk-Bræk, G. (2002). Alginate-polylysine-alginate microcapsules: effect of size reduction on capsule properties, *J. Microencapsul.*, **19**, pp. 615–630.

144. Strand, B. L., Mørch, Y. A., Espevik, T., and Skjåk-Bræk, G. (2003). Visualization of alginate-poly-L-lysine-alginate microcapsules by confocal laser scanning microscopy, *Biotechnol. Bioeng.*, **82**, pp. 386–394.

145. Strand, B. L., Ryan, L., Veld, P. I., Kulseng, B., Rokstad, A. M., Skjåk-Bræk, G., and Espevik, T. (2001). Poly-L-lysine induces fibrosis on alginate

microcapsules via the induction of cytokines, *Cell Transplant.*, **10**, pp. 263–275.

146. Strand, B. L., Mørch, Y. A., Syvertsen, K. R., Espevik, T., and Skjåk-Bræk, G. (2003). Microcapsules made by enzymatically tailored alginate, *J. Biomed. Mater. Res. A*, **64A**, pp. 540–550.
147. Svanem, B. I. G., Skjåk-Bræk, G., Ertesvåg, H., and Valla, S. (1999). Cloning and expression of three new *Azotobacter vinelandii* genes closely related to a previously described gene family encoding mannuronan C-5-epimerases, *J. Bacteriol.*, **181**, pp. 68–77.
148. Tao, S. L., and Desai, T. A. (2003). Microfabricated drug delivery systems: from particles to pores, *Adv. Drug Deliver. Rev.*, **55**, pp. 315–328.
149. Thu, B., Bruheim, P., Espevik, T., Smidsrød, O., Soon-Shiong, P., and Skjåk-Bræk, G. (1996). Alginate polycation microcapsules: I. Interaction between alginate and polycation, *Biomaterials*, **17**, pp. 1031–1040.
150. Thu, B., Gåserød, O., Paus, D., Mikkelsen, A., Skjåk-Bræk., G., Toffanin, R., Vittur, F., and Rizzo, R. (2000). Inhomogeneous alginate gel spheres: an assessment of the polymer gradients by synchrotron radiation-induced X-ray emission, magnetic resonance microimaging, and mathematical modeling, *Biopolymers*, **53**, pp. 60–71.
151. Thu, B., Bruheim, P., Espevik, T., Smidsrød, O., Soon-Shiong, P., and Skjåk-Bræk, G. (1996). Alginate polycation microcapsules: II. Some functional properties, *Biomaterials*, **17**, pp. 1069–1079.
152. Tielen, P., Strathmann, M., Jaeger, K. E., Flemming, H. C., and Wingender, J. (2005). Alginate acetylation influences initial surface colonization by mucoid *Pseudomonas aeruginosa*, *Microbiol. Res.*, **160**, pp. 165–176.
153. Tsui, H., Razavi, R., Chan, Y., Yantha, J., and Dosch, H. M. (2007). Sensing autoimmunity in type 1 diabetes, *Trends Mol. Med.*, **13**, pp. 405–413.
154. Turco, G., Donati, I., Grassi, M., Marchioli, G., Lapasin, R., and Paoletti, S. (2011). Mechanical spectroscopy and relaxometry on alginate hydrogels: a comparative analysis for structural characterization and network mesh size determination, *Biomacromolecules*, **12**, pp. 1272–1282.
155. Vandenbossche, G. M. R., Bracke, M. E., Cuvelier, C. A., Bortier, H. E., Mareel, M. M., and Remon, J. P. (1993). Host reaction against alginate-polylysine microcapsules containing living cells, *J. Pharm. Pharmacol.*, **45**, pp. 121–125.

156. Vandenbossche, G. M. R., Van Oostveldt, P., Demeester, J., and Remon, J. P. (1993). The molecular weight cut-off of microcapsules is determined by the reaction between alginate and polylysine, *Biotechnol. Bioeng.*, **42**, pp. 381–386.
157. Veliky, I. A., and Williams, R. E. (1981). The production of ethanol by *Saccharomyces cerevisiae* immobilized in polycation-stabilized calcium alginate gels, *Biotechnol. Lett.*, **3**, pp. 275–280.
158. Walvoort, M. T. C., van den Elst, H., Plante, O. J., Krock, L., Seeberger, P. H., Overkleeft, H. S., van der Marel, G. A., and Codée, J. D. C. (2012). Automated solid-Phase synthesis of β-mannuronic acid alginates, *Angew. Chem. Int. Ed.*, **51**, pp. 4393–4396.
159. Wang, T., Lacík, I., Brissová, M., Anilkumar, A. V., Prokop, A., Hunkeler, D., Green, R., Shahrokhi, K., and Powers, A. C. (1997). An encapsulation system for the immunoisolation of pancreatic islets, *Nat. Biotechnol.*, **15**, pp. 358–362.
160. Wong, T. Y., Preston, L. A., and Schiller, N. L. (2000). Alginate lyase: review of major sources and enzyme characteristics, structure-function analysis, biological roles, and applications, *Annu. Rev. Microbiol.*, **54**, pp. 289–340.
161. Yang, L., Hengzhuang, W., Wu, H., Damkiær, S., Jochumsen, N., Song, Z., Givskov, M., Høiby, N., and Molin, S. (2012). Polysaccharides serve as scaffold of biofilms formed by mucoid *Pseudomonas aeruginosa*, *FEMS Immunol. Med. Mic.*, **65**, pp. 366–376.
162. Zielinski, N. A., Roychoudhury, S., and Chakrabarty, A. M. (1994) Regulation of alginate gene expression in *Pseudomonas aeruginosa*, in *Methods in Enzymology-Bacterial Pathogenesis Part A: Identification and Regulation of Virulence Factors* Virginia, L. Clark, Academic Press, 493–502.

Chapter 14

Polysaccharide Hydrogels: The Present and the Future

Yeon Hee Yun,[a] Byung Kook Lee,[a] John Garner,[b] and Kinam Park[a]

[a] *Departments of Biomedical Engineering and Pharmaceutics, Purdue University, West Lafayette, IN 47907, USA*
[b] *Akina, Inc., West Lafayette, IN 47906, USA*
kpark@purdue.edu

14.1 Polysaccharides

Nature chose polysaccharides as one of the main materials for forming various structures including trees, plants, algae, and crustaceans. Polysaccharides, with about 40 different monosaccharide structures, different O-glycosidic linkage types, and various molecular weights, present extremely diverse physical, chemical, and biological properties [1, 2]. The monosaccharides (or sugar moieties) in homopolysaccharides and heteropolysaccharides can contain various substituents such as acyl groups, amino acids, or inorganic residues. It is the physicochemical properties of these monosaccharide repeating units and their sequences that determine their material properties in biological systems [3]. Polysaccharides have been used widely in pharmaceutical and

Polysaccharide Hydrogels: Characterization and Biomedical Applications
Edited by Pietro Matricardi, Franco Alhaique, and Tommasina Coviello

ISBN 978-981-4613-61-3 (Hardcover), 978-981-4613-62-0 (eBook)
www.panstanford.com

biomedical applications, particularly in drug delivery and tissue engineering, for their recognized biocompatibility (or nontoxicity), biodegradability, and hydrophilicity. Polysaccharides are known to have biochemical similarities with human extracellular matrix (ECM) components and, thus, may be readily accepted by the body [4]. Commonly used polysaccharides include agarose, alginate, carrageenan, cellulose, chitin, chitosan, chondroitin sulfate, dextran, gellan, guar gum, hyaluronic acid, pectin, pullulan, scleroglucan, starch, xanthan gum, xyloglucan, and other polysaccharides [4–8]. Polysaccharides form hydrogels when they are exposed to low amounts of water that are not sufficient to dissolve the polymer. Polysaccharides can also form hydrogels through crosslinking either by chemical or physical means [9].

14.2 The Big Picture

Various polysaccharides are used for a wide spectrum of applications. Some scientists focus on understanding the basic properties of polysaccharides, while others are interested in finding new applications and functions. For most biomedical and pharmaceutical scientists, however, the ultimate goal is to develop clinically useful devices and formulations. This requires approval by the Food and Drug Administration (FDA) after the safety and efficacy are proven by controlled clinical studies. Polysaccharides have also been frequently used for cell microencapsulation [10] and for tissue engineering [11]. Polysaccharides are expected to play important roles in developing tissue-engineered body components which are urgently needed because of the shortage of donated organs for organ transplantation. Figure 14.1 shows an example of development of a drug delivery system and its ultimate clinical application. The whole process requires concurrent considerations of multiple parameters and years of effort and resources. If a polysaccharide is chosen for a particular formulation, then it should provide the necessary physicochemical and biological properties. Thus, it is important to select a polysaccharide that is not expected to cause any unexpected and unwanted side effects. This limits the selection of polysaccharides to only a certain number of them.

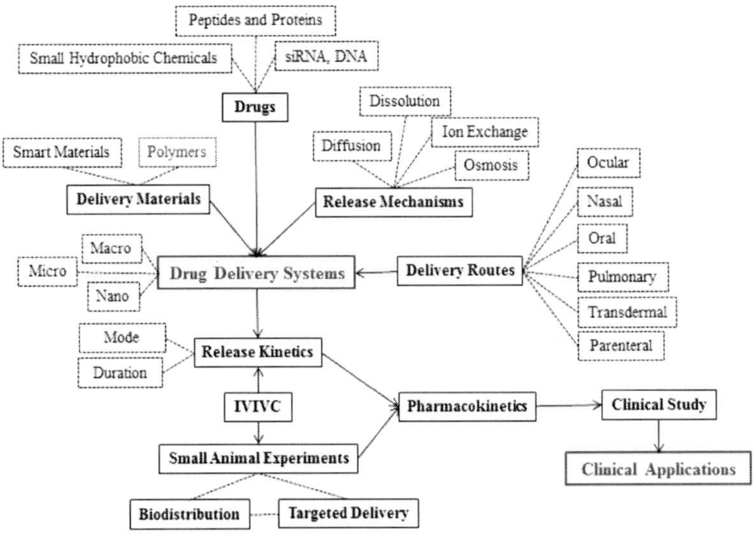

Figure 14.1 Overview of the development of drug delivery systems from basic research to clinical applications.

Polysaccharides commonly used in pharmaceutical formulations are found in the *Handbook of Pharmaceutical Excipients* [12]. It lists only a dozen polysaccharides, including alginic acid, cellulose (including microcrystalline cellulose, carboxymethylcellulose, ethylcellulose, hydroxyethyl cellulose, hydroxypropyl cellulose, and hydroxypropyl methylcellulose), carrageenan, guar gum, starch glycolate, and xanthan gum. Whenever a new polysaccharide, i.e., one that has not been used in products approved by the FDA, is used in biomedical devices or in pharmaceutical formulations, its safety has to be proven.

14.3 Uniqueness of Polysaccharides

The advantages of polysaccharides include natural abundance, relative ease of isolation, and the presence of functional groups for chemical modification to meet various technological needs [4]. Furthermore, polysaccharides undergo enzymatic and/or hydrolytic degradation in the biological environment without producing any toxic by-products. Polysaccharides are widely found in living organisms, and thus they can often be produced by recombinant DNA techniques. As renewable sources, polysaccharides have huge economic advantages over synthetic polymers. Polysaccharides show a number of unusual physicochemical properties that are useful for different applications in drug delivery systems [14].

Polysaccharides have drawbacks, too. The properties of polysaccharides are often varied based on the origin or even different batch and percentage of minor groups on the polymer backbone, causing different results even though seemingly the same polysaccharide is used [14]. Polysaccharides may have microbial contamination, different water uptake, poor mechanical strength, and an unpredictable degradation pattern [4]. The effective application of polysaccharides in biomedical and pharmaceutical applications requires more consistent properties. Obtaining polysaccharides with consistent properties is essential in reproducing the data and, thus, the functions of the final products. As a quality control, it is necessary to relate the physicochemical and the rheological properties to the polymer features and functions [14].

Semi-synthetic polymers have been prepared by modifying polysaccharides with synthetic functional groups, such as methacrylate [15] and *N*-isopropylacrylamide (NIPAM) [16]. Grafting NIPAM to a polysaccharide backbone affects the lower critical solution temperature and swelling behavior [16]. The swelling and mechanical properties can be controlled by making (semi-)interpenetrating polymer network polysaccharide hydrogels designed for specific drug delivery and tissue engineering applications [17]. The kinetics and extent of swelling can be adjusted by controlling the pore size of polysaccharide hydrogels. Polysaccharide-based superporous hydrogels possess superfast swelling and superabsorbent properties [18]. While many new polysaccharides with controlled properties have been prepared, more fundamental studies are necessary to establish a relationship between physicochemical properties and biological properties to fulfill unmet medical and pharmaceutical needs [17].

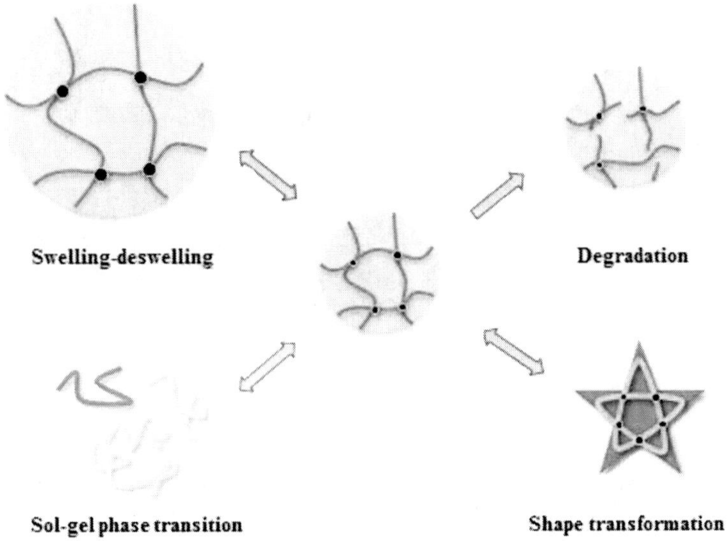

Figure 14.2 Smart hydrogels can undergo sol–gel phase transition, degradation, and shape changes in addition to swelling-deswelling.

14.4 Smart Hydrogels

Smart hydrogels are different from ordinary hydrogels which simply swell in the presence of an abundant amount of water. As shown in Fig. 14.2, smart hydrogels are able to do more than swell and deswell. They can undergo huge changes in physicochemical properties, degradation, sol–gel phase transition, and shape transformation in response to only a minute change in environmental conditions [19].

It is these additional properties that make smart hydrogels useful. It is important to note that drastic changes occur as a result of only a small change in environmental stimulus, e.g., a few degree changes in temperature and a change of 0.1 unit in pH. Various stimuli are shown in Fig. 14.3. Smart hydrogels have been applied in the areas of drug delivery, tissue engineering, bioseparation, and biosensors. One example of smart hydrogels is bioresponsive polysaccharide hydrogels [20]. Smart

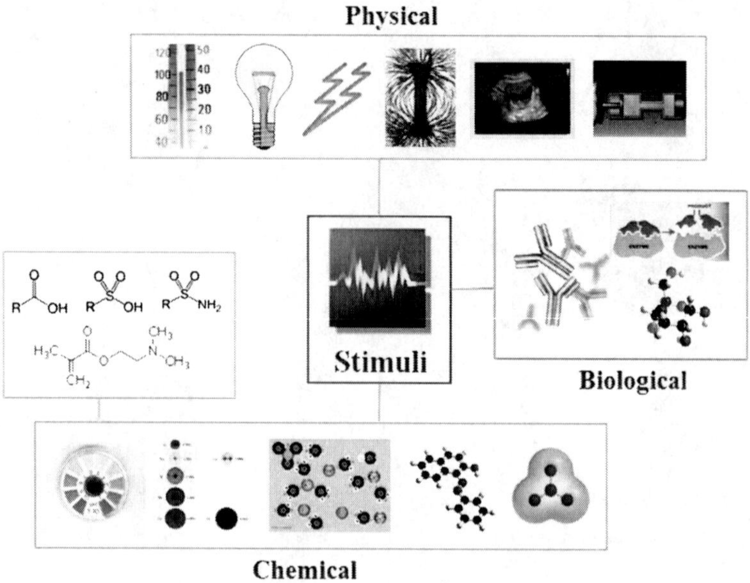

Figure 14.3 Environmental stimuli that cause smart hydrogels undergo unique changes include physical (temperature, light, electricity, magnetic field, ultrasound, and pressure), biological (antibody, enzyme, glucose), and chemical (pH, ion type, ionic strength, solvent, and hydrogen bonding) factors.

hydrogels may also have bioactive properties, e.g., antimicrobial properties [21].

A crosslinked carboxymethylcellulose (CMC) matrix as a diagnostic tool was designed to release a dye in response to pectinase and cellulase. The stability of CMC hydrogels was improved by introducing methacrylic groups as crosslinking molecules. These polysaccharide-based bioresponsive systems can be used in medical surfaces or packaging systems for detecting the presence of bacteria or fungi [20]. The ability of polysaccharides to be used as sensors for bacterial and fungal contamination is due to their biodegradability and selective enzymatic susceptibility. The enzymatic biodegradability is a unique property of polysaccharides, as most synthetic polymers are not degradable by this route other than being generally susceptible to hydrolysis. There has not been sufficient time for bacteria to evolve to a level to utilize synthetic polymers as a source of energy. In this respect, utilizing natural polysaccharides provide various benefits.

Although smart hydrogels have been used to develop hydrogels with new, unique functions, no smart hydrogels have proven themselves to be really smart. First of all, "smart" is a relative term. Smart hydrogels are smart only in comparison with conventional hydrogels which can undergo only swelling and deswelling depending on the availability of water. The question is whether the current smart hydrogels are "smart" enough to accomplish the intended job. Smart hydrogels still cannot achieve the modulated release of insulin, i.e., release the right amount of insulin only after an increase in glucose concentration in the blood. The current smart hydrogels have an IQ of only 2~4 [22]. They can, in addition to the basic function of swelling (IQ = 1), respond to environmental changes, such as physical, biological, and chemical changes, maybe at the same time. Smart hydrogels need to be a lot smarter in the future.

14.5 Future

The polysaccharide hydrogel technology has progressed substantially over the years. It, however, is still in its infancy stage. To propel it to the next level, new methods need to be developed so that the macroscopic properties, including biological properties, of polysaccharides can be predicted from the fundamental properties, such as physicochemical or rheological properties. Development of modeling tools for structure–property relationship is essential for effective biomedical and pharmaceutical applications. Such tools allow development of polysaccharide hydrogels with new functions and properties with pre-designed modification of polysaccharides. Chemical modification of polysaccharides may have to be done on specific saccharide units at a specific point of the polymer backbone. Polymer molecular weight can be optimized for the intended applications. The size of polysaccharide hydrogels can also be controlled from nanogels to macroscopic sizes [9, 23–25]. Understanding the structure-property relationship will provide new opportunities to make hydrogels that have not been possible previously. The term "bio-inspired" is still inspiring. This term was evolved from "bionic" in 1960 to "biomimetic" in 1969 [26]. Indeed, biology, through millions of years of trial and error, has produced remarkably adaptive structures with practical functions. Our job is to transfer such time-tested designs into engineering principles for developing better biomedical devices and pharmaceutical systems.

Currently, the most advanced hydrogels are known as "smart" hydrogels. The term "smart" has different meanings. In engineering, a smart system can sense, decide, and then respond autonomously or semi-autonomously to optimize its performance [26]. The three essential components, sensor, processor, and actuator, need to work flawlessly to achieve the intended functions. Utilizing the three-component smart system successfully, e.g., modulated insulin delivery, requires fast response time, too. Although the current approach to development of smart materials is based on structural design, it still more or less depends on the trial-and-error approach. Current hydrogel-based biomaterials still suffer from long-term failure due to the intrinsic limitations of inert, nonliving substances [27]. The so-called bioactive materials, or "really smart" biomaterials, have been developed to improve their functions [27]. Whether we call them "really smart" or not depends on what they can actually do. Smart materials will always become smarter, and calling them with proper names will be a challenge in the future. It may be time to distinguish smart materials on the basis of their IQ.

Plants have a highly modular design. Plants are adaptive, self-sensing, multifunctional, energy efficient, pollution free, and silent. These unique characteristics and behaviors of plants can be exploited to achieve the new properties necessary for developing future drug delivery systems and biomedical devices [28]. Suppose polysaccharide hydrogels can be used to make artificial jellyfish wandering around in water, artificial mimosa pudica that rapidly changes its shape upon a gentle touch, or an artificial virus that can self-reproduce. Making these artificial systems is not easy at all, but that is why we need to set a goal to make them. It will allow us to gather our brightest minds, figure out what technologies need to be developed, and create the future technologies.

Acknowledgment

This work was supported by the Showalter Research Trust Fund and the National Institute of Health through CA129287 and GM095879.

References

1. Puppi, D., Chiellini, F., Piras, A. M., and Chiellini, E. (2010). Polymeric materials for bone and cartilage repair. *Progress in Polymer Science* (**35**), pp. 403–440.
2. Dräger, G., Krause, A., Möller, L., and Dumitríu, S. (2011). Carbohydrates, in: A. Lendlein and A. Sisson (eds.), *Handbook of Biodegradable Polymers*, Wiley-VCH Verlag & Co., Weinheim, Germany, pp. 155–193.
3. Malafaya, P. B., Silva, G. A., and Reis, R. L. (2007). Natural-origin polymers as carriers and scaffolds for biomolecules and cell delivery in tissue engineering applications. *Advanced Drug Delivery Reviews* (**59**), pp. 207–233.
4. Aravamudhan, A., Ramos, D. M., Nada, A. A., and Kumbar, S. G. (2014). Natural polymers: polysaccharides and their derivatives for biomedical applications, in: S. G. Kumbar, C. T. Laurencin, and M. Deng (eds.), *Natural and Synthetic Biomedical Polymers*, Elsevier, Oxford, pp. 67–89 (chapter 4).
5. Brun-Graeppi, A. K. A. S., Richard, C., Bessodes, M. Scherman, D., and Merten, O. W. (2011). Cell microcarriers and microcapsules of stimuli-responsive polymers. *Journal of Controlled Release* (**149**), pp. 209–224.
6. Lee, K. Y., and Mooney, D. J. (2012). Alginate: properties and biomedical applications. *Progress in Polymer Science* (**37**), pp. 106–126.
7. Peniche, C., Argüelles-Monal, W., and Goycoolea, F. M. (2008). Chitin and chitosan: major sources, properties and applications, in: M. N. Belgacem and A. Gandini (eds.), *Monomers, Polymers and Composites from Renewable Resources*, Elsevier, Amsterdam, pp. 517–542 (chapter 525).
8. Dash, M., Chiellini, F., Ottenbrite, R. M., and Chiellini, E. (2011). Chitosan: a versatile semi-synthetic polymer in biomedical applications. *Progress in Polymer Science* (**36**), pp. 981–1014.
9. Varaprasad, K., Sadiku, E. R., Ramam, K., Jayaramudu, J., and Reddy, G. S. M. (2014). Significances of nanostructured hydrogels for valuable applications, in: S. Thomas, R. Shanks, and S. Chandrasekharakurup

(eds.) *Nanostructured Polymer Blends*, William Andrew Publishing, Oxford, pp. 273–298 (chapter 8).

10. Santos, E., Pedraz, J. L., Hernández, R. M., and Orive, G. (2013). Therapeutic cell encapsulation: ten steps towards clinical translation. *Journal of Controlled Release* (**170**), pp. 1–14.
11. Furth, M. E., and Atala, A. (2014). Tissue engineering: future perspectives, in: R. Lanza, R. Langer, and J. Vacanti (eds.), *Principles of Tissue Engineering* (4th ed.), Academic Press, Boston, pp. 83–123 (chapter 6).
12. Sheskey, R. C. R. P. J., and Quinn, M. E. (2009). *Handbook of Pharmaceutical Excipients*, Pharmaceutical Press, Grayslake, IL.
13. Park, K. (2014). The controlled drug delivery systems: past forward and future back. *Journal of Controlled Release* (**190**), pp. 3–8.
14. Coviello, T., Matricardi, P., Marianecci, C., and Alhaique, F. (2007). Polysaccharide hydrogels for modified release formulations. *Journal of Controlled Release* (**119**), pp. 5–24.
15. Corrente, F., Abu Amara, H. M., Pacelli, S., Paolicelli, P., and Casadei, M. A. (2013). Novel injectable and in situ cross-linkable hydrogels of dextran methacrylate and scleroglucan derivatives: preparation and characterization. *Carbohydrate Polymers* (**92**), pp. 1033–1039.
16. Hamcerencu, M., Desbrieres, J., Popa, M., and Riess, G. (2012). Original stimuli-sensitive polysaccharide derivatives/*N*-isopropylacrylamide hydrogels. Role of polysaccharide backbone. *Carbohydrate Polymers* (**89**), pp. 438–447.
17. Matricardi, P., Di Meo, C., Coviello, T., Hennink, W. E., and Alhaique, F. (2013). Interpenetrating polymer networks polysaccharide hydrogels for drug delivery and tissue engineering. *Advanced Drug Delivery Reviews* (**65**), pp. 1172–1187.
18. Kuang, J., Yuk, K. Y., and Huh, K. M. (2011). Polysaccharide-based superporous hydrogels with fast swelling and superabsorbent properties. *Carbohydrate Polymers* (**83**), pp. 284–290.
19. Lee, S. C., Kwon, I. K., and Park, K. (2013). Hydrogels for delivery of bioactive agents: a historical perspective. *Advanced Drug Delivery Reviews* (**65**), pp. 17–20.
20. Schneider, K. P., Wehrschuetz-Sigl, E., Eichhorn, S. J., Hasmann, A., Flock, T., Kaufmann, F., Shyng, Y.-T., and Guebitz, G. M. (2012). Bioresponsive systems based on crosslinked polysaccharide hydrogels. *Process Biochemistry* (**47**), pp. 305–311.

21. Kong, M., Chen, X. G., Xing, K., and Park, H. J. (2010). Antimicrobial properties of chitosan and mode of action: a state of the art review. *International Journal of Food Microbiology* (**144**), pp. 51–63.
22. Park, K. (2002). Preface. *Advanced Drug Delivery Reviews* (**54**), pp. 1.
23. Montoro, S. R., Medeiros, S. d. F., and Alves, G. M. (2014). Nanostructured hydrogels, in: S. Thomas, R. Shanks, and S. Chandrasekharakurup (eds.), *Nanostructured Polymer Blends*, William Andrew Publishing, Oxford, pp. 325–355 (chapter 10).
24. Moya-Ortega, M. D., Alvarez-Lorenzo, C., Concheiro, A., and Loftsson, T. (2012). Cyclodextrin-based nanogels for pharmaceutical and biomedical applications. *International Journal of Pharmaceutics* (**428**), pp. 152–163.
25. Fajardo, A. R., Pereira, A. G. B., Valente, A. J. M., Rubira, A. F., and Muniz, E. C. (2014). Stimuli responsive polysaccharide-based hydrogels, in: P. Matricardi, F. Alhaique, and T. Coviello (eds.), *Polysaccharide Hydrogels: Characterization and Biomedical Applications*, Pan Stanford Publishing, Singapore.
26. Jenkins, C. (2012). *Bio-Inspired Engineering*, Momentum Press, New York, NY.
27. van Noort, R., Hatton, P. V., and Haddow, D. B. (2003). Smart biomaterials: "out-smarting" the body's defense systems and other advances in materials for medicine, in: K. Worden, W. A. Bullough, and J. Haywood (eds.), *Smart Technologies*, World Scientific, River Edge, NJ.
28. Sater, J. M., Main, J. A., and Mchenry, R. (2012). Final discussion and conclusions, in: N. M. Wereleyand J. M. Sater (eds.), *Plants and Mechanical Motion: A Systematic Approach to Nastic Materials and Structures*, DEStech Publications, Lancaster, PA (chapter 10).

Index